普通高等院校双语教材

Engineering Surveying

工 程 测 量

陈学平 主编

中国建材工业出版社

图书在版编目（CIP）数据

工程测量 = Engineering Surveying/陈学平主编. —北京：中国建材工业出版社，2014. 3

ISBN 978-7-5160-0511-8

Ⅰ. ①工… Ⅱ. ①陈… Ⅲ. ①工程测量-教材
Ⅳ. ①TB22

中国版本图书馆 CIP 数据核字（2014）第 280878 号

内 容 简 介

本教材按学科系统性，并考虑教学与学生自学的方便组成新的体系。全书 12 章包括测绘的基础知识、基本理论与基本技能，以及若干工程领域的工程测量。第 1 ~ 5 章为绪论、误差理论、距离测量、水准测量和角度测量，第 7 ~ 8 章为控制测量与地形测量。前 8 章为普通测量学的主要内容。第 9 ~ 11 章包括公路曲线测量、面积与体积的计算以及施工测量，是工程领域的重要内容。第 6 章全站仪与第 12 章全球定位测量是已在生产中广泛使用的重要的测绘新技术。

本书可作为普通高等院校工程测量、土木建筑、公路交通等专业的双语教材、有关专业研究生的参考书，也可供从事测绘教学与生产人员作为英语学习资料使用。

Engineering Surveying

工程测量

陈学平 主编

出版发行：中国建材工业出版社
地　　址：北京市西城区车公庄大街 6 号
邮　　编：100044
经　　销：全国各地新华书店
印　　刷：北京雁林吉兆印刷有限公司
开　　本：787mm × 1092mm 1/16
印　　张：22. 75
字　　数：564 千字
版　　次：2014 年 3 月第 1 版
印　　次：2014 年 3 月第 1 次
定　　价：58. 00 元

本社网址：www. jccbs. com. cn　　微信公众号：zgjcgycbs

本书如出现印装质量问题，由我社发行部负责调换。联系电话：（010） 88386906

前　　言

目前国内英文版测绘教材和相关书籍较少，购买国外原版教材价格昂贵，且难以找到十分适用的版本。为了满足广大测绘科技工作者、生产单位作业员以及在校师生学习测绘专业英语的需求，本人尝试编写一本以反映测绘基础知识和基本理论与方法为主并能反映当代测绘科技特点的双语教材。为此，本人查阅国家图书馆近十年来四十多本英文测量学、工程测量及 GPS 教材，从中精选内容进行编辑。

本书选材与编写的要求是：

（1）明确章节的基本要求，从多本原著中选材，主要考虑科学性、系统性与实用性。内容深度适当，尽量避免因内容过于艰深而影响阅读理解。

（2）教材内容突出三基（基础知识、基本理论与基本技能），注意把阐述理论与讲解实践操作相结合。阐述概念与原理力求准确、简洁、清晰，叙述观测步骤注意条理性，算例尽可能表格化。

（3）测量精度要求符合国内规范，计算表格采用国内的通用格式。

（4）图文并茂。介绍仪器时，除照片外尽可能配以描述仪器主要结构的线划图，以便于读者理解与掌握。

（5）每节后列出单词与词组，遇有难以理解的句子做了注释。

全书分为两大部分，第一部分为英文课文，第二部分为课文参考译文。建议读者按英文课文顺序学习，独立阅读与翻译，不懂时再看参考译文。

本书的编写过程得到中国建材工业出版社教材教辅编辑部同志的大力支持与帮助，提出了许多宝贵意见，在此表示十分感谢。

编者水平有限，时间仓促，书中内容难免存在错误与漏洞，敬请读者批评指正。

陈学平

2014 年 1 月于北京

Part 1 TEXT

Contents

第二部分　课文参考译文

目　　录

Part 1 TEXT

Chapter 1 Introduction

1.1 Relevant definition

1.1.1 Surveying and Engineering surveying

*1 [Surveying may be defined as the science of determining the position, in three dimensions, of natural and man-made features on or beneath the surface of the Earth. These features may be represented in analogue form as a contoured map, plan or chart, or in digital form such as a digital ground model (DGM).]

Engineering surveying is defined as those activities involved in the planning and execution of surveys for the location, design, construction, maintenance, and operation of civil construction and other engineered projects. Such activities include all the survey work required before, during and after any engineering works.

(1) Before any works are started, large-scale topographical maps or plans are required as a basis for design. To produce up-to-date maps of the areas in which engineering projects are to be built. The scales of the maps are usually considerably larger than those produced in the other forms of land surveys. Civil engineering works commonly use scales 1:500, 1:1000, 1:2000 and 1:5000. Town planning commonly use scales 1:5000 and 1:10000. Preliminary survey map for route surveys commonly use scales 1:1000 to 1:10000. Cross sections and longitudinal sections are often drawn with exaggerated vertical scales.

(2) The proposed position of any new item of construction must then be marked out on the ground, both in plan and height, an operation generally termed setting out. Surveyors must ensure that the construction is built in its correct relative and absolute position on the ground. In land surveys and cadastral surveys, it is often required to calculate the areas and volumes of land.

(3) As-built surveys are made after a construction project is complete, to provide the positions and dimensions of the features of the projects as they were actually constructed. These surveys not only provide a record of what was constructed but also become a very important document that must be preserved for future maintenance, expansions and new construction.

(4) Providing permanent control points by which the future movement of structures such as dams and bridges can be monitored.

Engineering surveying is one of the most important areas of expertise in geomatics.

1.1.2 Geomatics

Where does the word Geomatics come from? GEODESY + GEOINFORMATICS = GEOMATICS or the combination of GEO—for Geoscience and—MATICS for informatics. The term geomatics emerged first in Canada and as an academic discipline; it has been introduced worldwide in a number of institutes of higher education during the past few years, *2 [mostly by renaming what was previously called "geodesy" or "surveying", and by adding a number of computer science and GIS-oriented courses.] Now the term includes the traditional surveying definition. Along with surveying steadily increased importance with the development of new technologies and the growing demand for a variety of spatially related types of information, particularly in measuring and monitoring our environment.

Geomatics is the science and technology of acquiring, storing, processing, managing, analyzing and presenting geographically referenced information (geo-spatial data). It integrates the following more specific disciplines and technologies including surveying and mapping, geodesy, satellite positioning, photogrammetry, remote sensing, geographic information systems (GIS), land management, computer systems, environmental visualization and computer graphics.

Geomatics not only covers the traditional work of the surveyor but also reflects the changing role of the surveyor in data management. This has arisen because of the advances made in surveying which make it possible to collect, process and display large amounts of spatial data with relative ease using digital technology. This in turn has created an enormous demand for this data from a wide variety of sources such as geology, geophysics, hydrology, forestry, transportation, government and human resources. For all of these, data is collected and processed by a computer in a Geographic Information System (GIS). These are databases that can integrate the spatial data provided by surveyors with environmental, geographic and social information layers (see Fig. 1-1) which can be combined, processed and displayed in any format according to the needs of the end user. Without any doubt, the most important part of a GIS is the spatial data on which all other information is based and the provision of this has been a huge growth area in surveying.

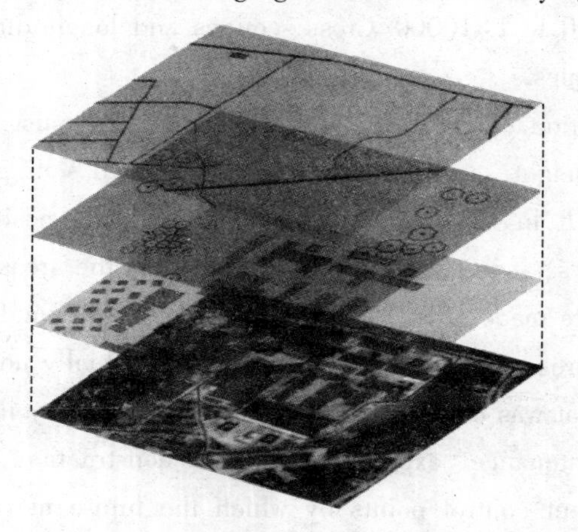

Fig. 1-1 Layers in a GIS

单词与词组

determine[di ˈtəmin] *vt.* 决定，确定，测定

position[pə ˈziʃən] *n.* 位置，地点；职位；观点

dimension[di ˈmenʃən] *n.* 尺寸，维(数)，量纲

man-made[mæn-meid] *a.* 人工的，人造的

feature[ˈfiːtʃə] *n.* 特点，性能，地物，地形(包含地物与地貌)特征，(复数)面貌，容貌

beneath[bi ˈniːθ] *prep.* 在……之下

represent[ˌrepri ˈzent] *vt.* 描述，描写，象征，表示

analogue[ˈænəˌlɔg] *n.* 相似体，模拟量(量，系统)

chart[tʃaːt] *n.* 图表，曲线图，略图，示意图

· DGM 即 Digital Ground Model 的缩写，数字地面模型

activity[æk ˈtiviti] *n.* 活动，活动的事物，领域，活动范围，功效，工作

planning[ˈplæniŋ] *n.* 规划，计划

execution[ˌeksi ˈkjuːʃən] *n.* 实行，实施，执行，完成

construction [kən ˈstrʌkʃən] *n.* 建筑，建设，工程

civil[ˈsivl] *a.* 公民的，市民的，民事的

· civil construction 土木建筑

· civil engineering 土建工程

location[ləu ˈkeiʃən] *n.* 定位，测位，勘定地界

operation[ˌɔpə ˈreiʃən] *n.* 操作，工作；活动；交易，经营；运算

· large-scale topographical map 大比例尺地形图

map[mæp] *n.* 地图，地形图

plan[plæn] *n.* 设计图，平面图，图样

· up-to-date 直到最近的，最新式的

· set out 放样，定线，布置，区划

preliminary [pri ˈliminəri] *a.* 初步的，预备的

· cross section 横断面

· longitudinal section 纵断面

exaggerate[igˈzædʒəreit] *vt.* 夸大，使增大

propose[prəˈpəuz] *vt.* 提议，建议，主张，推荐，提(名)，打算，计划

· as-built survey 竣工测量

preserve[priˈzəːv] *vt.* 保护，维护，保持，维持，支持，保管，保存

maintenance[ˈmeintinəns] *n.* 维持，保持，维修，保养

monitoring [ˈmɔnitəriŋ] *n.* 监视，控制，监测

dam [dæm] *n.* 坝，堤

bridge[bridʒ] *n.* 桥；桥梁

deformation [diːfɔːˈmeiʃən] *n.* 变形，畸形

movement[ˈmuːvmənt] *n.* 位移，运动

expertise [ˌekspə ˈtiːz]*n.* 专门知识或技能

geomatics[ˌdʒi:əu 'mætiks] n. 测绘学，地球空间信息学
geodesy [dʒi: 'ɔdisi] n. 大地测量学
geoinformatics[ˌdʒi:əuinfɔ'mætiks] n. 地球空间信息学
emerge[i'mə:dʒ] v. 出现，显出，暴露
academic [ˌædʒkə'demik] a. 大学的，学会的，学术的
discipline ['disiplin] n. 锻炼，训导，学科
worldwide ['wə:ldwaid] n. 全世界的，世界范围的
institute['institju:t] n. 学会，协会，学院
spatial['speiʃəl] a. 空间的，立体的，三维的
acquire [ə'kwaiə] vt. 得到，获取，招致
environment[in 'vaiərənmənt] n. 周围，围绕，环境
acquisition [ˌækwi 'ziʃən] n. 采集，收集，取得，获得
processing [prəu 'sesiŋ] n. （数据）处理，加工
present ['prezənt] vt. 提出，展示，显示；a. 现在的，目前的；n. 现在，礼物，瞄准
discipline['disiplin] n. 纪律，规定，学科
photogrammetry[ˌfəutəu 'grəmitri] n. 摄影测量术
· satellite positioning 卫星定位
· remote sensing 遥感
geographic[ˌdʒi:ə'græfik] a. 地理学的，地理的
· geographic information systems (GIS) 地理信息系统
visualization[vizjuəlai 'zeiʃən]n. 可视化，显象，目视
profession[prə'feʃən] n. 职业(尤指从事脑力劳动或受过专门训练的)
nature['neitʃə]n. 自然界，自然，性质，特性，生命力，精力，活力
environmental [enˌvaiərən 'mentl] a. 环境(产生)的

注释

[1] 句中两个介词短语：of natural and man-made features 作后置定语修饰 position，in three dimensions 作状语，而介词短语 on or beneath the surface of the Earth(在地球表面上面或下面)又修饰 features，该词还含有地貌含义。故此段译为：测量学可定义为研究测定地上或地下天然的和人工的地物与地貌特征在三维空间位置的科学。这些地物地貌可以模拟形式表示为地形图、平面图、曲线图，或以数字形式表示，如数字地面模型(DGM)。such as 词组意为“例如，正像……那样”。

[2] mostly by renaming what was previously called “geodesy” or “surveying” and by adding a number of computer science and GIS-oriented courses。句中 what 可用 that which 或 the thing which 替代。为了与上句衔接可加“其做法”，故此句译为：“其做法多半是通过对以前称为大地测量学或测量学进行改名以及通过增加许多计算机科学和 GIS 方向的课程”。

1.2 Types of survey

Surveys can be classified into two classes depending upon whether the curvature of the earth is taken into account or not.

Plane Survey In plane surveying, the curved nature of the earth's surface is neglected. A line on the earth's surface is taken to be a straight line. When three points lying on the surface of the earth are joined, they form a plane triangle. Consequently, all plumb lines are assumed to be parallel. The methods of plane surveying are used when the extent of the survey is small.

Geodetic Survey This type of survey is suited for large areas and long lines. The curvature of the earth is accounted for in the measurements. It is used to determine the precise location of basic point for establishing control for other surveys. In geodetic surveys, the stations are normally long distances apart, and more precise instruments and surveying methods are required for this type of surveying than for plane surveying.

Based upon their different purpose, surveys can be classified into the following types:

Control surveys *1 [Control surveys establish a network of horizontal and vertical monuments that serves as a reference framework for other surveys.] When the control survey is of small extent, plane surveying methods may be used, but generally geodetic methods are employed.

Topographic surveys Topographic surveys are used to prepare maps showing locations of natural and man-made features and elevations of points on the ground.

Construction surveys Construction surveys provide points and elevations for building civil engineering projects; they are often called engineering surveys.

Property surveys Property surveys establish property corners, boundaries, and areas of land parcels. They are also termed land surveys, cadastral surveys, and boundary surveys.

Route surveys *2 [Surveys of and for highways, railroads, pipelines, transmission lines, canals and other projects which do not close upon the starting point.]

Hydrographic surveys Hydrographic surveys are made to map shore-lines and bottom depths of lakes, streams, reservoirs, and other larger bodies of water. Sometimes the combination of topographic and hydrographic surveying is designated cartographic surveying.

Aerial surveys An aerial survey is done from aircraft, which take photographs of the surface of the earth in overlapping strips of land. Extensive areas can be covered by such surveys. This form of survey is also known as a aerophotogrammetry. The method is very expensive. It is, however, recommended in large projects in difficult terrain where ground surveys may be difficult or impossible.

单词与词组

applicable[ˈæplikəbl] *a.* 可适用的，能应用的，适当的，合适的

curvature[ˈkə:vətʃə] *n.* 弯曲(部分)，曲率，曲度

ignore[igˈnɔ:] *vt.* 忽视，不理睬；抹杀(建议)

neglect[niˈglekt] *vt.* 忽视，忽略，无视，不顾

join[dʒɔin] *vi.* 结合，联合，相遇(to; with)

plumb[plʌm] *n.* 铅锤，测锤，垂直

parallel[ˈpærəlel] *a.* 平行的，并行的，相似的，对应的

geodetic[ˌdʒi:əˈdetik] *a.* 大地测量的

· geodetic survey 大地测量
suite[swi:t] *n.* 随员，(房间、器具等)一套，一付，一组；*v.* 适合，适应，相配(称)
station['steiʃən] *n.* 台，所，站，车站，测点，测站
apart[ə'pa:t] *adv.* 相距，分隔
classify['klæsifai] *vt.* 分类，归类
control[kən'trəul] *n.* 控制
network['netwə:k] *n.* 网络
monument['mɔnjumənt] *n.* 标石，界碑，石(混凝土)桩，埋石点
framework['freimwə:k] *n.* 结构，框架，网格
extent[iks'tent] *n.* 广度，宽度，长度，大小，范围，限度，程度
employ[im'plɔi] *vt.* 使用，雇用
topographic[tɔpə'græfik] *a.* 地形学上的，地志的
· topographic survey 地形测量
· construction survey 建筑测量，施工测量
property['prɔpəti] *n.* 资产，房地产
· property survey 房地产测量
cadastral[kə'dæstrəl] *a.* 地籍的
boundary['baundəri] *n.* 分界线
route[ru:t] *n.* 路线
highway['haiwei] *n.* 公路
railroad['reilrəud] *n.* 铁路
pipeline['paipˌlain] *n.* 管道，管线
transmission[trænz'miʃən] *n.* 传输，输电，发射
canal[kə'næl] *n.* 运河，渠道，管道
hydrographic[ˌhaidrəu'græfik] *n.* 水文地理的
· hydrographic survey 水文测量
shore-line[ʃɔ:-lain] *n.* 岸线，海岸线
stream[stri:m] *n.* 小河
reservoir['rezəvwa:] *n.* 水库
cartographic[ka:'tɔgrəfik] *a.* 地形制图学的
aerial['ɛəriəl] *a.* 在空中的，空气的，大气的
· aerial survey 空中测量，航空测量
aerophotogrammetry[ˌɛərˌfəutəu'græmitri] *n.* 航空摄影测量

注释

[1] 此段译为：控制测量是建立水平和高程埋石点的控制网，这个控制网是作为其他测量的基准框架。句中 that serves as a reference framework for other surveys 是以 that 引导的定语从句修饰 network。句中 vertical monuments 直译为“垂直的埋石点”，为符合测绘术语改译为“高程埋石点”。

[2] 句中两个介词组合…of and for… 含意是“关于……的……”，翻译时可译成“关于公

路、铁路、管道、电力线、隧道以及不闭合至起始点的其他工程项目的测量。”

1.3 Survey reference surfaces

Surveying is concerned with the fixing of position whether it be control points or points of topographic detail and, *1[as such], requires some form of reference system. The physical surface of the earth, on which the actual survey measurements are carried out, is not mathematically definable. It cannot therefore be as a reference datum on which to compute position. We discover that a level surface is suitable, because the vertical axes of most surveying instruments are placed in the direction of gravity with the aid of level bubbles. Such a level surface would be normal to the gravity at each point. On the earth, a natural surface of this type would be the ocean surface, if it were in a mean position, free of other external influences, such as tides, currents, wind, etc. Such a surface, which can be defined uniquely for the whole earth, is considered a mathematical figure of the earth, and denoted as the "*geoid*" based on the Greek word for earth.

Indeed, the points surveyed on the physical surface of the earth are frequently reduced, initially, to their equivalent position on the geoid by projection along their gravity vectors.

The surface of the sea is formed according to the gravity. The gravity has irregularities due to mass distribution inside the earth, which causes the geoid to be an irregular surface and so cannot be used to locate position mathematically. The simplest mathematically definable figure which fits the shape of the geoid best is an *ellipsoid* formed by rotating an ellipse about its minor axis. By international agreement in 1967, its equatorial semi-diameter has been set at 6378160m with a shortening factor of 1:298.25 for the rotational axis in respect to the equatorial axis. This means that the rotational axis is only 3% shorter than the equatorial axis. *2[Where this shape is used by a country as the surface for its mapping system, it is termed the *reference ellipsoid*.]

Fig. 1-2 illustrates the relationship between these surfaces.

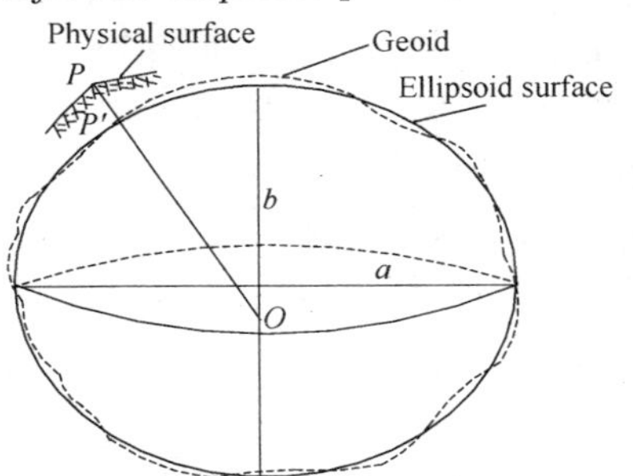

Fig. 1-2 Geoid, ellipsoid and physical surface

The majority of engineering surveys are carried out in areas of limited extent, in which case the reference surface may be taken as a tangent plane to the geoid and the principles of plane surveying applied. In other words, the curvature of the earth is ignored and all points on the physical surface are orthogonally projected onto a flat plane. For areas less than $10km^2$ the assumption of a flat earth is perfectly acceptable. The difference in length of an arc of approximately 20km on the earth's surface and its equivalent chord length is a mere 8mm.

The above assumptions of a flat earth, while acceptable for some positional applications, are not acceptable for finding elevations, as the geoid deviates from the tangent plane by about 80mm at 1km or 8m at 10km from the point of contact. Elevations are therefore referred at least theoretically, but usually to MSL practically.

单词与词组

· be concerned with 参与，涉及到，论述
fix [fiks] *v.* 使固定，决定，确定，规定，调整，安排
detail[ˈdiːteil] *n.* 小节，细节，碎部
reference[ˈrefrəns] *n.* 涉及，参考
· carry out 执行，实行，贯彻
datum[ˈdætəm] *n.* 数据，资料，基点，基面
· level surface 水准面
gravity[ˈɡræviti] *n.* 重量，重力，引力
external[eksˈtənəl] *adj.* 外面的，表面的，外部的
tide[taid] *n.* 潮汐，潮水
current[ˈkʌrənt] *n.* 水流，气流，电流
wind[wind] *n.* 风，气流
uniquely[juːˈniːkli] *adv.* 独特地，唯一地，罕有地
geoid[ˈdʒiːɔid] *n.* 大地水准面
frequently[ˈfriːkwəntli] *adv.* 时常，屡次，频繁地
reduce[riˈdjuːs] *vt.* 减少，简化，还原，归纳
initially [iˈniʃəli] *adv.* 开始，最初
equivalent[iˈkwivələnt] *a.* 相等的，相当的，等效的
projection[prəˈdʒekʃən] *n.* 预测，投影，投影图
vectors[ˈvektə] *n.* 向量
variation[ˌvɛvəriˈeiʃən] n. 变化，变异，变数
irregularity[iˌregjəˈlæriti] *n.* 不规则，不均匀，不对称，无规律
distribution[distriˈbjuːʃən] *n.* 分配，分布
ellipsoid[iˈlipsɔid] *n.* 椭圆体
rotate[rouˈteit] *v.* 旋转，回转，轮换
ellipse[iˈlips] *n.* 椭圆
· minor axis（椭圆）短轴
equatorial[ˌekwəˈtɔːriəl] *a.* 赤道的；赤道附近的
illustrate[ˈiləstreit] *v.* 给……加插图，说明
relationship[riˈleiʃənʃip] *n.* 关系，联系
majority[məˈdʒriti] *n.* 多数，半数以上
curvature[ˈkəːvətʃə] *n.* 弯曲(部分)，曲率，曲度，弧度
orthogonally[ɔːˈθɔgənlli] *adv.* 正交地，正射地
arc[aːk] *n.* 弧，弧线
equivalent[iˈkwivələnt] *a.* 相等的，相应的
chord[kɔːd] *n.* 乐弦，琴弦，弦
mere[miə] *a.* 仅仅的，只不过
deviate[ˈdiːvieit] *v.* 偏差，偏离

contact['kɔntækt] *n.* 联结，联系，相切
theoretical[ˌθiə 'retikəl] *a.* 理论上的
· at least 至少，不论怎样

注释

[1] 词组 as such 意为：像(作为)这样的人、事、物，以……资格、名义、身份，因而等。注意它与 such as 含义不同，后者意为：例如，正像……那样。

[2] Where this shape is used by a country as the surface for its mapping system, it is termed the reference ellipsoid. 这是主从复合句，where 引导地点或方式状语从句。从句中，介词短语 for its mapping system 作后置定语修饰 the surface。根据上下句的关系可知 this shape 就是指代 ellipsoid。主句的 it 就是指 this shape。整句译为：旋转椭球面被用于某一国家作为测图系统的面，该旋转椭球面被称为参考椭球面。

1.4 Basic measurements

An examination of Fig. 1-3 clearly shows the basic surveying measurements needed to locate points A, B and C and plot them orthogonally as A′, B′ and C′. Assuming the direction of B from A is known then the measured slope distance AB and the vertical angle to B from A will be needed to fix the position of B relative to A. The vertical angle to B from A is needed to reduce the slope distance AB to its equivalent horizontal distance A′B′ for the purposes of plotting. Whilst similar measurements will fix C relative to A, it also requires the horizontal angle at A measured from B to C (B′A′C′) to fix C relative to B. The vertical distance defining the relative elevation of the three points may also be obtained from the slope distance and vertical angle or by direct leveling relative to a specific reference datum.

The five measurements mentioned above comprise the basis of plane surveying and are illustrated in Fig. 1-4, i. e. AB is the slope distance, AA′ the horizontal distance, A′B the vertical distance. BAA′ the vertical angle(α) and A′AC the horizontal angle (θ).

It can be seen from the above that the only measurements needed in plane surveying are angle and distance. Nevertheless, the full impact of modern technology has been brought to bear in the acquisition and processing of this simple data. Angles may now be resolved with single-second accuracy using optical and electronic theodolites; electromagnetic distance measuring (EDM)

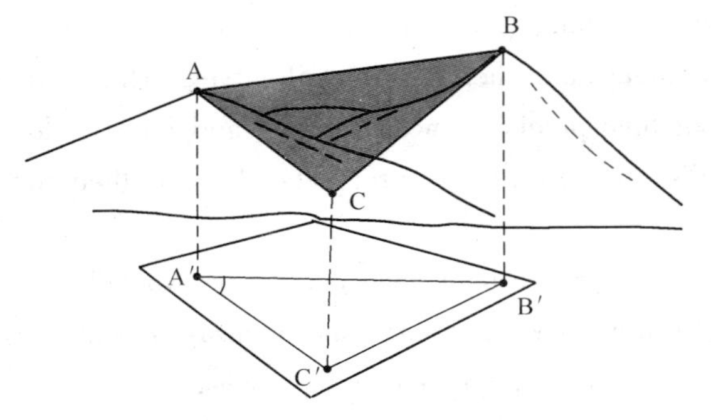

Fig. 1-3 Projection onto a plane surface

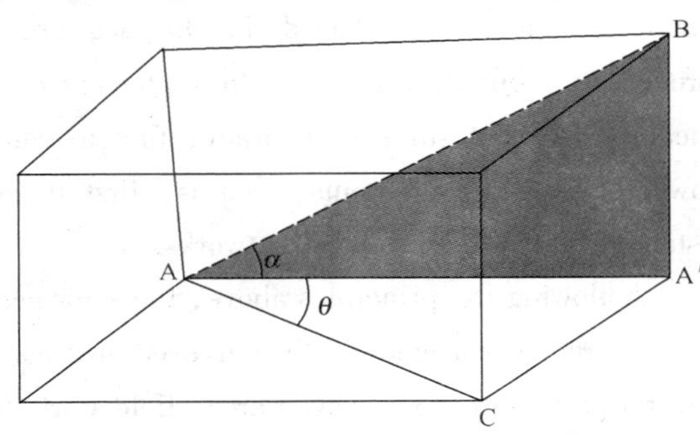

Fig. 1-4 Basic measurements

equipment can obtain distances up to several kilometers with millimeter precision, depending on the distance measured; lasers and north-seeking gyroscopes are virtually standard equipment for tunnel surveys; orbiting satellites are being used for position fixing offshore as well as on; continued improvement in aerial and terrestrial photogrammetric and scanning equipment makes mass data capture technology an invaluable surveying tool; finally, data loggers and computers enable the most sophisticated procedures to be adopted in the processing and automatic plotting of field data.

单词与词组

whilst[wailst] *conj.* whilst =while 同时，但是，可是，但另一方面
nevertheless[nevəð 'les] *adv.* 然而，尽管如此
impact['impækt] *n.* 影响，效力
acquisition[ˌækwi 'ziʃən] *n.* 取得，获得
electromagnetic[iˌlektrəmæg 'netik] *adj.* 电磁的
· north-seeking 指北的
laser['leizə] *n.* 激光，激光器
gyroscope['ʤaiərəskəup] *n.* 陀螺仪
tunnel['tʌnəl] *a.* 隧道的
orbit['ɔːbit] *n.*, *v.* 轨道，绕……作圆周运动，环绕
offshore['ɔ(ː)fʃɔː] *a.* 离开海岸的，向海面的，近海处的；*adv.* （指风）向海的，离岸的
terrestrial[tə 'restriːəl] *a.* 地球的，地上的，陆地的
scan[skæn] *v.*, *n.* 审视，扫描
invaluable[in 'væljuəbl] *a.* 无法估计的，无价的
logger['lɔgə] *n.* 数据记录器
sophisticated[sə'fistikeitid] *a.* 复杂的，高级的，尖端的，完善的

1.5 Two important principles of survey work

1.5.1 The organizing principle of survey work

The organizing principle of survey work is from the whole to the part, from the control surveys to detail surveys, and from a high-precision measurements to a low-precision measurements. The first meaning is about the general layout. First, what programs to adopt for the whole survey area must be confirmed and then what to do for the local area must be arranged. The second meaning is about the procedure of survey work, first to implement control surveys and then to do detail surveys. The third meaning is about surveys accuracy, first to carry out high-precision measurements and then to do low-precision measurements, that is, first to establish higher-order control networks and then to establish lower-order control networks.

Following the principles above, the error accumulation of measurement result can be avoided and the accuracy consistency in the surveyed area can be ensured. Owing to simultaneous measurements of several parties in the surveyed area, field work efficiency can be increased several times.

1. Control surveys

When a survey work is started up, surveyor should choose a limited number of points that have a bearing on the situation as a whole in the survey area. These points, which are established using highly precise equipment and methods, are called *control points*. The geometry composed of control points called *control networks*. The professional works of establishing control points and making the necessary measurements and calculations are called *control surveys*. For example, in Fig. 1-5, A, B, C, D, E and F are selected as control points. Using highly precise instrument, surveyors measure the distance between control points and the horizontal angle between two adjacent edges, and finally calculate the coordinates of each control point. In order to find elevation of points, the height differences between the control points to each other should also be measured. If the elevation of the point A is known, the elevation of the other control points can be obtained.

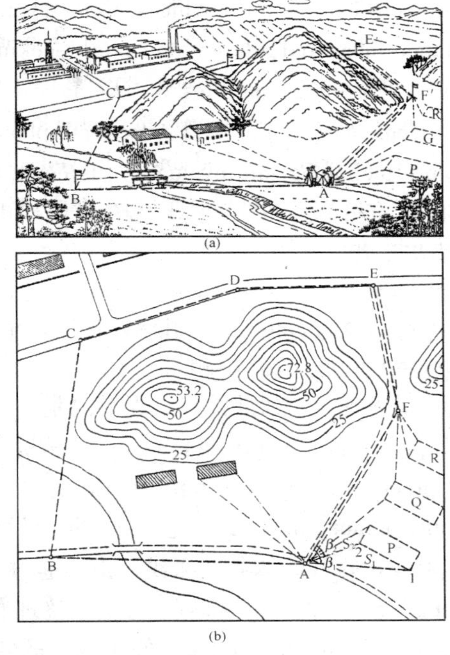

Fig. 1-5 Control survey and detail survey

2. Detail surveys

The purpose of detail surveys is to determine the position of characteristics of all features and relief, for example, highways, railroads, bridges, buildings, mountains and rivers etc. These features and reliefs can then be drawn to scale on a map. For instance in Fig. 1-5, to determine the house P, it is necessary to measure the detail points of the house 1, 2 etc. Therefore, measuring horizontal angle β_1 and side length S_1 at point A can determine point 1; similarly, measuring β_2 and S_2 can determine point 2. Finally, these detail points are drawn to scale on a map with polar coordinate method. Each detail points in turn connected to obtain house P.

1.5.2 The executive principle of survey work

It is inevitable that the errors or mistakes occur in the survey work. In order to avoid mistakes and to reduce errors, the following executive principles must be taken:

(1) *Repeated measurements* Many unknown quantities must be observed several times to avoid the occurrence of mistakes and improve the accuracy. For instance, measuring an angle or a distance several times and computing the average.

(2) *Redundant observations* The redundant observations are more observations than are absolutely necessary. It is also called the extra measurements. For example, suppose that we are interested in the shape of a plane triangle. Measuring two angles of the triangle, the shape of the triangle will be uniquely determined. However, measuring all three angles of the triangle will introduce one superfluous measurement. In data processing, least-squares adjustment is applied to solve some unknown parameters. The redundant observations are taken both to check measurements

and to improve the accuracy of the final results.

(3) *Step by step checked* As everyone knows, every project in the survey work involves several steps. Only after the first step have finished can the beginning of the second step be allowed. Every measurement must be immediately checked. The next step should never be allowed until the first step results have been checked and achieved all requirements. This is called as step by step checked.

Survey work includes a large number of field works, it is very important to follow the executive principle described above.

单词与词组

organize['ɔ:gənaiz] *vt.* 组织，编组，筹备，成立
· control survey 控制测量
· detail survey 细部测量，碎部测量
high-precision[ˌhaipri 'siʒən] *n.* 高精度
low-precision[ˌləupri 'siʒən] *n.* 低精度
layout['leiaut] *n.* 形式，轮廓，外形，设计图，草图，线路图，设备布置图，规划图
adopt[ə'dɔpt] *vt.* 采用，采纳，正式通过，选定
arrange[ə'reinʤ] *vt.* 排列，分类，整理商定，约定，准备，安排
implement['implimənt] *vt.* 使生效，贯彻，执行；*n.* 工具，器具，用具
· carry out 贯彻，执行，实施，落实
avoid[ə 'vɔid] *vt.* 避免，回避
accumulation[əˌkju:mjə 'leiʃən] *n.* 积累，积攒，积聚，堆积
consistency[kən 'sistənsi] *n.* 一致性，连贯性
efficiency[i 'fiʃənsi] *n.* 效率，效能，实力，生产能力，功率
multiply['mʌltiplai] *v.* 增殖，繁殖，增加，做乘法
bear[bɛə] *v.* 承担，负担，具有，怀有
· have a bearing on 关系到……，影响到……
adjacent[ə 'ʤeisənt] *a.* 接近的，附近的，毗连的，相邻的
edge[eʤ] *n.* 边，棱，边缘，端 边界，界线
characteristics[ˌkæriktə 'ristiks] *n.* 特性，特征，特色
relief[ri 'li:f] *n.* 地貌，地形，地势，浮雕
executive[ig 'zekjutiv] *a.* 实施的，实行的，执行的
inevitable[in 'evitəbl] *a.* 不可避免的，无法避免的，必然发生的
error['erə] *n.* 误差
mistake[mis 'teik] *vt.*，*vi.*（-took[-'tuk]；-taken[-'teikən]）弄错，估计错；*n.* 错误，过失，事故
redundant[ri 'dʌndənt] *a.* 多余的，过剩的，累赘的，冗长的，重复的，丰富的，众多的，备用的
superfluous[sju: 'pə:fluəs] *a.* 过多的，多余的
· least-squares adjustment 最小二乘法平差
· step by step 逐步地，一步一步地

Chapter 2 Error Theories

2.1 Errors in measurement

2.1.1 Definition, sources and types of errors

All measurements, no matter how carefully executed, will contain error, which is the difference between a measured value for any quantity and its true value. Therefore, an error can be defined as

$$\text{error} = \text{measured value} - \text{true value} \quad (2\text{-}1)$$

And a correction, which is the negative of the error, can be defined as

$$\text{correction} = \text{true value} - \text{measured value} \quad (2\text{-}2)$$

Because the true value of a measurement is never known, and the exact sizes of the errors present are always unknown. Even with the most sophisticated equipment, a measurement is only an estimate of the true size of a quantity. This is because the instruments, as well as the people using them are imperfect, and because changes in surrounding environment conditions cannot be fully predicted. However, measurements can approach their true values more closely as better equipment is developed, environmental conditions improve and observer ability increases, but they can never be exact. The sources of errors fall into three broad categories which are described as follows:

(1) Instrumental Errors. These errors are caused by imperfections in instrument construction or adjustment of the surveying instrument or the movement of individual parts.

(2) Natural Errors. These errors are caused by variation in the surrounding environment conditions, such as atmospheric pressure, temperatures, wind, refraction, gravitational fields, and magnetic fields, etc.

(3) Personal Errors. These errors arise due to limitations in human senses of sight, touch, and hearing, such as the ability to read a micrometer or to center a level bubble, etc.

*1 [The different types of error that can occur in surveying and other measurements are follows.]

Gross errors are often called mistakes or blunders, and they are usually much larger than the other categories of error. On construction and other sites, mistakes are often made by inexperienced engineers and surveyors who are unfamiliar with the equipment and methods they are using. As a result, gross errors are due to inexperience and also carelessness, and many examples can be given of these. Common mistakes include reading a levelling staff or tape graduation incorrectly or writing the wrong value in a field book by transposing numbers (for example 28. 342 is written as 28. 432). Failure to detect a gross error in a survey or in a setting out procedure can lead to serious problems on site.

Systematic errors are those which follow some mathematical law, and they will have the same magnitude and sign in a series of measurements that are repeated under the same conditions. In other words, if the instrument, observer and surroundings do not change, any systematic errors present in a measurement will not change. If the measuring conditions change, the size of the systematic errors will change accordingly. If an appropriate mathematical model can be derived for a systematic error, it can be eliminated from a measurement by using corrections. For example, the effects of a number of different factors in steel taping, such as temperature and tension, can be removed by calculation using simple formulae (each formula is a mathematical model in this case).

*2 [Another method of removing systematic errors is to calibrate the observing equipment and to quantify the error, allowing corrections to be made to further observations.] It is often necessary to perform an electronic calibration on a total station to measure a range of instrumental systematic errors that are present in the instrument. Calibration values are then applied automatically by the total station to any subsequent measurements taken with it. A good example of a systematic error in distance measurement with a total station is the prism constant, if this is ignored or applied incorrectly, it is an error that will be present in all readings.

Observational methods can also be selected to remove the effect of systematic errors; examples of these are to take the mean of face left and face right readings when measuring angles (to remove horizontal and vertical collimation errors from angles) and to keep the length of back sights and fore sights equal when leveling (to remove the collimation error in the level from height differences).

Random Errors (also know as *accident errors*) are those errors that remain after elimination of blunders and systematic errors. They are caused by limitations of the human senses, by imperfections of the measuring instruments, and by uncontrollable changes of the environment. Random errors are generally very small, but they can never be avoided in measurements. They must be dealt with according to the laws of probability.

单词与词组

equipment [iˈkwipmənt] *n.* 设备，配件，器材，仪器
estimate [ˈestimeit] *v.* 估计，预算，估价
quantity [ˈkwɔntiti] *n.* 量，数量，参量，程度
imperfect [imˈpəːfikt] *a.* 不完美的，有缺陷的
surrounding [səˈraundiŋ] *n.* 周围环境；*a.* 周围的，附近的
predict [priˈdikt] *v.* 预言，预测，预示
approach [əˈprəutʃ] *v.* 靠近，接近，临近
exact[igˈzækt] *a.* 准确的，确切的，精确的
source[sɔːs] *n.* 来源，源头，原因
broad [brɔːd] *a.* 宽的，广阔的，广泛的，显著的，主要的，粗略性的
category [ˈkætigəri] *n.* 种类，类别
variation [ˌɛəvriˈeiʃənl] *n.* 变化，改变，变量上
atmospheric [ˌætməˈsferik] *a.* 大气的，大气层的，大气引起的
gravitational [ˌgræviˈteiʃənl] *a.* 万有引力的，重力的

refraction [ri'frækʃən] *n.* 折射(程度)，折射作用
personal ['pə:sənl] *a.* 身体的，外貌的，能力的，气质的，关于个人的，涉及个人的
ability[ə'biliti] *n.* 能力，才能
micrometer [mai 'krɔmitə] *n.* 测微计，千分尺
level ['levl] *n.* 水平，水平线，水平面，水准仪
bubble ['bʌbl] *n.* 水泡，气泡
gross [grəus] *a.* 粗俗的，粗野的
· gross error 粗差，过失误差，错误，谬误，缺陷
blunder['blʌndə] *n.* (因无知、粗心等造成的)错误；*vi.* 犯错误
construction[kən 'strʌkʃən] *n.* 建造，建设；建筑业
site[sait] *n.* 位置，场所，地点
inexperience [ˌiniks 'piəriəns] *n.* 缺乏经验，不成熟
carelessness['keəlisnis] *n.* 粗心大意
graduation [ˌgrædʒu: 'eiʃən] *n.* 刻度，分度，校准，校正
transposing[træns 'pəusiŋ] *n.* 置换，更换
detect[di 'tekt] *v.* 发现；发觉，查明
vital['vaitəl] *a.* 极重要的，必不可少的
highlight ['hailait] *v.* 强调，突出，增强亮度
systematic [ˌsistə'mætik] *a.* 有系统的，有规则的
· systematic error 系统误差
law [lɔ:] *n.* 法，法律，法规
magnitude ['mægnitju:d] *n.* 大小，积，量，长(度)，尺寸
sign [sain] *n.* 标记，符号
observer [əb 'zə:və] *n.* 观察者，观察员；遵守者
surrounding [sə'raundiŋ] *n.* 环境，周围的事物；*a.* 周围的，附近的
present ['prezənt] *a.* 现在的，目前的，存在的，含有的
accordingly [ə'kɔ:diŋli] *adv.* 照着，相应地
· in addition 另外
appropriate [ə 'prəupriit] *a.* 适当的，恰当的
accumulate[ə'kju:mjuleit] *v.* 积累，存储，蓄积(财产等)，堆积
present [pri 'zənt] *v.* 显示，呈现出，引起，给予
derive[di'raiv] *v.* 获得，导出(from)，起源于，出自，推论，推究(from)
eliminate [i'limineit] *v.* 消除，排除
correction [kə'rekʃən] *v.* 修正，改正，校正，校准，订正错误
effect[i 'fekt] *n.* 结果，效果，影响
factor['fæktə] *n.* 因素，要素
temperature ['tempəritʃə] *n.* 温度，气温
tension ['tenʃən] *n.* 张力，拉力
formula['fɔ:mjulə] *n.* (复数为 formulae) 公式，方程式

calibrate [ˈkæləˌbreit] *v.* 校准(正)，检验，标定
quantify[ˈkwɔntifai] *v.* 确定……的数量
observation [ˌɔbzəːˈveiʃən] *n.* 注意，观察，言论，评论，意见
· total station 全站仪
measure [ˈmeʒə] *v.* 测量，估量，判断，比较
traverse [ˈtrævəs] *n.* 导线测量
alternative [ɔːl ˈtəːnətiv] *a.* 两者择一的，供替代的
specify[ˈspesifai]*v.* 规定，指定，确定；详细说明，具体说明
achieve [əˈtʃiːv] *v.* 实现，达到，完成
stated[ˈsteitid] *a.* 定期的，规定的，一定的
attempt [əˈtempt] *v. n.* 尝试，试图，企图
degree [di ˈgriː] *n.* 度数，度，程度，学位，地位，身份，次数
reliability [riˌlaiə ˈbiləti] *n.* 可靠性，安全性，可信赖性，确实(性)
unreliable[ˈʌnri ˈlaiəbəl] *a.* 不可靠的，不可信任的，不稳定的
knowing [ˈnəuiŋ] *a.* 知道的，有见识的，故意的
whereas [hwɛər ˈæz] *conj.* 但是，而
unlikely [ʌn ˈlaikliː] *a.* 未必的，不太可能的
subsequent [ˈsʌbsikwənt] *a.* 随后的，继……之后的
ignore [igˈnɔː] *v.* 不顾，不理，忽视
result[ri ˈzʌlt] *v.* 由……而造成(from)，结果，致使，导致(in)
collimation [kɔli ˈmeiʃən] *n.* 瞄准，视准

注释

[1] The different types of error that can occur in surveying and other measurements are follows. 此句中主语是 The different types of error，谓语是 are follows，其中 that 引导的定语从句是修饰 error，因此译为“在测量和其他量测中都会产生各种误差，其误差分类如下。”

[2] Another method of removing systematic errors is to calibrate the observing equipment and to quantify the error, allowing corrections to be made to further observations. 此句难点在于后面的分词短语的部分，它作为句子的补充说明，译为“允许对随后的观测加改正值”，所以全句译为：“消除系统误差的另一种方法是，检校测量仪器并确定误差大小，允许对随后的观测施加改正值”。

2.1.2 Characteristics of random errors

Fig. 2-1 represents the distribution of true errors Δ_i which were obtained from 160 observations of the same angle. According to their size, the Δ_i are, sorted into groups of 1″ as shown on the abscissa. The height of the rectangles plotted for each group is proportional to the number of errors within the respective group. *1[The resulting histogram shows that the frequency with which an error Δ occurs is a function of its magnitude.]

From above it ean be seen that random errors of a large number of repeated measurements of a quantity conform to *normal* or *Gaussian distribution*. *2[The plot of error sizes versus probabilities would approach a smooth curve of the characteristic bell-shape (see Fig. 2-1)]. This curve is known as the normal error distribution curve.

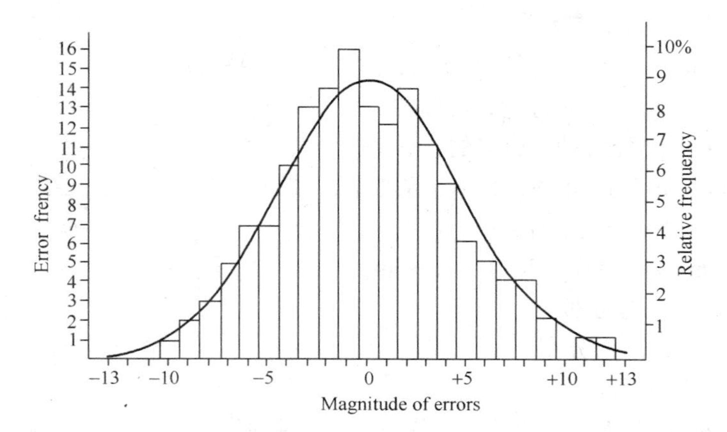

Fig. 2-1 The normal error distribution curve

The equation of the normal error distribution curve is

$$y = \frac{1}{\sigma\sqrt{2\pi}} e^{-\frac{\Delta^2}{2\sigma^2}} \tag{2-3}$$

Where y is the ordinate value of a point on the curve, i. e. probability of the occurrence of Δ, σ is referred to as the standard deviation or standard error of a group of measurement, e is the base of natural logarithms (2.718). The square of the standard deviation is called the variance (σ^2).

The general laws of probability of random errors are as follows:

(1) A random error will not exceed a certain amount.

(2) Positive and negative random errors may occur at the same frequency. Because the normal distribution curve is symmetrical, positive and negative random errors of the same size are equally probable and happen with equal frequency.

(3) *3[Errors that are small in magnitude are more likely to occur than those that are larger in magnitude.]

(4) The arithmetic mean of random errors tends to zero as the sample size tends to infinite.

In surveying, the true error (Δ_i) can be written by the observed value (l_i) minus the true value (X), i. e. $\Delta_i = l_i - X$. The true error removed systematic error is random error. When the number of observations is very large, we have

$$\lim_{n\to\infty} \frac{\Delta_1 + \Delta_2 + \cdots + \Delta_n}{n} = \lim_{n\to\infty} \frac{[\Delta]}{n} = 0$$

The variance (σ^2) is the mean of the sum of the squares of random errors, that is

$$\sigma^2 = \lim_{n\to\infty} \frac{\Delta_1^2 + \Delta_2^2 + \cdots + \Delta_n^2}{n} = \lim_{n\to\infty} \frac{[\Delta^2]}{n} = 0$$

Hence standard deviation σ:

$$\sigma = \lim_{n\to\infty} \sqrt{\frac{[\Delta\Delta]}{n}} \tag{2-4}$$

单词与词组

random ['rændəm] *a.* 任意的，无规则的，随机的

· random errors 随机误差，偶然误差

characteristic [kæriktə 'ristik] *n.* 特点，特性，性能； *a.* 特有的

symmetrical[si 'metrikəl] *a.* 对称的，匀称的

constant ['kɔnstənt] *a.* 经常的，稳定的，不变的

probability[ˌprɔbə'biliti] *n.* 几率，概率，或然率，(最)可能发生的事或结果

occur [ə'kə:] *v.* 发生；举行；存在

conform [kən 'fɔ:m] *vi.* 遵守，符合(to，with)

· normal distribution 正态分布

· Gaussian distribution 高斯分布

plot [plɔt] *n.* 标绘图，图表

versus ['və:səs] *prep.* 相对，相比，……与……的关系曲线

ordinate ['ɔ:dinit] *n.* 纵坐标，纵距

· standard deviation 标准偏差，标准差

· natural logarithm 自然对数

variance ['veəri:əns] *n.* 差异，变异，方差

demonstrate['demənstreit] *v.* 说明，演示

positive ['pɔzitiv] *a.* 正的

negative['negətiv] *a.* 负的

frequency ['fri:kwənsi] *n.* 频率

histogram['histəugræm]*n.* 矩形图，直方图

注释

[1] 此句是主从复合句，shows 要求的宾语是宾语从句(that…magnitude)，宾语从句中的 which 指代 frequency，宾语从句含义是“误差 Δ 出现的概率是误差大小的函数”。resulting 意为“作为结果的，因而发生的”，为使上下文连贯译为“可以看出”，全句译为：可以看出，形成的直方图显示误差 Δ 出现的概率是误差大小的函数。

[2] 此句中“the plot of error sizes versus probabilities”意为误差大小对应概率的关系图”作为句中主语。

[3] 句中 those 就是指句首的第一个词 errors，所以全句译为：数量级小的误差出现机率大于数量级大的误差。

2.1.3 Most probable value (MPV)

Most probable value (MPV) is the closest approximation to the true value that can be achieved from a set of measurements. If n times measurements with same precisions have been made to a quantity, the results are l_1, l_2, ……, l_n, and relevant true errors Δ_1, Δ_2, ……, Δ_n. Let the true value of the quantity is X, we have

$$\left.\begin{aligned}\Delta_1 &= l_1 - X \\ \Delta_2 &= l_2 - X \\ &\vdots \\ \Delta_n &= l_n - X\end{aligned}\right\} \tag{2-5}$$

Equation (2-5) are added to obtain

$$[\Delta] = [l] - nX \tag{2-6}$$

Where the symbol [ ] means "summation", $[\Delta] = \Delta_1 + \Delta_2 + \cdots + \Delta_n$, $[l] = l_1 + l_2 + \cdots + l_n$, n the number of observations.

Dividing both sides of equation (2-6) by n gives

$$\frac{[\Delta]}{n} = \frac{[l]}{n} - X \tag{2-7}$$

Use the symbol x to indicate the arithmetic mean, that is

$$x = \frac{[l]}{n}$$

Substituting the equation above into equation (2-7) gives following equation

$$x = X + \frac{[\Delta]}{n}$$

According the fourth characteristics of random errors, when the number of observation n tend infinite, we have

$$\lim_{n \to \infty} \frac{[\Delta]}{n} = 0$$

$$\therefore \qquad \lim_{n \to \infty} x = X \tag{2-8}$$

Therefore, it is concluded that the arithmetic mean is the most probable value (i. e. MPV) of measurements with same precisions. The MPV is the closest approximation to the true value.

单词与词组

· most probable value (MPV) 最或然值

symbol [ˈsimbəl] *n.* 记号，符号，象征，表征

summation [səˈmeiʃən] *n.* 总数，总和，加法，求和

substitute [ˈsʌbstitjuːt] *v.* 代替，替换，代入

divide [diˈvaid] *vt.* 除，除尽

· the arithmetic mean 算术平均值

2.2 Indices of precision

2.2.1 Standard deviation

The formula of standard deviation have mentioned by equation (2-4) in Section 2.1.3, that is

$$\sigma = \lim_{n \to \infty} \sqrt{\frac{[\Delta\Delta]}{n}}$$

Where σ is the standard deviation of a very large sample, and n is the very large sample size. In practice, however, the number of observation n is always finite number. For a group of observation values with same precisions, their standard deviation (also called root mean square error, m) can be written as

$$m = \pm\sqrt{\frac{[\Delta\Delta]}{n}} \tag{2-9}$$

The true value X replace by the arithmetic mean x in equation (2-5), we have

$$\left.\begin{aligned} v_1 &= l_1 - x \\ v_2 &= l_2 - x \\ &\vdots \\ v_n &= l_n - x \end{aligned}\right\} \tag{2-10}$$

Where v_i is the residuals of each measurement, or the deviation of each measurement from the mean x.

Subtracting equation (2-10) from equation (2-5) yields

$$\left.\begin{aligned} \Delta_1 &= v_1 + (x - X) \\ \Delta_2 &= v_2 + (x - X) \\ &\vdots \\ \Delta_n &= v_n + (x - X) \end{aligned}\right\} \tag{2-11}$$

Adding both sides of equation (2-11) respectively and considering $[v] = 0$ gives

$$[\Delta] = n(x - X)$$

$$x - X = \frac{[\Delta]}{n}$$

Squaring both sides of equation (2-11) and then adding gives

$$[\Delta\Delta] = [vv] + 2[v](x - X) + n(x - X)^2 \tag{2-12}$$

Where

$$(x - X)^2 = \frac{[\Delta]^2}{n^2} = \frac{\Delta_1^2 + \Delta_2^2 + \cdots + \Delta_n^2}{n^2} + \frac{2(\Delta_1\Delta_2 + \Delta_1\Delta_3 + \cdots + \Delta_{n-1}\Delta_n)}{n^2}$$

According to the fourth characteristics of random errors, the second term in equation above tend to zero, that is

$$\lim_{n\to\infty}\frac{2(\Delta_1\Delta_2 + \Delta_1\Delta_3 + \cdots + \Delta_{n-1}\Delta_n)}{n^2} = 0$$

Hence

$$(x - X)^2 = \frac{\Delta_1^2 + \Delta_2^2 + \cdots + \Delta_n^2}{n^2}$$

Substituting the equation above into equation (2-12) gives

$$[\Delta\Delta] = [vv] + n(x - X)^2 = [vv] + \frac{[\Delta\Delta]}{n}$$

$$\frac{[\Delta\Delta]}{n} = \frac{[vv]}{n-1} \tag{2-13}$$

Substituting equation (2-13) into equation (2-9) gives

$$m = \pm\sqrt{\frac{[vv]}{n-1}} \tag{2-14}$$

Formula (2-14) is called *Bessel Formula*, used to calculate the standard deviation of a group of observation values with same precision when its true value is unknown.

2.2.2 Relative precision

For many surveying measurements, the term relative precision (also called relative error) is usually used, and this is the ratio of the precision of a measurement to the measurement itself. For example, if the error of the measurement of a distance D is m_D (i. e, standard error), the relative precision expressed as 1 in $\frac{D}{m_D}$ (say 1 in 5000). In other words, the relative precision expressed as a fraction whose numerator is unity and denominator is rounded to the closet 100 units. For example, $D = 100\text{m}$, $m_D = \pm 0.019\text{m}$, the relative precisions expressed as

$$K = \frac{|\pm m_D|}{D} = \frac{1}{\frac{D}{|\pm m_D|}} = \frac{1}{5263} \approx \frac{1}{5300}$$

This is the method used to define precisions in taping and to define the precision (or accuracy) of a traverse. An alternative to this is to quote relative precision in parts per million or *ppm* (that is 1 in 1,000,000). This is also equivalent to a precision of 1mm per km, and both of these are used for distance measurement with total stations. *1 [The relative precision of a measurement is often calculated in surveying as soon as its precision is known, or it may be specified before starting a survey so that proper equipment and methods can be selected to achieve the stated relative precision.]

2.2.3 Confidence limits and allowable errors

Some of the useful features of normal distribution are:

(1) The probabitity of the occurence of Δ equals to zero as Δ goes to $\pm\infty$.

(2) *2 [The probability that Δ falls in interval between Δ_1 and Δ_2 is the area under the curve bounded by $\Delta = \Delta_1$ and $\Delta = \Delta_2$]. The shaded area in Fig. 2-2 represents the probability that Δ falls within $\pm\sigma$ of the mean μ (i. e. zero). The probabilities for deviation from the mean of the first three integral multiples of σ are

$$P(-\sigma < \Delta < +\sigma) = 0.6827$$
$$P(-2\sigma < \Delta < +2\sigma) = 0.9546$$

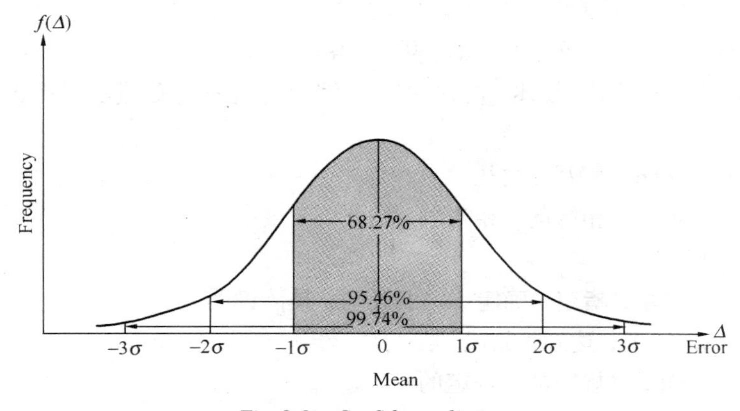

Fig. 2-2 Confidence limits

$$P(-3\sigma < \Delta < +3\sigma) = 0.9974$$

From some equations above can know the probability that random errors Δ falls in interval between -2σ and $+2\sigma$ is 95.5%, the probability that random errors Δ falls out side in this interval is only 5%; the probability that Δ falls in interval between -3σ and $+3\sigma$ is 99.7%, whereas the probability that Δ falls out side is only 0.3%.

Therefore, in practice surveying, allowable errors are specified two or three time of σ, that is

$$\left.\begin{array}{l} |\Delta_{\text{allowable}}| = 2|\sigma| \\ |\Delta_{\text{allowable}}| = 3|\sigma| \end{array}\right\} \tag{2-15}$$

If an error in practice is greater than $|\Delta_{\text{allowable}}|$, the survey data should be rejected or re-measured.

单词与词组

deviation [ˌdi:vi'eiʃən] *n.* 背离(from),偏向,偏差,(统计上的)误差

· standard deviation 标准偏差

· root mean square error 均方根误差,中误差

replace [ri'pleis] *vt.* 代替,取代(by, with);接替,替换

residual [ri'zidʒu:l] *a.* 剩余的,残余的[编者注]the residuals = residual errors(残余误差)

subtract[səb'trækt] *vt.* 减去,扣除,减少

consider [kən'sidə] *vt.* 考虑,照顾,顾及,认为

yield[ji:ld] *vt.* 产生,生出,出产

square [skwɛə] *n.* 平方,自乘,正方形,方形物;*v.* 使成方形,与……一致,自乘

substitute ['sʌbstitju:t] *v.* 代入,替换;*n.* 代用品,代入

· relative precision 相对精度

ratio ['reiʃiəu] *n.* (*pl.* ratios) 比率,比例,变换系数

fraction ['frækʃən] *n.* 分数

numerator['nju:məreitə] *n.* (分数的)分子,计算者,计数器

denominator[di'nɔməˌneitə] *n.* 分母,命名者

traverse ['trævəs] *n.* 导线测量

alternative [ɔ:l'tə:nətiv] *a.* 两者择一的,供替代的

quote [kwəut] *vt.* 摘引,引用,复述,提到,举出(例子),把……放进引号内,用引号把……括起来

· ppm = "parts per million" 百万分之几

equivalent[i'kwivələnt] *a.* 相等的,相当的

· total station 全站仪

specify['spesifai] *v.* 规定,指定,确定,详细说明,具体说明

achieve [ə'tʃi:v] *v.* 实现,达到,完成

stated['steitid] *a.* 定期的,规定的,一定的

interval['intəvəl] *n.* 间隔,空隙,区间

· allowable error 容许误差

approach [ə ˈprəutʃ] *vt.* 向……靠近，接近，趋于

bound[baund] *v.* 给……划界，限制

注释

[1] The relative precision of a measurement is often calculated in surveying as soon as its precision is known, or it may be specified before starting a survey so that proper equipment and methods can be selected to achieve the stated relative precision. 句中 it 是指代前面的 its precision, its precision is known 译为已知观测值的精度，句中 or 是"即"的意思。全句译为"在测量中，一旦已知观测值的精度，则观测值的相对精度就可计算，即开始测量之前预先规定观测值的精度，从而选择适当的仪器和方法，以便达到规定的相对精度"。

[2] 句中"the area under the curve bounded by $\Delta=\Delta_1$ and $\Delta=\Delta_2$"译为"在误差 $\Delta=\Delta_1$ 与 $\Delta=\Delta_2$ 限定范围曲线下的面积"。全句译为"误差落在 Δ_1 与 Δ_2 区间的概率就是在误差 $\Delta=\Delta_1$ 与 $\Delta=\Delta_2$ 限定范围的曲线下面的面积"。

2.3　Definition of precision and accuracy

The terms precision and accuracy are frequently used in engineering surveying both by manufacturers when giving specifications for their equipment and on site by surveyors and engineers to describe how good their fieldwork is.

Precision is the degree of refinement with which a given quantity is measured. In other words, it is the closeness of one measurement to another. If a quantity is measured several times and the values obtained are very close to each other, the precision is said to be high.

Accuracy refers to the degree of perfection obtained in measurements. It denotes how close a given measurement is to the true value of the quantity. The further a measurement is from its true value, the less accurate it is.

Precision represents the repeatability of a measurement and is concerned only with random errors. * [Good precision is obtained from a set of observations that are closely grouped together with small deviations from the sample mean $\bar{x}$ or MPV and have the normal distribution shown in Fig. 2-3a.] On the other hand, a set of observations which are widely spread and have poor precision will have the normal distribution shown in Fig. 2-3b.

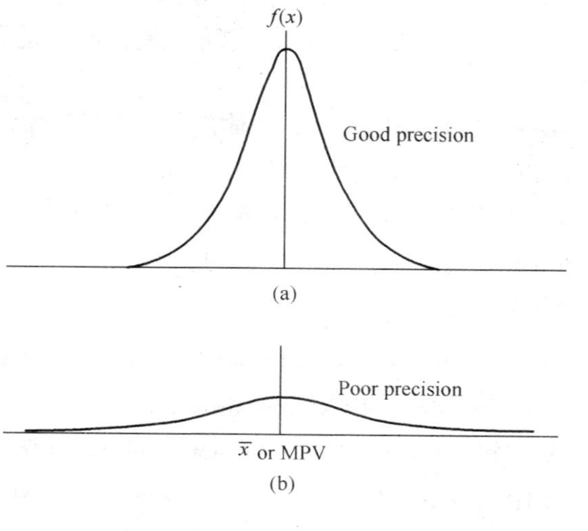

Fig. 2-3　Precision

In contrast, accuracy is considered to be an overall estimate of the errors present in measurements, including systematic effects. For a set of measurements to be considered accurate, the MPV or sample mean must have a value close to the true value, as shown in Fig. 2-4a. It is quite

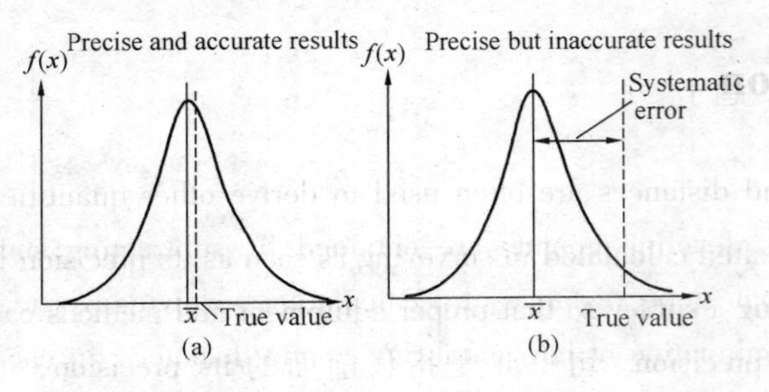

Fig. 2-4 Precision and accuracy

possible for a set of results to be precise but inaccurate, as in Fig. 2-4b, where the difference between the true value and the MPV is large and probably caused by one or more systematic errors. Since accuracy and precision are the same if all systematic errors are removed, In this case, precision can be regarded as an index of accuracy.

单词与词组

precision [pri ˈsiʒən] *n.* 精度，精密(性)
accuracy[ˈækjurəsi] *n.* 正确性，精确性，准确度
· good precision 精度好，较高精度
· poor precision 精度差，较低精度
manufacturer[ˌmænju ˈfæktʃərə] *n.* 制造商，制造厂
specification [ˌspesifi ˈkeiʃən] *n.* 说明书，详细的计划书
perfection [pəˈfekʃən] *n.* 完善，完美，完整性，极度
refinement [ri ˈfainmənt] *n.* 精密的程度，精致，精细
repeatability[ri ˈpi:təbiliti] *n.* 可重复性，反复性，再现性
· group together 聚集，把……归于一类
deviation[ˌdi:vi: ˈeiʃən] *n.* 偏离，背离(from)，偏向，偏差
· on the other hand 另一方面
widely[ˈwaidli] *a.* 广泛地，普遍地
spread [spred] *n.* 传播，散布，使扩大蔓延开
· in contrast 相比之下
overall [ˈəuvərɔ:l] *a.* 总体的，全面的，综合的
estimate [ˈestimeit] *v.* 估计，估价
accurate [ˈækjurit] *a.* 精确的，准确的
precise [pri ˈsais] *a.* 精密的，严谨的

注释

* Good precision is obtained from a set of observations that are closely grouped together with small deviations from the sample mean $\bar{x}$ or MPV and have the normal distribution shown in Fig. 2-3a. 这是复合句，主句是 Good precision is obtained from a set of observations，从 that 至句末是由 that 引导的限制性定语从句，that 指代 observations。从句含两个并列的简单句，即 that are closely grouped together with small deviations from the sample mean $\bar{x}$ or MPV 和 and have the normal distribution shown in Fig. 2-3a。第一从句中的第一简单句译为“这组观测值十分聚集与样本平均值 $\bar{x}$ 或最或然值 MPV 偏差小”；MPV 是 most probable value(最或然值)的缩写。第二简单句译为“这组观测值服从正态分布如图 2-3a 所示。全句译为：“精度好的一系列观测值紧密地聚集在一起，它与样本平均值 $\bar{x}$ 或最或然值 MPV 偏差小，并且具有正态分布，如图 2-3a 所示”。

2.4 The law of error propagation

Surveying measurements such as angles and distances are often used to derive other quantities using mathematical relationships: for instance, leveling heights are obtained by subtracting staff readings, horizontal distances are obtained from slope distances by calculation involving vertical angles, and coordinates are obtained from a combination of horizontal angles and distances. In each of these cases, the original measurements will be randomly distributed and will have errors, and it follows that any quantities derived from them will also have errors.

Let u be some function of independent measured quantities x, y, z, $\cdots q$; that is,

$$u = f(x, y, z, \cdots q) \tag{2-16}$$

If each independent variable (i. e. is not correlated with one another) is allowed to change by a small amount dx, dy, $\cdots dq$, then the quantity u will change by an amount du given by the expression from the calculus.

$$du = \frac{\partial u}{\partial x}dx + \frac{\partial u}{\partial y}dy + \cdots + \frac{\partial u}{\partial q}dq$$

* [Where $\frac{\partial u}{\partial x}$ denotes the partial derivative of u with respect to x, and likewise for the other variables.] Applying this to a set of measurements, and since the residuals are small quantities, let $\Delta x_i = dx_i$, $\Delta y_i = dy_i$, $\Delta z_i = dz_i \cdots \Delta q_i = dq_i$. Then, letting $\Delta u_i = du_i$,

$$\left.\begin{aligned}
\Delta u_1 &= \frac{\partial u}{\partial x}\Delta x_1 + \frac{\partial u}{\partial y}\Delta y_1 + \frac{\partial u}{\partial z}\Delta z_1 + \cdots + \frac{\partial u}{\partial q}\Delta q_1 \\
\Delta u_2 &= \frac{\partial u}{\partial x}\Delta x_2 + \frac{\partial u}{\partial y}\Delta y_2 + \frac{\partial u}{\partial z}\Delta z_2 + \cdots + \frac{\partial u}{\partial q}\Delta q_2 \\
&\vdots \\
\Delta u_n &= \frac{\partial u}{\partial x}\Delta x_n + \frac{\partial u}{\partial y}\Delta y_n + \frac{\partial u}{\partial z}\Delta z_n + \cdots + \frac{\partial u}{\partial q}\Delta q_n
\end{aligned}\right\} \tag{2-17}$$

Squaring both sides of equation (2-17) and adding gives

$$\begin{aligned}
\Delta u_1^2 &= \left(\frac{\partial u}{\partial x}\right)^2 \Delta x_1^2 + 2\left(\frac{\partial u}{\partial x}\right)\left(\frac{\partial u}{\partial y}\right)\Delta x_1 \Delta y_1 + \cdots + \left(\frac{\partial u}{\partial y}\right)^2 \Delta y_1^2 + \cdots + \left(\frac{\partial u}{\partial q}\right)^2 \Delta q_1^2 \\
\Delta u_2^2 &= \left(\frac{\partial u}{\partial x}\right)^2 \Delta x_2^2 + 2\left(\frac{\partial u}{\partial x}\right)\left(\frac{\partial u}{\partial y}\right)\Delta x_2 \Delta y_2 + \cdots + \left(\frac{\partial u}{\partial y}\right)^2 \Delta y_2^2 + \cdots + \left(\frac{\partial u}{\partial q}\right)^2 \Delta q_2^2 \\
&\vdots \\
\Delta u_n^2 &= \left(\frac{\partial u}{\partial x}\right)^2 \Delta x_n^2 + 2\left(\frac{\partial u}{\partial x}\right)\left(\frac{\partial u}{\partial y}\right)\Delta x_n \Delta y_n + \cdots + \left(\frac{\partial u}{\partial y}\right)^2 \Delta y_n^2 + \cdots + \left(\frac{\partial u}{\partial q}\right)^2 \Delta q_n^2
\end{aligned}$$

$$\sum \Delta u^2 = \left(\frac{\partial u}{\partial x}\right)^2 \sum \Delta x^2 + 2\left(\frac{\partial u}{\partial x}\right)\left(\frac{\partial u}{\partial y}\right)\sum \Delta x \Delta y + \cdots + \left(\frac{\partial u}{\partial y}\right)^2 \sum \Delta y^2 + \cdots + \left(\frac{\partial u}{\partial q}\right)^2 \sum \Delta q^2 \tag{2-18}$$

In which some of the squared and cross-product terms have been omitted for the sake of simplicity. If the measured quantities are independent and are not correlated with one another, the cross products tend toward zero. Dropping these cross products and then dividing both sides of equation (2-18) by (n-1) gives

$$\frac{\sum \Delta u^2}{n-1} = \left(\frac{\partial u}{\partial x}\right)^2 \frac{\sum \Delta x^2}{n-1} + \left(\frac{\partial u}{\partial y}\right)^2 \frac{\sum \Delta y^2}{n-1} + \left(\frac{\partial u}{\partial z}\right)^2 \frac{\sum \Delta z^2}{n-1} + \cdots + \left(\frac{\partial u}{\partial q}\right)^2 \frac{\sum \Delta q^2}{n-1}$$

$$m_u^2 = \left(\frac{\partial u}{\partial x}\right)^2 m_x^2 + \left(\frac{\partial u}{\partial y}\right)^2 m_y^2 + \left(\frac{\partial u}{\partial z}\right)^2 m_z^2 + \cdots + \left(\frac{\partial u}{\partial q}\right)^2 m_q^2$$

$$m_u = \sqrt{\left(\frac{\partial u}{\partial x}\right)^2 m_x^2 + \left(\frac{\partial u}{\partial y}\right)^2 m_y^2 + \left(\frac{\partial u}{\partial z}\right)^2 m_z^2 + \cdots + \left(\frac{\partial u}{\partial q}\right)^2 m_q^2} \qquad (2\text{-}19)$$

The general expression for propagation of error through any function expressed by equation (2-19) can then be applied to any special function.

(1) Let $u = x + y + z$, in which x, y and z are three independent measured quantities. Then from equation (2 – 19), We have

$$m_u = \pm\sqrt{m_x^2 + m_y^2 + m_z^2} \qquad (2\text{-}20)$$

(2) Let $u = xy$ in which x and y are two independent measured quantities. Then, from equation (2-19), We have

$$\because \quad \frac{\partial u}{\partial x} = y, \ \frac{\partial u}{\partial y} = x$$

$$\therefore \quad m_u = \pm\sqrt{y^2 m_x^2 + y^2 m_y^2} \qquad (2\text{-}21)$$

(3) Linear function $\qquad u = k_1 x_1 \pm k_2 x_2 \pm \cdots \pm k_n x_n$

where, x_l, x_2, $\cdots x_n$, are independent variables, and k_1, k_2, $\cdots k_n$, are constants. Then, from equation (2-19), We have

$$m_u = \pm\sqrt{k_1^2 m_1^2 + k_2^2 m_2^2 + \cdots + k_n^2 m_n^2} \qquad (2\text{-}22)$$

(4) Standard deviation of arithmetic mean of measured values with same precisions

The formula of arithmetic mean gives by

$$x = \frac{[l]}{n} = \frac{1}{n} l_1 + \frac{1}{n} l_2 + \cdots + \frac{1}{n} l_n$$

Where $\frac{1}{n}$ is a constant, the standard deviations of all independent measured values are m, Substituting these values into equation (2-22) gives following equation:

$$m_x = \pm\sqrt{\left(\frac{1}{n}\right)^2 m^2 + \left(\frac{1}{n}\right)^2 m^2 + \cdots + \left(\frac{1}{n}\right)^2 m^2}$$

Hence

$$m_x = \pm \frac{m}{\sqrt{n}} \qquad (2\text{-}23)$$

From equation above it can be seen that increasing observation time n in the formula (2-23) can decrease standard deviation of an arithmetic mean m_x. However, when the observation times become great enough, the decrease of the standard deviation is very limited.

Generally speaking, the law of propagation of error is used by the following three steps:

(1) Give a function equation: $u = f(x, y, z, \cdots q)$

For example, the dimensions of a rectangular field are measured, the side $a = 800.00 \pm 0.12$m, side $b = 550.00 \pm 0.07$m. What are the area of the field and the standard error of the area?

This example gives a function eguation:

$$A = a \cdot b$$

(2) Total differential of the formula above gives

$$dA = \frac{\partial A}{\partial a}da + \frac{\partial A}{\partial b}db = bda + adb$$

(3) Convert the differential expression of equation above into the expression in standard error, m_A is calculated according to equation (2-19), that is

$$m_A = \pm\sqrt{b^2 m_a^2 + a^2 m_b^2} = \sqrt{550^2 \times 0.12^2 + 800^2 \times 0.07^2} = \pm 86.6\text{m}^2$$

Example 2.1 In levelling, if the standard error of a backsight reading $m_a = \pm 0.002$m, and the standard error of the foresight reading m_b is also ± 0.002m, what is the standard error of the difference in elevation between the two points m_h?

Solution

The difference in elevation between the two points is the backsight reading minus the foresight reading, that is, $h_{AB} = a - b$. The standard error of the difference in elevation is given by

$$m_h = \pm\sqrt{m_a^2 + m_b^2}$$

$$= \pm\sqrt{0.002^2 + 0.002^2} = \pm 0.0028\text{m}$$

Example 2.2 An angle α is measured ten times with the same equipment by the same observer and the following results are obtained: 47°56′38″, 47°56′ 40″, 47°56′ 35″, 47°56′ 33″, 47°56′ 40″, 47°56′ 34″, 47°56′ 42″ 47°56′ 39″, 47°56′ 32″, 47°56′ 37″. Calculate the MPV of a set of measured values and its standard error.

Solution

The MPV is given by

$$\alpha_x = 47°56' + \frac{(38 + 40 + 35 + 33 + 40 + 34 + 42 + 39 + 32 + 37)''}{10}$$

$$= 47°56'37.0''$$

Based on the MPV, the residuals and squares of these are calculated and listed in Table 2-1:

Table 2-1 The tabular calculations for Example 2-2

i	α_i	$v_i = (\alpha_i - \alpha_x)$	v_i^2
1	47°56′38″	+1	1
2	47°56′40″	+3	9
3	47°56′35″	−2	4

Continued

i	α_i	$v_i = (\alpha_i - \alpha_x)$	v_i^2
4	47°56′33″	−4	16
5	47°56′40″	+3	9
6	47°56′34″	−3	9
7	47°56′42″	+5	25
8	47°56′39″	+2	4
9	47°56′32″	−5	25
10	47°56′37″	0	0
	α_x = 47°56′37″	$\sum v_i = 0$	$\sum v_i^2 = 102$

Using these values in Table 2-1, the equation (2-14) gives

$$m = \pm\sqrt{\frac{[vv]}{n-1}} = \pm\sqrt{\frac{102}{10-1}} = \pm 3.4''$$

The standard error of the MPV (m_x) is calculated as follows:

$$m_x = \frac{m}{\sqrt{n}} = \frac{\pm 3.4''}{\sqrt{10}} = \pm 1.1''$$

The MPV of angle α can be written as

$$\alpha_x = 47°56'37'' \pm 1.1''$$

Example 2.3 The two sides of the ground floor of a building were measured as $x = 32.00 \pm 0.01$m and $y = 18.00 \pm 0.02$m. Calculate the area of the ground floor of the building and its standard error.

Solution

The area of the ground floor of the building is given by $A = x \cdot y = 32.00 \times 18.00 = 576.00\text{m}^2$

Using the law of propagation of deviation error

$$m_A = \sqrt{\left(\frac{\partial A}{\partial x}\right)^2 m_x^2 + \left(\frac{\partial A}{\partial y}\right)^2 m_y^2}$$

$$= \sqrt{18.00^2 \times 0.01^2 + 32.00^2 \times 0.02^2}$$

$$= \pm 0.66\text{m}^2$$

单词与词组

propagation [ˌprɔpəˈgeiʃən] *n.* 传播，传导，普及

original[əˈridnəl] *a.* 最初的，原始的，固有的

likewise [ˈlaikˌwaiz] *adv.* 同样地，照样地，另外

cross-product [krɔs - ˈprɔdəkt] *n.* 交叉乘积

omit[əu ˈmit] *vt.* 省略，删除

· for the sake of 为了

simplicity [sim ˈplisiti] *n.* 简单，简易，简化

tend [tend] *v.* 往，朝向

drop[drɔp] *v.* 放弃，丢掉

· linear function 线性函数
manifest['mænifest] vt. 清楚表示，显露
decrease[di:'kri:s] v. 减小(少)，缩短
· less and less 愈来愈少(小)
· total differential 全微分
partial ['pɑ:ʃəl] a. 部分的，局部的，单独的
derivative[di 'rivətiv] a. 派生的，导出的
· partial derivative 偏导数，偏微商
respect [ris 'pekt] v. 考虑，关心，顾虑
· with respect to 关于，就……而论
backsight[bæksait] n. 后视
foresight['fɔ:sait] n. 前视
elevation[ˌelə 'veiʃən] n. 高地，高程，高度，海拔
minus ['mainəs] a. 负的，不利的；n. 减号，负号；prep. 减去，(表示否定)没有，缺少，差
area['ɛəriə] n. 面积，区域，范围，领域

注释

* "where $\frac{\partial u}{\partial x}$ denotes the partial derivative of u with respect to x, and likewise for the other variables."此句的前半句：式中$\frac{\partial u}{\partial x}$表示$u$对$x$的偏导数，后半句 and likewise for the other variables 意为"而其他变量也是这样"，即$\frac{\partial u}{\partial y}$表示$u$对$y$的偏导数，$\frac{\partial u}{\partial z}$表示$u$对$z$的偏导数等等。

2.5 Weight

2.5.1 Concept of weight

For a lot of survey works may be taken under different observing conditions with different equipment and observers. In this case, the reliability or different sets of measurements varies one to the other. Variations in precision are indicated by assigning different weights measurements: the higher weighted the number the more precise the measurement.

The unequal accuracy of observations possess different degree of reliability, which can be represented using a number, this numerical called weight and is usually denoted as p, the greater the weight, the less the corresponding standard error. For the measurements l_1, l_2, $\cdots l_n$, corresponding with the standard error m_1, m_2, $\cdots m_n$, the measurement weights p_1, p_2, $\cdots p_n$.

In order to compute the weight p_i of an observation l_i with a standard error m_i, the following equation is used.

From equation (2-23) in the previous section we know that weights were proportional to the number (n) of measurements. This means that the weights are inversely proportional to the square of the standard error. That is, generally,

$$p_i \propto \frac{1}{m_i^2} \tag{2-24}$$

Equation (2-24) can be written by

$$p_i = \frac{\mu^2}{m_i^2} \qquad i = 1, 2, \cdots, n \tag{2-25}$$

Where p_i = weight of the measurement l_i,

m_i = standard error of the measurement l_i,

μ = any constant.

For example, suppose that the measurements l_1, l_2, l_3, and with corresponding the standard error respectively $m_1 = \pm 2''$, $m_2 = \pm 4''$, $m_3 = \pm 6''$. The weight p_i of a measurement l_i is computed according equation (2-25) as bellow:

When $\mu = m_1$: $p_1 = 1 \quad p_2 = \frac{1}{4} \quad p_3 = \frac{1}{9}$

When $\mu = m_2$: $p_1 = 4 \quad p_2 = 1 \quad p_3 = \frac{4}{9}$

When $\mu = m_3$: $p_1 = 9 \quad p_2 = \frac{9}{4} \quad p_3 = 1$

Above discussion it can be seen that the weights are a group of proportional numbers, once μ value is determined, the weight of the measurements are also determined, μ can be taken different value, the weight of the measurements are also different, but proportional relationship of weight between the measurements are unchanged.

The measurements of weight being equal to 1 called *measurement of unit weight*, the standard error of weight being equal to 1 called *standard error of unit weight*, for example, $p_1 = 1$, we call l_1 measurement of unit weight, standard error m_1 is called standard error of unit weight.

In levelling, if the precisions of every mile's measurement are equal, the weights of elevation difference in levelling routes are inversely proportional to the lengths of leveling routes, i. e.

$$p_i = \frac{c}{l_i} \qquad i = 1, 2, \cdots n \tag{2-26}$$

Where p_i = the weights of elevation difference in the ith leveling route, c = any constant, l_i = the length of the ith leveling route.

2.5.2 Adjustment of measurements with unequal weights

1. The Weighted Mean

The MPV of a group of measurements with varying weights is called the *weighted mean* and is given by

$$x = \frac{p_1 l_1 + p_2 l_2 + \cdots + p_n l_n}{p_1 + p_2 + \cdots + p_n} = \frac{[pl]}{[p]} \tag{2-27}$$

or

$$x = x_0 + \frac{p_1 \delta_1 + p_2 \delta_2 + \cdots + p_n \delta_n}{p_1 + p_2 + \cdots + p_n} = x_0 + \frac{[p\delta]}{[p]} \tag{2-28}$$

Where measurement l_1 has weight p_1, measurement l_2 has weight p_2, …measurement l_n has weight p_n, $[P] = p_1 + p_2 + \cdots + p_n$, x_0 is the approximate value of x, and $\delta_i = l_i - x_0$.

If the deviation of each measurement from the mean x is v_i, then $v_i = l_i - x$. Substituting this equation into next equation gives

$$[pv] = [p(l_i - x)] = [pl] - [p]x$$

Considering equation (2-27), equation above becomes

$$[pv] = 0 \tag{2-29}$$

Therefore, the equation above can be used to check calculations.

2. Standard error of MPV of a group of measurements with unegual weights

Equation (2-25) can be written by

$$\mu^2 = p_1 m_1^2 = p_2 m_2^2 = \cdots = p_n m_n^2$$

$$n\mu^2 = [pm^2]$$

$$\mu = \sqrt{\frac{[pm^2]}{n}}$$

When the number of measurements n is very large, the standard error m can be substituted by true error.

$$\therefore \qquad \mu = \sqrt{\frac{[p\Delta\Delta]}{n}} \tag{2-30}$$

or

$$\mu = \pm\sqrt{\frac{[pvv]}{n-1}} \tag{2-31}$$

In order to obtain the standard error of MPV of a group of measurements with varying weights, the equation (2-27) can be written by

$$x = \frac{[pl]}{[p]} = \frac{p_1}{[p]}l_1 + \frac{p_2}{[p]}l_2 + \cdots + \frac{p_n}{[p]}l_n$$

Suppose that the standard error of the weighted mean is m_x, according to the law of error propagation, we have

$$m_x^2 = \frac{p_1^2}{[p]^2}m_1^2 + \frac{p_2^2}{[p]^2}m_2^2 + \cdots + \frac{p_n^2}{[p]^2}m_n^2$$

Considering the definition of weights, that is $p_i m_i^2 = \mu^2$, the equation above becomes

$$m_x^2 = \frac{p_1}{[p]^2}\mu^2 + \frac{p_2}{[p]^2}\mu^2 + \cdots + \frac{p_n}{[p]^2}\mu^2$$

$$= \frac{1}{[p]^2}(p_1 + p_2 + \cdots + p_n)\mu^2 = \frac{\mu^2}{[p]}$$

$$\therefore \quad m_x = \frac{\mu}{\sqrt{[p]}} \tag{2-32}$$

This is a formula for calculating the standard error of measurements with unequal weights.

Example 2.4 Fig. 2-5 shows levelling routes with one junction point, in order to get the elevation of point P, three elevations of the known points: $H_A = 50.148\text{m}$, $H_B = 54.032$, and $H_C = 49.895\text{m}$. Three levelling routes have been disposed and measured. Three lengths of the levelling routes are $l_{AP} = 2.4\text{km}$, $l_{BP} = 3.5\text{km}$ and $l_{CP} = 2.0\text{km}$. Three levelling measurements are $h_{AP} = +1.535\text{m}$, $h_{BP} = -2.332\text{m}$ and $h_{CP} = +1.780\text{m}$.

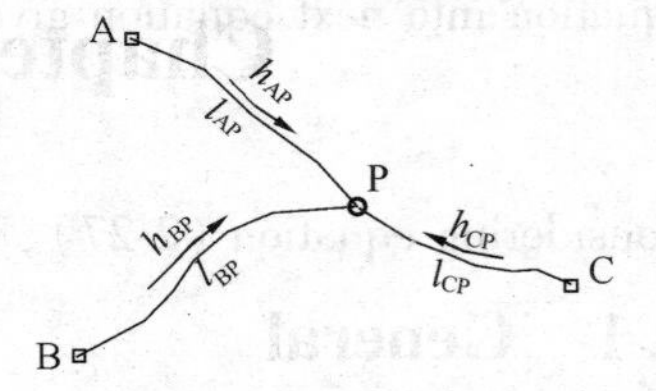

Fig. 2-5 Leveling routes with one junction point

Calculate: the most probable value of the elevation of point P (H_P) and its standard error m_p.

Solution:

Table 2-2 The tabular calculations for Example 2.4

Route	Elevation Hp (m)	Length of routes l_i (km)	Weight $P_i = 1/L_A$	v (mm)	Pv	Pvv
A - P	51.683	2.4	0.417	-0.7	-0.292	0.204
B - P	51.700	3.5	0.286	-16.3	+4.662	75.991
C - P	51.675	2.0	0.500	-8.7	-4.350	37.845
Σ	51.6837		1.203		0.02	114.040

The most probable value of the elevation of point P:

$$H_p = \frac{0.417 \times 51.683 + 0.286 \times 51.700 + 0.500 \times 51.675}{1.203} = 51.6837\text{m}$$

Standard error of unit weight μ:

$$\mu = \sqrt{\frac{[pvv]}{n-1}} = \sqrt{\frac{114.04}{2}} = \pm 7.6\text{mm}$$

Standard error of MPV of P point m_x:

$$m_x = \pm \frac{\mu}{\sqrt{[P]}} = \frac{\pm 7.6}{\sqrt{1.203}} = \pm 6.9\text{mm}$$

单词与词组

weight [weit] *n.* 重量，分量，权，加重值

reliability [riˌlaiə 'biliti] *n.* 可靠性，安全性，可信赖性

proportional [prə'pɔːʃənl] *a.* (成)适当比例的，(与……)相称的(to)

dispose [dis'pəuz] *v.* 布置，处置，安排

· measurement of unit weight 单位权观测值

· standard error of unit weight 单位权标准差

· junction point 结点

Chapter 3 Distance Measurement

3.1 General

Probably the most basic operation performed in surveying and perhaps the most difficult to do well is the measurement of horizontal distances. In plane surveying, whenever the length of a side of a parcel or the linear distance between two points on the earth's surface is given, it is understood that the distance is the horizontal distance. In general situation, measurements of lines in surveying are made by measuring either directly along a horizontal line or indirectly along an inclined one and then computing the corresponding horizontal distance. In the latter case the distance reported is, of course, the (computed) horizontal one.

There are a number of ways in which horizontal distances can be measured. Some common methods are as follows: (1) taping, (2) stadia, (3) electronic measurement.

3.2 Tape measurements

Taping is a common surveying method for measuring horizontal distance. When the length to be measured is less than that of the tape, measurements are carried out by unwinding and positioning the tape along the straight line between these points. The zero of the tape (or some convenient graduation) is held against one point, the tape is straightened and pulled taut, and the distance is read directly on the tape at the other point. This will be the normal procedure on construction sites, where short distances tend to be measured with a tape instead of a total station because it is a quicker and more convenient method of distance measurement.

In measuring a distance longer than one tape length, one must mark tape lengths at intermediate points; and if the total measurement is to be accurate, it is imperative that these intermediate points be established "on line". This is not much of a problem if the distance to be measured is relatively short and each end of the line can be seen from the other end. In this case, a range pole can be placed at each end of the line. When the first tape length is being marked off, the rear tapeperson, standing at behind the initial point, can sight between the range poles at each end of the line and signal the head tapeperson to move one way or the other until he or she is at the proper position to stick the pin on the straight line between the poles. * [When the second and succeeding tape lengths are being marked off, both the rear tapeperson, by sighting ahead to the range pole at the far end of the line, and the head tapeperson, by sighting backward to the range pole at the beginning of the line, can help keep intermediate points on line.]If extremely accurate work is required, a theodolite can be set up at one end of the line and used to set intermediate points on line.

When one end of a line cannot be seen from the other end, it is likely that the line has been run (established) prior to the time taping is done, and stakes (possibly with tacks on top) have been set on line at intermediate points along the line. These stakes (with range poles placed next to them, if necessary) may be used to keep the taping on line.

单词与词组

parcel[′pɑ:səl] *n.* 小包,(土地的)一块
tape[teip] *n.* 卷(皮,钢)尺;*v.* 用卷尺量
unwind[ʌn′waind] *v.* 解开,展开,摊开
graduation[′grædʒu:′eiʃən] *n.* 分段,分级;校准,校正;刻度,分度
straighten[′streitn] *v.* 使成直线,把……弄直
taut[tɔ:t] *a.* 拉紧的,绷紧的
tend[tend] *v.* 易于,倾向,往,朝向
intermediate[ρff′ntə′mi:djət] *a.* 中间的,中级的
imperative [im′perətiv] *a.* 绝对必要的,紧急的,迫切的,不可避免的
range[reindʒ] *v.* 整理,把……排成行,定线;*n.* 射程,距离
pole [pəul] *n.* 杆,柱,竿
tapeperson[teip′pə:sən] *n.* 尺手,掌尺者
· mark off 用标志区分,区分,标出
· rear tapeperson 后尺手
· head tapeperson 前尺手
· one way or the other 以这样或那样的方式(根据课文的含义译为"或左或右")
rod[rɔd] *n.* 杆,棒,标尺,测杆
stick [stik] *v.* 插入,刺入,卡住
pin[pin] *n.* 针,钎,销,钉,枢轴
backward [′bækwəd] *a.* 向后的,倒的
theodolite [θi′ɔdəlait] *n.* 经纬仪
· it is likely that… 很可能……
prior [′praiə] *a.* 在……之前,比……优先
stake[steik] *n.* 桩,柱;*vt.* 用桩支撑,用桩区分
tack[tæk] *n.* 大头钉,平头钉

注释

* "When the second and succeeding tape lengths are being marked off,"为时间状语从句, the second 即为 the second tape length 译为第二尺段,"……are being marked off,"为被动语态进行时,直译为"……被标志划分",或改译为主动语态,即"划分第二尺长段和后续尺长段标志"。从"both ……, and ……"为主句,既……又……,两者都……。整句译为:当第二尺长段和后续尺长段标志划分时,后尺手向前瞄准测线远端的定线杆,以及前尺手向后瞄准测线始端的定线杆,两者都能有助于使中间点保持在直线上。

3.3 Optical methods (Tachymetry)

3.3.1 Measurement by stadia for horizontal sights

Besides the center horizontal cross hair, a theodolite reticle equipped for stadia work has two shorter horizontal hairs spaced equidistant from the center one, called the stadia hairs. The stadia method is based upon the principle that in similar triangles corresponding sides are proportional. Thus in Fig. 3-1, which shows an external-focusing telescope, light rays from points A and B passing through the center of the lens form a pair of similar triangles AmB and amb. Here AB = l is the rod intercept (stadia interval) and *ab* is the interval between the stadia wires.

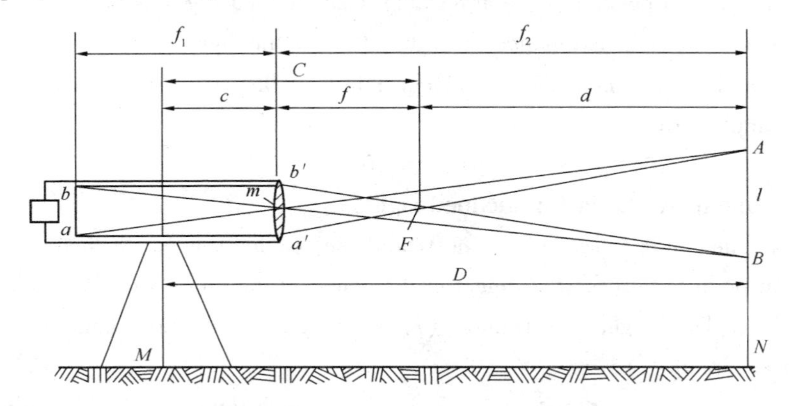

Fig. 3-1 Principle of stadia

Standard symbols used in stadia measurements, and their definitions, are as follows:

f = focal length of the lens (a constant for any particular compound objective lens). It can be determined by focusing upon a distant object and measuring the distance between the center (actually the nodal point) of the objective lens and the reticle.

f_1 = distance from the center (actually the nodal point) of the objective lens to the plane of the cross hairs when the telescope is focused on some definite point.

f_2 = distance from the center (actually the nodal point) of the objective lens to a definite point when the telescope is focused on that point. When f_2 is infinite, or very large, $f_1 = f$.

p = interval between the stadia wires (*ab* in Fig. 3-1).

$\frac{f}{p}$ = stadia interval factor, usually 100.

c = distance from the center of the instrument (spindle) to the center of the objective lens. It varies slightly as the objective lens moves in and out for different sight lengths but is generally considered to be a constant.

$C = c + f$ C is called the stadia constant although it varies slightly with c.

d = distance from the focal point in front of the telescope to the face of the rod.

$D = C + d$ distance from the center of the instrument to the face of the rod.

Then from similar triangles of Fig. 3-1,

$$\frac{d}{f} = \frac{l}{p} \quad \text{or} \quad d = \frac{f}{p}l$$

and

$$D = \frac{f}{p}l + C$$

Fixed stadia wires in theodolites, levels, and alidades are carefully spaced by instrument manufacturers to make the stadia interval factor $\frac{f}{p} = K$ and equal to 100. Then

$$D = Kl + C = 100l + C \tag{3-1}$$

When the internal-focusing instrument is employed, the objective lens of an internal-focusing telescope remains fixed in position while a movable negative focusing lens between the objective lens and the plane of the cross wires changes the direction of the light rays. By mean of appropriate selection of f, p and c can make the stadia constant $C \approx 0$. Therefore, equation (3-1) can be reduced to the simple form

$$D = 100l \tag{3-2}$$

3.3.2 Measurement by stadia for inclined sights

In Fig. 3-2, theodolite is set over point M and the rod held at N. With the horizontal hair of cross-hairs set on point O to read ON(height of the centre of the target v), the vertical angle (angle of inclination) is α. The height of instrument (i) is the height of the horizontal axis of the theodolite above the point occupied.

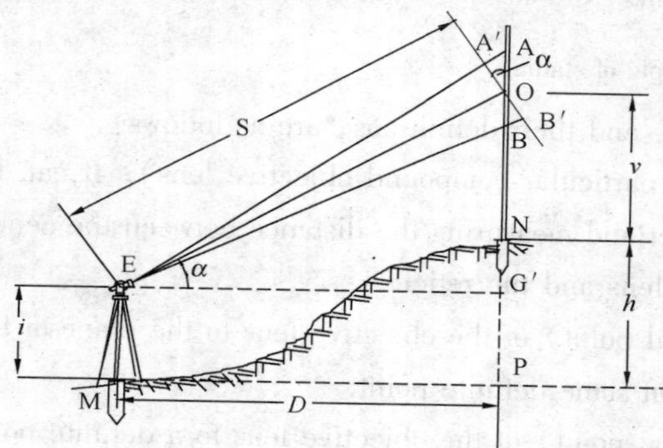

Fig. 3-2 Incilied stadia measurement

Let S represent the slope distance EO; D, the horizontal distance EG = MP; and h, the vertical distance NP. From Fig. 3-2

$$D = S \cos \alpha$$

In Fig. 3-2 where A, O and B are the readings given by the three lines with the staff held vertically. Asume AB = l.

* [If the rod could be held normal to the line of sight at point O, a reading $A'B'$, would be obtained.] Let $l' = A'B'$, according as equation (3-2), we yield

$$S = 100l'$$

It is consider that angle A A′O is a right angle, angle AOA′ equals α in Fig. 3-2. Therefore

$$l' = l \cos \alpha$$

$$S = 100l' = 100l\cos\alpha$$

$$D = S \cos\alpha = 100l\cos^2\alpha \tag{3-3}$$

The vertical distance h is found from Fig. 3-2

$$h = OG + GP - ON$$

$$\therefore \quad h = D \cdot \tan\alpha + i - v \tag{3-4}$$

单词与词组

stadia [steidjə] *n*. 视距，视距测量，视距仪

reticle['retikl] *n*. 十字线，刻线

equidistant [ˌi:kwi 'distənt] *a*. 距离相等的，等距的

corresponding[ˌkɔris 'pɔndiŋ] *a*. 相当的，对应的

proportional [prə 'pɔ:ʃənəl] *a*. 比例的，成比例的

· external-focusing telescope 外对光望远镜

· internal-focusing telescope 内对光望远镜

interval['intəvəl] *n*. 间隔，间歇

intercept[intə 'sept] *n*. 拦截，截距

· stadia interval 视距尺间隔

focal['fəukəl] *a*. 焦点(上)的，有焦点的

compound['kɔmpaund] *n*. 复合物

· objective lens 物镜

nodal[nəudəl] *a*. 结(节，交)点的，组合件的

spindle['spindl] *n*. (主)轴

alidade['ælideid] *n*. 照准仪，测高仪

· negative focusing len 凹透镜

· vertical angle(angle of inclination) 垂直角(倾斜角)

slope [sləup] *n*. 倾斜，斜坡，斜面

· horizontal axis (经纬仪的)水平轴

occupy['ɔkjupai] *vt*. 占用，使从事，把注意力集中于……，使用，占领

注释

* If the rod could be held normal to the line of sight at point O, a reading A′B′ would be obtained. If 引导的条件从句是一种非真实条件句,用了虚拟语气。虚拟语气表示动作或状态不是客观存在的事实,而是说话人一种主观愿望、假设或推测。此句属于下表的第一类。条件句的谓语动词用 could be held(过去式被动),主句语动词使用 would be obtained。因此翻译时条件句译为假设……。全句译为“假设能在 O 点置标尺垂直于视线，那么将得到读数 A′B′。”使用虚拟语气的规则看下表:

Table 3-1

类别	虚拟条件句	主 句	举 例
(1)与现在事实相反	If + 主语 + 动词的过去式(动词 be were)	should/would/could/might + 动词原形	If I had money now, I would lend it to you. 如果我有钱,我会借给你的。(实际上现在没有钱)
(2)与过去事实相反	If + 主语 + had + 过去分词	should/would/could/might + have + 过去分词	If you had come a few minutes earlier, you would have met her 如果你早来一会儿,你就会碰到她了。(实际上来晚了)

续表

类别	虚拟条件句	主句	举例
(3)与将来事实相反	If + 主语 + should + 动词原形或 were to + 动词原形	should/would/ could/ might + 动词原形	If it were to snow tomorrow, we would take photos. 如果明天下雪,我们就照相。(根据今天的天气状况推断,明天下雪的可能性很小)

3.4 Electromagnetic distance measurement (EDM)

3.4.1 Principles

There are basically two methods of EDM: the *pulse* method, in which pulses of radiation are used, and the *phase shift* method, which uses continuous electromagnetic waves.

1. Pulse method

A short, intensive pulse of radiation is transmitted to a reflector target, which immediately transmits it back, along a parallel path, to the receiver. The measured distance is computed from the velocity of the signal multiplied by the time it took to complete its journey, i. e.

$$2D = c \cdot \Delta t$$

$$D = c \cdot \Delta t/2 \tag{3-5}$$

If the time of departure of the pulse from gate A is t_A and the time of its reception at gate B is t_B (see Fig. 3-3), then $(t_B - t_A) = \Delta t$. In equation (3-5), c = the velocity of light in the medium through which it travelled. D = the distance between instrument and target.

It can be seen from equation (3-5) that the distance is dependent on the velocity of light in the medium and the accuracy of its transit time.

2. Phase shift method

This technique uses continuous electromagnetic waves for distance measurement. Although these are extremely complex in nature, electromagnetic waves can be represented in their simplest form as periodic sinusoidal waves, as shown in Fig. 3-3. The sinusoidal waves have following properties.

(1) *1 [The wave completes a cycle in moving from such identical points as A to E or D to H on the wave and the number of times in one second the wave completes a cycle is termed the frequency of the wave.] The frequency is represented by f hertz, 1 hertz (Hz) being 1 cycle per second.

(2) The wavelength of a wave is the distance which separates two identical points on the wave or is that length traversed in one cycle by the wave and is denoted by λ, in metres.

(3) The period is the time taken by the wave to travel through one cycle or one wavelength and is represented by T seconds.

(4) The velocity of the wave to travel through medium is denoted by v, in m/s or km/s.

The velocity of an electromagnetic wave in a vacuum is termed the speed of light and is given the symbol c. The value of c is known at the present time as 299792458 m/s.

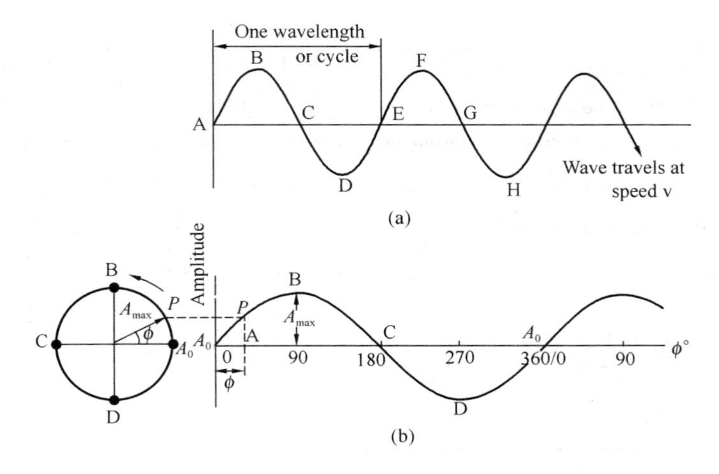

Fig. 3-3 Sinusoid wave motion

(a) As a function distance or time; (b) As a function phase angle ϕ

Interestingly, the definition of the meter is that it is the distance travelled by light in a vacuum during a time interval of 1/299792458 of a second, the inverse of the speed of light c.

The speed of electromagnetic radiation when propagated through the atmosphere will vary from the free space value according to the temperature, humidity and pressure at the time of measurement. The velocity (v) is given by

$$v = \frac{c}{n} \tag{3-6}$$

Where, n is the refractive index of the atmosphere. The value of n varies between 1.0001 and 1.0005 and is mainly a function of temperature and pressure.

All of the above properties of electromagnetic waves are related by

$$\lambda = \frac{v}{f} \tag{3-7}$$

A further term associated with periodic waves is the phase of the wave. As far as distance measurement is concerned, this is a convenient method of identifying fractions of a wavelength or cycle. A relationship that expresses the instantaneous amplitude of a sinusoidal wave is (see Fig. 3-3 b)

$$A = A_{max}\sin\phi + A_0 \tag{3-8}$$

where A_{max} is the maximum amplitude developed by the wave, A_0 is the reference amplitude and ϕ is the phase angle. Angular degrees are often used as units for a phase angle up to a maximum of 360° for a complete cycle.

The phase shift method determines distance by measuring the difference in phase angle between transmitted and reflected signals. This phase difference is usually expressed as a fraction of a cycle, which can be converted into distance when the frequency and velocity of the wave are known. The methods involved in measuring a distance by the phase shift method are as follows.

In Fig. 3-4a, a EDM instrument has been set up at A and a reflector at B so that distance AB

$= D$ can be measured.

Fig. 3-4b shows the same configuration as in Fig. 3-4a, but only the details of the electromagnetic wave path have been shown. The wave is continuously transmitted from A towards B, is instantly reflected at B (without change of phase angle) and received back at A. For clarity, the same sequence is shown in Fig. 3-4c but the return wave has been opened out. Points A and A′ are the same since the transmitter and receiver would be side by side in the instrument at A.

From Fig. 3-4c it is apparent that the distance covered by the wave in travelling from A to A′ is given by

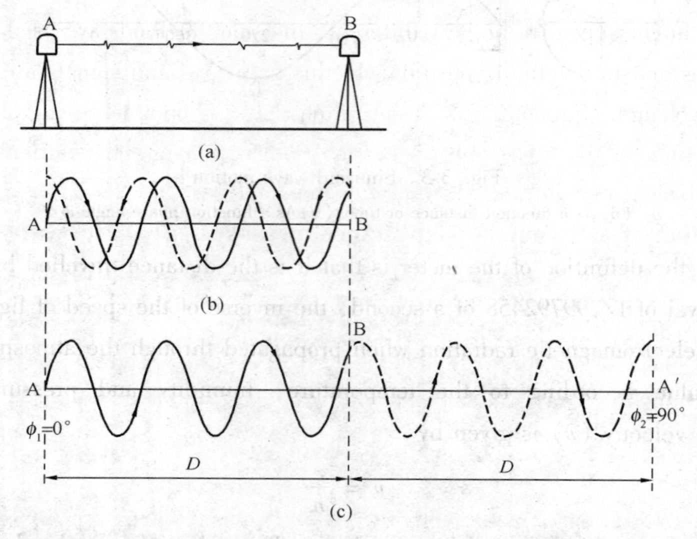

Fig. 3-4 Principle of phase comparison

$$2D = n\lambda_m + \Delta\lambda_m$$

$$D = n\frac{\lambda_m}{2} + \frac{\Delta\lambda}{2}$$

where D is the distance between A and B, λ_m the wavelength of the measuring wave, n the whole number of wavelengths travelled by the wave and $\Delta\lambda_m$ the fraction of a wavelength travelled by the wave. Since the double distance is measured, the effective measuring wavelength is $\lambda_m/2$.

The distance D is made up of two separate parts which are determined by two processes.

The phase shift or $\Delta\lambda_m$ measurement is carried out by measuring phase angles. An electronic phase meter or detector built into the total station at A measures the phase of the electromagnetic wave as it is transmitted. Let this be ϕ_1 degrees. Assume the same detector also measures the phase of the wave as it returns from the reflector at A′ (ϕ_2°). These two can be compared to give a measure of $\Delta\lambda_m$ using the relationship

$$\Delta\lambda_m = \frac{\lambda_m}{360^\circ} \times (\textit{phase difference in degrees}) = \frac{\lambda_m}{360^\circ} \times (\phi_2 - \phi_1)^\circ \qquad (3\text{-}9)$$

Since the phase value ϕ_2 can apply to any incoming wavelength at A′, the phase shift can only provide a means of determining by how much the wave travels in excess of a whole number of wavelengths. Therefore, some method of determining $n\lambda_m$, the other part of the unknown distance, is required. This is referred to as resolving the ambiguity of the phase shift and can be carried out by one of two methods. Either the measuring wavelength is increased in multiples of 10 until a coarse measurement of D is eventually made, or the distance can be found by measuring the line with different but related wavelengths to form simultaneous equations of the form $2D = n\lambda_m + \Delta\lambda_m$. These are solved to give a value for D.

Whatever techniques are used by a EDM instrument to carry out phase shift measurements and to resolve ambiguities, they are fully automated and the instrument will measure and display a distance at the press of a key—no calculations or any further action are required.

Although they might appear to be very different, the same methods are used by GPS equipment to measure distances and to determine position. In this case, L-Band signals are transmitted by the satellites with a wavelength of about 0.2m, from which a $\Delta\lambda_m$ is obtained using phase shift methods. Resolving the ambiguity to give $n\lambda_m$ is also necessary, but is a much more difficult process than with a EDM instrument because of the long distances to the satellites.

3.4.2 Classification of electromagnetic distance measurement equipment

The first electromagnetic distance device was the tellurometer which was a development from radar techniques. The instrument was designed in 1954 for geodetic work in South Africa. Since then there have been considerable developments in electronics and solid state design which have introduced electromagnetic distance devices for a range of purposes making use of different frequency ranges.

The electromagnetic spectrum is continuous from visible light with frequencies of the order of 10^{14} Hz, corresponding to wavelengths of the order of 10^{-6} m, to long radio waves with frequencies of 10^4 Hz, corresponding to wavelengths of the order of 10^4 m. The relationship between frequency and wavelength is shown in Fig. 3-5.

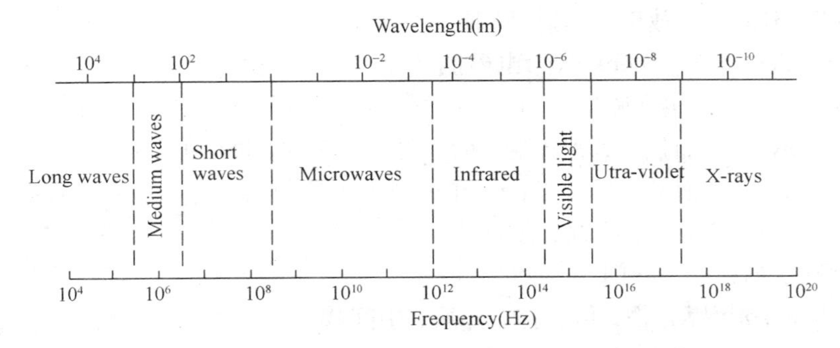

Fig. 3-5 The electromagnetic spectrum

1. Microwave instruments

Microwaves have frequencies between 10^8 and 10^{12} Hz. These instruments are built primarily for long range work. The master station transmits a signal which is received by the remote station. *2[As the signal is weak on arrival at the remote station, it amplifies the signal and returns the signal in exactly the same phase.]

This system employs an operator at each end and requires a communication system between the two. In order to dispel ambiguity in results, a range of four or five frequencies is employed. The method is appropriate for a range of 30 ~ 80 km and gives accuracies of ±(15 mm ±5 mm/km).

The microwave instrument is used for large trilaterations and triangulations where some sides are measured and for the control of very large civil engineering projects.

2. Visible light instruments

Visible light has frequencies between 10^{14} and 10^{16} Hz. Visible light instruments are for intermediate range. Here the remote station is replaced by a corner cube prism which reflects back the signal close to and parallel to the out-coming signal.

Light wave instruments normally measure the distance with three different wavelengths to ensure accuracy. The method is used up to 25 km and gives an accuracy of ±(10mm ±2mm/km).

3. Infra-red instruments

Infra-red waves have frequencies between 10^{12} and 10^{14} Hz. These are short wave instruments. The infra-red instrument is very popular due to its low costs. The carrier wave source is gallium arsenide which is an infra-red emitting diode. These diodes are easily amplitude modulated at high frequency and consequently provide a cheap and simple method of obtaining a modulated carrier wave. The simple design of this instrument makes it compact and light allowing theodolite mounting.

Because the infra-red wavelength is close to visible light it can be controlled by lenses in the same way as light. The main disadvantage of the diode is that power is low and the range limited to 2 or 3 km. The accuracy is normally ±(3mm ±2mm/km) within its range.

单词与词组

· pulse method 脉冲法

· phase shift method 相位移法

radiation [ˌreidi ˈeiʃən] *n.* 放射，辐射，发射

electromagnetic [iˌlektrəmæg ˈnetik] *a.* 电磁的

wave [weiv] *n.* (电，光，声)波

intensive [in ˈtensiv] *a.* 加强的，密集的，彻底的，精深的，透彻的

reflector [ri ˈflektə] *n* 反射器，反射光；*vt.* 在……上装反射器

receiver [ri ˈsiːvə] *n.* 接受者，听筒，接收器

gate [geit] *n.* 门，栅栏门，闸门

medium [ˈmiːdjəm] *n.* 媒介物，传导体，介质，中间物

periodic [ˌpiəri ˈɔdik] *a.* 周期(性)的，定期的，循环的

radiation [ˌreidi ˈeiʃən] *n.* 放射，辐射，发射，发光

sinusoidal [ˌsainə ˈsɔidəl] *a.* 正弦曲线的；*n.* 正弦曲线

cycle [ˈsaikl] *n.* 周期，循环，周而复始；一个操作过程

identical [ai ˈdentikəl] *a.* 同一的，恒等的，同卵的

frequency [ˈfriːkwənsi] *n.* 频率

Hertz [həːts] *n.* 赫兹(频率单位)

wavelength [ˈweivleŋθ] *n.* 波长

separate['sepəreit] *vt.* 分开，分离，切断，识别
vacuum ['vækjuəm] *n.* 真空，真空状态
universal[ˌju:ni 'və:səl] *a.* 宇宙的,绝对的，普遍的
atmosphere['ætməsfiə] *n.* 大气(层)，气圈，空气
temperature['tempəritʃə] *n.* 温度，气温
humidity [hju: 'miditi] *n.* 湿度，湿气，水分含量
well-established[wel-is 'tæbliʃt] *a.* 已为大家接受的，信誉卓著的，根深蒂固的
identify [ai 'dentifai] *vt.* 使等同于,认为一致(with);识别,鉴别,验明;认为与……有关系
instantaneous [ˌinstən 'teinjəs] *a.* 即刻的；瞬间的；立刻做成的，立即发生的
amplitude['æmplitju:d] *n.* 幅度，波幅，调幅
configuration [kənˌfigju 'reiʃən] *n.* 结构，构造，图形，组合,布置
sequence ['si:kwəns] *n.* 连续，一连串，前后次序，顺序
· open out 伸展开来，变宽
· side by side 肩并肩地，一起，并排，相互支持
multiple['mʌltipl] *a.* 复合的，复式的，多重，并联的，复接的;*n.* 倍数,并联
ambiguity [ˌæmbi 'gju:iti] *n.* 歧义，意义不明确，模糊
eventually[i 'ventjuəl] *adv.* 终于，最后
inverse [in 'və:s] *a.* 相反的，反向的
essential [i 'senʃəl] *a.* 必不可少的，非常重要的
incoming [ˌin 'kʌmiŋ] *a.* 正到达的，正来临的
tellurometer [ˌtelju 'rɔmitə] *n.* 用调制连续波的测距仪，无线电测距仪，微波测距仪
spectrum['spektrəm] *n.* (pl. spectra [–trə], spectrums) 谱，光谱，频谱
weak[wi:k] *a.* 弱的，虚弱的，微弱的；不牢固的，不稳固的
dispel[dis 'pel] *vt.* 驱散，消除(散，释)
out-coming ['aut-kəmiə] *a.* 输出口的
· corner cube prism 正立方角锥棱镜(详细解释见第 276 页)
infra-red ['infrə-red] *a.* 红外线的
gallium ['gæliəm] *n.* 镓
arsenide['ɑ:sənaid] *n.* 砷化物
· gallium arsenide 砷化镓
emit[i 'mit] *vt.* 发射，发放，喷，冒出
diode['daiəud] *n.* 二极管
modulate ['mɔdʒuəˌleit] *vt.* 调节,调制，使改变周波数
compact['kɔmpækt] *n.* 契约，合同,协定;*a.* 装填紧密的;*v.* 压紧
mounting['mauntiŋ] *n.* 座架，安装，装配，悬挂
lens [lenz] *n.* 透镜，一组透镜
disadvantage [ˌdisəd 'vɑ:ntidʒ] *n.* 缺点，缺陷，不利的情况

注释

[1] The wave completes a cycle in moving from such identical points as A to E or D to H on the wave and the number of times in one second the wave completes a cycle is termed the frequency of the wave。在 and 前为简单句，在 and 之后为一复合句。identical points 意为相同的点，即为相同波动状态的点，所以第一句译为电波从波上相同点 A 移到 E 或从 D 移到 H 称做完成一个循环（或称一整波）。and 后面部分为含主语从句的复合句，the number of times in one second the wave completes a cycle 是主语从句作主句的主语，译为一秒完成整波的次数，故整个复合句译为：一秒完成整波的次数称为频率。

[2] As the signal is weak on arrival at the remote station, it amplifies the signal and returns the signal in exactly the same phase。第一句为 as 引导的原因状语从句，其中 is 意为“变成，变为”，后面为主句，句中 it 指代前一整句的含义，amplifies（放大）与 returns（返回）为主句中并列的两个谓语。因此译为：由于信号到达远方站时变弱，所以必须放大信号，以精确相同相位返回信号。

Chapter 4 Levelling

4.1 General

There are several methods of elevation determination, i. e. , differential levelling, trigonometric levelling, barometric levelling, hydrostatic levelling and GPS etc.

Levelling, or the determination of the relative altitudes of points on the Earth's surface, is an operation of prime importance to the engineer, both in acquiring necessary data for the design of all classes of works, and during construction operations.

The basic terms in levelling defined below are illustrated in Fig. 4-1.

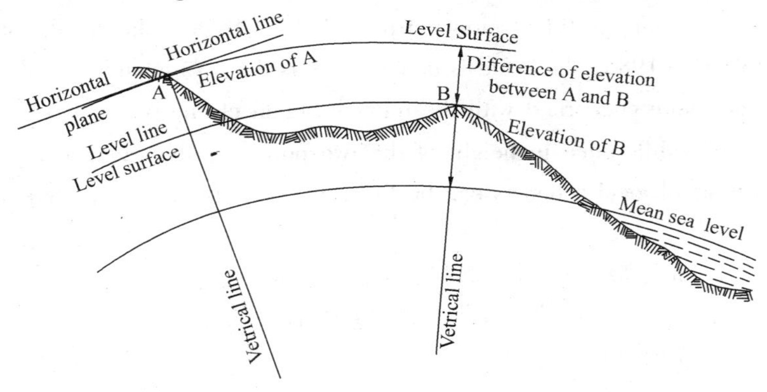

Fig. 4-1 Levelling terms

Level surface A level surface is a curved surface which is at all point normal to the plumb line (the direction in which gravity acts). Level surfaces are approximately spheroidal in shape. The surface of a still lake exemplifies a level surface.

Level line A level line is a line lying throughout on one level surface, and is therefore normal to the direction of gravity at all points.

Horizontal plane A horizontal plane passing through a point is the plane normal to the direction of gravity at the point.

Horizontal line A horizontal line is one that is normal to the direction of the force of gravity at a particular point. Fig. 4-1 shows a horizontal line through point A.

Vertical line A vertical line is a line perpendicular to the level surface and lies along the plumb line through that point.

Vertical plane A vertical plane is a plane containing the vertical line.

Elevation The elevation of a point is the height in a vertical line from the datum surface.

Reduced level The reduced level of a point is the height of the point obtained by adding the known or assumed datum surface elevation and the elevation of the point from the datum surface.

Difference in elevation The difference in elevation between two points is the vertical distance between the level surfaces through the two points.

Datum A datum is any reference surface to which the elevations of points are referred (for example, mean sea level), whose elevation is known. The most commonly used datum surface is that of mean sea level (MSL).

Benchmark (BM) BM is a permanently vertical control point of known elevation. Benchmarks are established using precise levelling techniques and instrumentation.

MSL is termed Geoid, which is the national geodetic vertical datum and assigned a vertical value (i. e. elevation) of 0.000 m or 0.000 ft. In the United States, the geoid is referred to as the "Sea Level Datum of 1929". In China, 7 years of observations at tidal station in Qingdao from 1950 to 1956 were deduced and adjusted to provide the height of leveling origin to be 72.289m. The geoid is referred to as the "Huanghai Vertical Datum of 1956". In the 1987, this datum was further refined to reflect long periodical ocean tide change to provide a new national vertical datum, according to the observations at tidal stations from 1952 to 1979. The height of levelling origin refined to be 72.260 m in 1985, the geoid is now termed as the "National Vertical Datum of 1985".

The engineer is, more concerned with the relative height of one point above or below another, in order to ascertain the difference in height of the two points, rather than a direct relationship to MSL. Therefore, on small local schemes can be to adopt a purely arbitrary reference datum.

单词与词组

elevation[ˌelə ˈveiʃən] *n*. 高程，高度，海拔，标高，仰角
differential[ˌdifə ˈrenəʃl] *a*. 差别的，微分的，与差别有关的
· differential levelling 几何水准测量
trigonometrical [ˌtrigənə ˈmetrikl] *a*. 三角学的
· trigonometric levelling 三角高程测量
barometric[ˌbærəu ˈmetrik] *a*. 大气压力的
· barometric levelling 气压高程测量
· hydrostatic levelling 流体静力高程测量
relative [ˈrelətiv] *a*. 相对的，比较的
altitude[ˈæltitjuːd] *n*. 高度，海拔高度
prime [praim] *a*. 首要的，主要的，基本的
operation [ˌɔpə ˈreiʃən] *n*. 操作，管理，经营；生产，施工，工作
acquire [ə ˈkwaiə] *v*. 得到，获得，招致
· level surface 水准面
normal [ˈnɔːməl] *a*. 正常的，常态的，普通的，垂直的，成直角的
plumb[plʌm] *n*. 铅锤，测锤，垂直;*a*. 垂直的
spheroidal [ˈsfiərɔidl] *a*. 椭球体的
· level line 水准线

lie [lai] *vi.* 撒谎，躺，卧，位于

throughout [θru(:) 'aut] *prep.* 遍及，贯穿从头到尾，到处，全面，完全，彻底

gravity ['græviti] *n.* 重力，(地心)引力

perpendicular[ˌpə:pən 'dikjələ] *a.* 成直角的，垂直的，正交的(to)

datum ['deitəm] *n.* 基点线，基准面

· reduced level 归化高程

datum ['deitəm] *n.* (*pl* data) 数据，基准(点、线、面)

· bench mark (BM) 水准点

permanently['pə:mənəntli] *adv.* 永久性地，固定不变地

instrumentation[ˌinstrumen 'teiʃən] *n.* 仪器，装备，手段

· mean sea level (MSL) 平均海水面

· Sea Level Datum of 1929　1929 年高程基准(美国国家高程系统)

tidal['taidl] *a.* 潮汐的，潮水般的

· Qingdao tidal station 青岛验潮站

shoreline ['ʃɔ:lain] *n.* 海岸线

deduce [di 'dju:s] *vt.* 推论，推断，演绎(from)，引出，推导

· Huanghai Vertical Datum of 1956　1956 年黄海高程系统(1985 年前中国国家高程系统)

· the height of leveling origin 水准原点高程

refine [ri 'fain] *vt.* 精炼，提炼，提纯，使(体态，语言等)文雅;*vi.* 提纯，变优雅，推敲，琢磨，改善，改进

· National Vertical Datum 1985　1985 国家高程基准(中国国家高程系统)

arbitrary ['ɑ:bitrəri] *a.* 任意的，随意的，主观的

ascertain[æsə 'tein] *vt.* 确定，调查，查明，断定

scheme [ski:m] *n.* 计划，规划，方案，设计，安排，办法

arbitrary ['ɑ:bitrəri] *a.* 随机的，任意的，专横的，独立的

4.2 Principle of levelling

One of the main surveying jobs is to determine the elevations of points on the earth. An elevation (also termed height) is a vertical distance above or below a referenced datum. The basic equipment required in levelling is:

(1) A device which can give a truly horizontal line (the level).

(2) A suitably graduated staff for reading vertical heights (the levelling staff).

The equipment used in the levelling process comprises levels and graduated staffs. The telescope fitted with a spirit bubble can furnish a truly horizontal line of sight. This is the reason that can determine the difference of elevation between two points on the Earth's surface.

The most precise and most commonly used method of determining elevation is differential levelling which means measuring the vertical distance directly. The differences in the readings on the vertically-held graduated staff where intersected by the horizontal line of sight is a direct measure of the difference in height between the two staff stations. For example, to determine the elevations of desired point B with respect to a point of known elevation A (see Fig. 4-2), the elevation of which is known to be H_A above Geoid or the arbitrary reference datum. The procedure of levelling is as follows.

The first, the level is set up approximately halfway between point A and point B. The graduated staffs are separately erected on A and B as shown in Fig. 4-2.

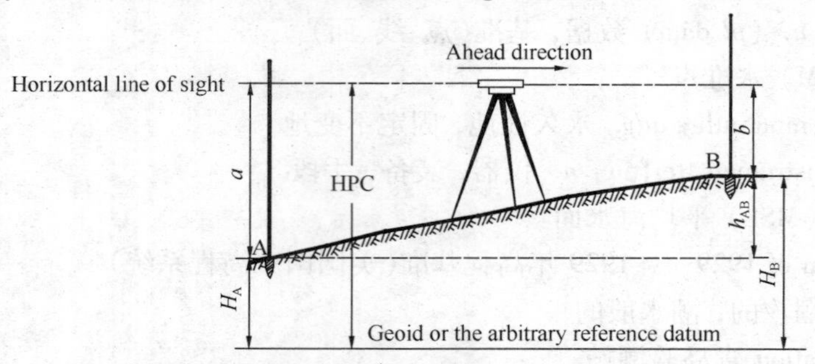

Fig. 4-2 Basic principle of levelling

The second, the horizontal line of sight through the telescope sights the vertical-held graduated staff at A and takes a reading *a*. In this case, the ahead direction of levelling is from A to B, the reading *a* taken with the level backsight, it is called backsight (BS).

The third, turning the telescope, the horizontal line of sight through the telescope sights the vertical-held graduated staff at B and take a reading *b*. Since *b* is taken with level foresight along the levelling direction, it is called foresight (FS).

From Fig. 4-2 can be seen, the difference of elevation between A and B (h_{AB}) is

$$h_{AB} = a - b$$

i. e., the difference of elevation between the two points = BS-FS

If the reduced level RL of point A is known, example $H_A = 100\text{m}$, therefore the RL (reduced level) of point B (H_B):

$$H_B = H_A + h_{AB}$$

$$H_B = H_A + a - b \tag{4-1}$$

From Fig. 4-2 can be seen:

$$\text{HPC} = H_A + a \tag{4-2}$$

where HPC is known as the *height of the plane of collimation or height of instrument*. The equation (4-1) becomes

$$H_B = \text{HPC} - b \tag{4-3}$$

The more general case occurs when the two points to be compared are in such situations as they are far apart or there are some obstacles to them. So it is impossible to set up the level only one station to determine the difference of elevation between the two points. For example, in Fig. 4-3, the point whose elevation is to be determined B is too far from the point of known elevation A. Such situation, the work is performed through a series of stations and the determination of the elevation of the desired point. The first station, the leveler sets up the level at a convenient point I_1, which is placed in such location that a clear rod reading is obtainable, but no attempt is made to keep on the direct line joining A and B. Back sight on the level rod erect held by the rear rodperson on A and then fore sight on the level rod erect held by the head rodperson on the turning point *TP*1. The turning point should be choose at some convenient spot that each foresight distance, such as I_1-*TP*1 be approximately equal to its corresponding backsight distance, such as I_1-*A*. The chief requirement is that the turning point should choose at stable object or using change plate (see Fig. 4-11c). Stability of the turning point is very important.

The second station, the level is moved to a convenient location (I_2) beyond *TP*1 and back sight on level rod erect held on *TP*1, and then forward on level rod erect held on new location *TP*2, which the level rod is moved by rear rodperson (see in Fig. 4-3). This procedure is from A to *TP*1, and then from *TP*1 to *TP*2, and from *TP*2 to *TP*3, and the end from *TPn* to B. At point *TP*i, both a FS and a BS have been recorded consecutively, each from a different instrument stations. This makes to relate the elevations of the lines of sight of two consecutive instrument stations in order to transfer elevation. Such point is called a turning point (*TP*) or change point (CP). At any change point the staff must be held on exactly the same spot for both foresight and backsight. A firm spot must be chosen and marked. If the ground is soft a change plate (see Fig. 4-11c) must be used.

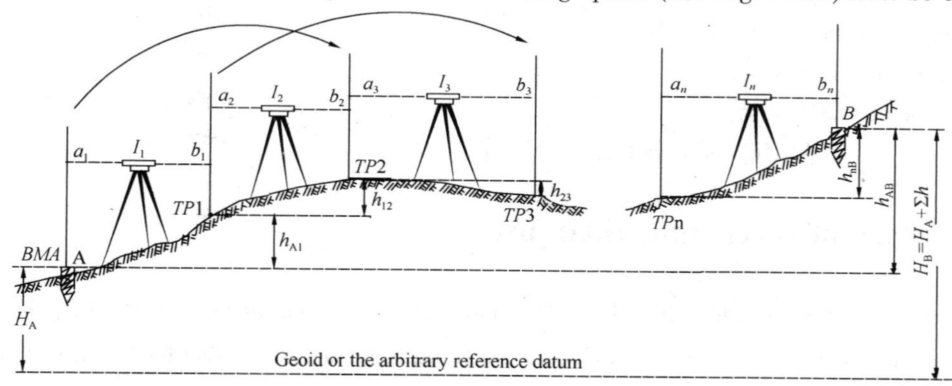

Fig. 4-3 Differential levelling

The readings on each station a_i and b_i are added up respectively, so we get Σa_i and Σb_i. Finally the difference of elevation between A and B is $h_{AB} = \Sigma a_i - \Sigma b_i$, therefore

$$H_B = H_A + h_{AB} \tag{4-4}$$

or

$$H_B = H_A + \Sigma a_i - \Sigma b_i$$

单词与词组

equipment [i 'kwipmənt] *n*. 设备；配件，器材，仪器，供给

device[di 'vais] *n*. 装置，设备，器具，方法，手段

suitably['su:təbli] *adv*. 适合地，适当地，相配地

graduated['grædʒueitid] *a*. 标有刻度的

staff [stɑ:f] *n*. 标杆(尺)

comprise[kəm 'praiz] *v*. 包含，由……组成

fit[fit] *v*. 安装，配备
spirit['spirit] *n*. 精神，心灵，灵魂，酒精
furnish['fə:niʃ] *vt*. 供给，提供，陈设，布置
· with respect to 关于，相对于，就……而论，根据
backsight['bæksait] *n*. 后视
foresight['fɔ:sait] *n*. 前视
forward['fɔ:wəd] *a*. 前部的，向前的
reduce[ri'dju:s] *vt*. 还原，折合，换算，归化
collimation[ˌkɔli'meiʃən] *n*. 视准，瞄准，准直
compare[kəm'pɛə] *vt*. 比较，对照(with)
· far apart 离得很远
obstacle ['ɔbstəkl] *n*. 障碍(物)，妨碍
situation [ˌsitju'eiʃən] n. 地点，场所，场合；形势，局势，局面，情况
leveller ['levələ] *n*. 使平者，校平机，司水准仪者
rodperson[rɔd'pə:sn] *n*. 扶尺者，持杆者
· change plate/levelling plate foot plate 尺垫
stability [stə'biliti] *n*. 稳定性，稳度
· instrument position 仪器站，测站
· turning point (TP) 转点
· change point (CP) 转点
respectively [ris'pektivli] *adv*. 各自地，分别地

4.3 Types of levels and their uses

Levels are divided into DS_{05}、DS_1、DS_3 and DS_{10} in China according to their instrument accuracy. "D" and "S" indicate the first letter of Chinese phonetic alphabet for "dadi celiang yiqi" and "shui zhun yi" respectively. Their subscripts indicate the standerd error of difference in elevation of 1km route levelling (i. e., ±0. 5mm, ±1mm, ±3mm and ±10mm.)

Levels are categorized into three groups according to instrument structure: the tilting level, the automatic level, and digital level.

4.3.1 Tilting levels

The basic purpose of a level is to provide a horizontal line of sight. The tilting level consists of four fundamental parts: (a) a telescope at the top; (b) tubular level fixed on the telescope; (c) a support mounted on vertical spindle (standing axis) in the middle; and (d) a tribrach at the base, the tribrach supports the remainder of the level. Its structure and appearance show in Fig. 4-4a and Fig. 4-4b.

1. Telescopes

The first telescope was invented by a Dutchman, Jan Lippershey, in 1608, and the great mathematician Johannes Kepler suggested how the device could be developed for use in surveying

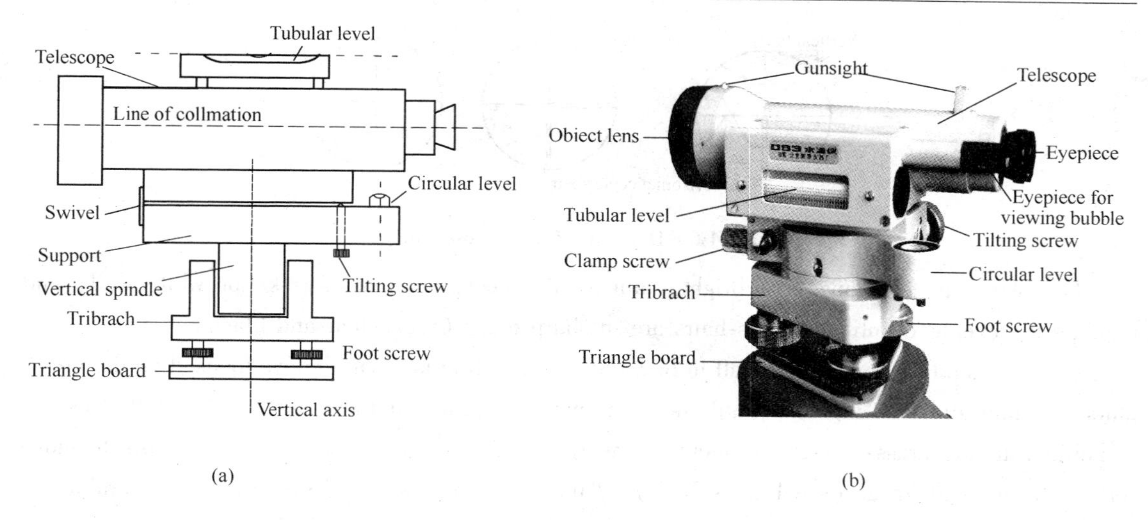

Fig. 4-4 Tilting level (Beijing survey factory DS_3)

instruments. This type of telescope is known as the Keplerian or astronomical telescope.

*[1] [On the tilting level, the telescope is not fixed to the base of the level, but swivel to a support and can be tilted a small amount in the vertical plane.] The amount of tilt is controlled by a tilting screw, which is usually directly underneath or next to the telescope eyepiece (see Fig. 4-4a). The foot screws are used to centre the circular bubble, thereby setting the telescope approximately in a horizontal plane. After the telescope has been focused on the staff, the line of sight is set more precisely to the horizontal using the highly sensitive tubular bubble and the tilting screw that raises or lowers one end of the telescope.

Tilting level have an internal-focusing telescope, in which an negative focusing lens has been introduced between the objective lens and the eyepiece. The telescope is focused by moving this lens within the tube by means of using focusing screw, which enables the length of the telescope to be kept constant and sealed against the entry of dirt and dust (see Fig. 4-5).

The double concave internal focusing lens is moved along the telescope tube by its focusing screw until the image of the staff is brought into focus on the cross-hairs. The cross-hairs are etched onto a circle of thin glass plate called a reticule. The eyepiece, with a magnification of about 35 diameters, is then used to view the image in the plane of the cross-hairs.

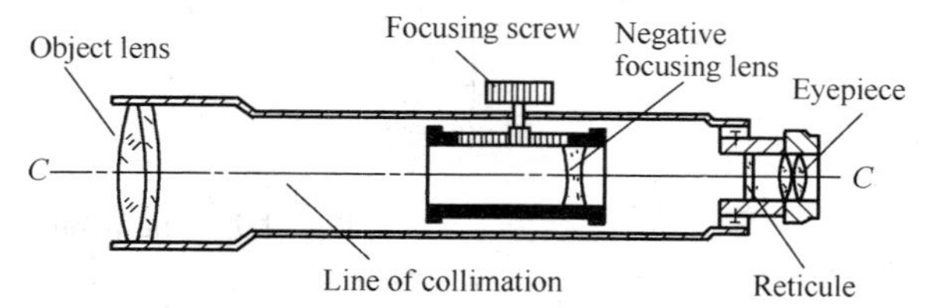

Fig. 4-5 Internal-focusing telescope

Different types of cross-hairs are shown in Fig. 4-6. A line from the centre of the cross-hairs and passing through the centre of the object lens is the line of sight or line of collimation of the telescope.

Prior to the actual observation, three steps are required to focus a telescopic sight for greatest accuracy.

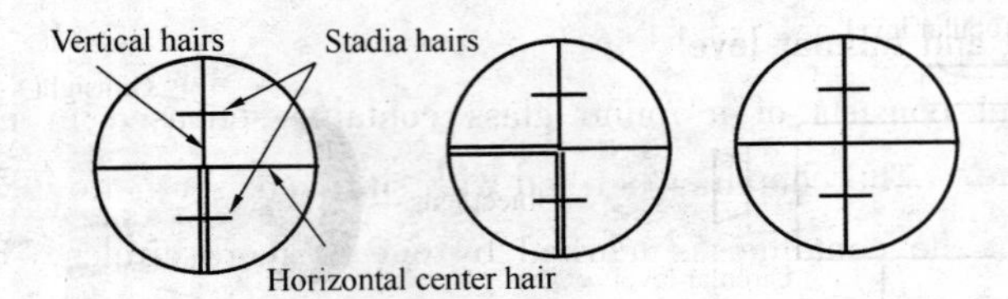

Fig. 4-6 Different types of cross-hairs

(1) Aim the telescope at a bright, unmarked object, such as the sky or white wall, and regulate the eyepiece until the cross-hairs are in sharp focus (i. e. clear and black).

(2) Aim the telescope at the staff to be viewed and, while keeping the eye focused on the cross-hairs, regulate the focusing lens until the staff image is clear. If the image of the staff does not coincide with the cross-hairs, movement of the observer's eye will cause the cross-hairs to move relative to the staff image (see Fig. 4-7a,b). This phenomena is called the cross-hairs parallax.

(3) In order to eliminate parallax, move observer's eye up and down. If the cross-hairs appear to move with respect to the object sighted, change the focus of the objective until the apparent motion is discarded. Continue focusing back and forth, reducing the apparent motion each time until it is eliminated. Finally, it may be necessary to adjust the eyepiece slightly to make the image and the cross-hairs appear clear and distinct. At the time, the plane of the image and the plane of the cross-hairs must be very nearly coincident (see Fig. 4-7c).

Practically, the eliminating parallax need not be every time step by step. When the eyepiece has been set for a particular observer after the parallax has been once eliminated, it is common practice to keep the eyepiece in this position throughout the work and to rely on focusing the objective so that both the cross hairs and the object are in sharp.

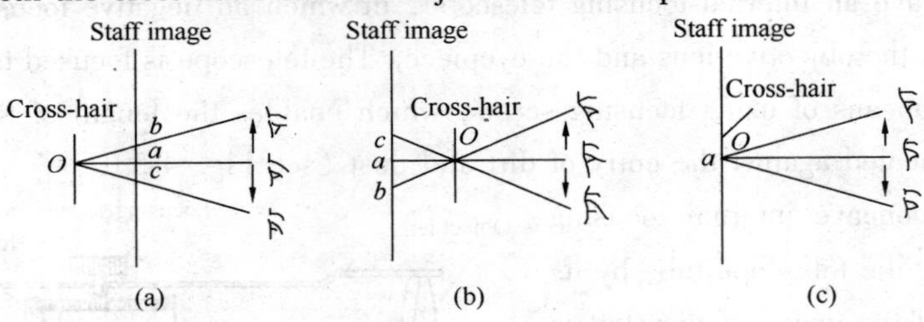

Fig. 4-7 The parallax phenomenon

(a) Presence parallax; (b) Presence parallax; (c) No parallax

The main characteristics defining the quality of the telescope are its powers of magnification, the size of its field of view, the brightness of the image formed and the resolution quality when reading the staff.

Magnification is the ratio of the size of the object viewed through the telescope to its apparent size when viewed by the naked eye. The resolution quality or resolving power of the telescope is its ability to define detail. Surveying telescopes are limited in their magnification in order to retain their powers of resolution and field of view.

2. Circular level and tubular level

The circular level consists of a round glass container encased in metal. Its lid is milled spherically on the inside. The container is filled with sufficient synthetic alcohol to leave a small air bubble. The centre of the container is marked by one or more circles (Fig. 4-8). The axis of circular level ($L'L'$) is the imaginary sphere radius through the circular center of upper surface of the circular level.

The tubular level consists of a cylindrical glass tube encased in metal, its inside surface is milled in such a way that its longitudinal section represents a circular arc. *2[The axis of tubular level (LL) is the imaginary longitudinal line tangent to the upper inside surface at the midpoint.] When the bubble is in the center of its run, the axis should be a horizontal line, as in Fig. 4-9.

The tube is graduated generally in intervals of 2mm. The sensitivity of the tubular spirit bubble is chiefly determined by its radius of curvature (R); *3[the larger the radius, the more sensitive the bubble.]

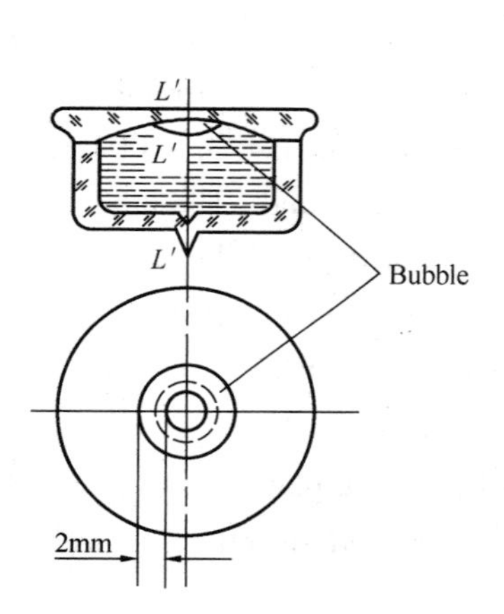

Fig. 4-8　Circular level

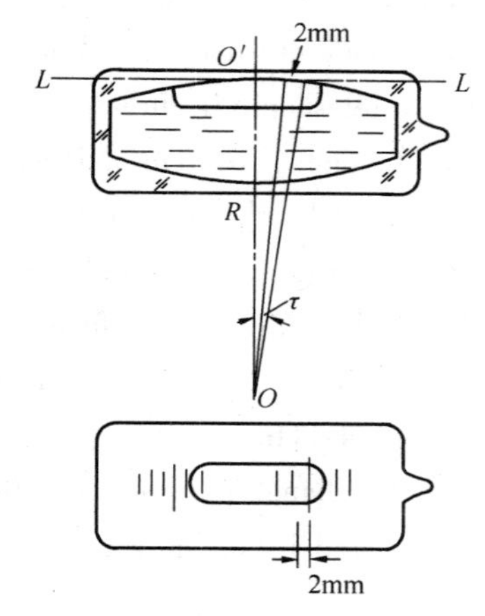

Fig. 4-9　Tubular level

The sensitivity of the tubular level is the central angle (τ) subtended arc length 2mm as shown in Fig. 4-9, hence

$$\tau = \frac{2\text{mm}}{R} \times \rho'' \tag{4-5}$$

Where R = the tube's radias of curvature, $\rho'' = 206265''$。

If $R = 20.63$m $\tau = 20''$. In this case, the bubble moves off centre by one interval it represents an angular tilt of the axis of tubular level of 20 seconds of arc.

The bubble attached to the tilting level may be viewed directly or by means of a coincidence reading system (Fig. 4-10a). In this latter system the two ends of the bubble are viewed and appear as shown at Fig. 4-10b and Fig. 4-10c; Fig. 4-10b shows the image when the bubble is centered by means of the tilting screw; Fig. 4-10c shows the image when the bubble is off centre. This method of viewing the bubble is three or four times more accurate than direct viewing.

The circular level is for rough settings, and tubular levels for precision measurements. The

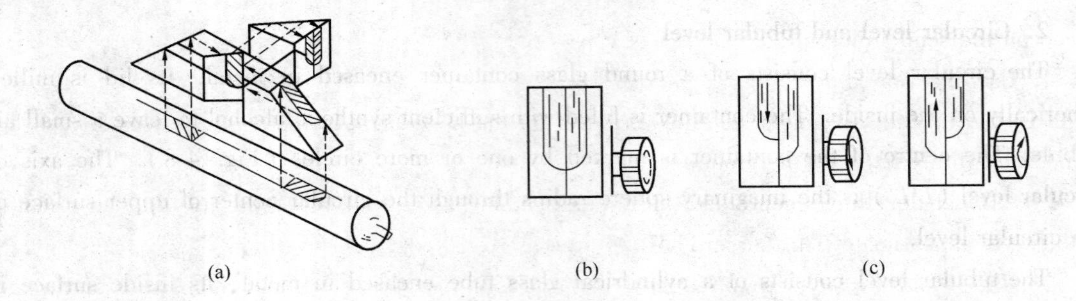

Fig. 4-10 Bubble coincidence reading system

(a) Principle of bubble coincidence reading system

(b) Image when bubble is centre; (c) Image when bubble is off centre

tilting level has only one critical condition, which is that the line of collimation and the bubble axis must be parallel.

3. Levelling staffs and change plates

Levelling staffs are made of wood, metal or glass fiber and graduated in meters and centimeters. The majority of staffs are telescopic in three or four sections for easy carrying (see Fig. 4-11a). The smallest graduation on the staff is 0.01m and readings are estimated to 0.001m. The alternate meter lengths are usually shown in black and red on a white background. The other staff is solid staff with two faces of different graduations. The graduations of one face of the staff are colored black on a white background (generally called black face), with the other face showing red graduations (red face). The bottom of the staff is zero in black face, while 4.687m (NO.1) or 4.787m (NO.2) in red face as shown in Fig. 4-11b.

Fig. 4-11c shows the change plate (also called levelling plate or foot plate) which is made from a solid piece of cast iron and its top is a semicircle ball and smooth. In leveling a staff is upright

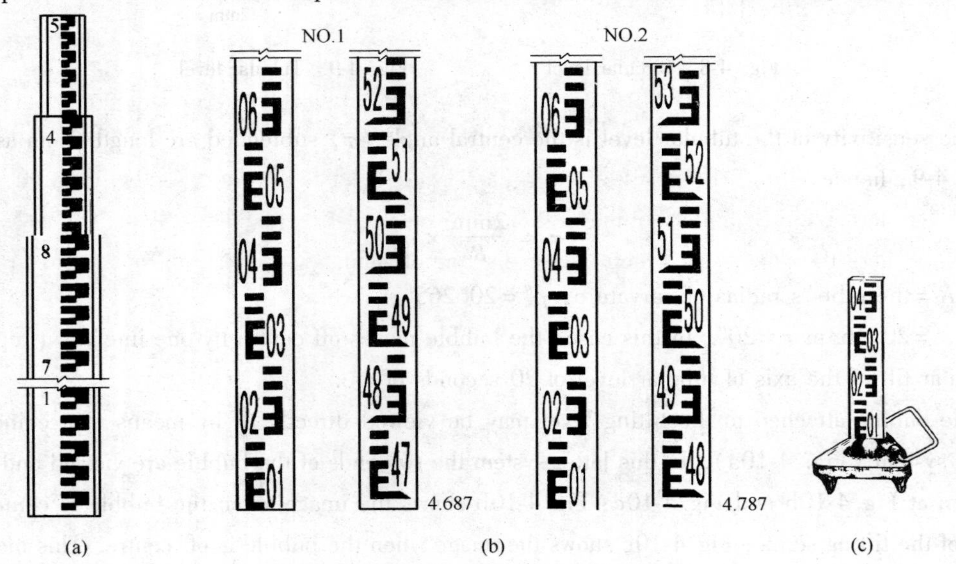

Fig. 4-11 Levelling staffs and change plate

(a) Telescopic staff; (b) Solid staff; (c) Change plate

placed on the semicircle ball.

4. Use of levels

(1) Set up the instrument—set up the level on the head of the tripod, whose legs have been opened and firmly fixed on the ground.

(2) Rough levelling—centralize the circular bubble using the foot-screws in order to circular bubble axis approximately in the vertical line, this is so-called rough levelling.

For this, hold the foot screws by the thumb and forefinger of the left and right hands and rotate the screws either inwards or outwards simultaneously. Also note that the bubble moves in the direction of movement of the left thumb during this operation. First by rotating foot screws 1 and 2 in opposite directions simultaneously until the bubble stays approximately in the line *ab* (see Fig. 4-12a). And then, by rotating foot screw 3 only, the bubble can be centred in the target ring (see Fig. 4-12b).

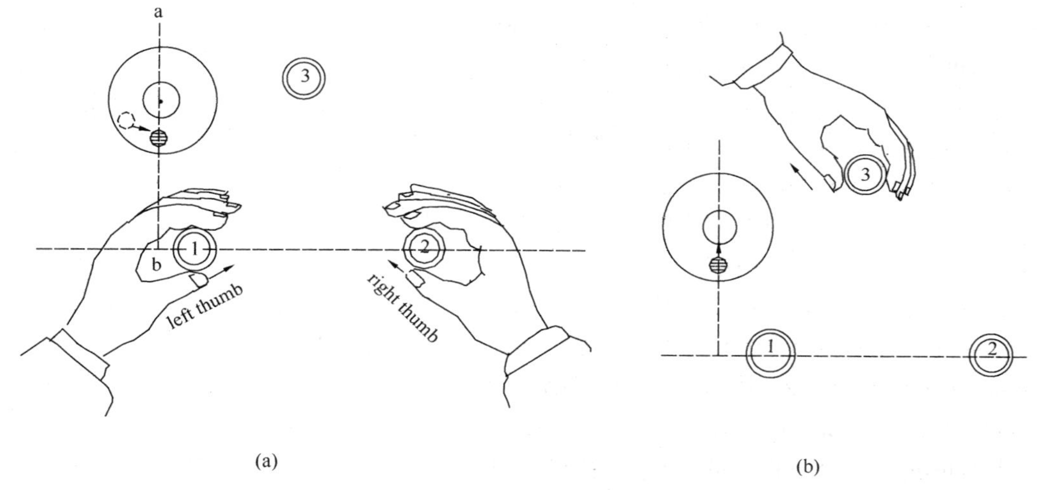

Fig. 4-12 Centralize the circular bubble

(3) Eliminate parallax—the method of removing paralax has been described in Part one above (i. e., "1. telescopes").

(4) Precise levelling—because the level is not exactly leveled, consequently, the tilting screw must be reset to make the tubular bubble in the centre before every reading is taken, this is so-called precise levelling, the perpose of which is that the line of sight becomes a truly horizontal line.

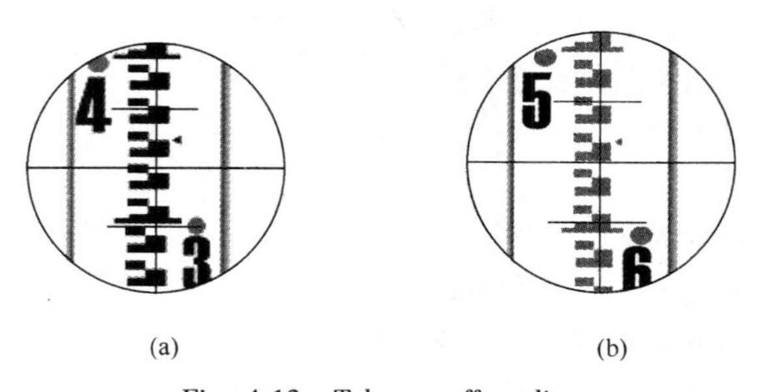

Fig. 4-13 Take a staff reading

(a) The staff figures increase upward; (e) The staff figures increase downward

(5) Take a reading—take the reading of horizontal center hair on the staff. The staff figures in the telescope field of view, regardless of increase upward or downward increase, take reading should be from small to large number. The Fig. 4-13a notes the staff reading 1.334m, Fig. 4-13b the staff reading 1.560m.

单词与词组

phonetic[fəu'netik] *a.* 语音的,音形一致的
alphabet[ælfəbit] *n.* 字母,字母表
tilt ['tilt] *v. n.* 倾斜,倾卸
· tilting level 微倾水准仪
categorize['kætigə'raiz] *vt.* 把……归类, 把……列作
· tubular level 管形水准器, 水准管
fix[fiks] *vt.* 使固定, 安装
support [sə 'pɔ:t] *n.* 支撑物, 支柱, 支座, 支架; *v.* 支撑, 托住, 支持
mount[maunt] *vt.* 上演, 登上, 配有……
spindle['spindl] *n.* 轴, 转轴
tribrach['tribræk] *n.* 有三支脚的物体, 基座
remainder [ri 'meində] *n.* 剩余部分
structure ['strʌktʃə] *n.* 结构, 构造
invent[in 'vent] *vt.* 发明, 创造
Dutchman['dʌtʃmən] *n.* (=[美] Hollander) 荷兰人
mathematician [ˌmæθəmə 'tiʃən] *n.* 数学家
Keplerian[kep 'liəriən] *a.* 开普勒的, 开普勒定律的
astronomical[ˌæstrə 'nɔmikəl] *a.* 天文学的, 天体的
internal-focusing[in 'tə:nəl 'fəukəsiŋ] *a.* 内对光的
negative['negətiv] *a.* 否定的, 负的, 阴极的
lens[lenz] *n.* 透镜, (凹、凸)镜片, 一组透镜
· objective lens 物镜
eyepiece['aipi:s] *n.* 目镜
sealed [si:ld] *a.* 密封的, 未知的
swivel['swivəl] *v.* 用转节连结, 用铰链连结; *n.* 旋转节, 铰接部
underneath ['ʌndə 'ni:θ] *prep.* 在……下面, 在……底下
sensitive ['sensitiv] *a.* 有感觉的, 敏感(锐)的
concave[ˌkɔn 'keiv] *a.* 凹(面)的
etch [etʃ] *vt.* 蚀刻
reticule['retikju:l] *n.* (=reticle)十字线
magnification[ˌmægnəfi 'keiʃən] *n.* 放大率, 扩大
regulate ['regjuleit] *vt.* 管理, 控制, 使整齐, 调节, 调整
sharp[ʃɑ:p] *a.* 急转的, 陡峭的, 突然的, 急剧的, 明显的, 清晰的
parallax['pærəˌlæks] *n.* 视差, 视差角度

distinct [dis 'tiŋkt] *a.* 独特的，清晰的，明显的

· with respect to 关于，就……而论，相对于

characteristics[ˌkæriktə 'ristiks] *n.* 规格参数，特性

diameter[dai 'æmitə] *n.* 直径，放大倍数

brightness['braitnis] *n.* 亮度，白度

resolution[ˌrezə 'lu:ʃən] *n.* 分辨力

apparent[ə'pærənt] *a.* 明显的，表面上的，外观上的

naked['neikid] *a.* 裸体的，无遮盖的，无罩的，赤裸裸的

detail['di:teil, di 'teil] *n.* 细节，细目，影像的细节，清晰度

· circular level 圆水准器

container [kən 'teinə] *n.* 容器，箱，盒

encase[en 'keis]*vt.* 把…放入盒内，插(镶)在……内

lid[lid] *n.* 盖，罩，帽

mill[mil] *v.* 磨，铣，粉碎;*n.* 磨坊，碾磨厂

synthetic [sin 'θetic] *a.*综合(性)的，合成的

alcohol['ælkəhɔl] *n.* 乙醇，酒精

imaginary [i 'mædʒinəri] *a.* 想象中的，假想的，虚构的

cylindrical[sə 'lindrikəl] *a.* 圆柱形的;圆筒状的

graduate['grædʒueit] *vt.* 授予学位，刻(分)度

interval['intəvəl] *n.* 间隔，间歇

sensitivity[ˌsensi 'tiviti] *n.* 灵敏度

coincide[ˌkəuin 'said] *vi.* 与……相重合;与……相同[同时发生]，与……相符合(with)

prior['praiə] *a.* (用作前置定语)在前的，更早的，优先的;(与 to 连用)在……之前，比……优先

magnification[ˌməgnifi 'keiʃən] *n.* 放大(率，倍数)，倍率

curvature['kə:vətʃə] *n.* 曲率，曲度，弧度

subtend[səb 'tend] *vt.* 对着，(弦、边)对(弧、角)

tripod['traiˌpɔd] *n.* 三脚架(台)

firmly [fə:mli] *adv.* 坚固地，稳固地

fiber['faibə] *n.* 纤维，光纤

alternate['ɔ:ltə:neit] *v.* (使)交替，(使)轮换

majority[mə 'dʒɔriti] *n.* 多数，大半

telescopic[ˌteli 'skɔpik] *a.* 望远镜的，用望远镜的；伸缩的，套叠的

solid['sɔlid] *a.* 固体(态)的，实心的，紧密的，结实的

inwards['inwədz] *adv.* 向内地，向内部地

outwards['autwədz] *adv.* 向外地，外表上，向国外

opposite['ɔpəzit] *a.* 对面的，对立的，相反的

target['tɑ:git] *n.* 目标，靶子

ring [riŋ] *n.* 圈，环

reset [ri 'set] *vt.* 重新安放或安置
regardless[ri 'gɑ:dlis] *adv.* 不管怎样地，无论如何
increased[in 'kri:st] *a.* 增加的，增强的

注释

[1] On the tilting level, the telescope is not fixed to the base of the level, but swivel to a support and can be tilted a small amount in the vertical plane. 句中 support 是名词，意为支撑物或支撑托扳，swivel to support 意为“与支撑托板铰接”。全句译为“微倾水准仪望远镜不是固定在水准仪的基座上，而是与支撑托板铰接并能在垂直面内作微小倾斜”。

[2] The axis of tubular level (*LL*) is the imaginary longitudinal line tangent to the upper inside surface at the midpoint. 句中 tangent to the upper inside surface at the midpoint 可看为 longitudinal line(纵向的线)的后置定语，其意为“通过水准管中点并切于水准管上部内表面的切线”，这样切线有多条，特指水准管纵向的一条，所以译为“水准管轴可想象为通过中点切于其上内表面的纵向切线”。

[3] the larger the radius, the more sensitive the bubble. “the more……, the more……,”这种结构英语句子非常常见，意为“越……，越……”所以这句话译为“半径越大，气泡的灵敏度越高”。

4.3.2 Automatic Levels

1. Features of the automatic level

In the automatic level, examples of which are shown in Fig. 4-14, a tubular level is no longer used to set a horizontal line of collimation. *1[Instead, the line of collimation is directed through a system of compensators which ensure that the line of sight viewed through the telescope takes horizontal reading even if the optical axis of the telescope tube itself is not horizontal.]

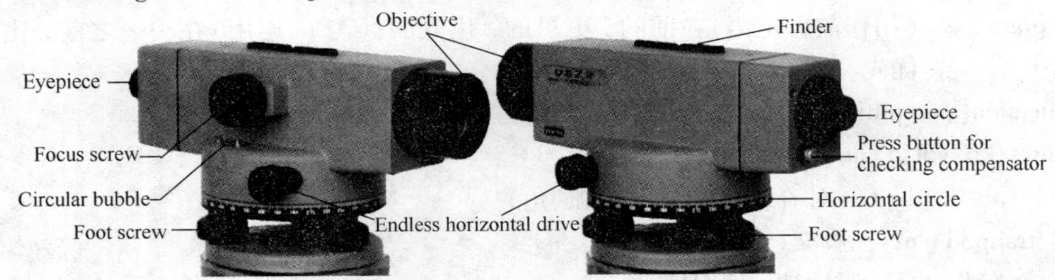

Fig. 4-14 Automatic level (Suzhou SOIF Co., Ltd DSZ2)

*2[As the automatic level is only approximately leveled by means of its low-sensitivity circular bubble, the collimation axis of the instrument will be inclined to the horizontal by a small angle α (see Fig. 4-15b) so the entering ray would strike the plane of the cross-hairs at *b* with a displacement of *Zb* equal to fα.] The compensator situated at K would need to redirect the ray to pass through the cross-hair at *Z*. Thus

$$f\alpha = Zb = s\beta$$

$$\beta = \frac{f\alpha}{s} = n\alpha \tag{4-6}$$

It can be seen from this that the positioning of the compensator is a significant feature of the compensation process. For instance, if the compensator is fixed halfway along the telescope, then $s \approx f/2$ and $n = 2$, giving $\beta = 2\alpha$. There is a limit to the working range of the compensator, about 15′. This is achieved by using the three footscrews together with the circular bubble.

In order to compensate for the slight residual tilts of the telescope, the compensator requires a reflecting surface fixed to the telescope, movable surfaces influenced by the force of gravity and a dampening device (air or magnetic) to swiftly bring the moving surfaces to rest and rapid viewing of the staff. Such an arrangement is illustrated in Fig. 4-15a. When the telescope is horizontal, the line of sight is as shown in Fig. 4-15a, the suspended prisms take up a position under the influence of gravity. If the telescope is not horizontal, the line of collimation tilts through angle α from the horizontal at O (see Fig. 4-15b). In this case, the fixed prism also tilt angle α, while the suspended prisms take up original position under the influence of gravity. A horizontal ray of light entering the compensator will be occur multiple refraction and refraction inside the compensator, and finally, it is deflected by angle β to pass through the centre of the cross hair Z. Therefore, the system of compensators can ensure that the line of sight viewed through the telescope takes horizontal reading a_0.

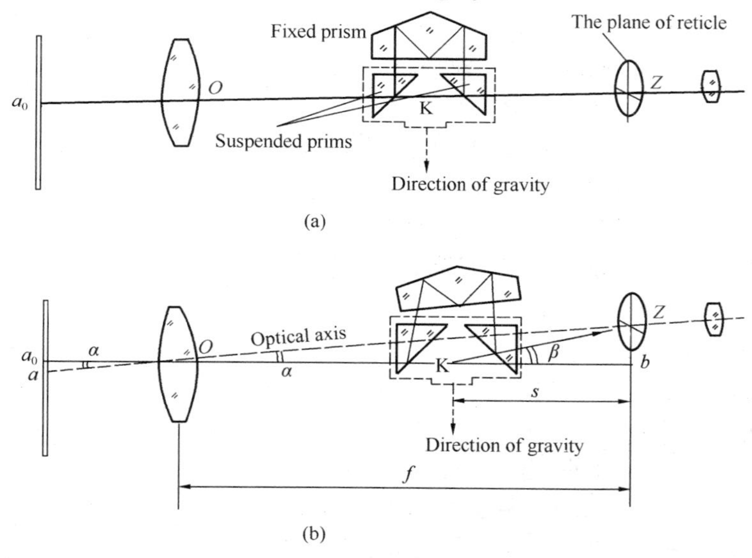

Fig. 4-15 Principle of compensator

(a) Telescope is horizontal; (b) Telescope is tilted

2. Use of the automatic level

When an automatic level has been roughly levelled, the compensator function can automatically to take reading of a horizontal line of sight. Therefore, no further levelling is required after the initial levelling.

As with all types of level, parallax must be removed before any readings are taken. In addition to the levelling procedure and parallax removal, a test should be made to see if the compensator is functioning before readings commence. Ordinary automatic levels have a button, which when pressed moves the compensator to prevent it sticking. If the compensator is working, the horizontal hair is seen to move and then return immediately to the horizontal line of sight. Some levels incorporate a warning device that gives a visual indication to an observer, in the field of view of the telescope, when the instrument is not level.

The particular advantage of the automatic level is the greater speed with which accurate levelling may be carried out compared with using conventional levels. This is attributable to the fact that there is no main bubble to centralize. In addition, the compensating system eliminates errors caused by either forgetting to set the bubble or setting it inaccurately.

单词与词组

· automatic level 自动安平水准仪
· endless horizontal drive 无限位水平驱动螺旋
compensator [ˈkɔmpenseitə] *n.* 补偿器
residual[riˈzidjuəl] *a.* 残留(渣)的；剩(余)的，未加说明的
reflect [riˈflekt] *vt.* 反射，映出(形象)
movable [ˈmu:vəbl] *a.* 活动的，可移动的
influence[ˈinfluəns] *n.* 影响；感化(力)；*vt.* 影响，改变
· force of gravity 重力
dampen[ˈdæmpən] *vt.* 阻尼，减震，缓冲
swift [swift] *a.* 迅速的，突然的，短促的
arrangement [əˈreindʒmənt] *n.* 整顿，布置，排列，安装，装配
suspend[səsˈpend] *vt.* 吊起，悬挂
original[əˈridʒnəl] *a.* 最初的，最早的，原始的
deflect [diˈflekt] *v.* (使)偏折，(使)偏离，(使)转向
refraction [riˈfrækʃən] *n.* 折射作用，折射度
· as with 正如……情况一样
function[ˈfʌkʃən] *n.* 功能，机能，作用，函数；*v.* 运行，起作用
sticking[ˈstikiŋ] *a.* 粘的，胶粘的
incorporate[ˈinkɔpərit] *vt.* (使)合并，并入，加上
indication[ˌindiˈkeiʃən] *n.* 指示，标记，信号(设备)
advantage[ədˈvɑ:ntidʒ] *n.* 利益，益处，优越性，优点
attributable[əˈtribjutəbl] *a.* 可归于……的
inaccurately[inˈækjuritli] *adv.* 不精密地，不准确地

注释

[1] Instead, the line of collimation is directed through a system of compensators which ensure that the line of sight viewed through the telescope takes horizontal reading even if the optical axis of the telescope tube itself is not horizontal. 理解本句应搞清句中关系代词的所指，which 指代 compensators，从 which 到句末是较长的定语从句，其中由 that 引导的宾语从句(到句末)，宾语从句中又含 even if 引导的让步状语从句。因此译为："取代之，视准线是直接通过补偿器系统，即使望远镜光轴本身不水平，补偿器系统确保通过望远镜观察的视线读取水平读数"。

[2] As the automatic level is only approximately leveled by means of its low-sensitivity circular

bubble, the collimation axis of the instrument will be inclined to the horizontal by a small angle α (see Fig. 4-15b) so the entering ray would strike the plane of the cross-hairs at *b* with a displacement of *Zb* equal to *f* α. 此句中第1句是由 as 引导的原因状语从句;从 the collimation axis 到 (see Fig. 4-15b)为结果状语从句;从 so 到句末才是主句。全句译为:"因为自动水准仪借助于低灵敏度的圆水准器仅能做到粗略整平,所以视准轴对水平面倾斜小角度α(图 4-15b),因而进入的光线射在十字丝环平面上 *b* 点,位移 $Zb = f\alpha$"。

4.3.3 Digital levels

Shown in Fig. 4-16, the digital level is similar in appearance to an automatic or tilting level. In use, it is set up in the same way as an automatic level by attaching it to a tripod and centralizing a circular bubble using the footscrews. A horizontal line of sight is then established by a compensator and readings could be taken from levelling staff, all readings are taken and recorded completely automatically.

This instrument has been designed to carry out all reading and data processing automatically via an on-board computer which is accessed through a display and keyboard. When leveling, a special bar-coded staff is sighted (see Fig. 4-17), the focus is adjusted and a measuring key is pressed. There is no need to read the staff as the display will show the staff reading about two or three seconds after the measuring key has been pressed.

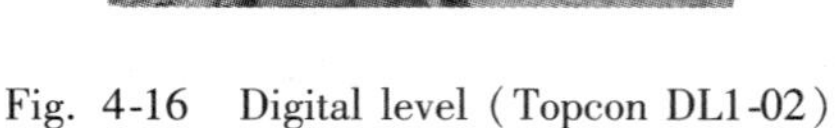

Fig. 4-16 Digital level (Topcon DL1-02)

Fig. 4-17 Bar-coded staff

[*1] [The instrument has a beam splitter in the optical path, which arranges for infra-red radiation to reach a photodiode array while visible light passes through a reticule and eyepiece.] When being used in scanning mode the photodiodes effectively replace the eye of the observer when the bar-coded staff is scanned. The photodiode array converts the bar-code image into a video signal, which is then amplified, digitized electronically and passed to a microprocessor.

The captured bar-coded image of the staff is compared by the on-board computer to the bar codes stored in the memory for the staff and when a match is found, this is the displayed staff reading. In addition to staff readings, it is also to display the horizontal distance to the staff with a precision of about 20 ~ 25mm. All readings can be coded using the keyboard and as levelving proceeds, each staff reading and subsequently all calculations are stored in the level's internal memory.

In good conditions, a digital level has a range of about 100m. The power supply for the digital level is standard AA or rechargeable batteries, which are capable of providing enough power for a complete day's levelling. If it is not possible to take electronic staff readings (because of poor lighting, obstructions such as foliage preventing a bar code from being imaged or loss of battery power), the reverse side of the bar-coded staff has a normal E type face and optical readings can be taken and entered manually into the instrument instead.

The digital level has many advantages over conventional levels since observations are taken quickly over longer distances without the need to read a staff or record anything by hand. This eliminates two of the worst sources of error from levelling—reading the staff incorrectly and writing the wrong value for a reading in the field book.

The data stored in a digital level can also be transferred to a removable memory card and then to a computer, where it can be processed further.

单词与词组

· digital level　数字水准仪

appearance [ə'piərəns] *n.* 出现，出场，登台；外观(表)，容貌，体面

identify [ai 'dentifai] *vt.* 使等同于，认出，识别，鉴别，验明认为与……有关系；参与

centralize ['sentrəlaiz] *v.* 使(国家等)实行中央集权制，成为……的中心，把……集中起来，集中，形成中心

via [ˈvaiə, 'vi:ə] *prep.* 经过，经由，取道，以……作媒介，通过

· on-board computer　机载电脑，机内电脑

· beam splitter　电子射线分离器

arrange[ə'reindʒ] *v.* 安排，准备，处理

· infra-red radiation 红外辐射

photodiode[ˌfəutəu 'daiəud] *n.* 光电二极管

array[ə'rei] *n.* 展示，陈列，数组，阵列

convert[kən 'və:t] *v.* 转变为，转化

microprocessor['maikrəˌprɔsesə] *n.* 微处理机

encoder[in 'kəudə] *n.* 译码器，编码器

detector[di 'tektə] *n.* 探测器，检波器，指示器

foliage['fəuliidʒ] n. 树叶

prevent[pri 'vent] *vt.* 防止，预防，阻挡，妨碍(from)

reverse[ri 'və:s] *a.* 相反的，颠倒的，反向的；*vt.* 颠倒，翻转

transfer[træns 'fə:] *vt.* 转(迁)移，传送

注释

* The instrument has a beam splitter in the optical path, which arranges for infra-red radiation to reach a photodiode array while visible light passes through a reticule and eyepiece. 此句后半句为 which 引导的非限制性定从句，which 指主句中 beam splitter(电子射束分离器)，arranges for(安排)，which arranges for infra-red radiation to reach a photodiode 直译：电子射束分离器安排红外射线进入光电二极管阵列，改译为：电子射束分离器分离出红外射线到达光电二极管阵列。

4.4 Field procedure for levelling

When a level has been correctly set up, the line or plane of collimation generated by the instrument coincides with or is very close to a horizontal plane. If the height of this plane is known, the heights of ground points can be found from it by reading a vertically held leveling staff.

In Fig 4-18, a level has been set up at point $I_1$and readings R_1 and R_2 have been taken with the staff placed vertically in turn at ground points A and B. If the reduced level of A (RL_A) is known then, by adding staff reading R_1 to RL_A, the reduced level of the line of collimation at instrument position I_1 is obtained. This is known as the *height of the plane of collimation* (HPC) or the *collimation level*. This is given by

$$\text{collimation level at } I_1 = RL_A + R_1 \quad (4\text{-}7)$$

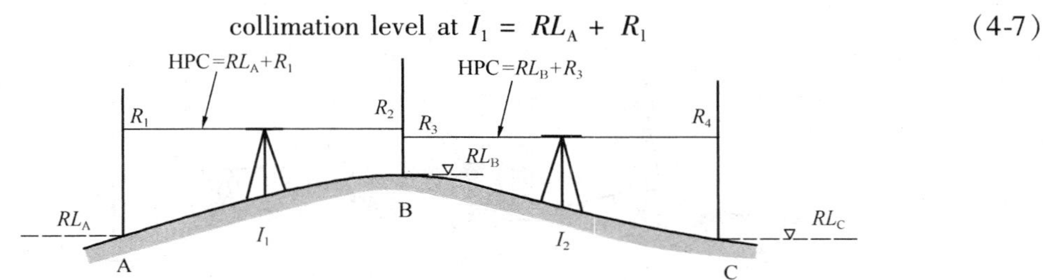

Fig. 4-18 Levelling from A to C

In order to obtain the reduced level of point B (RL_B), staff reading R_2 must be subtracted from the collimation level to give

$$RL_B = \text{collimation level} - R_2 = (RL_A + R_I) - R_2 = RL_A + (R_1 - R_2) \quad (4\text{-}8)$$

The direction of levelling in this case is from A to B and R_1 is taken with the level facing in the opposite direction to this. For this reason it is known as a backsight (BS). Since reading R_2 is taken with the level facing in the direction from A to B, it is called a foresight (FS). The height change between A and B, both in magnitude and sign, is given by the difference of the staff readings taken at A and B. Since R_1 is greater than R_2in this case, ($R_l - R_2$) is positive and the base of the staff has risen in moving from A to B. Because ($R_1 - R_2$) is positive it is known as a rise.

The level is now moved to a new position I_2 so that the reduced level of C can be found. Reading $R_3$is first taken with the staff still at point B but with its face turned towards I_2. This will be the back sight at position I_2 and the fore sight R_4 is taken with the staff at C. At point B, both an FS and a BS have been recorded consecutively, each from a different instrument position and this is called a turning point (TP).

From the staff readings taken at I_2, the reduced level of C (RL_C) is calculated from

$$RL_C = RL_B + (R_3 - R_4)$$

The height difference between B and C is given both in magnitude and sign by ($R_3 - R_4$). In this case, ($R_3 - R_4$) is negative since the base of the staff has fallen from B to C. This time, the difference of the staff readings is known as a fall.

From the above, it can be seen that when calculating a rise or fall, this is always given by (back sight-fore sight). If this is positive, a rise is obtained and if negative, a fall is obtained.

In practice, a BS is the first reading taken after the instrument has been set up and is always to a bench mark or calculated reduced level. Conversely, a FS is the last reading taken at an instrument position. Any readings taken between the BS and FS from the same instrument position are known as *intermediate sights* (IS).

A more complicated leveling sequence is shown in cross-section and plan in Fig 4-19, in which an engineer has leveled between two *TBM*s to find the reduced levels of points A to E. The readings could have been taken with any type of level and the field procedure followed to determine the reduced levels is as follows.

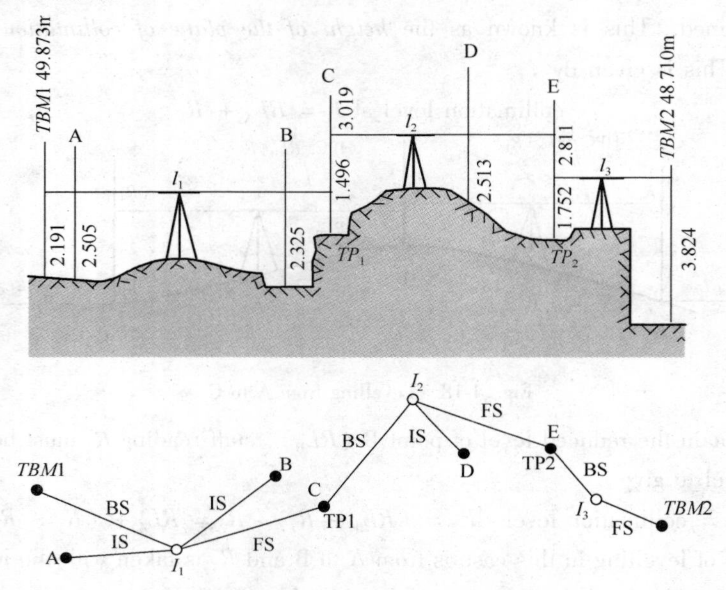

Fig. 4-19 A more complicated levelling

(1) The level is set up at some convenient position I_1, and a BS of 2.191m is taken to *TBM*1, the foot of the staff being held on the *TBM* and the staff held vertically.

(2) The staff is then moved to points A and B in turn and readings are taken. These are intermediate sights of 2.505m and 2.325m respectively.

(3) A change point must be used in order to reach D owing to the nature of the ground. Therefore, a change point is chosen at C (*TP*1) and the staff is moved to C and the staff is moved to C and a FS of 1.496m taken.

(4) While the staff remains at C, the instrument is moved to another position I_2. A BS of 3.019m is taken from the new level position to the staff at turning point C.

(5) The staff is moved to D and E in turn and readings of 2.513m (IS) and 2.811m (FS) are taken where E is another *TP*2.

(6) Finally, the level is moved to I_3, a BS of 1.752m taken to E and a FS of 3.824m taken to TBM2.

The final staff position is at a *TBM*, it is most important that all levelling field work must start and finish at a bench mark, otherwise it is not possible to detect errors in the levelling.

单词与词组

coincide [ˌkəuin 'said] *vi.* 与……相重合，与……一致(相符合)(with)

· in turn 依次，轮流地

· reduced level (RL) 归化高程

· height of the plane of collimation (HPC) 视准面高程，又称视线高程

subtract[səb 'trækt] *vt.* 减去，扣除

face[feis] *n.* 脸，面孔，面貌;*vt.* 朝，临，面向，面对，应付

opposite['ɔpəzit] *a.* 对面的，对立的，相反的

· back sight (BS) 后视读数

· fore sight (FS) 前视读数

magnitude['mægnitju:d] *n.* 大小，积，量，长(度)，尺寸，幅度，宽狭

positive ['pɔzətiv] *a.* 正的，阳性的；确实的，明确的，确定的，确信的

rise[raiz] *vi.* 升起，上升

consecutively [kən 'sekjutivli] *adv.* 连续地,依顺序地

fall[fɔ:l] *vi.* (fell[fel], fallen['fɔ:lən]) 下降，落下，降落

· bench mark(BM) 水准点

· temporary bench mark(TBM) 临时水准点

conversely ['kɔnvə:sli] *adv.* 相反地，颠倒地;*a.* 复杂的,麻烦的

complicate ['kɔmplikeit] *vi.* 变复杂

· intermediate sights (IS) 中间视线

cross-section ['krɔ:s 'sekʃən] *n.* 横断面

otherwise ['ɔðəwaiz] *adv.* 另外，否则

detect[di 'tekt] *vt.* 发觉，探测，检定

4.5 Assessment of levelling precision

There are three types of levelling routes which are frequently used in levelling, they are:

1. A connecting levelling route

Levelling is carried out from a starting point BM*A* of known along the elevation points to be determined connecting into another BMB of known, it is called a connecting levelling route as shown in Fig. 4-20.

Theoretically, the sum of measuring every section height differences in a route (Σh_i) should equal to difference of elevations between two bench marks of known, that is

$$\sum h_i = H_B - H_A$$

Because of measuring errors, so $\sum h_i \neq H_B - H_A$, the misclosure for levelling is given by

$$f_h = \sum h_i - (H_B - H_A) \tag{4-9}$$

Where f_h is called *misclosure for levelling*.

2. A loop levelling route

Levelling is carried out from a starting point BM*A* of known along the elevation points to be determined back to the BM*A* from which it started, it is called a loop levelling route as shown in Fig. 4-21. Obviously, the misclosure in levelling (f_h) can be written by

$$f_h = \sum h_i \tag{4-10}$$

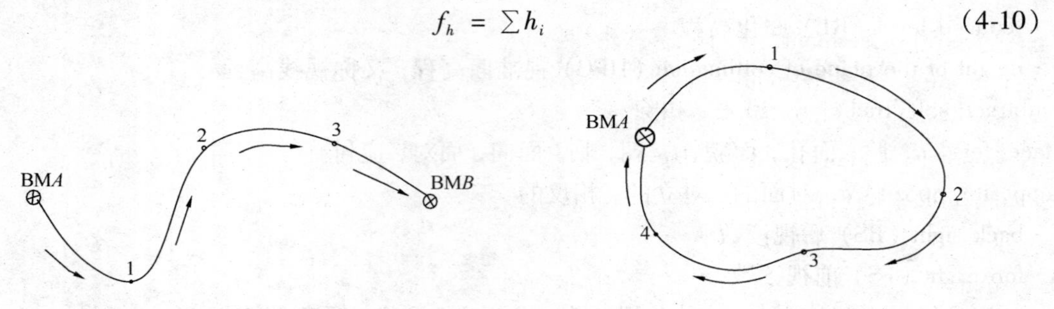

Fig. 4-20 A connecting levelling route

Fig. 4-21 A loop levelling route

3. An open-ended levelling route

Since lack of inner checking condition for an open-ended levelling route, it is necessary to measure and return to measure point P of unknown (See Fig. 4 – 22). Therefore, the misclosure for levelling is given by

$$f_h = \sum h_{to} + \sum h_{return} \tag{4-11}$$

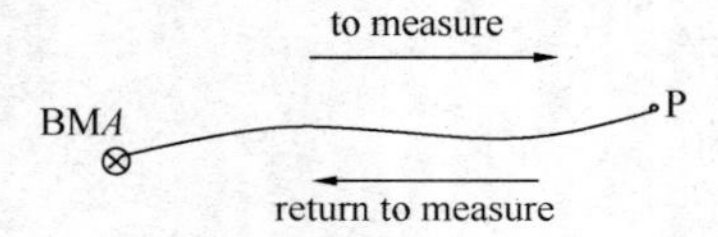

Fig. 4 – 22 An open-ended levelling route

In the section 4. 6. 1, a misclosure of – 6mm was obtained for the levelling by comparing the reduced level of the closing bench mark TBM2 (48. 710m) with its initial *RL* obtained from staff readings (48. 704m). By comparing the two reduced levels for TBM2 in this way an assessment of the quality or precision of the levelling can be made and it is usual to check that the misclosure obtained must be less than some specified value called the allowable misclosure.

On construction sites and other engineering projects, levelling is usually carried out over short distances and it can include a lot of instrument positions. For this type of work, the allowable misclosure for levelling is given by

$$f_{h-allowable} = \pm m\sqrt{n} \tag{4-12}$$

Where $f_{h-allowable}$ is the allowable misclosure in millimetres, m is a constant with units of millimetres and n is the number of instrument positions. For construction levelling, the value of m frequently used is ±5 mm.

When the misclosure obtained from staff readings is compared to the allowable misclosure and it is found that the misclosure is greater than the allowable value, the levelling is rejected and has to be repeated. If the misclosure is less than the allowable value, the misclosure should be equal distributed to each station positions, it is detail discussed in the next section.

When assessing the precision of any levelling by this method, it may be possible for a site engineer to use a value of m based on site conditions. For example, if the reduced levels found are to be used to set out earthwork excavations, the value of m might be ±30 mm but for setting out steel and concrete structures, the value of m might be ±3 mm. Values of m may be specified as tolerances in contract documents or where they are not given, may simply be chosen by an engineer based on experience.

单词与词组

assessment [ə 'sesmənt] *n.* 评估,评价,评定

· connecting levelling route　附合水准路线

· loop levelling route　闭合水准路线

open-ended ['opən 'endid] *a.* 没有固定限度的, 广泛的,无限度的, 可以变更的

· open-ended levelling route　无约束水准路线,支水准路线

misclosure [mis 'kləuʒə] *n.* 闭合差

allowable [ə 'lauəbl] *a.* 可允许的, 可承认的, 可原谅的

· allowable misclosure　容许闭合差

reject [ri 'ʤekt] *vt.* 拒绝, 抵制,丢弃,否决, 否认

distribute [dis 'tribjut] *vt.* 分配, 分给, 分送, 区分,分类, 分布, 散布

excavation [ekskə 'veiʃən] *n.* 挖掘, 开凿

tolerance ['tɔlərəns] *n.* 宽容, 容忍

earthwork ['ə:θwə:k] *n.* 土方(工程), 土木工事

concrete ['kɔnkri:t] *a.* 实际的,具体的, 特定的; 固结成的, 混凝土制的, 水泥的

contract [kən 'trækt] *n.* 契约, 合同

4.6　Calculations of levelling

4.6.1　Height of collimation method

In using the height of collimation method, the height of collimation is calculated from the reading taken to a staff station of known elevation. When the instrument is shifted, the height of collimation will be changed. Table 4-1 shows the field book for the levelling shown in Fig. 4-19. The calculations in Table 4-1 are using the height of collimation method. The method is based on the HPC calculated for each instrument position. The computational steps are as follows.

Table4-1　Height of collimation method

Point No	Staff reading			HPC (m)	Initial RL (m)	Adjustment (m)	Adjusted RL (m)
	BS (m)	IS (m)	FS (m)				
TBM 1	2.191			52.064	49.873		49.873
A		2.505			49.559	+0.002	49.561
B		2.325			49.739	+0.002	49.741
C (*TP*1)	3.019		1.496	53.587	50.568	+0.002	49.570
D		2.513			51.074	+0.004	51.078
E (*TP*2)	1.752		2.811		50.776	+0.004	49.741

Continued

Point No	Staff reading			HPC	Initial RL	Adjustment	Adjusted RL
	BS (m)	IS (m)	FS (m)	(m)	(m)	(m)	(m)
TBM 2			3.824	52.528	48.704	+0.006	48.710
Checks	6.962		8.131				
	f_h = Last Initial RL − Last known RL = 48.704 − 48.710 = −0.006m						

Note: The date, observer and booker (if not the observer), the survey title, level number, weather conditions and anything else relevant should be recorded as well as the staff readings

(1) If the BS reading taken to TBM1 is added to the RL of this bench mark, then the HPC for the instrument position I_1 will be obtained. This will be 49.873 + 2.191 = 52.064m and this is entered in the HPC column on the same line as the BS (2.191).

(2) To obtain the initial reduced levels of A, B and C the staff readings to those points are now subtracted from the HPC. The relevant calculations are

RL of A = 52.064 − 2.505 = 49.559m

RL of B = 52.064 − 2.525 = 49.739m

RL of C = 52.064 − 1.496 = 50.568m

(3) At point C, a change point, the instrument is moved to position I_2 and a new HPC is established. This collimation level is obtained by adding the BS at C to the RL found for C from I_1. For position I_2, the HPC is 50.568 + 3.019 = 53.587m. The staff readings to D and E are now subtracted from this to obtain their reduced levels.

(4) The procedure continues until the Initial RL of TBM2 is calculated. With the initial RL column in the table completed, the following check can be applied.

$$\sum BS - \sum FS = \text{LAST initial RL} - \text{FIRST RL}$$

that is $$6.962 - 8.131 = 48.704 - 49.873 = -1.169$$

Obviously, in this case the misclosure of levelling is the difference between last initial RL obtained from observations (48.704m) and its known value (48.710m). Therefore the height misclosure f_h can be written

$$f_h = \text{Last initial RL} - \text{Last known RL} = 48.704 - 48.710 = -0.006\text{m}$$

After the checking, the misclosure is distributed according to apply an equal, but cumulative, amount of the misclosure to each instrument position, the sign of the adjustment being opposite to that of the misclosure.

Since there is a misclosure of −0.006m in this example a total adjustment of +0.006m must be distributed. As there are three instrument positions, +0.002m is added to the reduced levels found from each instrument position. The distribution is shown in the adjustment column, in which the following cumulative adjustments have been applied: Levels A, B and C +0.002 m, levels D and E + (0.002 + 0.002) = +0.004m and TBM2 + (0.002 + 0.002 + 0.002) = +0.006m. No adjustment is applied to *TBM*1 since this is temporary bench mark. The adjustments are applied to the Initial RL values to give the Adj (adjusted) RL values in Table 4 −1.

4.6.2 The rise and fall method

If the level of one of the two points is known, the level of the other can be found by adding the level of known point and the height difference of the two points. In levelling of one instrument position, when the backsight reading greater than the foresight reading, the height difference is positive, it indicates a rise; but when the backsight reading less than the foresight reading, the height difference is negative, it indicates a fall. The reduced level of all points can be obtained by continuously adding a rise (or a fall), this is called a rise and fall method.

Table 4-2 shows all the records for the levelling shown in Fig. 4-19. All calculations will be carried out by a rise and fall method.

Table 4-2 Rise and fall method

Point NO.	Staff readings			Height differences h_i		Adjusted height differences h_{i-adj} (m)	Adjusted reduced levels RL (m)
	BS (m)	IS (m)	FS (m)	Rise (m)	Fall (m)		
TBM 1	2.191						49.873
A		2.505			-0.314	-0.313	49.560
B		2.325		0.180		+0.181	49.741
TP1	3.019		1.496	0.829		+0.830	50.570
D		2.513		0.506		+0.507	51.078
TP2	1.752		2.811		-0.298	-0.297	50.780
TBM 2			3.824		-2.072	-2.071	48.710
$\sum$	6.962m		8.131m	1.515m	-2.684m	-1.163	-1.163
Checks	$\sum$BS − $\sum$FS = 6.962 − 8.131 = −1.169m $\sum h$ = $\sum$Rise + $\sum$Fall = 1.515 − 2.684 = −1.169m Misclosure $f_h = \Sigma h - (H_B - H_A)$ Allowable misclosure $f_{t-allowable} = \pm m\sqrt{n} = \pm 5\sqrt{3} = \pm 9$mm = −1.169 − (48.710 − 49.873) = −0.006m						

Note: The date, observer and booker (if not the observer), the survey title, level number, weather conditions and anything else relevant should be recorded as well as the staff readings.

(1) From the TBM1 to A there is a small fall. A BS of 2.191m has been recorded at TBM1 and an IS of 2.505m at A. So, for the fall from TBM1 to A, the height difference is given by (2.192 − 2.505) = −0.314m. The negative sign indicates a fall. −0.314m must be filled in the relevant position of column named height difference.

(2) The procedure is repeated and the height difference from A to B is given by (2.505 − 2.325) = +0.180m. The positive sign indicates a rise. +0.180m must be filled in the relevant position of column named height difference.

(3) The rise from B to C up to TP1 is (2.325 − 1.496) = +0.829m. The height change from C to D is (3.019 − 2.513) = +0.506m. These data must be filled in the relevant position of column named height difference.

(4) Such calculation is repeated until the height differences of every segment are calculated, and then check on the arithmetic must be applied. This check is

$$\Sigma BS - \Sigma FS = \Sigma Rises + \Sigma Falls$$

thal is $6.962 - 8.131 = 1.515 + (-2.684) = -1.169\text{m}$

(5) The misclosure of the levelling is calculated by formula (4 – 9). That is

$$f_h = \Sigma h - (H_B - H_A) = -1.169 - (48.710 - 49.873) = -0.006\text{m}$$

The allowable misclosure $f_{h-allowable}$ is calculated by

$$f_{h-allowable} = \pm m\sqrt{n} = \pm 5\sqrt{3} = \pm 9\text{mm}$$

Since there is the misclosure of −0.006m less than the allowable value so the misclosure with opposite sign should be equal distributed to height difference of every section, and then the adjusted height differences are calculated by

$$h_{1-adj} = -0.314 + 0.001 = -0.313\text{m}$$
$$h_{2-adj} = +0.180 + 0.001 = +0.181\text{m}$$
$$h_{3-adj} = +0.829 + 0.001 = +0.830\text{m}$$
$$\vdots$$
$$h_{6-adj} = -2.072 + 0.001 = -2.071\text{m}$$

The adjusted height differences fill in the relevant position in Table 4-2. At last column, adjusted reduced level of points A to TP2 are computed point-by-point from starting point TBM1.

Example 4 – 1 It is gives a practical example for a connecting levelling route as shown in Fig. 4-23. The point A and point E are the two temporary bench marks, their elevations: $H_A = 89.763\text{m}$, $H_E = 93.504\text{m}$. The point B, C and D are three unknown points to determine their elevations, the measuring height differences and numbers of instrument set-ups in every section note in Fig. 4 – 23. It is now desired to compute elevations of unknown points B, C and D.

Solution

The results in the field book have been arranged and filled in Table 4-3, and then calculations should be carried out according to the columns in the table.

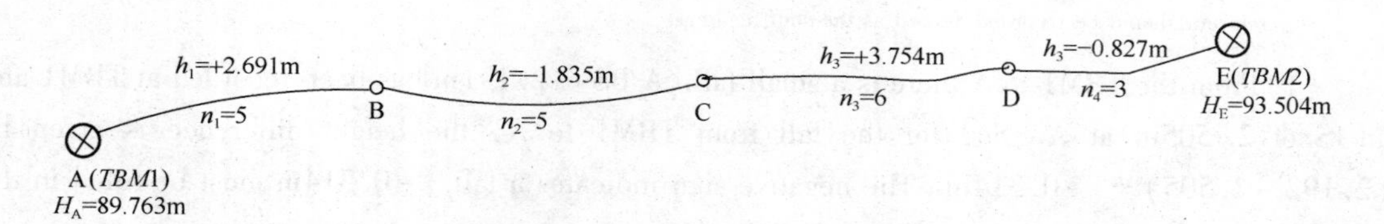

Fig. 4-23 A practical example for a connecting levelling route

(1) The sum of measuring height difference of each section is calculated by

$$\sum h_{mea} = +2.691 - 1.835 + 3.754 - 0.827 = +3.783$$

(2) The height difference of two known points is $H_E - H_A = 95.504 - 89.763 = +3.741\text{m}$

(3) The misclosure of levelling is calculated by formula (4 – 9), that is

$$f_h = \sum h_{mea.} - (H_E - H_A) = +0.042\text{m}$$

(4) The allowable misclosure in railway surveying specification in China is:

$$f_{h\text{-}allowable} = \pm 12\sqrt{n} = \pm 12\sqrt{19} = \pm 52\text{mm}$$

(5) The corrections of height difference (v_i) are calculated by

$$v_i = \frac{-f_h}{n} \times n_i \tag{4-13}$$

Table 4-3 The computations for a connecting levelling route

Point NO	Numbers of station setup	Measuring height difference h_i (m)	Corrections of height difference v_i(m)	Adjusted height difference h_{i-adj} (m)	Adjusted elevations H (m)
A(TBM1)					87.763
	5	+2.691	−0.011	+2.680	
B					92.443
	5	−1.835	−0.011	−1.846	
C					90.597
	6	+3.754	−0.013	+3.741	
D					94.338
	3	−0.827	−0.017	−0.834	
E(TBM2)					93.504
Σ	19	+3.783	−0.042	−3.741	+3.741
checks	misclosure: $f_h = \sum h_{mea.} - (H_E - H_A) = +3.783 - (95.504 - 89.763) = +0.042$m Allowable misclosure: $f_{h-allowbclc} = \pm 12\sqrt{n} = \pm 12\sqrt{19} = 52 \pm$ mm Corrections of height difference: $v_1 = -0.0022 \times 5 = -0.011$m; $v_2 = -0.011$m; $v_3 = -0.0022 \times 6 = -0.013$m; $v_4 = -0.0022 \times 3 = -0.007$m					

where n = total numbers of instrument set-up for whole levelling route,

n_i = numbers of instrument set-up for a certain section.

The corrections of height difference of every section are

$v_1 = -0.0022 \times 5 = -0.011$m $v_2 = -0.011$m $v_3 = -0. .0022 \times 6 = -0.013$m $v_4 = -0.0022 \times 3 = -0.007$m

(6) The adjusted height difference (h_{i-adj}) are calculated by

$$h_{i-adj} = h_{i-mea} + v_i$$

$$\therefore \quad h_{1-adj} = h_{1-mea} + v_1 = +2.691 - 0.011 = +2.680\text{m}$$

$$H_{2-adj} = -1.846\text{m}; H_{3-adj} = +3.741\text{m}; H_{4-adj} = -0.834\text{m}$$

(7) The adjusted elevation of points B to D are computed point-by-point from starting point A.

单词与词组

· height of collimation method 视高法

· rise and fall method 高差法

column ['kclәm] *n.* 列，栏

reduction[ri 'dʌkʃәn] *n.* 归纳，整理，缩减，简化

reduce[ri 'dju:s] *vt.* 减少，使成为，把……分类，简化

relevant['relivәnt] *a.* 有关的，恰当的，贴切的，切题的，中肯的(to)，成比例的，相应的

set-up [setʌp] *n.* 建立,装配,(仪器安置)位置,计划,方案

· instrument set-ups 仪器的安置站,仪器测站

注释

* The usual method of correction is to apply an equal, but cumulative, amount of the misclosure to each instrument position, the sign of the adjustment being opposite to that of the misclosure。此句要分成两部分:句首至 position 为第一部分,其后为第二部分。句中“The usual method of correction is to apply an equal, but cumulative”,意思是改正的常用方法是执行相等(翻译改为平均分配)并要考虑积累,amount of the misclosure to each instrument position(每个仪器站的闭合差的数量)短语成分为对 cumulative 的具体解释说明,即每个仪器站闭合差会积累。最后的部分容易理解,但应注意句中 that 指代其前面的 sign。所以全句译为:“进行改正常用的方法是对每个测站平均分配闭合差,但应注意仪器站闭合差的积累,调整数的符号反其闭合差的符号”。

4.7 Instrument adjustment

For equipment to give the best possible results it should be frequently tested and, if necessary, adjusted. Surveying equipment receives continuous and often brutal use on construction sites. In all such cases a calibration base should be established to permit weekly checks on the equipment.

4.7.1 Circular level adjustment

Although the circular level is relatively insensitive, nevertheless it plays an important part in the efficient functioning of the compensator.

The compensator has a limited working range. If the circular level is out of adjustment, thereby resulting in excessive tilt of the line of collimation and the vertical axis, the compensator may not function efficiently or, as it attempts to compensate, the large swing of the pendulum system may cause it to stick in the telescope tube. Therefore, it is very necessary to adjust circular level.

From the above it can be seen that not only must the circular level be tested or adjusted but it should also be accurately centered when in use.

The adjustment purpose is to ensure that the axis of circular level is parallel to the vertical axis.

To adjust the circular level, firstly bring it exactly to centre using the footscrews, and then rotate the instrument through 180° about the vertical axis. If the bubble moves off centre, bring it halfway back to centre with the footscrews and then exactly back to the centre using its adjusting screws.

4.7.2 Adjustment for collimation axis

1. Tilting level adjustment

The tilting level requires adjustment for collimation error. Collimation error occurs if the line of sight is not truly horizontal when the tubular bubble is centered, i. e. the line of sight is inclined up or down from the horizontal plane. The adjustment purpose is to ensure that the axis of tubular level is parallel to the line of collimation.

The usual method of testing and adjusting a level is to carry out a two-peg test. The two-peg test is carried out as follows, with reference to Fig. 4-24.

(1) On the ground, hammer in two pegs A and B about 60 ~ 80m apart. Let this distance be D

meters.

(2) Set up the level exactly midway between the pegs at point M and level carefully. Place a levelling staff at each peg in turn and obtain readings a_1 and b_1, as shown Fig. 4-24.

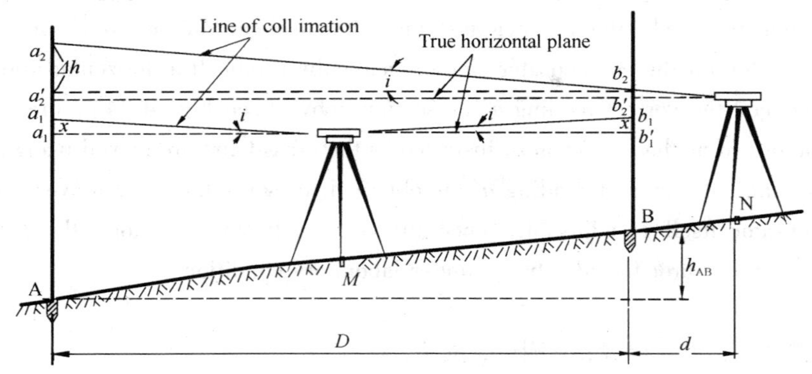

Fig. 4-24 Two-peg test

Since $AM = MB$, the error x in the readings a_1 and b_1 will be the same. This error is due to the collimation error, the effect of which is to incline the line of collimation by angle i. It can be seen from the figure:

$a_1 - b_2 = (a_1' + x) - (b_2' + x) = a_1' - b_1' = h_{AB}$ (i. e., true difference in height between A and B)

(3) Move the level about 3 ~ 5m from peg B at N and take readings a_2 at A and b_2 at B. Compute the *apparent difference* in height between A and B from $(a_2 - b_2)$, i. e. $h'_{AB} = (a_2 - b_2)$.

(4) Compute the collimation error that the line of collimation is not horizontal but inclined by angle i. Let us assume

$$\Delta h = h'_{AB} - h_{AB} = (a_2 - b_2) - h_{AB} = [a'_2 + a_2a'_2 - (b'_2 + b_2b'_2)] - h_{AB}$$

Considercing $a'_2 - b'_2 = h_{AB}$, the equation above becomes

$$\Delta h = a_2a'_2 - b_2b'_2$$

From Fig. 4-24 can be obtain

$$i = \frac{\Delta h}{D} \times \rho'' = \frac{|h'_{AB} - h_{AB}|}{D} \times \rho'' \tag{4-14}$$

If the collimation error $i < 20''$, then the level need not adjust, otherwise the level should be adjusted.

(5) Because of the instrument sets still at N, we can first deduce to observe the staff correct reading a'_2 at A by the following formula

$$a'_2 = a_2 - \frac{(D + d) \times i}{\rho''}$$

Then the telescope aims staff at A, the tilting screw is adjusted until the line of collimation takes a reading a_2'. This time, the line of collimation would be on the true horizontal plane, but the tubular bubble moves from the centre of its run, so it is brought back to the centre by adjusting the bubble capstan screws.

The test should be repeated to ensure that the adjustment has been successful.

2. Automatical level adjustment

On the automatic level, the two-peg test must be also carried out to ensure that once the circular bubble is central the compensator can automatically establish a horizontal line of sight.

Having deduced the correct reading a_2' as section above (see Fig. 4-24), the adjustment can be made by one of two methods. For most instruments the cross hairs are moved using the diaphragm adjusting screws until the correct reading a'_2 is obtained. In some levels, however, it is necessary that the compensator itself is adjusted. Since this is a delicate operation, the level should be returned to the manufacturer for adjustment under laboratory conditions.

单词与词组

test[test] *n.* 测试，试验，检验；*v.* 测试，试验，检验
adjust[ə'ʤʌst] *vt.* 调准，校正，调整，整理
brutal ['bru:tl] *a.* 残忍的，野蛮的，粗暴的，令人难受的
nevertheless [ˌnevəðə'les] *conj.* (尽管如此)还是，然而，不过
insensitive[in'sensitiv] *a.* 感觉迟钝的，不机敏的，低灵敏度的
efficient [i'fiʃənt] *a.* 有效率的，最经济的，有能力的，能胜任的
excessive [ik'sesiv] *a.* 过度的，份外的，额外的
swing [swiŋ] *vt.* (swung [swʌŋ]) 摆动，使旋转，使转向悬挂，吊运；*n.* 摆动，摇摆
pendulum ['pendjuləm] *n.* 摆，振动体
stick [stik] *v.* 插入，刺入，卡住；容忍，忍受
· collimation error 视准轴误差
parallel['pærəlel] *a.* 平行的，并行的
· two-peg test 双桩检测
hammer ['hæmə] *vi.* 接连锤打；延伸，拔长，推敲；埋头工作；重申，一再强调(away)
cause [kɔ:z] *vt.* 成为……的原因，导致；*n.* 原因，起因，理由，缘故
· bubble capstan screw 水准管的校正螺旋
apparent [ə'pærənt] *a.* 明显的，明白的，表面上的，外观上的，貌似的，表观的，视在的
· apparent difference 貌似高差，非真正高差
capstan['kæpstən] *n.* 绞盘，六角刀架
· capstan screws 校正螺丝
delicate['delikit] *a.* 精美的，精致的，精巧的

4.8 Curvature error and refraction

Over short distances a level line and a horizontal line (the line of sight through the telescope of a surveyors' level) are taken to coincide; but over long distances a correction for their divergence becomes necessary. When considering the divergence between level and horizontal lines, one must also account for the fact that all sight lines are refracted downward by the earth's atmosphere.

Although the magnitude of the refraction error is dependent on atmospheric conditions, it is generally considered to be about one-seventh of the curvature error.

* [It is seen in Fig. 4-25 that the refraction error of AB compensates for part of the curvature error AE, resulting in a net error due to curvature and refraction (c − r) of *BE*.]

From Fig. 4-25, the curvature error can be computed by

Line of Sight
Horizontal
K
A
r
B
Level
d
C
E
c+r
R=Mean Radius of Earth =6371km
Vertical
R
Vertical
R
0

Fig. 4-25 Effects of curvature and refraction

$$(R+c)^2 = R^2 + D^2$$

$$R^2 + 2Rc + c^2 = R^2 + D^2$$

$$c(2R+c) = D^2$$

$$c = \frac{D^2}{2R+c} \approx \frac{D^2}{2R} \tag{4-15}$$

where: c is curvature error, D is the length of sight (D = KA, see Fig. 4-25), R is mean radius of Earth(i. e. ,6371km).

Refraction (r) is affected by atmospheric pressure, temperature, and geographic location, but, as noted earlier, it is usually expressed as being roughly equal to one-seventh of curvature error (c). The combined effects of curvature and refraction ($c-r$) can be determined from the following formula:

$$f = c - r = \frac{D^2}{2R} - \frac{1}{7} \times \frac{D^2}{2R} = \frac{3D^2}{7R} = 0.43\frac{D^2}{R} \tag{4-16}$$

It can be seen from the figures in Table 4 − 4 that f errors are relatively insignificant for differential levelling. Even for precise levelling, where distances of rod readings are seldom in excess of 60m, it would seem that this error is of only marginal importance. As a matter of fact, if successive backsight and foresight are made at equal distances from level to rod, it effectively cancels out this type of error.

Table 4-4 The combined effect (f) of curvature and refraction for different distance (D)

Distance (m)	30	60	100	120	150	200	300	350	400	500	1000
f (m)	0.0001	0.0002	0.0007	0.001	0.002	0.003	0.006	0.008	0.011	0.017	0.068

单词与词组

divergence[dai ˈvəːdʒəns] *n*. 分叉，分歧，偏离，偏差，发散性

compensate[ˈkɔmpənseit] *vt*. 补偿,偿还,酬报(for)

· curvature error 曲率误差

net[net] *a*. 净的，纯的

geographic[ˈʤiːə ˈgræfik] *a*. 地理学的,地理的

marginal[ˈmɑːʤinəl] *a*. 边缘的，边界的，不重要的，微小的，少量的

· as a matter of fact 事实上，实际上

注释

* “It is seen in Fig. 4-25 that the refraction error of AB compensates for part of the curvature error AE, resulting in a net error due to curvature and refraction ($c-r$) of BE.”此句后半句是由分词 resulting 引导的结果状语从句，该从句译为：由于曲率及折光差的影响，其结果纯误差为 BE 即（$c-r$）。全句译为：图 4-25 可看出折光差 AB 抵偿了部分曲率误差 AE，其结果纯误差为 BE 即（$c-r$）。

上述句型（由分词引导结果状语从句，进一步说明与补充主句）在科技英语中十分常见，现另举一例说明：

例：In a liquid or solid, the molecules are much closer together, resulting in much more material in a given volume. 在液体或固体中，分子结合得紧密得多，其结果在一定体积内的物质也多得多。

4.9 Trigonometrical levelling

Trigonometrical levelling is used where difficult terrain, such as mountainous areas, precludes the use of conventional differential levelling. It may also be used where the height difference is large but the horizontal distance is short such as heighting up a cliff or a tall building. The vertical angle and the slope distance between the two points concerned are measured. Slope distance is measured using electromagneitc distance measurers (EDM) and the vertical (or zenith) angle using a theodolite.

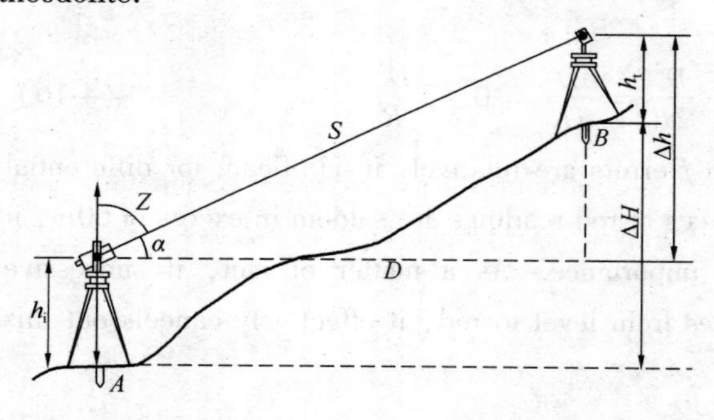

Fig. 4-26 Trigonometrical leveling-shot lines

Total stations contain algorithms that calculate and display the horizontal distance and vertical height. This latter facility has resulted in trigonometrical levelling being used for a wide variety of heighting porocedures, including contouring.

4.9.1 Short lines

From Fig. 4-26 it can be seen that when measuring the angle

$$\Delta h = S\sin\alpha \tag{4-17}$$

When using the zenith angle z

$$\Delta h = S\cos z \tag{4-18}$$

If the horizontal distance is used

$$\Delta h = D\tan\alpha = D\cot z \tag{4-19}$$

The difference in elevation (ΔH) between ground points A and B is therefore

$$\Delta H = h_i + \Delta h - h_t = \Delta h + h_i - h_t \tag{4-20}$$

where h_i = vertical height of the measuring centre of the instrument above A, h_t = vertical height of the centre of the target above B.

This is the basic concept of trigonometrical levelling. The vertical angles are positive for angles of elevation and negative for angles of depression. The zenith angles are always positive, but naturally when greater than 90° they will produce a negative result.

What constitutes a short line may be derived by considering the effect of curvature and refraction compared with the accuracy expected. The combined effect of curvature and refraction over 100m equal 0.7mm, over 200m egual 3 mm, over 300m egual 6mm, over 400 m egual 11mm and over 500m equal 17mm (see Table 4-4).

The basic equation(4-20) of trigonometrical levelling is written:

$$\Delta H = S\sin a + h_i - h_t$$

Take differential

$$d(\Delta H) = \sin\alpha dS + S\cos\alpha\, d\alpha + dh_i - dh_t$$

and taking standard errors:

$$\sigma_{\Delta H}^2 = (\sin\alpha \cdot \sigma_s)^2 + \left(S\cos\alpha \cdot \frac{\sigma_\alpha}{\rho''}\right)^2 + \sigma_i^2 + \sigma_t^2$$

Consider a vertical angle of $\alpha = 5°$, with $\sigma_\alpha = \pm 5''$ (= 0.000024 radians), $S = 300$m with $\sigma_s = \pm 10$mm and $\sigma_i = \sigma_t = \pm 2$mm. Substituting in the above equation gives:

$$\sigma_{\Delta H}^2 = 0.87^2 + 7.2^2 + 2^2 + 2^2$$

$$\sigma_{\Delta H} = \pm 7.8\text{mm}$$

This value is similar in size to the effect of curvature and refraction over this distance and indicates that short sights should never be greater than 300m (refer to Table 4-4). It also indicates that the accuracy of distance S is not critical when the vertical angle is small. However, the accuracy of measuring the vertical angle is very critical and requires the use of theodolite, with more than one measurement on each face.

4.9.2 Long lines

For long lines the effect of curvature (c) and refraction (r) must be considered. From Fig. 4-27, it can be seen that the difference in elevation (ΔH) between A and B is:

$$\begin{aligned}\Delta H &= GB = GF + FE + EH - HD - DB \\ &= h_i + c + \Delta h - r - h_t \\ &= \Delta h + h_i - h_t + (c - r)\end{aligned} \qquad (4\text{-}21)$$

Thus it can be seen that the only difference from the basic equation for short lines is the correction for curvature and refraction ($c-r$).

Although the line of sight is refracted to the target at D, the telescope is pointing to H, thereby measuring the angle α from the horizontal. It follows that S $\sin\alpha = \Delta h = EH$ and requires a correction for refraction equal to HD.

4.9.3 Reciprocal observations

Reciprocal observations are observations taken from A and B, the arithmetic mean result being accepted. If one assumes a symmetrical line of sight from each end and the observations are taken simultaneously, then the effect of curvature and refraction is cancelled out. For instance, for elevated sights, ($c - r$) is added to a positive value of vertical angle to increase the height

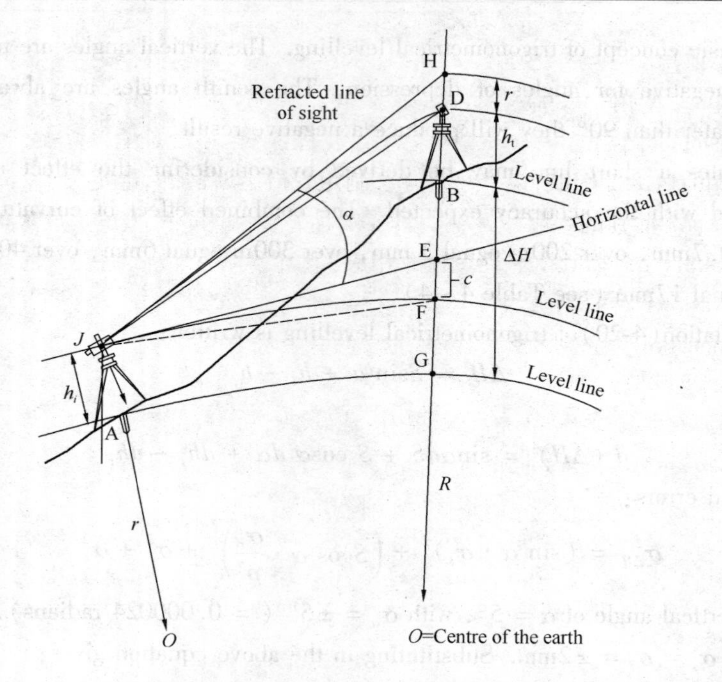

Fig. 4-27 Trigonometric leveling-long lines

difference. For depressed or downhill sights, $(c - r)$ is added to a negative value of vertical angle and decreases the height difference. Thus the average of the two values is free from the effects of curvature and refraction.

Observing from A to B, equation (4-10) gives $\Delta H_{AB} = \Delta h_{AB} + h_{Ai} - h_{Bt} + (c - r)$

Observing from B to A, equation (4-10) gives $\Delta H_{BA} = \Delta h_{BA} + h_{Bi} - h_{At} + (c - r)$

These two measurements of the reciprocal observations are averaged to give

$$\Delta H_{AB} = \frac{1}{2}(\Delta H_{AB} - \Delta H_{BA}) = \frac{1}{2}(\Delta h_{AB} - \Delta h_{BA}) + \frac{1}{2}(h_{Ai} - h_{Bi}) + \frac{1}{2}(h_{At} - h_{Bt}) \tag{4-22}$$

As can be seen from equation above, $(c - r)$ have been cancelled out. This statement is not entirely true as the assumption of symmetrical lines of sight from each end is dependent on uniform ground and atmospheric conditions at each end, at the instant of simultaneous observation.

单词与词组

terrain ['terein] *n*. (= terrane) 地域，地带，地势，地形

mountainous ['mauntinəs] *a*. 山多的，山似的，巨大的

preclude [pri 'klu:d] *vt*. 排(消)除，预防，阻止，妨碍

cliff [klif] *n*. 悬崖，峭壁

algorithms ['ælgərieðmδ] *n*. (algorithm 的复数)算法，算法式

· trigonometrical levelling 三角高程测量

facility [fə 'siliti] *n*. 设备，设施

curvature ['kə:vətʃə] *n*. 弯曲(部分)，曲率，曲度，弧度

refraction [ri 'frækʃən] *n*. 折射作用，折射度，折光差(度)

· zenith angle 天顶角
· angles of elevation 仰角
· angles of depression 俯角
constitute['kɔnstitju:t] *vt.* 构成,设立(机构),制定,指定,任命,选定
critical ['kritikəl] *a.* 批判的,批评性的,紧要的,关键性的,严重的,危险的,危急的
reciprocal[ri 'siprəkəl] *a.* 相互的,互惠的,相应的,相对的,彼此相反
· reciprocal observations 对向观测
symmetrical [si 'metrikəl] *a.* 对称的,匀称的
· cancel out 取消
elevate ['eliveit] *vt.* 举起,提高,抬高,提升,提拔
depressed[di 'prest] *a.* 压下的,降低的,低于标准的;忧伤的,抑郁的,消沉的
downhill[daun 'hil] *adv.* 向山下;*a.* 向下的,下降的

Chapter 5 Angle Measurement

5.1 Definition of horizontal and vertical angles

The measurement of angles is one of the most important required in surveying and construction. Angles are usually measured using either a theodolite or a total station.

Fig. 5-1 shows two points A and B and a theodolite or total station T set up on a tripod above a ground point G. Point A is higher than the instrument and is above the horizontal plane through T, whereas B is lower and below the horizontal plane. *1 [At T, the instrument is mounted a vertical distance h above G on its tripod.]

The horizontal angle at T between A and B is not the angle in the sloping plane containing A, T and B, but the angle θ on the horizontal plane through T between the vertical planes containing the lines of sight TA and TB. The vertical angles to A and B from T are α_A (an angle of elevation) and α_B (an angle of depression).

Another angle often referred to is the zenith angle. This is defined as the angle in the vertical plane between the direction vertically above the instrument and the line of sight, for example Z_A in Fig. 5-1.

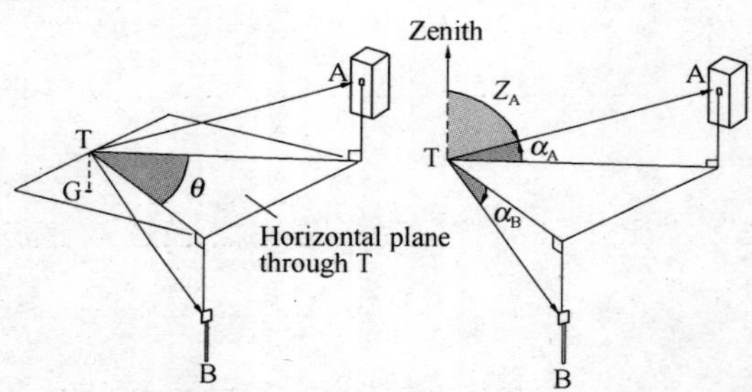

Fig. 5-1 Horizontal, vertical and zenith angles

Horizontal angles are used to determine bearings and directions in control surveys. They are used for locating detail when mapping and are essential for setting out all types of structure.

Vertical angles are used when determining the heights of points by trigonometrical methods, and can be used to calculate slope corrections for horizontal distances.

In order to measure horizontal and vertical angles, a theodolite or total station must be centered over point G and must be leveled to bring the angle reading systems of the instrument into the horizontal and vertical planes. *2 [Although centering and levelling ensure that horizontal angles measured at T are the same as those that would have been measured if the instrument had been set on the ground at point G, the vertical angles from T are not the same as those from G, and the value of h, the height of the instrument, must be taken into account when height differences are being calculated.]

单词与词组

horizontal [ˌhɔriˈzɔntl] *a.* 地平的，地平线的，水平的；平(坦)的

vertical ['və:tikəl] *a.* 垂直的，铅垂的，竖式的，直立的，纵的
theodolite [θi 'ɔdəlait] *n.* 经纬仪
· total station 全站仪
tripod ['traipɔd] *n.* 三脚架
mount[maunt] *v.* 安装，固定
· angle of elevation 仰角
· angle of depression 俯角
zenith['zeniθ] *n.* 天顶
· zenith angle 天顶角，天顶距
bearing ['beəriŋ] *n.* 方位，方位角
direction [di 'rekʃən] *n.* 方向
detail['di:teil, di 'teil] *n.* 细节，细目，详细，详情
· set out 放样,定线
center['sentə] *vt.* 集中，使聚集在一点，定中心，居中；*vi.* 居中，有中心，被置于中心

注释

[1] At T, the instrument is mounted a vertical distance *h* above G on its tripod。句中 a vertical distance *h* above G 这个词组是状语，修饰谓语(is mounted 被安装)的状态，说明仪器的安置状况(高于 G 点上方垂直距离 *h*)，并非说明三脚架状况，因此译成:在 T 点,仪器被安置在三脚架上,与 G 点垂直距离为 *h*。

[2] 此句的翻译应搞请两个 those 指代什么，第 1 个 those(…those that would have been measured…)指代前面的 horizontal angles，第 2 个 those(…those from G,…)指代前面的 vertical angles。最后一句中 the height of the instrument 是 *h* 的同位语。

5.2 Introduction to theodolites

Theodolites are precision instruments. There are basically two types of theodolite, the optical mechanical type or the electronic digital type, both of which may be capable of reading directly to 1′, 20″, 1″or 0.1″of arc, depending upon the precision of the instrument. The electronic theodolite is capable of displaying angle readings automatically. The BOIF optical theodolite and electronic theodolite are shown in Fig. 5-2.

(a)

(b)

Fig. 5-2 BOIF optical theodolite and electronic theodolite
(a) Optical theodolite; (b) Electronic theodolite

The electronic theodolite is the predominant instruments for angle measurement on site and elsewhere, but optical theodolites are still in use. Both optical and electronic theodolite will be described in this chapter, but it focuses on the former.

Theodolites in china are divided into five levels, i. e. Dj_{07} Dj_1, Dj_2, Dj_6 and Dj_{15}, which are according to instrument accuracy. "D" and "J" indicate the first letter of Chinese phonetic alphabet for "dadi celiang yiqi" and "jing wei yi" respectively. Their subscripts indicate the accuracy of measuring direction with the instrument, for example, DJ_6 indicates 6″ level theodolite.

5.2.1 Optical theodolites

1. Structure of an optical theodolites

An optical theodolite consists of three fundamental parts: the tribrach at the base, the horizontal circle (the lower plate) in the middle and the alidade (the upper plate) at the top. Its structure is shown in Fig. 5-3a.

(1) The *tribrach* supports the remainder of the instrument and is supported in turn by the levelling foot-screws. The tribrach can, therefore, be leveled independently the top of the tripod.

The optical theodolite has the facility for detaching the upper part of the theodolite from the tribrach. A special target or other piece of equipment can then be centered in exactly the same position occupied by the theodolite. This ensures that angular and linear measurements are carried out between the same positions, thereby reducing errors, particularly centering errors.

(2) The *horizontal circle* is mounted on outer spindle, which is concentric to the inner center (i. e. the vertical axis of the instrument). The inner spindle of most theodolites is hollow to provide a line of sight for the optical plummet, which takes the place of the plumb bob used to center the theodolite over the point to be occupied (see Fig. 5-3a).

The horizontal and vertical circles on which the angle graduations are etched are made of glass. Many types of glass arc theodolite are available, varying in reading precision from 1′to 1″, although 20″and 1″ reading theodolites are most commonly used in engineering surveying.

Most modern theodolites do not have a lower plate clamp and tangent screw. There is a facility for altering the position of the horizontal circle within the instrument and this is achieved using one control only, called the horizontal circle setting knob or screw which can rotate the horizontal circle to any reading required (see Fig. 5-3b).

(3) *1 [The *alidade* is carried on an inner spindle, which fits down into and can be free to rotate hollow outer spindle of the circle assembly(see Fig. 5-3a).] The circle assembly includes the horizontal circle that is covered by the alidade plate. The outer spindle is inserted into and can be free to rotate within the hollow socket of the tribrach.

The alidade includes the standards, telescope, vertical circle, horizontal circle, circle reading system, plate level (i. e. tubular level), upper plate clamp and tangent screw etc.

2. Sighting devices

The alidade involves a telescopic sighting device mounted on a horizontal axis (the trunnion axis) whose bearings are in turn mounted on standards. Thus the telescope can be pointed freely in any direction. In particular, the telescope may be rotated through 180° about the horizontal axis, the telescope points in the opposite direction after transiting. This process is also known as plunging or reversing.

The telescope consists of four parts: (1) an objective lens at the forward end of the telescope, which produces a reversed and reduced object image; (2) an eyepiece, which magnifies the cross-

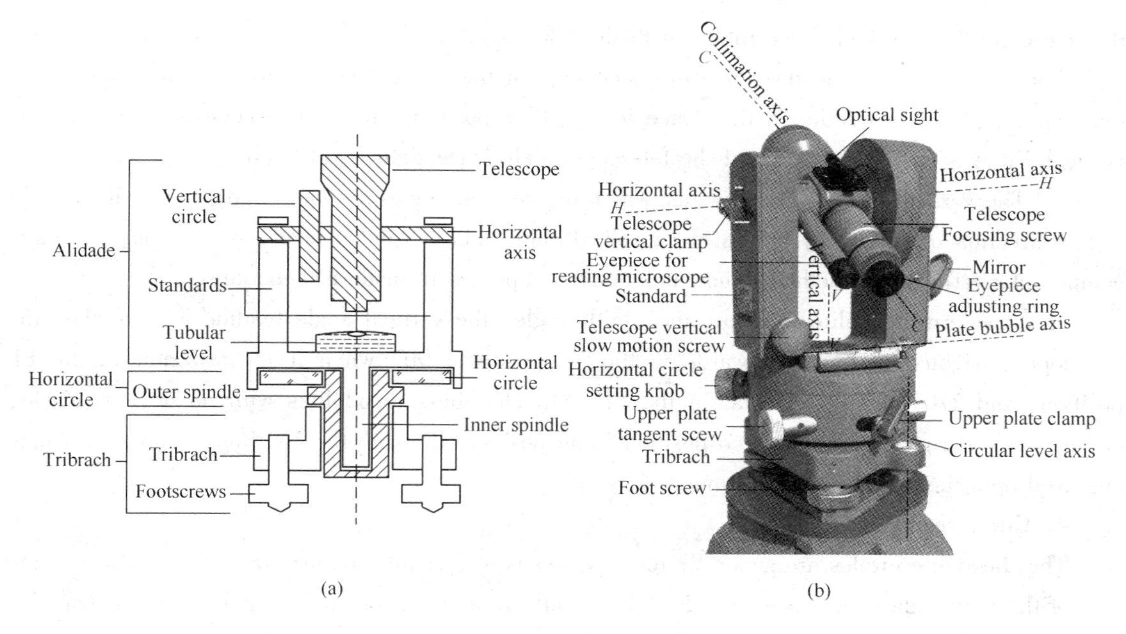

Fig. 5-3 An optical theodolite

hairs and must be focused on them according to the eyesight of the observer; (3) a reticule, which provides the cross-hairs near the rear of the telescope tube; and (4) a focusing lens, which can be moved back and forth to focus the image of object.

Since the image formed by the objective lens is inverted, the eyepieces of most theodolites are designed to erect the image. Telescopes which erect the image are called erecting telescopes; the others are called inverting telescopes.

The exact form of the reticle lines (often called cross hairs) engraved on the diaphragm varies, but most instruments have a single horizontal center line and two shorter horizontal lines, called the stadia lines, one above and one below the center line (see Fig. 5-4). For vertical angle readings, the central horizontal line is used.

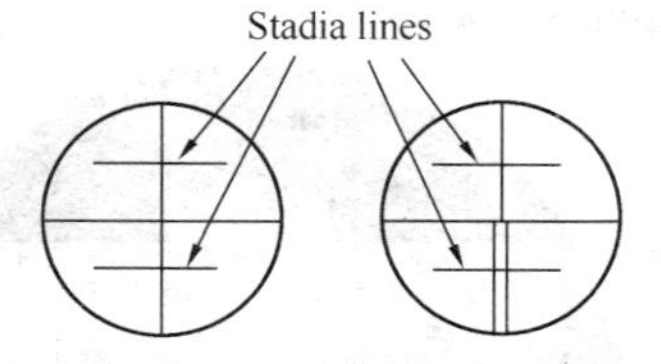

Fig. 5-4 Reticle pattern

The line of collimation is the line joining the intersection of the cross-hairs to the optical centre of the object glass and its continuation. This is also called the line of sight.

The cross-hairs must be brought into sharp focus by the eyepiece focusing screw prior to commencing observations. This process is necessary to remove any cross-hairs parallax caused by the image of the target being brought to a focus in front of or behind the cross-hairs. The presence of parallax can be checked by moving the observer's eye from side to side or up and down when looking through the telescope. If the image of the target does not coincide with the cross-hairs, movement of the observer's eye will cause the cross-hairs to move relative to the target image.

3. Vertical circle

A vertical circle is mounted on a plane perpendicular to the horizontal axis and to one side of the telescope. The vertical circle rotates with the telescope.

A surveyor faces to the telescope eyepiece end, if the vertical circle is on the left side of the telescope, call the theodolite in the "face left" (FL) position; if the telescope is transited, the vertical circle is on the right side of the telescope, call "face right" (FR) position.

*2 [The vertical circle index, against which the vertical angles are measured, is set with respect to the direction of gravity by means of (a) an altitude bubble (index level) or (b) an automatic compensator. The latter method is now universally employed in modern theodolites.]

Most modern theodolites measure the zenith angle; the vertical angle reading is zero when the telescope is sighted vertically upwards in the FL position, 90° when it is horizontal in the FL position, and 270° when horizontal in the FR. On electronic theodolites with no obvious circle, circles left and right are often called position I and position Ⅱ, with the two Roman numerals being engraved onto the body of the instrument.

4. Circle reading systems

The theodolite circles are generally read by means of a small auxiliary reading telescope at the side of the main telescope (see Fig. 5-3b). Small circular mirrors reflect light into the complex system of lenses and prisms used to read the circles.

There are basically two types of reading system: optical microscope reading and optical micrometer reading system.

(1) The *optical microscope reading system* is generally used on theodolites with a resolution of 20″ or less. Both horizontal and vertical scales are simultaneously displayed and are read directly with the aid of the auxiliary telescope.

The auxiliary telescope used to give the direct reading may be a "line microscope" or a "scale microscope".

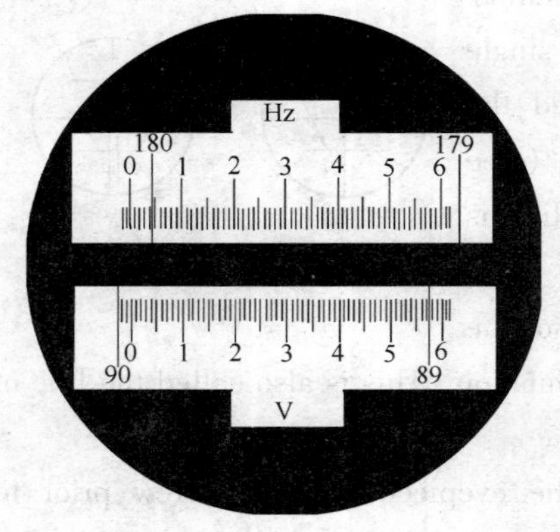

Fig. 5-5 The readings of DJ6 theodolite

The line microscope uses a fine line etched onto the graticule as an index against which to read the circle.

The scale microscope has a scale in its image plane, whose length corresponds to the line separation of the graduated circle. Fig. 5-5 illustrates this type of reading system and shows the scale from 0′ to 60′ equal in scale of one degree on the circle. This type of instrument is frequently referred to as a direct-reading theodolite and, at best, can be read, by estimation, to 6″. In the figure, the horizontal one (Hz) reads 180° 04′24″, whereas the vertical circle scale (V) reads 89°57′30″.

(2) The *optical micrometer system* generally uses a line microscope, combined with an optical micrometer using exactly the same principle as the parallel plate micrometer on a precise level.

On more precise theodolites, reading to 1″ of arc and a coincidence microscope is used. This enables diametrically opposite sides of the circle to be combined and a single mean reading taken.

This mean reading is therefore free from circle eccentricity error.

Fig. 5-6 shows the diametrically opposite scales brought into coincidence by means of the optical micrometer screw. The number of divisions on the main scale between 94°and 95° is three; therefore each division represents 20′. Therefore, the horizontal circle reads 94°10′, the micrometer scale reads 2′44″, giving a total reading of 94°12′44″. All improved version of this instrument is shown in Fig. 5-7.

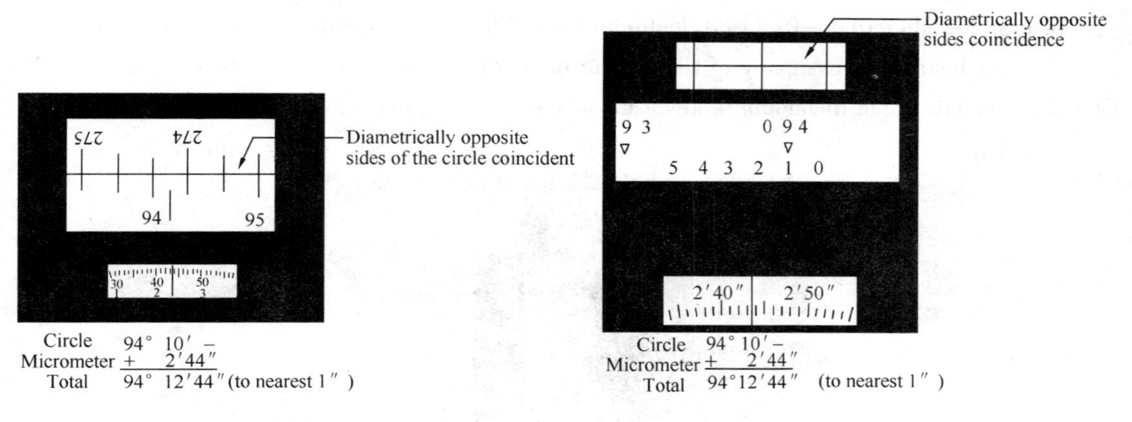

Fig. 5-6 Wild T2 (old pattern) theodolite reading system

Fig. 5-7 Wild T2 (new pattern) theodolite reading system

The above process is achieved using two parallel plates rotating in opposite directions until the diametrically opposite sides of the circle coincide.

5.2.2 Electronic theodolites

The electronic theodolites are composed of a precision optical devices , mechanical devices , electronic scanning dial, electronics sensors and microprocessor, and automatic digital display angle values (horizontal and vertical right angle) under the microprocessor control according to dial location information.

The electronic theodolites contain encoders that sense the rotations of the spindles and the telescope, convert these rotations into horizontal and vertical (or zenith) angles electronically, and display the values of the angles on liquid crystal displays (LCDs) or light emitting diode displays (LEDs). These readouts can be recorded in a conventional field book or can be stored in data collector for future printout or computations. The instrument contains a pendulum compensator or some other provision for indexing the vertical circle readings to an absolute vertical direction. The circles can be set to zero readings by a simple press of a button on the instrument. The horizontal circle readings of some of the electronic theodolites can be preset to any desired angle before a backsight is taken. The horizontal circle can be switched to measure either clockwise or anticlockwise and the scales give a near continuous display as the instrument is turned.

Electronic theodolite contains digitized scales which replace glass circles in an optical theodolite. It's horizontal and vertical circles are actually encoders scanned by a photodiode. This scan produces electrical pulses that can be converted to angles in digital form to be stored or displayed. The encoders have two types——the incremental encoding system and the absolute encoding system. Most electronic theodolites use incremental decoding system.

Angle readouts can be to 1″, with precision ranging from 0.5″ to 20″. The surveyor should check the specifications of new instruments to determine their precision, some instruments with 1″ readouts may be capable of only 5″precision.

In Fig. 5-8 shows ET－02 laser-electronic theodolite which is produced by the South Surveying and Mapping Instrument Company of China. The name of the instrument parts note on Fig5-8. The ET－02 laser electronic theodolite's keyboard is shown as the Fig 5-9.

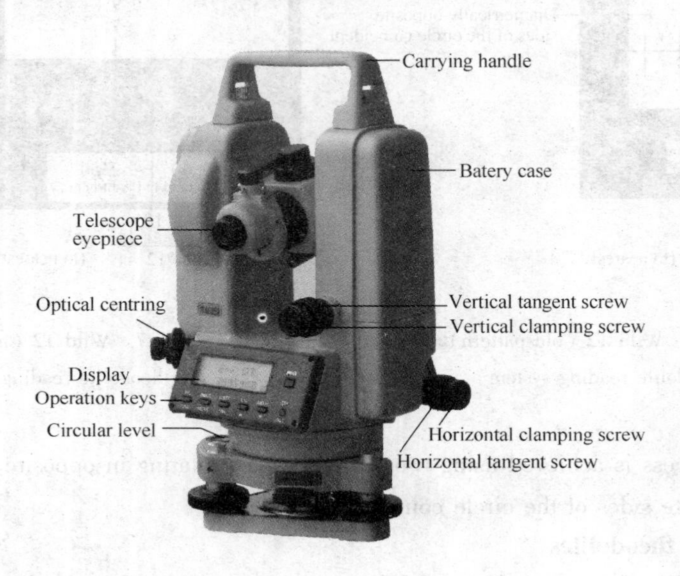

Fig. 5-8 ET－02 laser-electronic theodolite

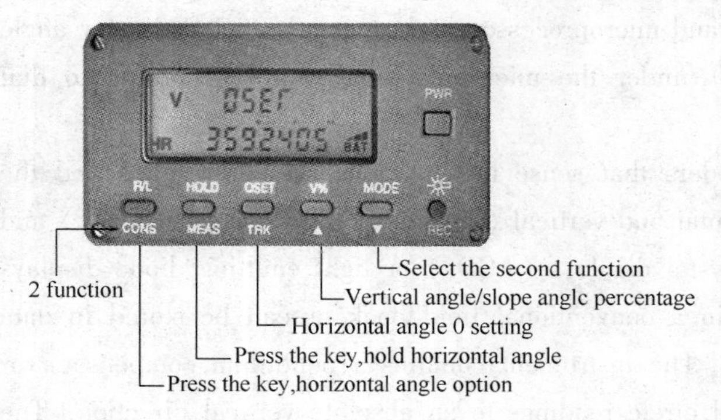

Fig. 5-9 The keyboard of ET－02/05 theodolite

【R/L】key: Left and right rotating to measure angle shift key each other, when right turning to measure angle i. e. the horizontal scale is clockwise to increase, the display shows HR; when left turning to measure angle i. e. the horizontal scale is anti-clockwise to increase, the display shows HL.

【OSET】key: Reading to return zero keys, continuous press 2 times, the horizontal scale reading is set to 0°00′00″.

Other key names and function are detailed to see explication in Fig. 5-9.

单词与词组

· the optical mechanical theodolite 光学机械经纬仪
· the electronic digital theodolite 电子数字经纬仪

· Chinese phonetic alphabet 汉语拼音字母
predominant [pri 'dɔminənt] *a.* 主要的；突出的；最显著的
tribrach['tribræk] *n.* 三脚台，三角基座
remainder[ri 'meində] *n.* 剩余部分，其余的人
tripod[ˌtrai 'pɔd] *n.* 三脚架
facility[fə 'siliti] *n.* （常用复）便利，设备，器材，工具，装置
detach[di 'tætʃ] *vt.* 分开，解开，分离
fit [fit] *v.* 安装，配备，（使）适合
spindle['spindl] *n.* 主轴，门锁的转轴，锭子，纺锤
concentric[kən 'sentrik] *a.* 同一中心的，同轴的
hollow['hɔləu] *n.* 穴；孔，洞坑，洼地
plummet ['plʌmit] *n.* 铅锤，测锤，准绳，垂球
· optical plummet 光学对中器
assembly [ə'sembli] *n.* 集合，装配，组装，组合件
socket ['sɔkit] *n.* 孔，洞，窝，凹处，插口，轴孔，管座
graduation['grædju 'eiʃən] *n.* 刻度，分度，（大学毕业时）授学位，获学位
etched['etʃid] *a.* 被侵蚀的，被蚀刻的，风化的
glass [glɑːs] *n.* 玻璃，玻璃制品，玻璃器皿，玻璃杯
· lower plate clamp 下盘固定夹
· upper plate clamp 上盘固定夹
· tangent screw 微动螺旋
alter['ɔːltə] *vt.* 修改，改动，改变，改建
· setting knob 安置钮
standard['stændəd] *n.* 标准，规范，直立机架（框架）
plate[pleit] *n.* 盘子，盆子，金属板
· plate level （上）盘水准器（编者注：plate level = tubular level 长水准管）
bearing['bɛəriŋ] *n.* 轴承，关系，影响，方面，意义，（常用复数）方位，方向，方位角
· trunnion axis 横抽
plunge[plʌndʒ] *vt.* 使投入，使插入，使陷入，倒转
reverse[ri 'vəːs] *vt.* 颠倒，反转，翻转
· erecting telescope 正像望远镜
· inverting telescope 倒像望远镜
reticle[retikl] *n.* 分划板，十字线
diaphragm[daiəfræm] *n.* 光阑，光圈
hair[hɛə] *n.* 头发，毛发
· stadia lines 视距丝
presence ['prezns]*n.* 出席，到场，在场；参加，列席，（存）在
· face left （FL） 盘左
· face righ （FR） 盘右

· with respect to 关于，相对于，根据
· altitude bubble(=index level) 高度水准器，(垂直度盘)指标水准管
auxiliary[ɔ:g 'ziljəri]a. 辅助的，补助的
mirror['mirə]n. 镜子，反射镜
microscope ['maikrəskəup] n. 显微镜
micrometer[mai 'krɔmitə] n. 测微计，千分尺
graticule['grætikju:l] n. (光学仪器目镜)网格，分度线，标线，十字线，分划线，交叉丝，方格图
correspond[kɔris 'pɔnd] v. 相当于，相关(to)，对应，符合(to，with)
separation[sepə 'reiʃən] n. 间距
· parallel plate 平行玻璃板
sensor['sensə] n. 传感器，感受器
microprocessor['maikrəuˌprɔsesə] n. 微处理器
crystal['kristl] n. 水晶，石英，结晶，晶体
· liquid crystal display (LCD) 液晶显示
· light emitting diode display (LED) 发光二极管显示
printout['printˌaut] n. 输出数据，用打字机印出，印刷输出，印出
provision[prə 'viʒən] n. 供给，准备，(预防)措施，设备，装置，构造
preset['pri: 'set] vt. 预先装置，预调，调整
· laser-electronic theodolite 激光电子经纬仪
readout['ri:daut] n. 读出，读数器
uncertainty[ʌn 'sə:tnti] n. 变化无常，靠不住，不确知，未确定
interpolation[inˌtə:pəu 'leiʃən] n. 插入(值)，内插
provision [prə 'viʒən] n. 供给，准备，措施，设备，装置，构造

注释

[1] The *alidade* is carried on an inner spindle, which fits down into and can be free to rotate hollow outer spindle of the circle assembly(see Fig. 5-3a)。句中 which 引导的从句中并列两个谓语 fits down into 与 can be free to rotate，第一个谓语 fits down into 意为“向下装配到”或“向下插入到”，故该从句译为：照准部安装在内轴上，内轴向下装入度盘组合件的外轴内，并能在空心外轴内自由转动(见图 5-3a)。

[2] 此句语法并不复杂，但需要一定的专业知识。即测量水平角，度盘不能动，指标随照准部而动；测量垂直角正相反，垂直度盘随望远镜而动，指标固定不动。指标设置由重力来控制，自动补偿器受重力的影响使指标部件始终处于重力方向。因此本句译为：测量垂直角是依赖垂直度盘指标，指标被设置在重力方向而言可借助于(a)垂直度盘指标水准管(指标水准器)或(b)自动补偿器自动安置，后一种方法在现代的经纬仪中普遍使用。

5.3 Setting up the theodolite

The perpose of setting up a theodolite is that the vertical axis of a theodolite is in the plumb line direction and passes through the centre of ground mark. The process of setting up a theodolite is carried out in two stages: centering the theodolite and levelling the theodolite.

5.3.1 Centering the theodolite

The perpose of centering is that the horizontal circle centre of the theodolite is in the plumb line through the centre of ground mark.

1. Setting up the tripod

(1) The tripod is first set up over the ground mark. The legs of the tripod are extended to suit the height of the observer. Standing back a few paces from the tripod, the centre of the tripod head is checked to see if it is vertically above the ground mark. The tripod head should be made as level as possible by eye.

(2) The theodolite is taken out of its case, it is securely attached to the tripod head. Three footscrews are adjusted to achieve approximately the same height. (i.e. the tribrach of the instrument is approximately parallel to the tripod head)

2. Rough centering—using a plumb bob

Hang plumb bob and note deviation between it and the ground mark, and then centering by moving the tripod legs. If the plumb bob deviated from the ground mark is not large, the tripod legs are pushed firmly into the ground. At last, slightly loosening centering clamping screw and slightly shifting the instrument on the tripod, make the plumb bob just above the ground mark. The centering in this way is known as rough centering, its error should be within 5mm.

3. Precise centering—using optical plummet

The instrument should be roughly leveled by adjusting three foot-screws, the observer slightly loosens the center clamping screw, and shifts the instrument on the tripod until the image of the ground mark seen through the plummet coincides with the reference mark (either cross hairs or a circle). Since the top of the tripod may not be perfectly level, it is worth noting that the instrument must be shifted parallel with itself. If it is rotated while being shifted, then the line of sight of the optical plummet will no longer be vertical. Both the leveling and the centering must be checked after this shifting operation.

At this stage, the theodolite is almost level. To level the instrument exactly, the plate level has to be used.

5.3.2 Levelling the theodolite

The perpose of levelling the theodolite is that makes the horizontal circle of the theodolite in a truly horizontal plane. The procedure of levelling the theodolite consists of rough levelling and precise levelling.

1. Rough levelling

Centralize the circular bubble using the foot-screws, referring to section 4.3.1 (see chapter 4 levelling).

2. Precise levelling

Centralize the plate level bubble using the foot-screws, the procedures are as follows:

(1) The theodolite is rotated until the plate level axis is approximately parallel to the line through any two foot-screws. These two footscrews are turned until the plate level bubble is brought to the centre of its run. The levelling footscrews should be turned in opposite directions simultaneously, remembering that the bubble will move in a direction corresponding to the movement of the left thumb (see Fig. 5-10a).

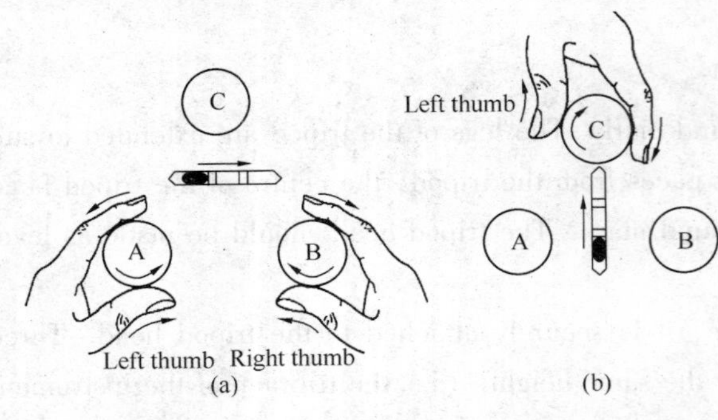

Fig. 5-10 Levelling the theodolite

(2) The instrument is turned through 90° and the bubble centered again, but using the third footscrew only (see Fig. 5-10b).

(3) This process is repeated until the plate level bubble is central in both positions.

(4) The instrument is now turned so that the plate level is in a position 180° from the first. If the plate level bubble is still in the centre of its run, the theodolite is level and no further adjustment is needed. If the bubble is not central, it has an error equal to half the amount the bubble has run off centre. If, for example, the bubble moves off centre by two plate level divisions to the left, the error in the bubble is one division to the left.

It may be possible to level a theodolite electronically after the rough levelling has been carried out. In this case, the position of an electronic bubble is shown on the display and the three levelling foot screws are used to centre it as shown in Fig. 5-11. All instruments capable of electronic levelling will give a warning if they are not levelled properly and will not function until they have been re-levelled.

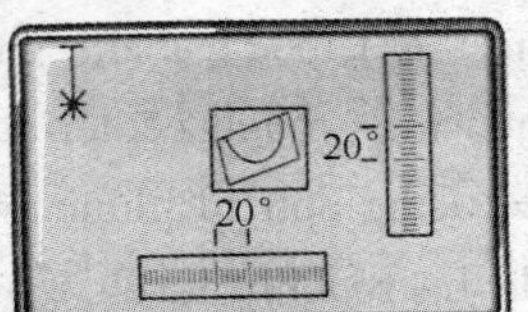

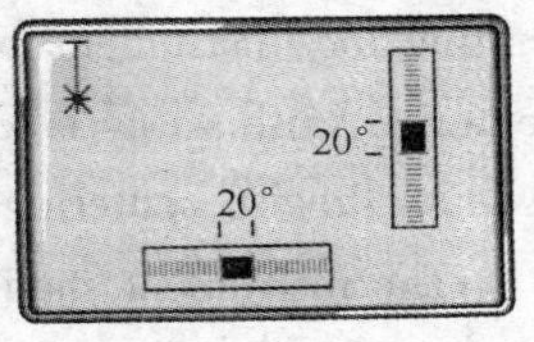

Fig. 5-11 Electronic levelling

When using the optical or laser plummet for centering, it is essential that the theodolite is properly levelled before this is done. If the theodolite is not levelled, the axis of the plummet will not be vertical and even though it may appear to be centered, the theodolite will be miscentered.

If any point is occupied for a long time, it is necessary to check the levelling and centering at frequent intervals, especially when working on soft ground or in hot sunshine.

单词与词组

· setting up 安置

centre[ˈsentə] *vt.* 对中(点, 心), 定中心 centering[ˈsentəriŋ, ˈsentriŋ] = centring; *n.* 置于中心, 定中心

involve[in ˈvɔlv] *vt.* 包括, 包含, 涉及, 使参与, 即是使

tripod['trai͵pɔd] *n.* 三脚架
extend[iks 'tend] *v.* 延长，扩展
· rough centring 粗对中
· fine centring 精对中
plumb[plʌm] *n.* 铅锤，测锤，垂直
bob [bɔb] *n.* 钟摆，秤锤
deviation[͵di:vi 'eiʃən] *n.* 偏离，偏向，偏差
loosen['lu:sn] *vt.* 使松动，放松
slightly['slaitli] *adv.* 轻微地，有一点
· centering clamping screw 中心固紧螺旋
parallel['pærəlel] *a.* 平行的，并行的，相似的，对应的，并联的
thumb [θʌm] *n.* 拇指
· laser plummet 激光对中器
miscentered[mis 'sentəd] *a.* 偏离中心的

5.4 Measuring horizontal angles

It is assumed that the theodolite has been set up over station O, accurately centering and levelling in section above. The horizontal angle to two distant points A and B are to be measured. Targets must be set up at these points. The procedure of measuring a horizontal angle is as follows:

(1) The theodolite is set face left (FL) position. The horizontal circle is set to is zero or near zero. The modern optical theodolites have a horizontal circle setting screw which can rotate the horizontal circle to any reading required. Having been set the horizontal circle, make the alidade and horizontal circle disjoin by click the horizontal circle setting screw (the specific operation depends on the type of instrument used).

(2) Loosen the Upper plate clamp and use the optical sight to find first target A. And then the Upper plate clamp is locked and the vertical hair of the telescope is accurately sighted onto the target A (usually at the bottom of target) using the upper plate tangent screw. Attention must be paid to check and remove parallax. Finally record the reading of the horizontal circle "a_1".

(3) Rotate the instrument is to sight second target B in a clockwise manner and record the other reading "b_1" similar to the step above. Therefore, the horizontal angle β_1 between A and B is $\beta_1 = b_1 - a_1$,

The procedure above, first half round of angles has been completed.

(4) Transit the telescope so that the theodolite is now in face right (FR) to start other half round. But remember to begin with target B and then rotate to target A in an anticlockwise manner. The difference of the two readings is the second half round of angle β_2. Take average of β_1 and β_2 to get finally horizontal angle.

Above these procedures, one round of angles has been completed.

(5) Take as many rounds as possible if it is necessary, starting with different pre-set initial

readings. It is customary to distribute the different initial readings around the circle to minimize the effect of circle graduation distortions. If measuring n rounds the initial reading is altered by $\frac{180°}{n}$ each time. Therefore, starting a second round of angles, set the horizontal circle reading to 90° when sighting A.

Repeat steps (2) to (4) inclusive to complete a second round of angles.

At least two rounds of angles should be taken at each station in order to detect errors when the angles are computed. Since each round is independently observed, both rounds must be computed and compared before the instrument and tripod are moved. It is worth noting that every time sighting target must use the central same part on the vertical hair in order to reduce the collimation error.

Table 5-1 Horizontal angle measurement and calculation

Stations	Objects	Face	Horizontal circle readings (° ′ ″)	Half round of horizontal angles (° ′ ″)	One round of horizontal angles (° ′ ″)	Average angles (° ′ ″)
O (1st round)	A	L	0 00 24	91 55 42	91 56 00	91 55 50
	B		91 56 06			
	B	R	271 56 54	91 56 18		
	A		180 00 36			
O (2nd round)	A	L	90 00 12	91 55 42	91 55 40	
	B		181 55 54			
	B	R	1 56 30	91 55 38		
	A		270 00 52			

单词与词组

disjoin[dis ′dʒɔin] *vt.* 把……分开

click[klik] *vt.* 点按，单击(快速按一下键钮，马上松开)

sight[sait] *n.* *v.* 瞄准器，瞄准，察看

· upper plate clamp 上盘固定夹(螺旋)，照准部制动螺旋

· upper plate tangent screw 上盘微动螺旋，照准部微动螺旋

clockwise[′klɔkwaiz] *a.*, *adv.* 顺时针方向转动的(地)

manner[′mænə] *n.* 方法，方式

anticlockwise [′ænti͵klɔkwaiz] *a.*, *adv.* (= counterclockwise) 逆时针方向的(地)

initial[i ′niʃəl] *a.* 最初的，开始的，初期的

distortion[dis ′tɔːʃən] *n.* 扭曲；歪曲，曲解，变形

5.5 Measuring vertical angles

5.5.1 The procedure of measuring vertical angles

It is assumed that the theodolite has been set up over station, accurately centering and levelling described in section 5. 3.

(1) The theodolite is set in face left (FL) position. Turn the horizontal or vertical slow motion screws to make horizontal hair of cross-hairs sight target A precisely.

(2) The altitude bubble (if fitted) must be brought to the middle of its run before every reading is taken. Then get the reading of vertical circle "L".

* [Modern instrument possesses an automatic index to the vertical circle, the reading will give the angle relative to the instrument zero. This may be a (+) or (−) angle related to the horizontal plane or, on most modern instruments, the angle related to zero vertically upwards.]

(3) Transit the telescope so that the theodolite is now in face right (FR) position to start the other half round. Turn the horizontal or vertical slow motion screws to make horizontal hair of cross-hairs sight target A precisely.

(4) The altitude bubble (if fitted) must be brought to the middle of its run again. Then get the reading of vertical circle "R".

The procedure above, one round of vertical angles has been completed, all records and calculations are as shown Table 5-2.

Table 5-2 Vertical angle measurement and calculation

Stations	Objects	Face	Vertical circle readings (° ′ ″)			Vertical angles: Half round (° ′ ″)			Vertical angles: One round (° ′ ″)			Index errors (″)
O	A	L	78	18	18	11	41	42	11	41	51	+9
		R	281	42	00	11	42	00				
O	B	L	96	32	48	−6	32	48	−6	32	34	+14
		R	263	27	40	−6	32	20				

5.5.2 The calculations of vertical angle and index error

There are two types of vertical circle, one is the circle graduation increasing in clockwise direction from 0° to 360° as shown Fig. 5-12a, the other is the circle graduation increasing in anticlockwise as shown Fig. 5-12b.

When the line of sight is horizontal and the altitude bubble (if fitted) is central, the theoretical vertical circle reading is 90° or some multiple of 90° depending on vertical circle type of a theodolite. The difference between practical vertical circle reading and its theoretical value is termed index error x (see Fig. 5-12).

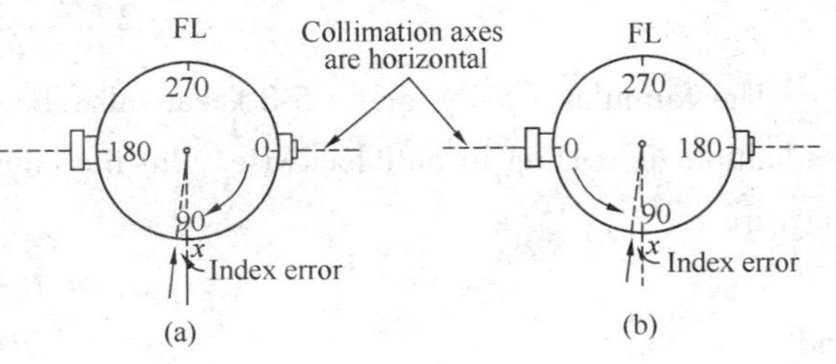

Fig. 5-12 Two types of the vertical circle graduation

(a) Increasing in clockwise direction;

(b) Increasing in anticlockwise direction

The theodolite with the circle graduation increasing in clockwise direction (Fig. 5-12a) will be discussed as follows. Assume L means reading of vertical circle, x index error when theodolite is set in FL. From Fig. 5-13b can obtain:

$$\alpha = 90° + x - L \tag{5-1}$$

Assume

$$\alpha_L = 90° - L \tag{5-2}$$

Hence

$$\alpha = \alpha_L + x \tag{5-3}$$

Where α = vertical angle, α_L = FL half round of vertical angle.

Similar, when theodolite is set in the FR, From Fig 5-14b can obtain:

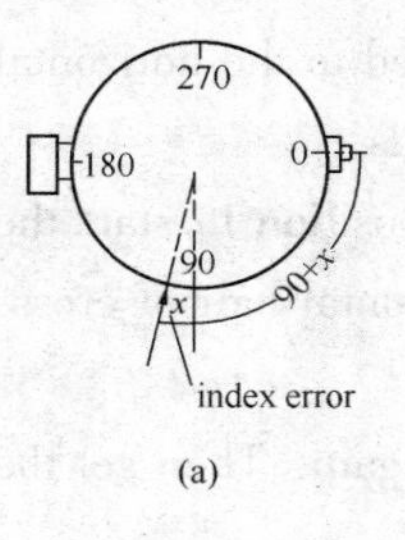

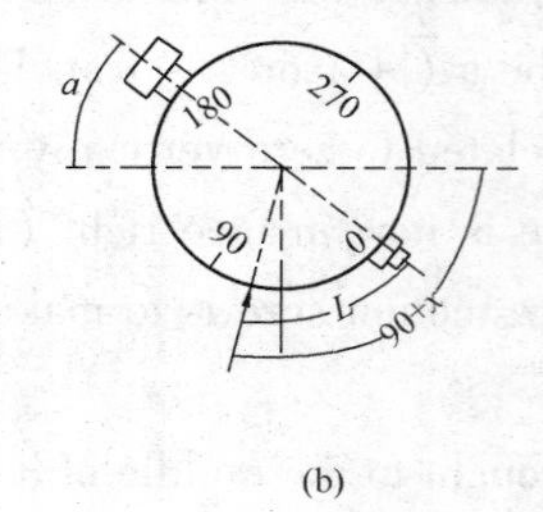

Fig. 5-13 When FL, calculation of vertical angle
(a) When FL, the line of sight is horizontal;
(b) When FL, the linr of sight is in clined

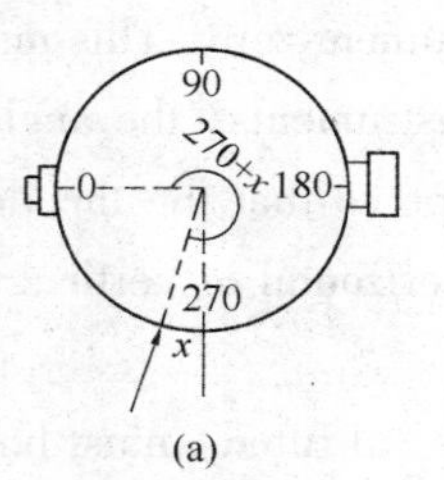

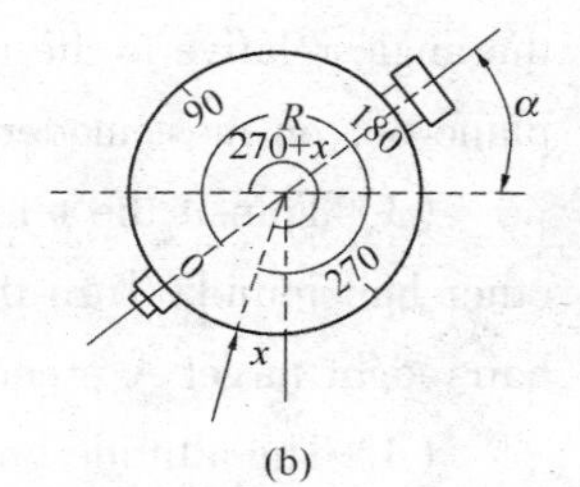

Fig. 5-14 When FR, calculation of vertical angle
(a) When FR, the line of sight is horizontal;
(b) When FR, the line of sight is in clined

$$\alpha = R - (270° + x) \tag{5-4}$$

Assume

$$\alpha_R = R - 270° \tag{5-5}$$

Hence

$$\alpha = \alpha_R - x \tag{5-6}$$

Where α = vertical angle, α_R = FR half round of vertical angle.

Adding equation (5-3) and equation (5-6) gives

$$\alpha = \frac{1}{2}(\alpha_L + \alpha_R) \tag{5-7}$$

Subtracting equation (5-6) from equation (5-3) gives

$$x = \frac{1}{2}(\alpha_R - \alpha_L)$$

Substituting equations (5-2) and (5-5) into equation above gives

$$x = \frac{1}{2}(L + R - 360°) \tag{5-8}$$

The formulas (5-7) and (5-8) can also be the same with Fig. 5-12b, which is the circle graduation increasing in anticlockwise. But the calculating formulas of α_L and α_R are different only, they are:

$$\alpha_L = L - 90°$$

and

$$\alpha_R = 270° - R$$

All records and calculations are shown Table 5-2.

单词与词组

possess[pə 'zes] *vt.* 有，具有，拥有，占有，掌握(知识)

multiple['mʌltipl] *n.* 倍数

· index error (垂直度盘)指标差

represent[ˌrepri 'zent] *vt.* 表示，表现，代表

· be the same with 适用于，与……一样

注释

* 此段有两个句子:① Modern instrument possesses an automatic index to the vertical circle, the reading will give the angle relative to the instrument zero. ② This may be a (+) or (-) angle related to the horizontal plane or, on most modern instruments, the angle related to zero vertically upwards.

第一句 ① 后半句中 relative to the instrument zero 短语是修饰 angle，意为相对于仪器零位的角度，故第一句译为：现代的经纬仪（句中 instrument 就是指 theodolite）具有对垂直度盘的自动指标，读数给出（或译显示）相对于仪器零位的角度。

第二句 ② 前半部分译为：这可能给出相对于水平面的“ + 角度”（注：即仰角）或“ - 角度”（即俯角）。编者说明：此仪器当望远镜视线水平时，竖盘读数指标读为 0°；zero vertically upwards 意为垂直向上为 0°，即当望远镜视线垂直向上时，竖盘读数指标读为 0°。所以第二句后半句译为：大多数现代的仪器给出相对垂直向上为 0°的角度。

5.6 Adjustments of the theodolite

5.6.1 The geometrical relationships of the axes of the theodolite

The arrangement of the axes of the theodolite is shown in Fig 5-15. The most important relationships are as follows:

(1) The axis of the plate bubble should be in a plane perpendicular to the vertical axis (the rotational axis of a theodolite), i. e. $LL \perp VV$.

(2) The line of sight (collimation axis) should be perpendicular to the horizontal axis, i. e. $CC \perp HH$.

(3) The vertical cross-hair should be perpendicular to the horizontal axis.

(4) The horizontal axis should be perpendicular to the vertical axis, i. e. $HH \perp VV$.

(5) The axis of circular level should be parallel to the vertical axis, i. e. $L'L' // VV$.

Fig. 5-15 Axes of the theodolite

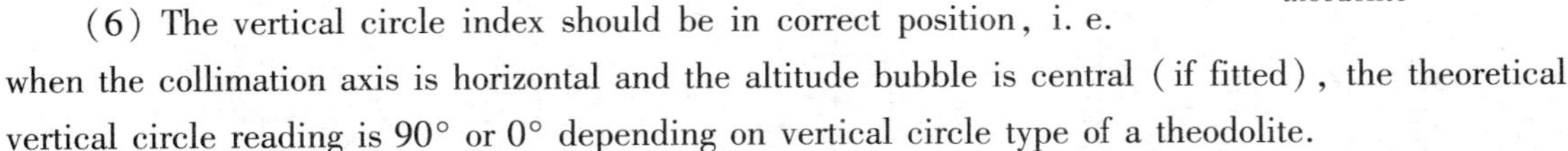

(6) The vertical circle index should be in correct position, i. e. when the collimation axis is horizontal and the altitude bubble is central (if fitted), the theoretical vertical circle reading is 90° or 0° depending on vertical circle type of a theodolite.

The purpose of the adjustments is to make the primary axes of the theodolite in their correct geometrical relationships.

5.6.2 Plate bubble adjustment

The plate bubble is set on the alidade of the thodolite. It is high sensitive tubular level, so also usually call tubular level. The instrument vertical axis must be truly vertical when the plate bubble is centralized. The vertical axis of the instrument constructed by the instrument manufacturer is strictly perpendicular to the horizontal plate which carries the plate bubble. In order to ensure that the vertical axis of the instrument is truly vertical, it is necessary to align the bubble axis parallel to the horizontal plate.

Test: Assume the bubble axis is not parallel to the horizontal plate but is in error by angle e. The plate bubble is set parallel to a pair of foots crews, leveled approximately, then turned through 90° and leveled again using the third foot screw only. It is now returned to its former position, accurately leveled using the pair of footscrews, and will appear as in Fig. 5-16a. The instrument is now turned through 180° and will appear as in Fig. 5-16b, i. e. the bubble will move off centre by an amount representing twice the error in the instrument ($2e$).

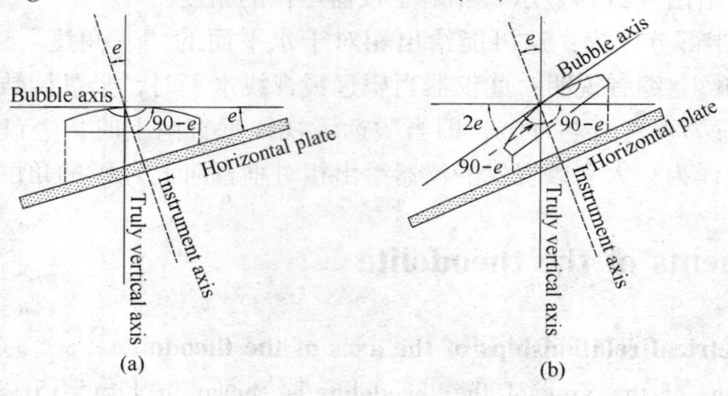

Fig 5-16 Principle of plate level adjustment

Adjustment: The bubble is brought half-way back to the centre using the pair of footscrews which are turned by a strictly equal and opposite amount. This will cause the instrument axis to move through e, thereby making it truly vertical and, in the event of there being no adjusting tools available, the instrument may be used at this stage. The bubble will still be off centre by an amount proportional to e, and should now be centralized by raising or lowering one end of the bubble using its capstan adjusting screws.

5.6.3 Collimation axis adjustment

The purpose of this test is to ensure that the line of sight is perpendicular to the horizontal axis. If the collimation axis is not setting of its correct position, then the line of sight will trace a cone and

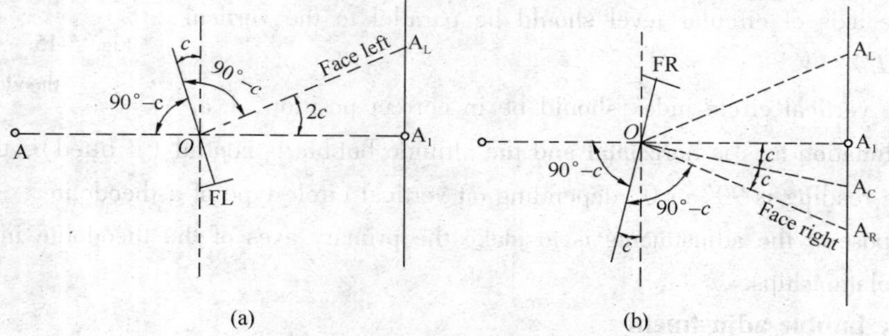

Fig. 5-17 Principle of the collimation axis adjustment

(a) Face left; (b) Face right

not a vertical plane when the telescope is rotated around the horizontal axis.

Test: *1 [The instrument is set up and accurately leveled, and the telescope directed to bisect a fine mark at A, situated at instrument height about 50m away (Fig. 5-17). If the line of sight is perpendicular to the horizontal axis, thus when the telescope is rotated vertically through 180°, we give mark A_1 on horizontal rod set at about same instrument height and about 50m on the line of sight.] However, assume that the line of sight makes an angle of $(90° - c)$ with the horizontal axis, then when in the FL position the telescope sight to a fine mark at A, transit the telescope and give mark A_L on horizontal rod (Fig. 5-17a). This instrument is now face right, re-bisect point A, transit the telescope and give mark A_R on horizontal rod (Fig. 5-17b). From the sketch it is obvious that distance A_LA_R represents four times the error in the instrument $(4c)$.

Adjustment: The cross-hairs are now moved in azimuth using the two-side diaphragm adjusting screws, from A_R to a point (A_C) midway between A_R and A_1, this is one-quarter of the distance A_LA_R.

This movement of the reticule carrying the cross-hair may cause the position of the vertical hair to be disturbed in relation to the horizontal axis; i. e. it should be perpendicular to the horizontal axis. It can be tested by traversing the telescope vertically over a fine dot. If the vertical cross-hair moves off the dot then it is not at right angles to the horizontal axis and is corrected with the adjusting screws.

5.6.4 Diaphragm adjustment

In carrying out the collimation axis adjustment, the diaphragm is moved. This may upset the setting of the vertical hair in a plane perpendicular to the horizontal axis.

Assuming that a collimation axis adjustment has just been completed, the following procedure should be adopted.

(1) Relevel the instrument carefully and sight A on either face.

(2) Move the telescope up and down while observing A. If the vertical hair stays on point A then it is set correctly.

(3) If adjustment is necessary, the diaphragm is rotated until the vertical hair remains on point A while moving the telescope in altitude.

Tests 5.6.3 and 5.6.4 are interdependent and both tests are undertaken consecutively until a satisfactory result is obtained for each.

The diaphragm is constructed by the instrument manufacturer so that the horizontal and vertical hairs are perpendicular. Setting the vertical hair vertical therefore sets the horizontal hair in a horizontal plane.

5.6.5 Trunnion axis test

The purpose of this test is to set the horizontal axis perpendicular to the vertical axis. The horizontal axis will then be horizontal when the instrument is levelled. If the horizontal axis is not horizontal the telescope will not define a vertical plane and this will give rise to incorrect vertical and horizontal angles.

It should be noted that in modern instruments this adjustment cannot be carried out owing to their excellent construction and, consequently, most do not provide for this adjustment. Manufacturers claim that this error does not occur in modern equipment. However, *2 [satisfactory results will be obtained in practical by meaning FL and FR readings.]

5.6.6 Circular level adjustment

The purpose of this adjustment is to ensure that the axis of circular bubble parallel to the vertical axis. Otherwise, the vertical axis of the instrument isn't in vertical position approximately even if circular bubble is centralized. Besides, if the circular level is not leveled properly, the electronic level will not function.

The circular level adjustment is very simple. After the plate bubble has been set and adjusted, as described in Section 5.6.2. The instrument must be accurately leveled. This time, if the circular bubble is centre off, it can be adjusted (centered) by adjusting one or more of the three capstan screws beneath the circular level.

5.6.7 Vertical circle index adjustment

The aim of this adjustment is to ensure that when the line of sight is horizontal and the altitude bubble central (if fitted), the vertical circle reads 90° or some multiple of 90° depending on the type of instrument.

Test: Assume that the type of instrument is the circle graduation increasing in clockwise direction, and that the vertical circle reading is 90° when the theodolite is set in FL position and the line of sight is horigontal. The test procedure is as follows:

(1) The theodolaite is set in face left (FL), carefully level the instrument.

(2) The horizontal hair of cross-hair of the telescope sight precisely a fine point.

(3) Adjust the altitude bubble to the middle of its run and take a vertical circle reading L.

(4) Transite the telescope the theodolite in face right (FR), repeat step (2). Adjust the altitude bubble again and take a vertical circle reading R.

For example: test results has been obtained, FL vertical circle reading $L = 78°18'18''$, FR vertical circle reading $R = 281°42'00''$. According formula (5-8), we obtain:

$$x = \frac{1}{2}(78°18'18'' + 281°42'00'' - 360°) = +9''$$

Adjustment: The theodolite is in the FR position, the line of sight of telescope still sight a fine point.

(1) Equation (5-4) indicates that the theodolite in the FR position vertical circle correct reading should be to equal $(R - x)$, i. e., $(R - x) = 281°42'00'' - 09'' = 281°41'51''$. This reading is set by rotating the altitude bubble levelling screw, which causes the altitude bubble to move off centre.

(2) The altitude bubble is re-centralized by adjusting its capstan adjusting screws.

For a theodolite with an automatic vertical index, the manufacturer's handbook should be consulted for the correct adjustment procedure.

5.6.8 Adjustment of the optical plummet

The line of collimation of an optical plummet must coincide with the vertical axis of the theodolite. Two tests are possible, depending on the type of instrument used.

(1) If the optical plummet is on the alidade and can be rotated about the vertical axis (Fig. 5-18a)

Firstly, the instrument is set up and accurately leveled. And then secure a piece of paper on the ground below the instrument, the plummet axis make a mark where the optical plummet intersects it. Rotate the alidade through 180° in azimuth and make a second mark. If two marks coincide, the plummet is in adjustment. If not, the correct position of the plummet axis is given by a point midway between the two marks. Consult the instrument handbook and adjust either the diaphragm (cross hairs) or objective lens on the optical plummet.

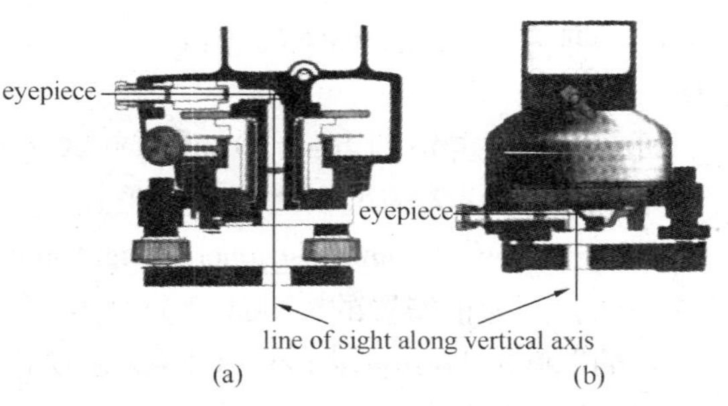

Fig. 5-18 Two types of optical plummets

(a) Optical plummet mounted on alidade;

(b) Optical plummet mounted on tribrach

(2) If the optical plummet is on the tribrach it cannot be rotated (Fig. 5-18b).

Set the theodolite on its side on a bench with its base facing a wall and mark the point on the wall intersected by the optical plummet. Rotate the tribrach through 180° and again mark the wall. If both marks coincide, the plummet is in adjustment. If not, the plummet diaphragm should be adjusted to intersect a point midway between the two marks.

单词与词组

geometrical[ʤiə 'metrikəl] *a.* 几何的,几何学的, 几何图案的

relationship[ri 'leiʃənʃip] *n.* 关系, 联系

· in the event of 即使,万一

· capstan adjusting screws 校正调节螺丝

cone[kəun] *n.* 圆锥体(面,形),锥形物

· transit axis (=horizontal axis = trunnion axis) 横轴

undertake[ˌʌndə 'teik] *vt.* (-took [-'tuk]; -taken [-'teikən]) 承担, 接受, 承办

equivalent [i 'kwivələnt] *a.* 相等的, 相当的, 相同的

bisect[bai 'sekt] *vt.* 把…一分为二,二等分

consult[kən 'sʌlt] *v.* 商议(量),考虑, 翻阅, 查阅, 考虑

traverse['trævəs] *vt.* 横越,横贯, 通过; *vi.* 来回移动, 旋转, 横越

diaphragm['daiəfræm] *n.* 光阑, 光圈, 十字丝环

adopt[ə'dɔpt] *vt.* 采用, 采纳, 正式通过

orientation[ˌɔ(:)rien 'teiʃən] *n.* 向东, 定向, 辨向作用

upset[ʌp 'set] *vt.* 推翻, 颠覆, 打翻, 打乱, 弄糟

beneath [bi 'ni:θ] *prep.* 在……之下; 在……的(正)下方, 紧挨着……的底下

centralize['sentrəˌlaiz] *vt.* 使居中, 成为……的中心

consult[kən 'sʌlt] *vi.* 商量,考虑,咨询, 查阅

secure[si 'kjuə] *a.* 安全的，牢固的；*vt.* 保护，(使)获得，固定；*vi.* 获得安全，安全，保险，担保

disturb[dis 'tə:b] *vt.* 打扰，扰乱，妨碍，妨害

bench[bentʃ] *n.* 长凳，船上的坐板，座，架，(工作，试验)台

注释

[1] 理解此段要弄清两点：(1) situated at instrument height about 50 m away (Fig. 5-17)。这个词组意为"大约50m外仪器高位置的"修饰前句的A。(2) we give mark A_1 on horizontal rod set at about same instrument height and about 50m on the line of sight。set 为 set 的过去分词，意为被安置的，划线部分意为"在视线上大约50m并与仪器高大约同高度位置安置的" 修饰 horizontal rod(水平尺)。所以课文中这两句译为"望远镜瞄准约50m外与仪器高同高位置A点精细标志(图 5-17)。如果视线垂直于水平轴，那么当望远镜倒转180°，在视线的方向大约50m且与仪器高大约同高位置放置的水平尺得记号A_1。"

[2] satisfactory results will be obtained in practical by meaning FL and FR readings. 句中 meaning 为动名词，意为取平均值。全句释为"在实践中通过取盘左和盘右的平均读数将获得满意的结果。"

Chapter 6 Total Station

6.1 General

Today in surveying there is an increasing use of global positioning systems, however, the most commonly used surveying instrument is the total station.

The total station is an electronic instrument integrated with an electronic theodolite and with an electronic distance meter (EDM). A typical total station is shown in Fig. 6-1. The appearance of total station is very similar to an electronic theodolite. The total station combines both angle and distance measurement in the same unit. Measurements are carried out easier, faster and more accurately with total station.

Total stations are capable of performing a number of different survey tasks and they can store relativity large amounts of data. It can average multiple angle measurements, average multiple distances, and determine coordinates, elevations, areas, and calculate atmospheric and instrument corrections, etc. All the functions of a total station are controlled by its microprocessor (or computer), which is accessed through a keyboard and display.

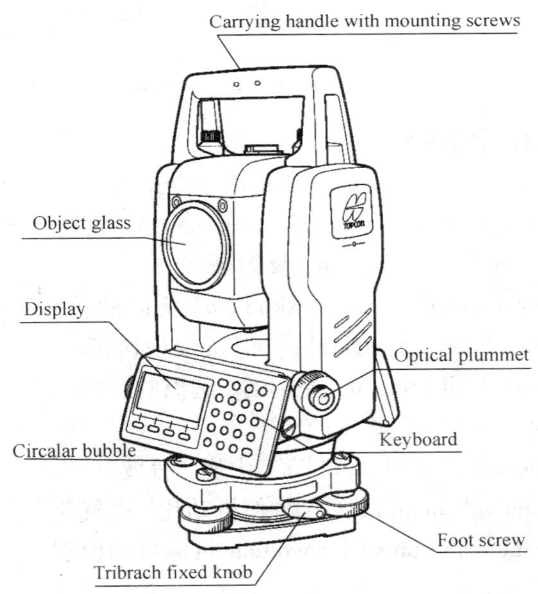

Fig. 6-1 Typical total station

To use a total station, it is set over one end of the line to be measured and some form of reflector is positioned at the other end such that the line of sight between the instrument and the reflector is unobstructed. This is shown in Fig. 6-2, where the reflector is a prism attached to a detail pole. The telescope is aligned and pointed at the prism, the measuring sequence is initiated and a signal is transmitted from the instrument towards the reflector, where part of it is returned to the instrument. This is processed, in a few seconds, to give the slope distance together with the horizontal and vertical angles.

A total station can also be used in reflectorless mode, it is possible to survey areas from a distant location or cliff. Even areas of possible danger such as landslide areas can safely and efficiently be surveyed with this method.

Some instruments have motorized drives and can use automatic target recognition to search and lock onto a prism—this process is fully automated and does not require an operator. Some total stations can be controlled from the detail pole, enabling surveys to be carried out by one person, as shown in Fig. 6-3.

Fig. 6-2 Measuring with total station

Fig. 6-3 Robotic total station in use

单词与词组

· integrated total station 集成化全站仪
microprocessor ['maikrəuˌprɔsesə] *n.* 微处理器
reflector[ri 'flektə] *n.* 反射器
unobstructed[ʌnəb 'strʌktid] *a.* 无阻的，不受阻拦的
reflectorless[ri 'flektəlis] *n.* 无反射棱镜
distant['distənt] *a.* 遥远的，远离的，稀疏的
cliff [klif] *n.* 悬崖，绝壁
landslide['lændˌslaid] *n.* 山崩，滑坡
motorize['məutəraiz] *vt.* 使机动化，电气化，汽车化；给……安装发动机
· automatic target recognition 目标自动识别

6.2 Features of total stations

1. Telescope sighting axis and the EDM ranging emission optical axis are coaxial

The total station system consists of electronic theodolite and a compact EDM device so that the EDM ranging emission optical axis and the sighting axis are coaxial. Thus, measurement of sight target prism center can be simultaneously to determine the horizontal angle, vertical angle and slope distance of the target point.

2. Total stations can perform a variety of operations and can monitor the instrument's status

and have a wide variety of built-in programs

Total stations can read and record horizontal and vertical angles together with slope distances. The microprocessors in the total stations can perform a variety of operations: for example, averaging multiple angle measurements; averaging multiple distance measurements; X, Y, Z coordinate determination; atmospheric and instrumental corrections and so on.

Total stations can automatically monitors the instrument's status (e. g., battery status, horizontal and vertical axes status, return signal strength, etc.).

Typical total station built-in programs include point location, set up a free station, azimuth calculations, remote object elevation measurement, offset measurements, layout or setting-out positions, and area measurement etc. All these topics are discussed in the sections 6.5.

3. Total stations have large capability for the data store and data transfer

Many modern total stations have the data stored onboard, thus eliminating handheld data collectors. For example, Topcon GTS 300 total station, data are stored onboard in internal memory capacity of which is about 1300 points. The data can be directly transferred to the computer from the total station via a RS – 232 cable.

Some instruments have memory cards (about 2,000 points per card). Some manufacturers use PCMCIA cards, which can be read directly into a computer through a PCMCIA reader. Other total stations can be downloaded by connecting the instrument (or its keyboard) directly to the computer.

Fig. 6-4 shows a Nikon DTM 750 Total Station having card readers for applications program cards (upper reader), and for data storage cards (lower reader). The data storage cards can be removed when full and read into a computer using PCMCIA card readers—now standard on most notebook computers. Total Station operating software is MS-DOS compatible, permitting user-defined additions to applications software.

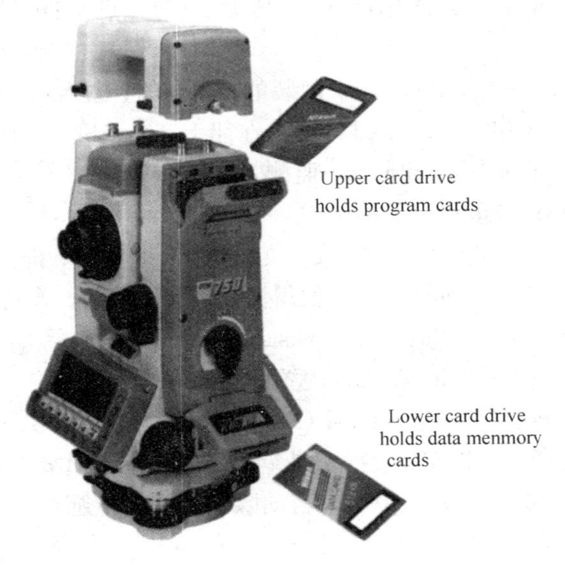

Fig. 6-4 Nikon DTM 750 Total Station

4. Total stations have dual-axis liquid compensator

The total station incorporates a dual-axis liquid compensator. The built-in dual-axis tilt sensor constantly monitors the inclination of the vertical axis in two directions, that is the collimation direction (i.e., x-axis) and the horizontal axis direction (i.e., y-axis). It calculates the compensation value and automatically corrects the horizontal and vertical angles. The dual-axis compensation illustration in Sokkia SET 1000 total station is shown in Fig. 6-5.

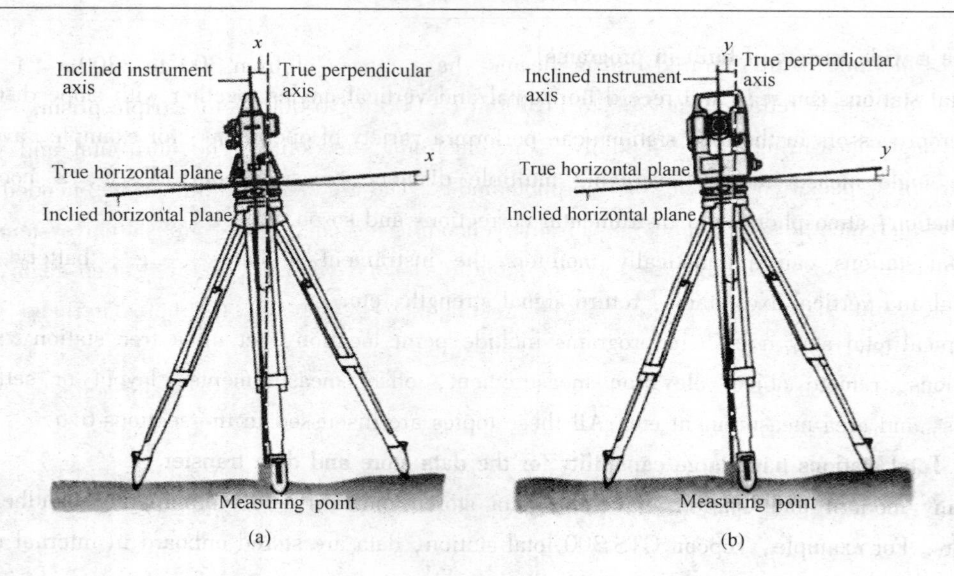

Fig. 6-5 Dual-axis compensation illustration

(a) Side view; (b) Front view

单词与词组

coaxial[kəu 'æksəl] *a.* 同轴的，共轴的

compact [kəm 'pækt] *v.* 压紧，（使）坚实；*a.* 装填紧密的，整齐填满的，小巧的，袖珍的；*n.* ['kɔmpækt] 契约，合同，协定

built-in[ˌbilt 'in] *a.* 内装的，固定的，嵌入的，固有的，机内的

onboard ['ɔn 'bɔ:d] *adv.* 在船上，在飞机上，在运载工具上

monitor ['mɔnitə] *n.* 监测，检测

· free station 自由站

offset['ɔ:fset] *n.* 分支，支脉，支族，后裔，抵消，补偿，横距，支距

notable['nəutəbl] *a.* 值得注意的，显著的，著名的

dual-axis['dju(:)əl- 'æksis] *a.* 双轴的

built-in[ˌbilt 'in] *a.* 内置的，机内的

· PCMCIA 是Personal Computer Memory Card International Association PC 机内存卡国际联合会的缩写，PCMCIA 插槽是笔记本电脑上最重要的设备扩展接口，可以用来插入传真卡/网卡/存储卡/声霸卡等等，目前除了笔记型电脑可使用 PCMCIA 规格的卡片外，还有 PDA、数字相机、数字电视、机顶盒（set-top boxes）等等也都有对应的产品可以使用 PCMCIA 规格的卡片。

6.3 The configuration of total stations and their components

6.3.1 The configuration of total stations

The configuration of total stations contains three components: the distance measuring unit or EDM, the angle measuring device or electronic theodolite, and an onboard microprocessor. The most of EDM units, with infrared carrier beams, have ranges of from 300 to 3300m (1 000 to 11 000ft) to a single reflector and 900 to 11 000m (3 000 to 36 000ft) to a triple prism.

The angle measuring unit essentially is an electronic theodolite. The horizontal and vertical circles are in a special code to be read by photodiodes. The graduated circle can be encoded in two ways—the incremental system and the absolute encoding system. Most total station systems have incremental decoding. In either case, it is possible to assign 0 degrees, or any desired angle, to the pointing after the sight has been taken. The displayed resolutions of the horizontal and vertical circles vary from 0.5 to 20″.

The Total Station has an on-board microprocessor that monitors the instrument status and makes corrections to measured data. In addition, the microprocessor controls the acquisition of angles and distances and then computes horizontal distances, vertical distances, coordinates, and the like.

The axis configuration comprises the vertical axis, the tilting axis and line of sight (or collimation), and these should all be mutually perpendicular. Other parts include the tribrach with levelling foot-screws, the keyboard with display and the telescope which is mounted on the standards and which rotates around the tilting axis. Levelling is carried out in the same way as for a theodolite by adjusting the footscrews to centralize a plate level or electronic bubble. The telescope can be transited and used in the face left (or face Ⅰ) and face right (or face Ⅱ) positions. Horizontal rotation of the total station about the vertical axis is controlled by a horizontal clamp and tangent screw and rotation of the telescope about the tilting axis is controlled by a vertical clamp and tangent screw. These can be replaced by endless friction drives that do not require a clamp, and some total stations incorporate dual-speed drives-coarse for rapid target location and fine for exact target location. All total stations will have a horizontal and vertical circle for measurement of angles, which are measured in digital form and displayed as degrees-minutes-seconds or gons. The angular accuracy of a total station varies from instrument to instrument but this will be in the range 1 ~ 10″ for most of the instruments that are likely to be used in construction surveying. All of total stations have either an optical or laser plummet.

Three broad categories of total station are available. Instruments intended for use in building and construction have a shorter measuring range and lower angular specification than others, but they are made to be more robust and will resist water and dust penetration to a higher degree. It has been shown that 95% of all site distance measurements are under 500m and that a 10″ angular accuracy is adequate for most setting out procedures. Examples of instruments in this category are Sokkia's Series 10 and the Topcon GPT 3005 shown in Fig. 6-6. The next category of totalstation covers those intended for surveying applications. These will have better angle and distance, specifications more functions, and better data storage and processing capabilities. Examples of instruments in this category are Sokkia's 030R Series and Trimble's 3600 DR total stations are shown in Fig. 6-7. The remaining category covers motorized total stations, which tend to have the best specifications but are the most expensive. Examples of instruments in this category are Leica TPS1200 and Timble S6 total stations shown in Fig. 6-8. These total stations use technology of automatic target recognition (ATR), which requires the total station to be fitted with an ATR sensor.

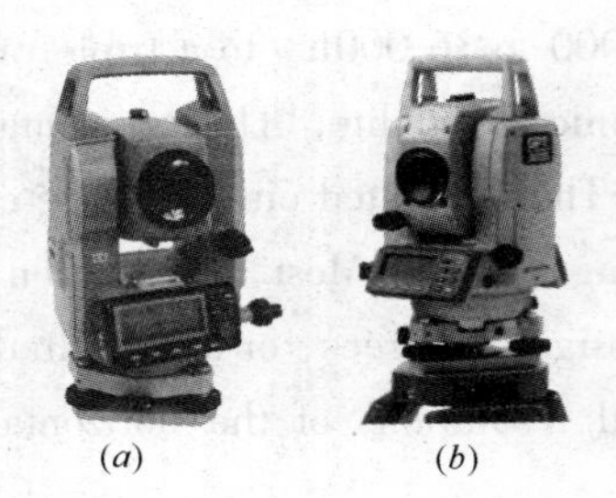

Fig. 6-6 Total station for building anconstruction
(a) Sokkia Serics 10; (b) Topcon GPT3005

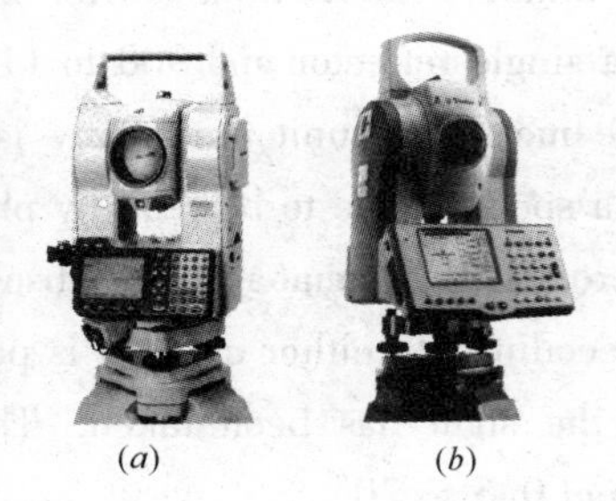

Fig. 6-7 Total station for surveying applications
(a) Sokkia 030R; (b) Trimble 3600DR

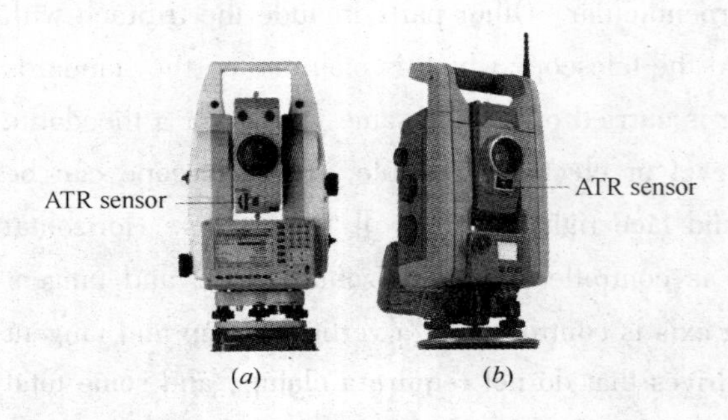

Fig. 6-8 Motorized total station
(a) Leica TPS1200; (b) Trimble S6

Although angles and distances can be measured and used separately, the most common applications for total stations occur when these are combined to define position in control surveys, mapping and setting out.

As well as total station, site surveying is increasingly being carried out using GPS equipment. Some predictions have been made that this trend will continue, although the use of GPS is increasing, total stations are one of the predominant instruments used on site for surveying and will be for some time. Eventually, the two will find applications that complement rather than compete with each other.

6.3.2 The components and accessories of total stations

1. Keyboard and display

A total station is activated through its control panel, which consists of a keyboard and multiple line liquid crystal display (LCD). Total station displays are moisture-proof and can be illuminated; some incorporate contrast controls to accommodate different viewing angles. A number of instruments have two control panels, which makes them easier to use. The keyboard enables the user to select and implement different measurement modes, enables instrument parameters to be changed and allows special software functions to be accessed. Some keyboards incorporate multifunction keys that carry out specific tasks, whereas others use keys to activate and display menu systems which enable the total station to be used as a computer might be.

Angles and distances are usually recorded electronically by a total station in digital form as raw data (slope distance, vertical angle and horizontal angle). For mapping, if a code is entered from the keyboard to define the feature being observed, the data can be processed much more quickly when it is downloaded into and processed by an office-based computer and plotter. On numerical keyboards, codes are represented by numbers only, whereas on alphanumeric keyboards, codes can be represented by numbers and/or letters, which give greater versatility and scope. Many total stations now have large graphic screens that make it possible for data to be edited on site.

Some examples of keyboards and displays are given in Fig. 6-9. On some total stations, the keyboard and display can be detached from the instrument and interchanged with other total stations or with GPS receivers, in what is known as integrated surveying. This enables data to be shared between different instruments and systems using a single interface. There combined keyboards and displays not only control the instrument, but are also data storage devices.

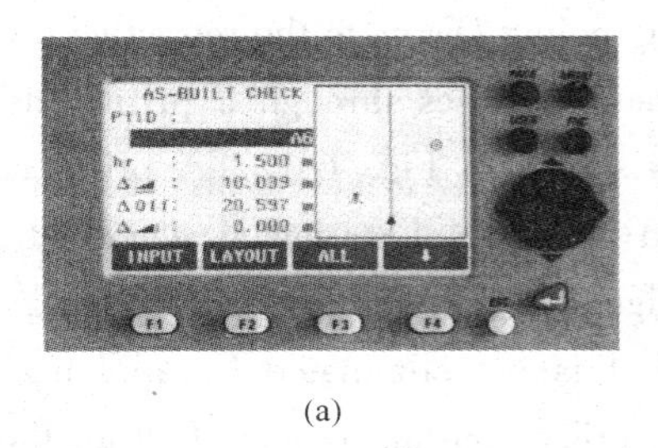

(a)

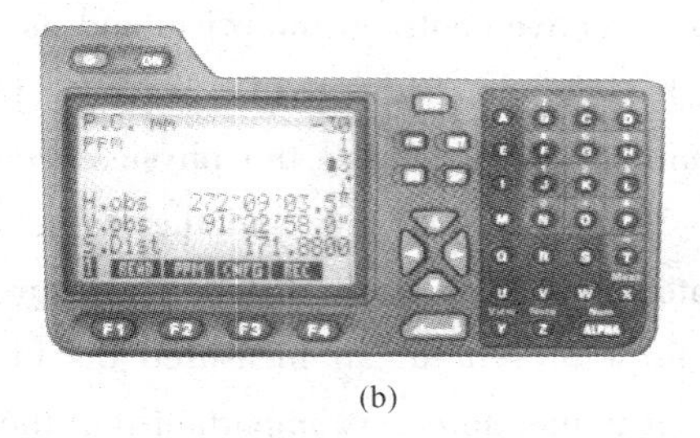

(b)

Fig. 6-9 Examples of total station keyboards and displays

(a) Leica TCR410C with basic functions required on site; (b) Sokkia 030R with greater functionality required for mapping and other surveys

2. Power supply

Three types of rechargeable battery are used in surveying instruments: these are Nickel Metal hydride (NiMh), Nickel Cadmium (NiCad) and Lithium-ion batteries. The NiMh battery is the most popular because it is compatible with standard camcorder batteries and has a better capacity than the NiCad battery. However, the NiCad battery has been available for many years, is still in widespread use and has more charging cycles than the NiMh battery. Lithium-ion batteries have the advantage of being easy to charge and maintain. As an alternative to rechargeable batteries, some instruments will accept AA size alkaline batteries. Most total stations are capable of giving a battery power indication and some have an auto power save feature which switches the instrument off or into some standby mode after it has not been used for a specific time.

3. Reflectors

Since the waves or pulses transmitted by a total station are either visible or infrared (which behaves like light but is invisible), a plane mirror could be used to reflect them. Unfortunately, this would require very accurate alignment of the mirror, because the transmitted wave or pulses have a narrow spread. To overcome this problem, special reflecting prism known as a *corner cube prism*, is always used. *1 [It is constructed from the corners of cubes of glass which have been cut away in a plane making an angle of 45° with the faces of the cube, as shown in Fig 6-10a.] *2 [Such a prism will always return a wave or pulse along a path exactly parallel to the incident path, but over a range of angles of incidence of about 20° to the normal of the front face of the prism, as shown in Fig. 6-10b.] As a result, the alignment of the prism is not critical and it is quickly set when on site.

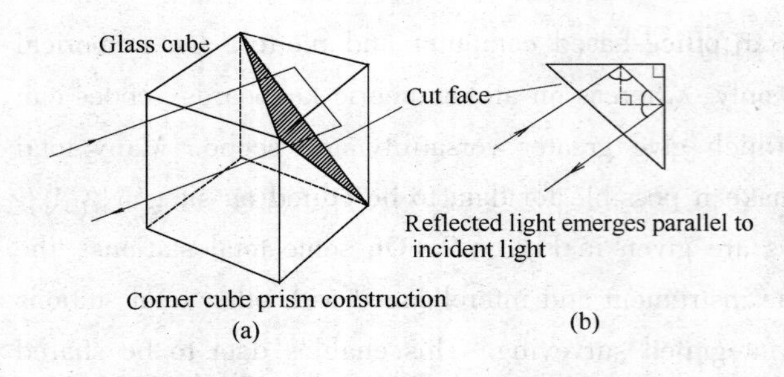

Fig. 6-10 Corner cube reflecting prism

A wide range of reflecting prisms are available to suit short range measurements (small prisms) and long range measurements (large and multiple prisms). Single and triple prism sets (a combination of a prism with an optical target) for tripod mounting are shown in Fig. 6-11.

Associated with all reflecting prisms is a prism constant—this is the distance between the effective centre of the prism and its plumbing point. Owing to the refractive properties of glass, which slows down the carrier wave or pulse when it passes through a prism, its effective centre is normally well behind the physical centre, as shown in Fig. 6-12. A prism constant is typically −30mm or −40 mm and this value is keyed into a total station as a correction that is applied automatically to each distance measured. If ignored, or applied incorrectly, this is a systematic error present in all measured distances and it is not eliminated by applying any field procedure. It is therefore very important that the correct prism constant is identified for the prism in use with a total station and that it is applied to all measurements. It is also necessary to enter a constant even when measuring to reflecting foil and in reflectorless mode.

Fig. 6-11 Single and triple prism sets

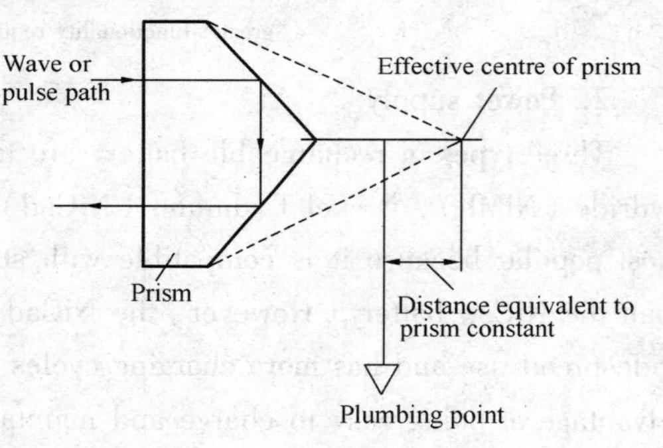

Fig. 6-12 Prism constant

单词与词组

configuration[kən ˌfigju ˈreiʃən] *n.* 结构，构造，外形，布置
component [kəm ˈpəunənt] *n.* 成分，组成部分，部件，元件
endless [ˈendlis] *a.* 无限的，无穷的，无止境的
friction[ˈfrikʃən] *n.* 摩擦，摩擦力
incorporate [in ˈkɔ:pərit] *vt.* (使)合并，并入，合编，具体表现
dual-speed [ˈdju:əl – spi:d] *a.* 双速的
drives-coarse[draivs – kɔ:s] *n.* 粗驱动装置
gon[gɔn] *n.* 哥恩，百进位度(圆周 =400g，1g =100C，1C =100CC)
dual-axis[ˈdju:əl – ˈæksis] *a.* 双轴的
broad[brɔ:d] *a.* 宽的，广阔的，主要的，粗略性的，大概的

robust [rəu ′bʌst] *a.* 健全的，坚固的，耐用的
penetration [ˌpeni ′treiʃən] *n.* 穿透，渗透，侵入
intend[in ′tend] *vt.* 想要，打算，企图，设计，计划
format [′fɔːmæt, －mɑːt] *n.*（出版物的）开本，版式，格式，大小尺寸，幅度；*vt.* 使格式化，设计，安排
robotic [rəu′bɔtik] *a.* 机器人的，自动的
· automatic target recognition（ATR）目标自动识别
complement[′kɔmplimənt] *n.* 补足物，补充量
activate[′æktiveit] *v.* 使活动，起动，触发
panel[′pænl] *n.* 面，板，控制板，仪表盘
moisture-proof[′mɔistʃə－pruːf] *a.* 防潮气的
illuminated [i ′luːmineitid] *a.* 被照明的，发光的
accommodate[ə ′kɔmədeit] *vt.* 给方便，帮助，使适应
multifunction[ˌmʌlti ′fʌŋkʃən] *n.* 多功能
versatility[vəː sə ′tiliti] *n.* 多才多艺，多用途，可转动性，易变
integrate[′intigreit] *v.* 与……结合起来（with），使一体化
nickel[′nikl] *n.* 镍（28 号元素，符号 Ni）
hydride[′haidraid] *n.* 氢化物
cadmium[′kædmiəm] *n.* 镉（48 号元素，符号 Cd）
lithium-ion [′liθiəm － ′aiən] *n.* 锂离子
compatible[kəm ′pætəbl] *a.* 适宜的，符合的
camcorder [kæm′kɔːdə] *n.* 摄录像机，便携式摄像机
widespread[′waidspred] *a.* 分布广的，普遍的，广泛的
alkaline[′ælkəlain] *a.* 碱（性）的，含（强）碱的
standby[′stændˌbai] *n.* 备用的人或物，待命状态
narrow[′nærəu] *a.* 狭窄的，严格的，精细的，仔细的
cube[kjuːb] *n.* 立方体，立方形物，正六面体，立方晶系（格）

注释

[1] It is constructed from the corners of cubes of glass <u>which have been cut away in a plane making an angle of 45° with the faces of the cube,</u> as shown in Fig 6-10a. 划线部分为定语从句修饰 cubes of glass（玻璃正立方体），从 in a plane making...... 到从句末是方式状语说明切割的方法。have been cut away……现在完成时被动语态，意为“已被切开”，这里不是切掉弃去。例如，He cut the collar away.（例句选自参考文献[34]）意为他裁出领子而不是裁去领子。所以全句译为“它是用一个与立方体面夹角为 45°的平面切割正立方体玻璃而构成的立方角锥，如图 6-10a 所示。”

[2] 此句前部分不难翻译，仅解释从 but 到句末：这里的 but 不是连接词，而是介词，除……之外，除非。over a range 意为超出范围。angles of incidence 意为入射角。the normal 这里指垂直线，to the normal of the front face of the prism（棱镜前面的垂直线）用来修饰说明 20°，入射角就是指入射光线与这条垂直线的夹角。全句译为：这样的棱镜

返回的电波总是精确地平行于入射波的路径，除非对于棱镜前面的垂直线形成的入射角大约超过 20°范围，如图 6-10b 所示。

6.4 Use of total station

The usual requirement at the start of any work with a total station is to level and centre it over a point, center with the optical plummet, and level using the plate level bubble or electronic bubble on the instrument's display. Most total stations allow the user to view the amount by which the instrument is off level at any time, as shown in Fig. 6-13, and all will issue a warning and stop working when the instrument is not levelled within specified limits.

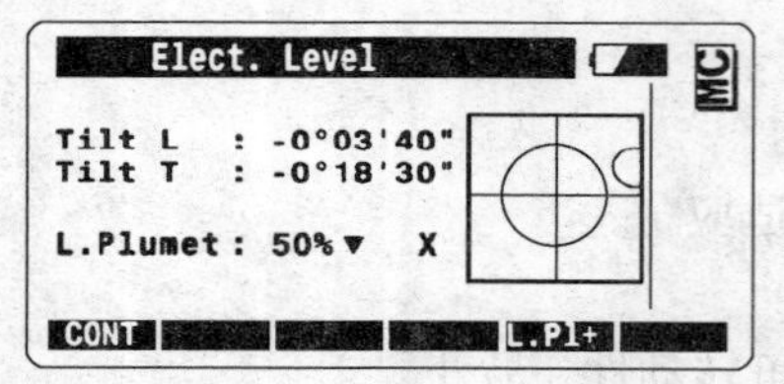

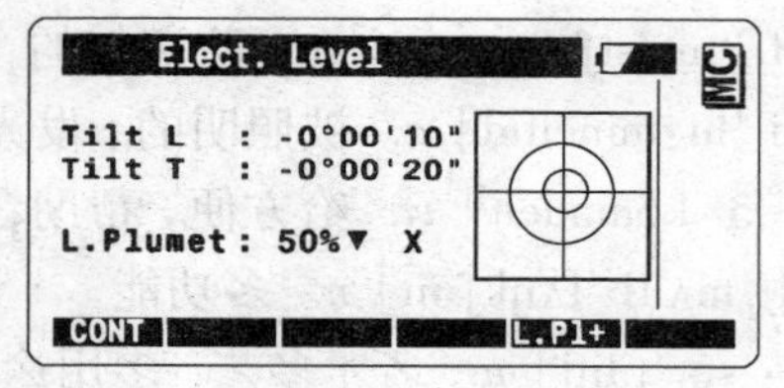

Fig. 6-13 Tilt errors displayed by total station (courtesy Leica Geosystems)

When the instrument has been leveled and centered over the station, turn on the power and initialize the system, usually by passing the telescope through a zenith angle of 90°. Next, some of the parameters that can be entered through the control panel, these parameters mainly are as bellows:

(1) Units of distance and angular measurement in meters or feet and degrees or gons, respectively;

(2) Temperature, ℉ or ℃, and atmospheric pressure, millimeters or inches of mercury (some systems have built-in sensors for temperature and atmospheric pressure);

(3) Prism constants;

(4) Curvature and refraction settings;

(5) Number of measurement repetitions-angle or distance (the average value is computed);

After the total station is set up and initialized with all necessary parameters entered. we can begin all kinds of measurements.

For example, to measure a horizontal angle, the telescope is sighted on the reflector and target at the backsight station and the zero-set key is pressed to set the horizontal circle to 0° or a preselected azimuth is entered via the keyboard. And then, the instrument is rotated and pointed to the second point, the horizontal angle is automatically displayed on the register and recorded in the field book or stored in the onboard or data-collector.

The total station measures the slope distance from the instrument to the reflector along with the vertical and horizontal angles. The microprocessor in the instrument then computes the horizontal and vertical components of the slope distance. Furthermore the microprocessor, using these computed components and the azimuth of the line, determines by trigonometry the north-south and east-west components of the line and the coordinates of the new point. These new coordinates are stored in memory.

单词与词组

issue['isju:] *v.* 出版，发行，发表，发布；冒出，流出
warning['wɔ:niŋ] *n.* 警告，告诫，前兆警报
initialize[i 'niʃəlaiz] *vt.* 预置(初始状态)，初始化
parameter[pə 'ræmitə] *n.* 参数，变数，特性
mercury['mə:kjuri] *n.* 汞，水银
preselect[ˌpri:si 'lekt] *vt.* 预选
memory ['meməri] *n.* 回忆，存储，存储器

6.5 Total station basic operations

Total stations and/or their attached data collectors have been programmed to perform a wide variety of surveying functions. All total station programs require that the instrument station and at least one reference station be identified so that all subsequent tied-in stations can be defined in X, Y and elevation (Z) coordinates. Because survey programs require that the proposed instrument station and backsight station be identified before data collection can commence, the proposed instrument station's coordinates – X, Y, and elevation (Z) – must be uploaded or manually entered into the total station's microprocessor, in addition, the backsight's station identity must be uploaded or manually entered into the total station's microprocessor. At a minimum, the azimuth to the reference station must be entered; however, if the coordinates of the backsight station were previously uploaded or manually entered, the needed azimuth to that station is easily inversed using the stored data. After setup, and before the instrument has been oriented for surveying as described above, the instrument and prism heights must be measured and recorded.

6.5.1 Point location

After the instrument has been properly oriented, the coordinates (northing, easting, and elevation) of any sighted point can be determined, displayed, and recorded in the following format: N E Z or E N Z (see Fig. 6-14). The format chosen reflects needs the format of the software program chosen to process the field data. At this time, the sighted point is numbered and coded for attribute data (point description)—all of which is recorded with the location data. This program is extensively used in topographic surveys. It should be noted here that when total stations with dual-axis compensation are used to determine elevations, the results compare favorably with elevations determined using automatic and digital levels.

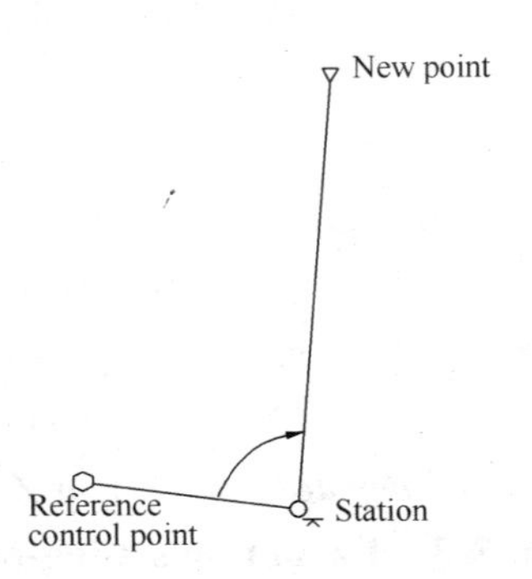

Fig. 6-14 Point location

Point location program:

Known: (1) N, E, and Z coordinates of the instrument station.

(2) N, E, and Z coordinates of reference control point,

or at least the azimuth of the line joining the instrument station and the control point.

The above station/control point information can be uploaded from the computer prior to going into the field or manually entered in the field.

Measured: (1) Angles, or azimuths, from the control point.

(2) Distances to the new point from the instrument station.

Computed: (1) N, E, and Z coordinates of new points.

(2) Azimuth of the line (and its distance) joining the instrument station to the new point.

6.5.2 Free station

This technique permits the surveyor to set up the total station at any convenient position referred to as a free station and then to determine the coordinates and elevation of that instrument position by sighting previously coordinated reference stations (see Fig. 6-15).

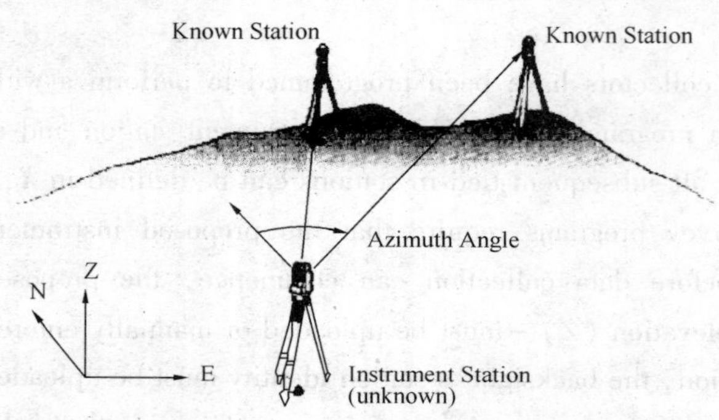

Fig. 6-15 Free station

When sighting only two points of known position, it is necessary to measure and record both the distances and the angle between the reference points. When sighting several points (three or more) of known position, it is necessary only to measure the angles between the points. *1 [It is important to stress that most surveyors take more readings than are minimally necessary to obtain a solution]; these redundant measurements increase the survey precision and check on the results. Once the instrument station's coordinates have been determined, the instrument is now oriented, and the surveyor can continue to survey using any or the other techniques described in this section.

Resection program:

Known: (1) N, E and Z coordinates of control point #1.

(2) N, E and Z coordinates of control point #2.

(3) N, E and Z coordinates of additional sighted control stations (up to a total of 10 control points). These can be entered manually or up-loaded from the computer depending on the instrument's capabilities.

Measured: (1) Angles between the sighted control points.

(2) Distances are also required from the instrument station to the control points, if only two control points are sighted. The more measurements (angles and distances), the better solution to the free station.

Computed: N, E and Z (elevation) coordinates of the instrument station.

6.5.3 Layout or setting-out positions

After the coordinates and elevations of the layout points have been uploaded into the total station, the layout/setting-out software will enable the surveyor to locate any layout point by simply entering that point's number when prompted by the layout software. *2[The instrument's display shows the left/right, forward/back, and up/down movements needed to place the prism in each of the desired position locations (see Fig. 6-16).] This capability is a great aid in property and construction layouts.

Setting out program:

Known: (1) N, E and Z coordinates of the instrument station.

(2) N, E and Z coordinates of a reference control point, or at least the azimuth of the line joining the instrument station and the control point.

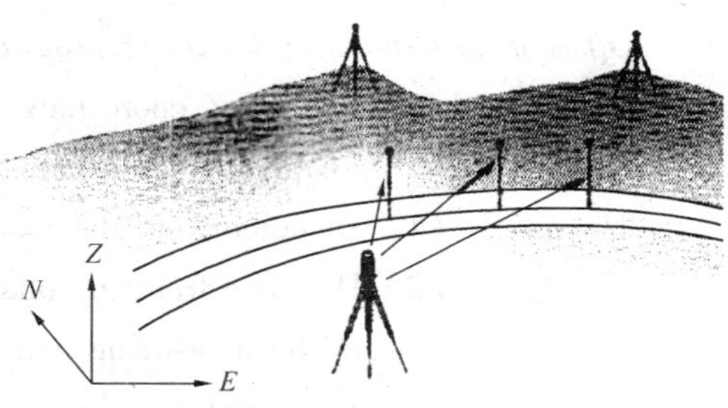

Fig. 6-16 Layout or setting-out positions

(3) N, E and Z coordinates of the proposed layout points (Entered manually, or previously uploaded from the computer).

Measured: (1) Indicated angles (on the instrument display), or azimuths, and distances from the control to each layout point. The angles may be turned manually or automatically if a servomotor-driven instrument is being used.

(2) The distances (horizontal and vertical) are continually re-measured as the prism is eventually moved to the required layout position.

6.5.4 Remote object elevation measurements

The surveyor can determine the heights of inaccessible points (e. g., electrical conductors, bridge components, etc.) by simply sighting the pole-mounted prism while it is held directly under the object. When the object itself is then sighted, the object height can be promptly displayed (the prism height must first be entered into the total station). See Fig. 6-17.

Remote object elevation measurement program:

Known: (1) N, E and Z coordinates of the instrument station.

(2) N, E and Z coordinates or a reference control point, or at least the azimuth of the line joining the instrument station and the control point.

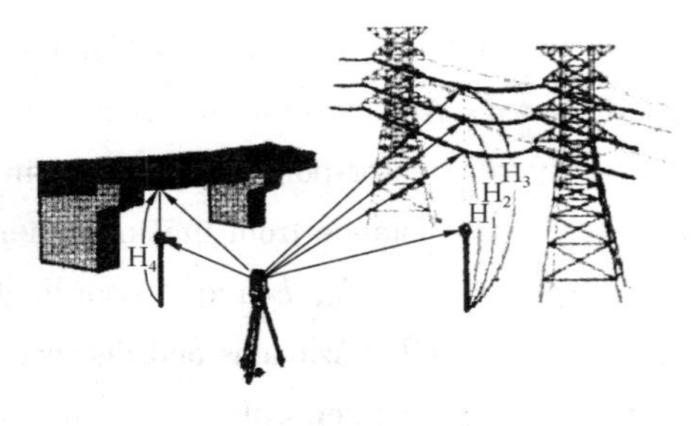

Fig. 6-17 Remote object elevation measurement

Measured: (1) Horizontal angle, or azimuth, from the reference control point (optional) and distance from the instrument station to the prism held directly below (or above) the target point.

(2) Vertical angles to the prism and to the target point.

(3) h_i height of the instrument station.

Computed: Distance from the ground to the target point (and its coordinates if required).

6.5.5 Offset measurements

When an object is hidden from the total station, the offset measurements must be used. The prism is placed appropriately at a point (offset point) for the position of an object is determined. There are two cases described below:

Offset measurements (distance) programs (see Fig. 6-18):

Known: (1) N, E and Z coordinates of the instrument station.

(2) N, E and Z coordinates (or the azimuth) for the reference control point.

Measured: (1) Distance from the instrument station to the offset point.

(2) Distance from the offset point (at a right angle to the instrument line-of-sight) to the measuring point.

Computed: (1) N, E and Z (elevation) coordinates of the "hidden" measuring point.

(2) Azimuths and distance from the instrument station to the "hidden" measuring point.

Offset measurements (angles) program (see Fig. 6-19):

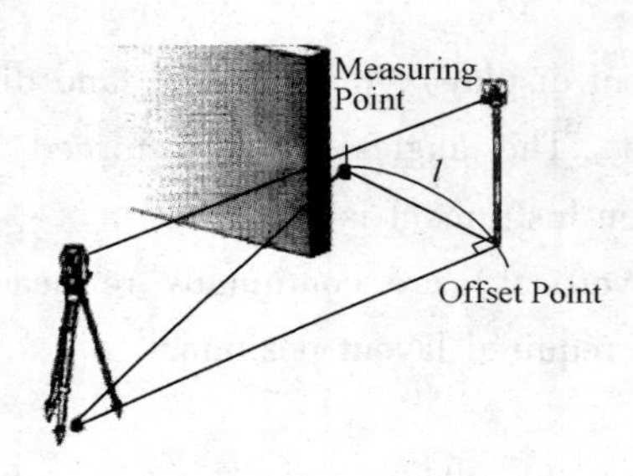

Fig. 6-18 Offset measurements (distance)

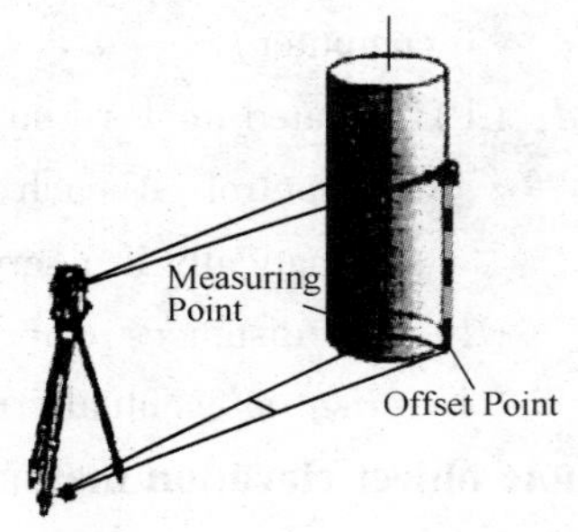

Fig. 6-19 Offset measurements (angles)

Known: (1) N, E and Z coordinates of the instrument station.

(2) N, E and Z coordinates (or the azimuth) of the reference control point.

Measured: Angles from the prism being held on either side of the measuring point, to the target center-point. (The prism must be held such that both readings are the same distance from the instrument station—if both sides are measured).

Computed: (1) N, E and Z coordinates of the "hidden" measuring points.

(2) Azimuths and distance from the instrument station to the "hidden" measuring point.

6.5.6 Area measurements

While this program has been selected, the processor will compute the area enclosed by a series of measured points. The processor first determines the coordinates of each station as described earlier and then, computes the areas using those coordinates (see Fig. 6-20).

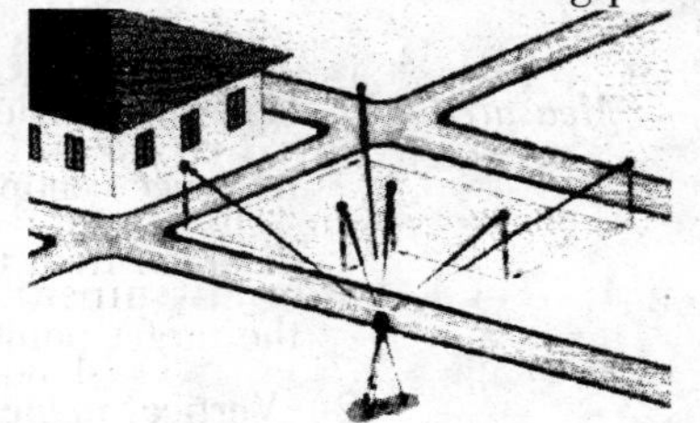

Fig. 6-20 Area measurement

Area measuremeut program:

Known: (1) N, E and Z coordinates of the instrument station.

(2) N, E and Z coordinates of a reference control point, or at least the azimuth of the line joining the instrument station and the control point.

Measured: (1) Angles, or azimuths, from the control point to the new points.

(2) Distances from the instrument station to the new points.

Computed: (1) N, E and Z coordinates of the area boundary points.

(2) Area enclosed by the coordinated points.

单词与词组

identify[ai 'dentifai] *vt.* 使等同于，认为一致(with)认出，识别，鉴别，验明认为，与……有关系，参与

tied-in[taid-in] *a.* 打结的，结合成整体的，相配的

propose[prə 'pəuz] *vt.* 提议，建议，主张推荐，打算，计划

previous['pri:vjəs] *a.* 先前的,事前的,以前的

redundant[ri 'dʌndənt] *a.* 多余的

orient['ɔ:riənt] *vt.* 使熟悉，使适应，定向

prompt[prɔmpt] *a.* 敏捷的，迅速的；干脆的，即时的，马上的

servomotor['sə:vəu,məutə] *n.* 伺服电动机(马达)，辅助电动机

· remote object elevation measurement 弧高目标的高程测量,悬高测量

· offset measurements 偏心测量

enclose[in 'kləuz] *vt.* 把……围起来，封入，附寄

注释

[1] It is important to stress that most surveyors take more readings than are minimally necessary to obtain a solution；此句难点在于 that 引起的从句，属于 more…than…的句型，翻译时根据句子含义强调 more 后面的部分，削弱、减少甚至否定 then 后面的部分。此句译为“强调这一点很重要，大多数测量采用多次读数，而不是采用求解的最小必要读数”。类似地再举一例：He is more witty than wise. （与其说他聪明，不如说他机智。）

[2] The instrument's display shows the left/right, forward/back, and up/down movements needed to place the prism in each of the desired position locations (see Fig. 6-16)。此句中 needed to place the prism in each of the desired position locations 作为 movements 后置定语，意为“在每一个所要求的点位上放置棱镜必需的”。全句译为：仪器显示屏上显示每一个所要求的点位上放置棱镜必须左右、前后、上下移动的量(见图 6-16)。

6.6 Total station instrument errors and accuracy of distance measurement

6.6.1 Total station instrument errors

In chapter 5, Fig. 5-15 shows the correct arrangements of the axes of a theodolite, these are the same for a total stations. This arrangement is seldom achieved in practice and gives rise to instrumental errors in total stations which are the same as those for a theodolite, some of which have already been described in Section 5.6. The most sophisticated can measure and correct for horizontal and vertical collimation, tilting axis error, and compensator index error, as described in the following sections.

1. Horizontal collimation error (or line-of-sight error)

This axial error is caused when the line of sight is not perpendicular to the tilting axis. It affects all horizontal circle readings and increases with steep sightings, but is eliminated by observing on two faces. For single face measurements, an onboard calibration function in the total station is used to determine c (see Fig. 6-21), the deviation between the actual line of sight and a line perpendicular to the tilting axis. A correction is then applied automatically for this to all horizontal circle readings. If c exceeds a specified limit, the total station should be returned to the manufacturer for adjustment.

2. Tilting axis error

This axial error occurs when the tilting axis of the total station is not perpendicular to its vertical axis. This has no effect on sightings taken when the telescope is horizontal, but introduces errors into horizontal circle readings when the telescope is tilted, especially for steep sightings. As with the collimation error, this error is eliminated by two face measurements, or the tilting axis error a (see Fig. 6-22) is measured in a calibration procedure and a correction applied for this to all horizontal circle readings. As before, if α is too big, the instrument should be returned to the manufacturer for adjustment.

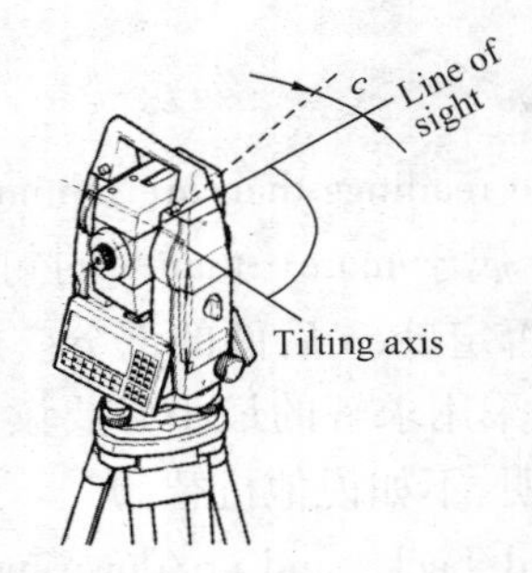

Fig. 6-21 Horizontal collimation error

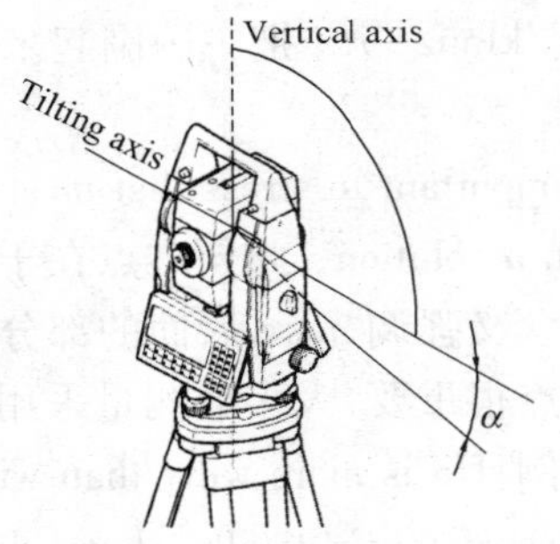

Fig. 6-22 Tilting axis error

3. Compensator index error

Any errors caused by not levelling a theodolite or total station carefully (sometimes called the vertical axis error) cannot be eliminated by observing on two faces. If the instrument is fitted with a compensator and this is switched on, it will measure any residual tilts of the instrument and will apply corrections to horizontal and vertical angles for these. However, all compensators have a longitudinal error l in the direction of the line of sight of the total station and a transverse error t perpendicular to this (see Fig. 6-23). These are known as zero point errors. For l and t can be determined using the onboard calibration function so

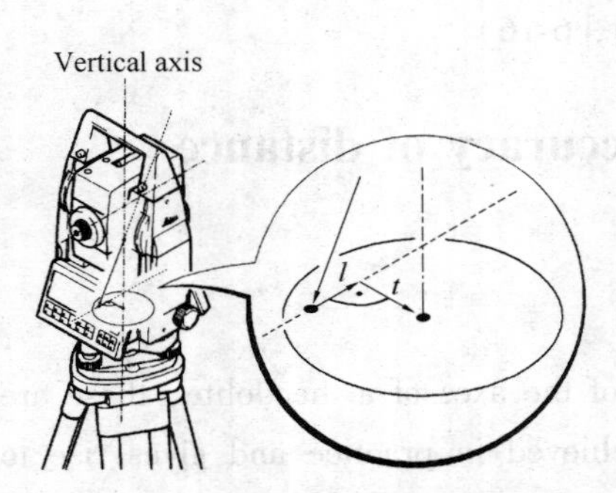

Fig. 6-23 Compensator index error

that measured angles can be corrected.

4. Vertical collimation error (or vertical index error)

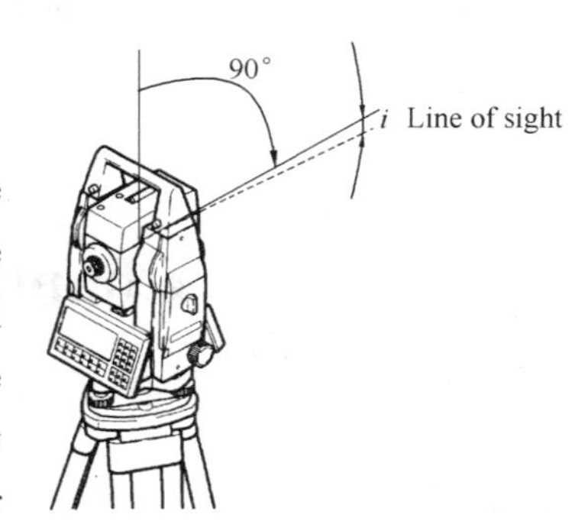

Fig. 6-24 Vertical index error

A vertical index error exists in a total station if the 0° ~ 180° line of the vertical circle is not perpendicular to the vertical axis when the line of sight is horizontal. This vertical index error is present in all vertical circle readings, and like the vertical index horizontal collimation error, it is eliminated by taking two face readings or by determining i, the index error shown in Fig. 6-24, in another calibration procedure. In this case, corrections for i are applied directly to vertical angles.

6.6.2 Accuracy of distance measurement

Accuracies of distance measurement of all total stations are stated following form:

$$\pm (a \text{ mm} + b \text{ ppm} \times D)$$

Where constant a is independent of the length being measured and is made up of internal sources within the instrument that are normally beyond the control of the user. It mainly consists of zero error (i. e., differences in the mechanical, electrical and optical centres of the total station and reflectors) and cyclic error (caused by unwanted interference between signals generated and processed inside the total station).

The systematic error b is proportional to the distance being measured, where 1ppm (part per million) is equivalent to an additional error of 1mm for every kilometre measured. It depends on the atmospheric conditions at the time of measurement and on any frequency drift in the oscillator. At short distances, this part of the distance error is small in comparison to instrumental and centring errors and it can be ignored for most survey work. However, at longer ranges, atmospheric conditions can be the worst source of error for electronic distance measurements, and since these are proportional to distance, extra care should be taken in the recording of meteorological conditions when these are used to calculate corrections that are applied to long lines.

Typical specifications for a total station vary from $\pm$ (2mm + 2ppm × D) to $\pm$ (5 mm + 5ppm × D). Taking the $\pm$ (2mm + 2ppm × D) specification as an example, at 100 m the error in distance measurement will be $\pm$ 2mm but at 1.5 km, the error will be $\pm$ (2mm + [2mm/km × 1.5km]) = $\pm$5mm.

单词与词组

steep[sti:p] *a.* 陡的，急剧升降的
longitudinal [lɔnʤi 'tju:dinl] *a.* 经度的，纵向的
transverse ['trænzvə:s] *a.* 横向的，横断的，横切的
unwanted[ˌʌn 'wɔntid] *a.* 不需要的，多余的，有缺点的
interference [ˌintə 'fiərəns] *n.* 干扰；干涉，干预，冲突，抵触，妨碍，打扰，阻碍物
component[kəm 'pəunənt] *n.* 成分，组成部分，部件，元件
oscillator ['ɔsileitə] *n.* 振荡器，发生器，加速器
meteorological[ˌmi:tiərə 'lɔʤikəl] *a.* 与气象学有关的，气象的

Chapter 7 Control Surveys

7.1 General

A control survey is the establishment of control networks in the survey area, in order to obtain coordinates of the control points. The control survey mainly includes reconnaissance, measurement and calculation of control points. These have been described in section 1.5. For all engineering projects, a control survey is carried out to fix the positions of reference point required for mapping, setting out and other dimensional work.

In fairly extensive surveys, such as those of a large industrial estate or of a town, the first thing to be done is to establish a control networks. Adopt the organizing principle of working from the whole to the part, that is, choose a small number of primary (first-order) control points which form a high-order precision of network covering the whole area and break these down into smaller networks of figures, as necessary, covering particular parts inside the main area by establishing secondary (second-order) and, if necessary, tertiary (third-order) control points. Although the areas covered in construction are usually quite small, the accuracy may be required to a very high order. Therefore, as surveys in the large area, the surveys in a building site are also to establish higher-order control network.

Control surveys can be divided into two general types: horizontal and vertical control systems. They are laid out in the form of nets covering the areas to be surveyed. The control points are established at locations where other surveys can be conveniently and accurately tied into them.

Horizontal control can be carried out by precise traversing, triangulation, trilateration and some combination of these methods. The exact methods used depend on the terrain, equipment available, information needed and economic factors.

For the majority of engineering work, the positions of horizontal control points are specified as plane rectangular coordinates (equivalent to X and Y coordinates used in mathematics). This is normal practice for construaion sites as survey work is greatly simplified and fewer mistakes are made when using rectangular coordinates for setting out and other dimensional work.

* [Using the technique of traversing, the relative position of the control points is fixed by measuring the horizontal angle at each point, between the adjacent stations, and the horizontal distance between consecutive pairs of stations.] All of the field procedures and calculations required for traversing are described in this chapter.

Alongside horizontal control, all sites will have some form of vertical control. The method used to provide vertical control on most construction sites is levelling, which is used to establish a series of benchmarks around a site. However, trigonometrical heighting with a total station can also be used for this. On large construction and engineering projects, the position of control points must be defined taking into account the curvature of the Earth.

单词与词组

reconnaissance [ri 'kɔnisəns] *n.* 侦察，搜索，勘测，踏勘，草测，选点
tie [tai] *n.* 带子，线，绳 *v.* 打结，系上，连结
trilateration[traiˌlætə 'reiʃən] *n.* 三边测量
· plane rectangular coordinates 平面直角坐标
equivalent [i 'kwivələnt] *a.* 相等的，相当的
dimensional[di 'menʃənəl] *a.* 尺寸的，空间的，……维的
traversing['trævəsiŋ] *n.* 导线测量
adjacent[ə'dʒeisənt] *a.* 接近的，附近的，毗连的，相邻的
consecutive[kən 'sekjutiv] *a.* 连续的，依顺序的，连贯的，相邻的
alongside[ə'lɔŋ 'said] *adv.* 在……的侧面，傍，沿着边靠，与……并排
· trigonometrical heighting 三角高程测量

注释

* Using the technique of traversing, the relative position of the control points is fixed by measuring the horizontal angle at each point, between the adjacent stations, and the horizontal distance between consecutive pairs of stations. 此句难点在后半部分，at each point（在每一点上）是修饰horizontal angle，而 between the adjacent stations(相邻点之间)也是修饰 horizontal angle，故译为“在每一点相邻点之间的水平角”。the horizontal distance between consecutive pairs of stations 其中 pairs of 意为一对，译为两个，此短句译为“相邻两个站之间的水平距离”。

7.2 Direction conception

7.2.1 Meridians

In surveying, the direction of a line is described by the horizontal angle that it makes with a reference line or direction. Usually this is done by referring to fixed line of reference which is called a meridian. The reference line usually selected is either the true meridian or the magnetic meridian. A *true meridian* is an intersection line that one plane passing through the geographic north and south poles and the observer's position with the earth's surface as shown in Fig. 7-1. True directions are obtained by sighting on the sun or one of the numerous stars whose astronomical position is known (the North Star, or Polaris, is the most common).

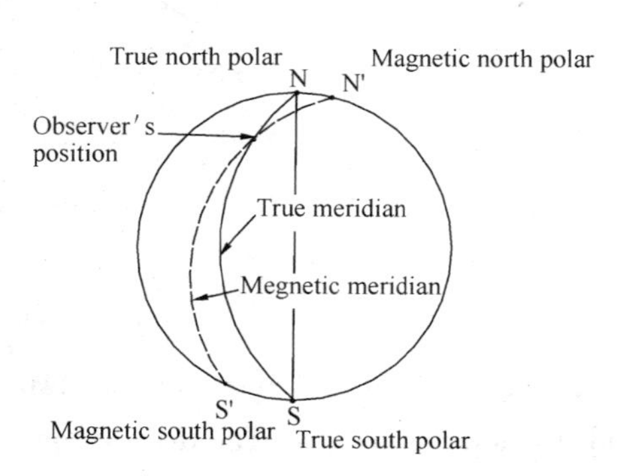

Fig. 7-1 True meridian and magnetic meridian

A *magnetic meridian* is an intersection line which one plane passing through the magnetic north and south poles and the observer's position with the earth's surface. The direction of magnetic meridian is the direction taken by the magnetized needle of compass at the observer's position. If neither a true meridian nor a magnetic meridian is readily available, the surveyor may use an assumed meridian, which just an arbitrary direction taken for convenience.

True meridians should be used for all surveys of large extent and, in fact are desirable for all surveys of land boundaries. They do not change with time and can be reestablished decades later. Magnetic meridians have the disadvantage of being affected by many factors some of which vary with time.

Sometimes for surveys of limited extent another type of meridian is used. A line through one point of a particular area is selected as a reference meridian and all other meridians in the area are assumed to be parallel to this, the so-called *grid meridian.* It could be aligned along some feature to suit site conditions (such as the centreline of a long bridge or along a building line). The use of a grid meridian eliminates the need for considering the convergence of meridians at different points in the area.

7.2.2 Azimuths and bearings

1. Azimuths

A common term used for designating the direction of a line is the azimuth. An azimuth is the direction of a line as given by an angle measured clockwise from the north end of a meridian. Azimuths range in magnitude from 0° to 360°. The azimuths of several lines AB, AC, AD are shown in Fig. 7-2a, the values respectively 60°, 152°, and 294°.

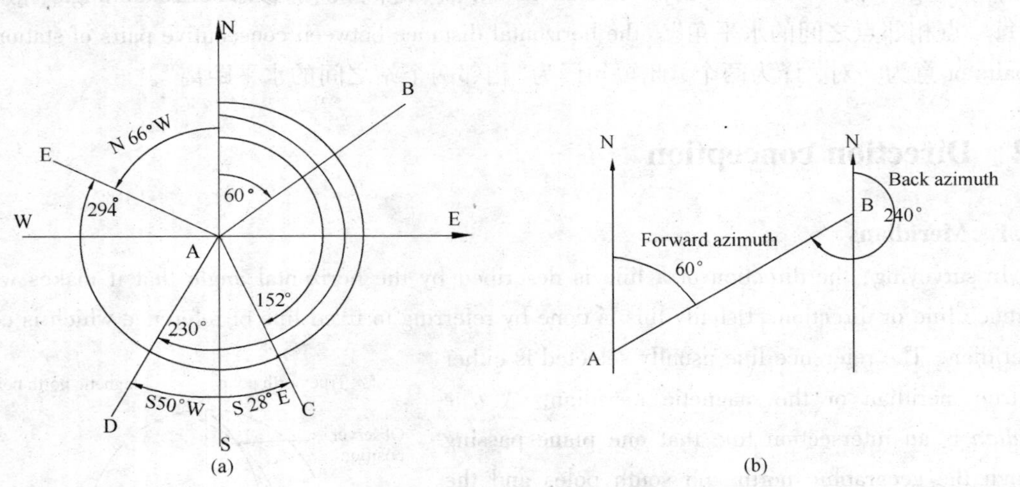

Fig 7-2 Azimuths and bearings

(a) Azimuths and bearings; (b) Forward and back azimuths

It can be said that every line has two directions. The line shown in Fig. 7-2b has direction AB and it has direction BA. The designations of forward and back are often arbitrarily chosen. Suppose that the direction of from point A to a point B is forward direction and from point B to a point A is back direction, the azimuth A_{AB} from A to B is called *forward azimuth*, the azimuth A_{BA} from B to A is called *back azimuth* of AB. For instance, the forward azimuth (A_{AB}) of line AB is 60° and its back azimuth (A_{BA}), which may be obtained by adding 180°, equals 240° (see Fig. 7-2b). Relationship between forward azimuth ($A_{forward}$) and back azimuth (A_{back}) can be written as

$$A_{back} = A_{forward} \pm 180° \tag{7-1}$$

Where if $A_{forward} < 180°$ it is to add 180°, and if $A_{forward} > 180°$ it is to subtract 180°.

Azimuths are classified as true, magnetic, and grid azimuth, depending on the meridian used. The type of meridian being used should be clearly indicated.

2. Bearings

Another method of describing the direction of a line is to give its bearing. * [The *bearing* of a line is indicated by the quadrant in which the line falls and the acute angle that the line makes with the meridian in that quadrant, and is measured from the north or south ends of the meridian.] In all cases, values of bearing angles lie between 0° and 90°. It cannot be greater than 90°. Therefore, in Fig. 7-2a, the bearing of the line AB is read north 60° east and written N60° E, The bearings of AC, AD, and AE are, respectively, S28°E, S50°W, and N66°W. If the direction of the line is parallel to the meridian and north, it is written as N0° or due North; if perpendicular to the meridian and east, it is written as N90°E or due East.

单词与词组

direction[di 'rekʃən, dai 'rekʃən] *n.* 指导，指挥，导演，管理；方向，方位
meridian[mə 'ridiən] *n.* 子午圈(线)
· true meridian 真子午线
· magnetic meridian 磁子午线
intersection [intə 'sekʃən] *n.* 横断，交叉，交点，交叉线
geographic[ˌdʒiːə 'græfik] *a.* 地理学的，地理的
polaris[pəu 'lɛəris] *n.* 北极星
compass['kʌmpəs] *n.* 罗盘,指南针
magnetize['mægnitaiz] *vt.* 使有磁性，使磁化
magnetic[mæg 'netik] *a.* 有磁性的，(可)磁化的
· assumed meridian 假定子午线
arbitrary['ɑːbitrəri] *a.* 任意的
azimuth['æziməθ] *n.* 方位角
· reference meridian 基准子午线
extent[iks 'tent] *n.* 广度，宽度，长度，大小
boundary['baundəri] *n.* 界线，边界，境界，范围
reestablish[riis 'tæbliʃ] *vt.* 重新建立
decade['dekeid] *n.* 十，十个一组，十年，十年间
disadvantage[ˌdisəd 'vɑːntidʒ] *n.* 不便，不利，不利的情况
· grid meridian 网格子午线，坐标子午线
designation[ˌdezig 'neiʃən] *n.* 指出，指明，任命，选派，称呼，名称
· forward azimuth 正方位角

· back azimuth 反方位角
bearing[′bɛəriŋ] *n.* 关系，影响，方面，方位，象限角
quadrant[′kwɔdrənt] *n.* 象限，圆周的四分之一
acute[ə ′kju:t] *a.* 尖的，锐的，尖锐的
due[dju:] *a.* 适当的，应有的，应做的，应得的，正当的，充分的

注释

* The *bearing* of a line is indicated by the quadrant in which the line falls and the acute angle that the line makes with the meridian in that quadrant, and is measured from the north or south ends of the meridian. 句中 which 指的是 quadrant,它又作 fall 的宾语。由 that 引导的定语从句“that the line…quadrant”修饰 angle,所以前一句译为“某线段的象限角是指该线段所处的象限并且与那个象限内与子午线形成的锐角”。后一句为简单句，省略了主语 bearing，译为“象限角由子午线的北端或南端量算”。

7.3 Traversing

One of the principles of engineering surveying is that horizontal and vertical control must be established for surveying detail and for setting out engineering projects. A traverse is a series of consecutive lines whose lengths and angles between adjoining lines have been determined from field measurements. Traversing is the professional work of establishing traverse points and making the necessary measurements, which mainly involve measuring angles and measuring distances. Angles are measured by theodolites or total stations; the distances can be measured by electronic distance measurement (EDM) instruments or total stations, sometimes by steel tapes.

There are three basic types of traverses: closed traverse, connecting traverse and open traverses.

7.3.1 Closed traverse

In Fig. 7-3, a traverse starts at station 1 and returns to the same point 1 via stations 2, 3 and 4. Station 1 can be of known position or can have an assumed position. In this case the traverse is called a *closed or polygon traverse* since it closes back on itself.

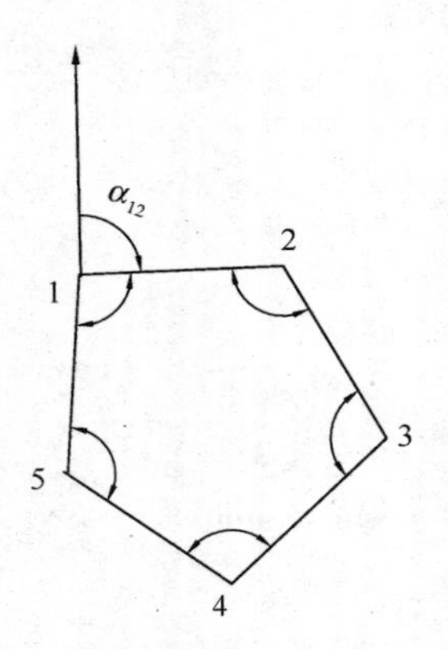

Fig. 7-3 Closed traverse

In the field works, the interior angle is measured at each station. For a closed polygon of n sides, the sum of interior angles should equal $(n-2)\times 180°$. Therefore, the observed angles can be checked by comparing the sum of the observed angles with theoretical values in the field, this is a mainly advantage of the closed traverse.

In Fig. 7-3, given the coordinates of the first station and the azimuth of the first line the coordinates of all successive points can be calculated.

7.3.2 Connecting traverse

Fig. 7-4 illustrates a typical connecting traverse commencing from the precisely coordinated point B and closing onto the precisely

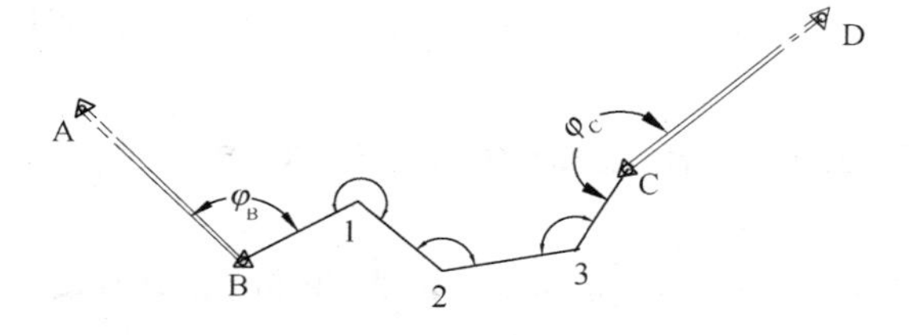

Fig. 7-4　Connecting traverse

coordinated point C. Generally, points A, B, C and D would be control points of an existing precisely coordinated control network. The traverse B123C is called a *connecting* or *link traverse*. * [In field work, traverse angles 1, 2, 3 should be measured and two attachment angles (i. e., φ_B and φ_C) between traverse line and high-grade line should be measured at same time.]

In the connecting traverse there are external checks on the observations since the traverses start and finish on points.

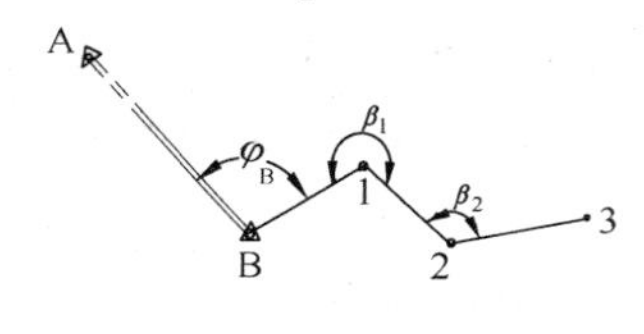

Fig. 7-5　Open Traverse

7.3.3　Open traverses

The open traverse is a "free" or "open" traverse, which commences at a known point and finishes at an unknown point and does not close back onto any known point as shown Fig. 7-5. Therefore, the measuring technique must be refined and repeated to guard against mistakes, at a minimum, distances are measured twice (once in each direction), and angles are doubled. The open traverses usually are used to route surveys.

单词与词组

traverse['trævə: s] *v.* 横越，横切，横贯；*n.* 导线，横梁

traversing['trævə: siŋ] *n.* 导线测量

· professional work 业务

· closed traverse 闭合导线

· connecting traverse 附合导线

· open traverses 支导线

polygon['pɔli:gɔn] *n.* 多边形

interior[in 'tiəriə] *a.* 内(部)的

attachment[ə 'tætʃmənt] *n.* 附属(物)，附件，连结

· attachment angle 连结角

refine[ri 'fain] *vt.* 精炼，提炼，提纯，提去(杂质等)

verification[verify'keiʃən] *n.* 证实,验证,核对,确认,确定

detect[di'tekt] *vt.* 发现,探测,检定

quantify['kwɔntifai] *vt.* 确定……的数量

coordinate[kəu'ɔ:dinit] *n.* 坐标

注释

* In field work, traverse angles 1, 2, 3 should be measured and two attachment angles (i. e., φ_B and φ_C) between traverse line and high-grade line should be also measured at same time。句中 traverse line 意为导线边，high-grade line 意为高等级的边。整句译为:在野外应测量导线 1、2、

3 的角度,并同时测量导线边与高级边的连结角(即 φ_B 与 φ_C)。

7.4 Traversing fieldwork

7.4.1 Reconnaissance and station marking

Reconnaissance is a vitally important part of any survery project, its purpose is to decide the best location for the traverse points. Reconnaissance is summarized as follows.

(1) Successive points in the traverse must be intervisible to make observations possible.

(2) Stations should be placed in firm, level ground so that the theodolite and tripod are supported adequately when observing angles at the stations.

(3) If the purpose of the control network is the location of topographic detail only, then the survey points should be positioned to afford the best view of the terrain, thereby ensuring that the maximum amount of detail can be surveyed from each point.

(4) Traverses for road works and pipelines generally require a link traverse since these sites tend to be long and narrow. The shape of the road or pipeline dictates the shape of the traverse.

(5) If the traverse is to be used for setting out, say, the centre-line of a road, then the stations should be sited to afford the best positions for setting out *1 [the intersection points and tangent points.]

When a reconnaissance is completed, the stations have to be marked for the duration, or longer, of the survey.

The construction and type of station depends on the requirements of the survey. For general purpose traverses, wooden pegs are used which are hammered into the ground until the top of the peg is almost flush with ground level. A nail should be tapped into the top of the peg to define the exact position of the station. A more permanent station would normally require marks set in concrete.

7.4.2 Angular measurement

Once the traverse stations have been placed in the ground, the next stage in the field procedure is to use a theodolite to measure the included angles between the lines.

In most cases it will be necessary to provide a signal at the observed stations since the station marks may not be directly visible. The signals has to be erected perpendicularly above the station marks, otherwise centering errors will result.

Signals should be perfectly straight objects set up vertically and centered exactly over the station mark. If a signal is not vertical then a centering error will be introduced, even though the base of the signal may be centered accurately over the mark. This is demonstrated by Fig. 7-6a.

From Fig. 7-6a, the lower the point of observation on the signal, the smaller will be the centering error. For this reason, the lowest visible point on any signal should always be observed when measuring angles.

Some simple types of signal are shown in Fig. 7-6b. Suggestions for their uses are as follows.

(1) When traverse line is very short the signal can use a nail tapped into the top of a

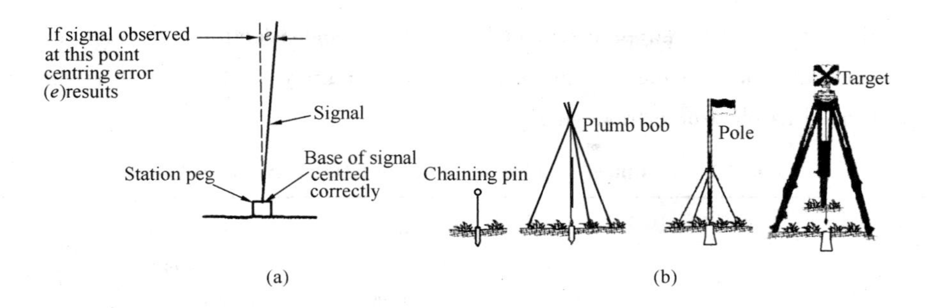

Fig. 7-6 Signals

(a) Signal not vertical; (b) Miscellaneous signals

wooden peg.

(2) If the station mark cannot be seen directly, but visibility condition is good, then a chining pin inserted into top of a wooden peg.

(3) *2[Three bamboo poles make to a support so that a plumb bob can be suspended from it.] The plumb line can then be observed.

(4) For larger line, it is preferable to the pole or target held tripod.

In the field works, either the left-hand angle or the right-hand is measured at each station. Measuring left-hand angle sequence is (see Fig. 7-7):

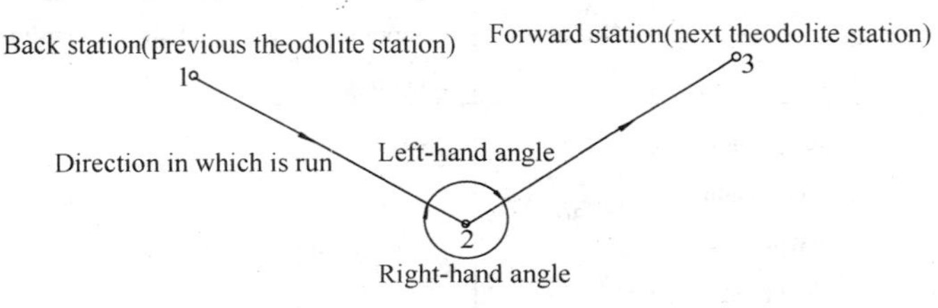

Fig. 7-7 Left-hand angle and right-hand angle

(1) Observe back station, Face Left.

(2) Observe fore station, Face Left.

(3) Observe fore station, Face Right.

(4) Observe back station, Face Right.

The sequence above may be term "B – F – F – B". On the contrary, measuring right-hand angle sequence is: "F – B – B – F".

The left-hand angle or the right-hand is calculated by

(1) left-hand angle = forward circle reading-back circle reading, or (2) right-hand angle = back circle reading-forward circle reading.

7.4.3 Distance measurement

Distance measurement of the traverse line is normally undertaken using steel taping or EDM, Technical requirements of distance measurement for traverse in China are listed in Table 7-1 and Table 7-2.

The length of each traverse is usually obtained by the simplest and most economical method capable of satisfying the required precision on a given project. Electronic devices and taping are used more frequently and provide the highest order of accuracy. When EDM is employed, the procedure is termed as electronic traversing.

Table 7-1 Technical requirements in the EDM traverse in China

Classification	Total length of traverse (km)	Mean line (m)	Mean square error of measuring distance (mm)	Mean square error of measuring angles (″)	Fractional linear misclosure of traverse
3rd level	15	3000	≤ ±18	≤ ±1.5	≤1/60000
4th level	10	1600	≤ ±18	≤ ±2.5	≤1/40000
1st class	3.6	300	≤ ±15	≤ ±5	≤1/14000
2nd class	2.4	200	≤ ±15	≤ ±8	≤1/10000
3rd class	1.5	120	≤ ±15	≤ ±12	≤1/6000

Table 7-2 Technical requirements in the mapping control in China

Traverse length	Precision ratio of the traverse	Line length	Mean square error of measuring angles		DJ6 observation set numbers	Misclosure in azimuth	
			Ordinarily	First class mapping control		Ordinarily	First class mapping control
≤1.0M	≤1/2000	≤1.5 the maximum sighting distance of mapping	30″	20″	1	$\pm 60''\sqrt{n}$	$\pm 40''\sqrt{n}$

Notes: M is the denominator of map scales. *n* is angle numbers of the traverse.

单词与词组

reconnaissance[ri ˈkɔnəsəns] *n.* 侦察，搜索，勘测，踏勘，草测

vital[ˈvaitl] *a.* 严重的，极其重要的，必不可少的

summarize[ˈsʌməraiz] *vt.* 概括，总结，摘要

successive[sək ˈsesiv] *a.* 连续的，相继的

intervisible[ˌintə ˈvizəbl] *a.* 可通视的

firm[fəːm] *a.* 结实的，坚固的，稳固的

pipeline[ˈpaipˌlain] *n.* 管道，输油管，输送管

flush [flʌʃ] *a.* 直接的，齐平的，同平面的，同高的(with)

tap [tæp] *vt.* 轻打，安接;*n.* 塞子,开关,龙头

concrete[ˈkɔnkriːt] *a.* 混凝土制的,水泥的

· station mark 测站标志

signal [ˈsignl] *n.* 信号，信号器，瞄准标志(编者注:signal = sight mark)

bamboo[bæm ˈbuː] *n.* 竹，竹竿

bob [bɔb] *n.* 钟摆，秤锤，垂球
suspend[sɔs 'pend] *vt.* 吊起，悬挂
mix-ups [miks-ʌps] *n.* 混乱，混淆，混合，搅匀
fractional['frækʃənl] *a.* 分数的，小数的
· mean square error of measuring angles 测角中误差
· mean square error of measuring distance 测距中误差
· fractional linear misclosure of traverse 导线全长相对闭合差
· misclosure in azimuth 方位角闭合差
economical[ˌiːkə 'nɔmikəl] *a.* 经济的，经济学的
· electronic traversing 电子导线测量

注释

[1] the intersection points and tangent points. 组词中 the intersection point 意为交点,参见第 9 章 9.1 节图 9-2 中的 P. I.（point intersection）路线的交点桩。tangent point 就是指图9-2 中的切线上曲线起点 B. C. 与曲线终点 E. C.。

[2] Three bamboo poles make to a support so that a plumb bob can be suspended from it. 句中 Three bamboo poles make to a support 直译 3 根竹竿做成一个支撑物，意译为“用 3 根竹竿捆扎成三脚架”。

7.5 Plane rectangular coordinate system

A plane rectangular coordinate system is as defined in Fig. 7-8. *[It is split into four quadrants with the typical mathematical convention of the axis to the north and east being positive and to the south and west, negative.] The four quadrants Ⅰ, Ⅱ, Ⅲ and Ⅳ are arranged in clockwise.

In pure mathematics, the axes are defined as x and y, with angles measured anticlockwise from the x-axis. In surveying, the x-axis is defined as the east-axis (E) and the y-axis as north-axis (N), with angles (α) measured clockwise from the y-axis (north-axis).

Another definition of method: the north-south direction is defined as the x-axis, the east-west direction perpendicular to north-south direction is defined as the y-axis, with angles (α) measured clockwise from the x-axis.

For all types of survey and engineering works, the origin is taken at the extreme south and west of the area so that all coordinates are positive.

From Fig. 7-8, it can be seen that to obtain the coordinates of point B, the coordinates of point A and the difference in coordinates between the ends of the line AB will be required. We have

$$\left.\begin{aligned} E_B &= E_A + \Delta E_{AB} \\ N_B &= N_A + \Delta N_{AB} \end{aligned}\right\} \tag{7-2}$$

or

$$\left.\begin{aligned} Y_N &= Y_A + \Delta Y_{AB} \\ X_B &= X_A + \Delta X_{AB} \end{aligned}\right\} \tag{7-3}$$

where ΔE_{AB} or ΔY_{AB} can call departure (or east coordinate difference), ΔN_{AB} or ΔX_{AB} can call latitude (or north coordinate difference).

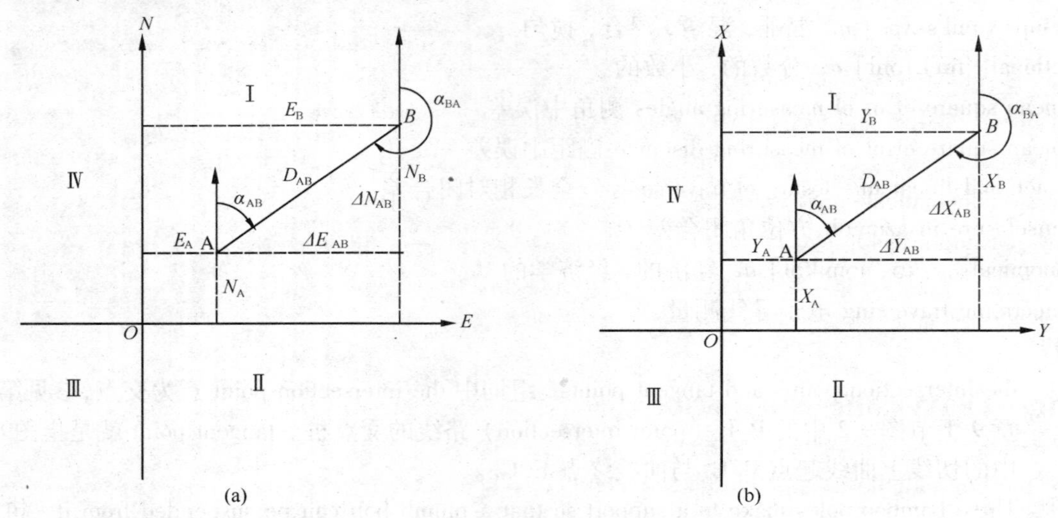

Fig. 7-8 Plane rectangular coordinate system

From Fig. 7-8, it is easy to get formula of coordinate differences:

$$\left.\begin{aligned}\Delta X_{AB} &= D_{AB}\cos\alpha_{AB}\\ \Delta Y_{AB} &= D_{AB}\sin\alpha_{AB}\end{aligned}\right\} \tag{7-4}$$

In the formula (7-4), the sign of ΔX_{AB} and ΔY_{AB} are dependent α_{AB} on its quadrant.

If the coordinates of point A(x_A, y_A) and point B (x_B, y_B) are given, from Fig. 7-8, it is easy to get

$$\tan\alpha_{AB} = \frac{\Delta Y_{AB}}{\Delta X_{AB}} \tag{7-5}$$

$$D_{AB} = \frac{\Delta Y_{AB}}{\sin\alpha_{AB}} = \frac{\Delta X_{AB}}{\cos\alpha_{AB}} \tag{7-6}$$

$$D_{AB} = \sqrt{\Delta X^2 + \Delta Y^2} \tag{7-7}$$

By formulae (7-5), (7-6) and (7-7), azimuths and side lengths can be obtained. But pay attention to formula (7-5), it usually can't give a correct azimuth directly. A judgment process should be done in the light of the signs of Δx_{AB} and Δy_{AB}, to determine which quadrant the azimuth belongs to.

单词与词组

· plane rectangular coordinate system 平面直角坐标系

convention [kən ˈvenʃən] *n.* 会议，大会，常规，习俗，惯例

extreme [iks ˈtriːm] *a.* 尽头的，末端的，最远的

departure [di ˈpɑːtʃə] *n.* 离开，出发，东西距离，横距，横坐标增量

latitude ['lætitju:d] *n.* 纬度(线)，纵距，纵坐标增量

· in the light of 鉴于，由于，按照

注释

* It is split into four quadrants with the typical mathematical convention of the axis to the north and east being positive and to the south and west, negative. 此句下划线的部分是主干,只是简单句,下划线的部分意为"它分成4个象限",但附加成分(从 with…到句末)很复杂,其中 with the typical mathematical convention 又是主要部分，意为"用典型的数学习惯",从 of …negative 又是修饰 convention,向北向东的轴是正的，"是"只能用分词 being。句末在 west 与 negative 之间用了逗号，省略 being。附加成分较长可单独译成两句。整句译成:"它分成4个象限,用典型的数学习惯,即向北向东的轴是正的，向南向西为负的"。

7.6　Traversing calculations

7.6.1　The calculations of closed traverse

1. Data preparation

When all the traverse fieldworks have been completed, mean angles observed and mean side lengths measured should be prepared. It is preferable to show all the data on a sketch of the traverse as this helps in the following calculations and can minimize the chance of a mistake.

The field data is shown in Fig. 7-9, the angles and lengths being entered on to a traverse figure. All the coordinates of points and surveyed data must fill in a table, see Table 7-1.

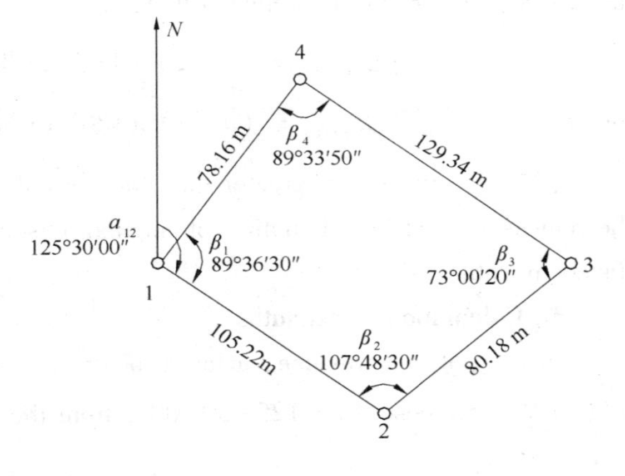

Fig. 7-9　The diagram of a closed traverse

2. The calculations of angular misclosure and adjustment

The observed angles of a polygon traverse can be either the internal or external angles. The angular misclosures are found by comparing the sum of the observed angles with one of the following theoretical values.

(1) Sum of internal angles = $(n-2)\times180°$

or　(2) Sum of external angles = $(n+2)\times180°$

where n is the number of angles or sides of the polygon.

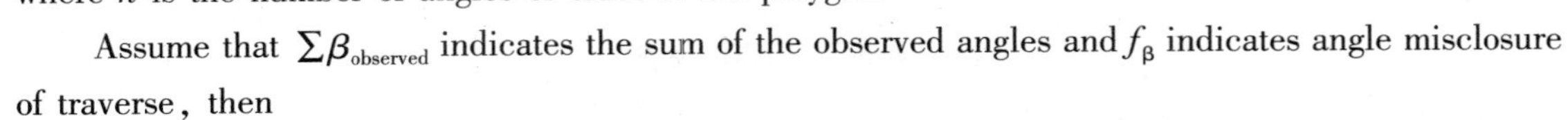

Assume that $\sum\beta_{\text{observed}}$ indicates the sum of the observed angles and f_β indicates angle misclosure of traverse, then

(1) $f_\beta = \sum\beta_{\text{observed}} - (n-2)\times180°$

or　(2) $f_\beta = \sum\beta_{\text{observed}} - (n+2)\times180°$

$f_{\beta_{\text{allowable}}}$ indicates allowable angle misclosure, it is different for different class of traverses. For mapping control survey, allowable angle misclosure is shown below:

$$f_{\beta\text{-allowable}} = \pm 60'' \sqrt{n} \tag{7-8}$$

where n is the number of angles or lines of the polygon.

When the actual angle misclosure (f_β) is compared to its allowable value ($f_{\beta\text{-allowable}}$) two cases may arise:

(1) The misclosure less than the allowable this result is acceptable.

For angular misclosure adjustment, *1 [an equal distribution is acceptable method since each angle is measured in the same way and there is an equal chance of the misclosure having occurred in any of the angles.] The angular misclosure (f_β) with opposite sign should be equal distributed to each obsereved angles, thus the corrected value for each angle is

$$v_\beta = \frac{-f_\beta}{n} \tag{7-9}$$

This example has $v_\beta = \dfrac{-(-50'')}{4} = 12.5''$, the corrections rounded to second, that is the two $-12''$, the other two $-13''$.

$$\text{corrected angles } \beta_{\text{corrected}} = \beta_{\text{observed}} + v_\beta \tag{7-10}$$

After all corrected angles are calculated, it should be to check if the sum of the corrected angles is equal to the theoretical value, i. e. :

$$\sum\beta_{\text{corrected}} = (n-2) \times 180° \text{ (when internal angles are observed)}$$

or

$$\sum\beta_{\text{corrected}} = (n+2) \times 180° \text{ (when external angles are observed)}$$

(2) The misclosure greater than the allowable this result is not acceptable. The angles should be remeasured. The calculation of angular misclosure and adjustment for the polygon traverse is listed in Table 7-1.

3. Calculation of azimuths

In the polygon traverse, azimuth of the first line is known or assumed, it indicates α_{12} in the Fig. 7-9a. Suppose $\alpha_{12} = 125°\ 30'\ 00''$, from the figure can obtain

$$\left.\begin{aligned} \alpha_{23} &= \alpha_{12} + 180° \pm \beta_2 \\ \alpha_{34} &= \alpha_{23} + 180° \pm \beta_3 \\ \alpha_{41} &= \alpha_{34} + 180° \pm \beta_4 \\ \alpha_{12} &= \alpha_{41} + 180° \pm \beta_1 \end{aligned}\right\} \tag{7-11}$$

Where β_1, β_2, β_3, β_4 indicate traverse angles of point 1, 2, 3, 4 respectively. In using the equation above, attention must be paid to the sign. When left-hand angles are measured, sign of traverse angle should be positive, when right-hand angles are measured, sign of traverse angle negative. It is special attentive to that the corrected traverse angle should be used in the above equations, that is corrected angles ($\beta_{\text{corrected}}$). Finally, check calculation should be done, i. e.

$$\alpha_{41} + 180° \pm \beta_1 = \alpha_{12} \text{ (checked)}$$

This example: $215°53'17'' + 180° + 89°36'43'' = 125°30'00''$

4. Calculate the latitudes (Δx) and departures (Δy)

Δx and Δy of each traverse side can be calculated in the light of formula (7-4), and then they should be filled in Table 7 - 1. After this, the sum of latitudes ($\sum\Delta x$) and departures ($\sum\Delta y$) can be calculated. From Fig. 7-10a can show $\sum\Delta x = 0$ and $\sum\Delta y = 0$.

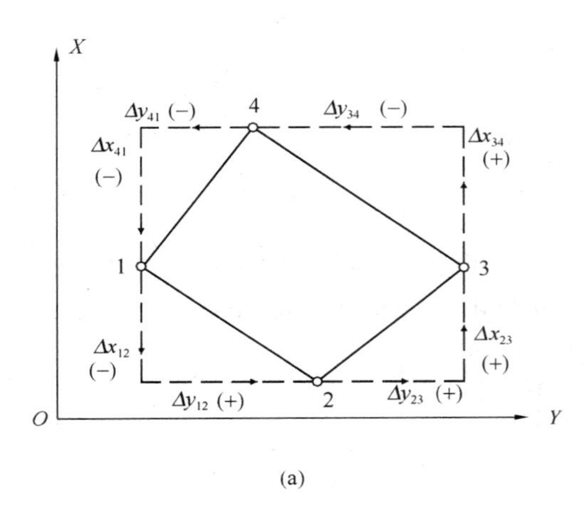

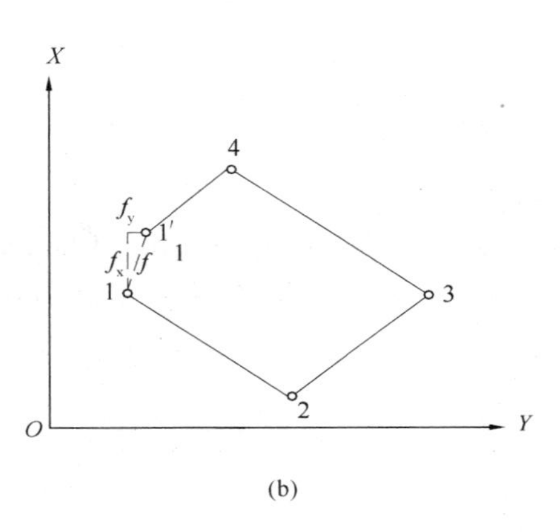

Fig. 7-10 Adjustment of the closed traverse
(a) Theoretical traverse figure; (b) Actual traverse figure

Owing to measuring angles and measuring lines occur errors, lead $\sum\Delta x \neq 0$, $\sum\Delta y \neq 0$. Let

$$\left.\begin{aligned} f_x &= \sum\Delta x \\ f_y &= \sum\Delta y \end{aligned}\right\} \tag{7-12}$$

where f_x can be called latitude misclosure, f_y departure misclosure.

Theoretically speaking, $f_x = 0$, $f_y = 0$, from Fig. 7-10a can obviously obtains

$$f_x = \sum\Delta x = (-\Delta x_{12}) + (-\Delta x_{23}) + \Delta x_{34} + \Delta x_{41} = 0$$

$$f_y = \sum\Delta y = \Delta x_{12} + \Delta x_{23} + (-\Delta x_{34}) + (-\Delta x_{41}) = 0$$

Owing to field errors, the traverse ends at 1′ instead of 1 (see Fig. 7. 10b). The total length misclosure f is given by

$$f = \sqrt{f_x^2 + f_y^2} \tag{7-13}$$

This example have $f = \sqrt{0.09^2 + (-0.07)^2} = \pm 0.11\text{m}$

To obtain a measure of the accuracy of the traverse, this misclosure is compared with the total length of the traverse lines ($\sum D$), to give the relative length misclosure of traverse K, i. e.

$$K = \frac{f}{\sum D} \tag{7-14}$$

This example: $$K = \frac{0.11}{392.90} = \frac{1}{3571}$$

We know that the allowable relative length misclosure $= \frac{1}{2000}$. The actual relative linear misclosure is less than the permissible value, the traverse fieldwork is satisfactory.

The values of the adjustment of coordinate difference are found directly proportional to the length of the individual traverse lines, i. e.:

$$\left.\begin{aligned} v_{x_i} &= \frac{-f_x}{\sum D} \times D_i \\ v_{y_i} &= \frac{-f_y}{\sum D} \times D_i \end{aligned}\right\} \qquad (7\text{-}15)$$

Where, v_{x_i} = the corrected value of latitude of *2 i th side, v_{y_i} = the corrected value of departure of ith side. It is obvious that the sum of all corrections of coordinate difference should satisfy next formula:

$$\left.\begin{aligned} \sum v &= -f_x \\ \sum v_y &= -f_y \end{aligned}\right\} \qquad (7\text{-}16)$$

Now commence to calculate corrected latitudes Δx and corrected departures Δy, i. e.

$$\left.\begin{aligned} \Delta x_{i_{\text{correted}}} &= \Delta x_i + v_{x_i} \\ \Delta y_{i_{\text{correted}}} &= \Delta y_i + v_{x_i} \end{aligned}\right\} \qquad (7\text{-}17)$$

After all corrected different of coordinates are calculated, it is to check that the sum of the corrected different of coordinates should be to zero, i. e.:

$$\left.\begin{aligned} \sum \Delta x_{i_{\text{corrected}}} &= 0 \\ \sum \Delta y_{i_{\text{corrected}}} &= 0 \end{aligned}\right\} \qquad (7\text{-}18)$$

5. Calculate the coordinates of traverse stations

For closed traverses, the coordinates of the starting point have to be known, which can also be assumed. The process of calculate the coordinates is as follows:

$$\left.\begin{aligned} x_2 &= x_1 + \Delta x_{12} \\ x_3 &= x_1 + \Delta x_{23} \\ &\vdots \\ x_1 &= x_n + \Delta x_{n,1} \end{aligned}\right\} (\text{checked}) \qquad (7\text{-}19)$$

Table 7-3 The computation of a closed traverse

Point NO	Observed horizontal angles (° ′ ″)	Corrections v (″)	Corrected horizontal angles (° ′ ″)	Azimuths α (° ′ ″)	Distances D (m)	Coordinate differences		Corrected coord. differ.		Coordinates	
						Δx (m)	Δy (m)	Δx (m)	Δy (m)	X (m)	Y (m)
1	2	3	4 = 2 + 3	5	6	7	8	9	10	11	12
1										500.00	500.00
				125 30 00	105.22	-2 -61.10	+2 +85.66	-61.12	+85.68		
2	107 48 30	+13	107 48 43							438.88	585.68
				53 18 43	80.18	-2 +47.90	+2 +64.30	+47.88	+64.32		
3	73 00 20	+12	73 00 32							486.76	650.00
				306 19 15	129.34	-3 +76.61	+2 -104.21	+76.58	-104.19		
4	89 33 50	+12	89 34 02							563.34	545.81
				215 53 17	78.16	-2 -63.32	+1 -45.82	-63.34	-45.81		
1	89 36 30	+13	89 36 43							500.00	500.00
				125 30 00							
Σ	359 59 10	+50	360 00 00		392.90	-0.09	+0.07	0.00	0.00		

$f_\beta = -50''$

$f_x = +0.09 \quad f_y = -0.07$

$f_{\beta\text{-allowable}} = \pm 60''\sqrt{n} = \pm 60''\sqrt{4} = \pm 120''$

total length misclosure of traverse $f = \sqrt{0.09^2 + (-0.07)^2} = \pm 0.11\text{m}$

allowable relative length misclosure $K_{\text{allowable}} = \dfrac{1}{2000}$

relative length misclosure of traverse $K = \dfrac{0.11}{392.90} = \dfrac{1}{3571}$

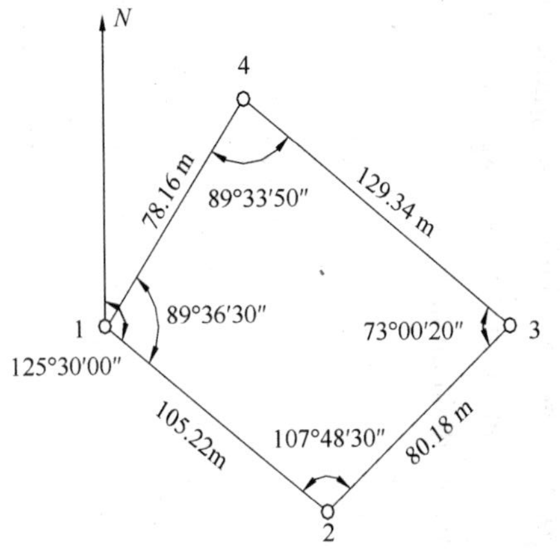

7.6.2 The calculations of connecting traverse

A connecting traverse was run between the high-grade control stations B and C as shown in the traverse diagram of Fig. 7-11.

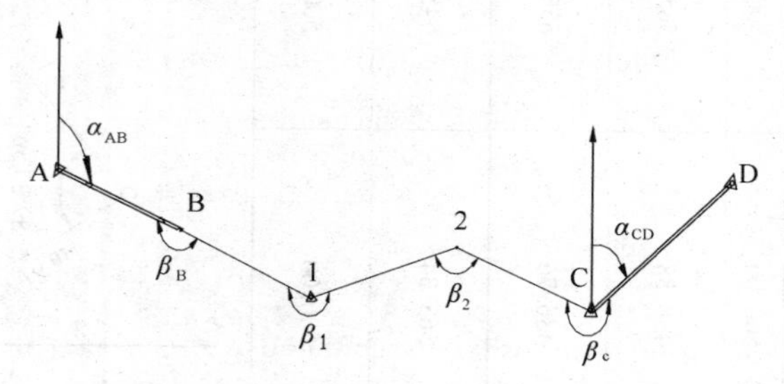

Fig. 7-11 The diagram of a connecting traverse

The precise coordinates of the controlling stations at the ends of the traverse are as follows:

The coordinates of the point B: $X_B = 509.580\text{m}$ $Y_B = 675.890\text{m}$

The coordinates of the point C: $X_C = 529.000\text{m}$ $Y_C = 801.540\text{m}$

The initial azimuth have been given, $\alpha_{AB} = 127°20'30''$, the final azimuth have been given, $\alpha_{CD} = 34°26'00''$. These data are listed in Table 7-4.

The calculation process of connecting traverse is much the same as the closed traverse, but the calculations of angular misclosure and coordinate differences misclosure are different since the two ends of connecting traverse connect high-grade points. The two items calculations will be discussed as follow.

1. The calculations of angular misclosure and adjustment

In Fig. 7-11, A, B, C, D are the high-grade known control points, β_B, β_C are attachment angles measured, β_1, β_2 are right-hand angles measured. From figure can obtain:

$$\left.\begin{aligned} \alpha_{B1} &= \alpha_{AB} + 180° - \beta_B \\ \alpha_{12} &= \alpha_{B1} + 180° - \beta_1 \\ \alpha_{2C} &= \alpha_{12} + 180° - \beta_2 \\ \alpha_{CD} &= \alpha_{2C} + 180° - \beta_C \end{aligned}\right\} \tag{7-20}$$

The sum of above four equations obtain

$$\alpha_{CD} = \alpha_{AB} + 4 \times 180° - \Sigma(\beta_B + \beta_1 + \beta_2 + \beta_C)$$

The above equation can be written universal formula as follows

$$\alpha_{final} = \alpha_{initial} + n \times 180° - \Sigma\beta_{right} \tag{7-21}$$

or

$$\alpha_{final} = \alpha_{initial} + n \times 180° + \Sigma\beta_{left} \tag{7-22}$$

We have the measured right-hand angles ($\beta_{right_{measured}}$) instead of the theoretical right-hand angles (β_{right}), and calculate the finally azimuth symbolizing as α'_{final}, then equation (7-21) changes

$$\alpha'_{final} = \alpha_{initial} + n \times 180^{\circ} - \Sigma\beta_{right_{measured}} \quad (7\text{-}23)$$

The same reason
$$\alpha'_{final} = \alpha_{initial} + n \times 180^{\circ} + \Sigma\beta_{left_{measured}} \quad (7\text{-}24)$$

This example has $\alpha'_{CD} = 127°20'30'' + 4 \times 180° - 822°54'06'' = 34°26'24''$

The angular misclosure (f_β) is difference of measured value and theoretical value, that is

$$f_\beta = \alpha'_{final} - \alpha_{final}$$

This example has $f_\beta = \alpha'_{CD} - \alpha_{CD} = 34°26'24'' - 34°26'00'' = +24''$

The allowable angle misclosure can still be calculated with formula (7-8). If $f_\beta < f_{\beta\text{-allowable}}$, then measuring angles are acceptable. The angular misclosure (f_β) should be equal distributed to each measuring angles, the corrected value for each angle is

$$v_\beta = \frac{f_\beta}{n} \quad (7\text{-}25)$$

The care must be taken to ensure that the correctness of calculating angles. For the connecting traverse measuring left-hand angle and right-hand angle are different in the calculate corrected value (v_β). When right-hand angles are measured, sign of corrected value v_β is the same to angle misclosure (f_β); but when left-hand angles are measured, sign of corrected value v_β is not same to angle misclosure (f_β).

This example is measuring right-hand angle, therefore sign of corrected value v_β should be same to angle misclosure (f_β). If $f_\beta = +24''$, then $v_\beta = \frac{+24''}{4} = +6''$, see Table 7-4.

2. Calculate the latitudes (Δx) and departures (Δy)

For the connecting traverse, the sum of coordinate difference of each line should equal coordinate difference between initial point and final point, i. e.

$$\Sigma\Delta x_{theoretical} = x_{final} - x_{initial} \quad (7\text{-}26)$$

$$\Sigma\Delta y_{theoretical} = y_{final} - y_{initial} \quad (7\text{-}27)$$

There are some errors in measuring angles and measuring lines, therefore $\Sigma\Delta x_{measure}$ and $\Sigma\Delta y_{measure}$ are not equal to theoretical values, and result in misclosure of coordinate difference, i. e.

$$f_x = \Sigma x_{measure} - \Sigma x_{theoretical} \quad (7\text{-}28)$$

$$f_y = \Sigma y_{measure} - \Sigma y_{theoretical} \quad (7\text{-}29)$$

$$f = \sqrt{f_x^2 + f_y^2} \quad (7\text{-}30)$$

This example have $f_x = -0.031\text{m}$, $f_y = -0.033\text{m}$, $f = +0.045\text{m}$, $K = \frac{0.045}{178.670} = \frac{1}{3953}$. The relative length misclosure of traverse less than the allowable value, therefore all observed values can be acceptable. The adjustment of misclosure of coordinate difference is the same with closed traverse. The process is listed in Table 7-4.

Table 7-4 The computation of a connecting traverse

Point NO	Observed horizontal angles (° ′ ″)	Corrections (″)	Corrected horizontal angles (° ′ ″)	Azimuths α (° ′ ″)	Distances D (m)	Coordinate differences ΔX (m)	Coordinate differences ΔY (m)	Corrected values ΔX (m)	Corrected values ΔY (m)	Coordinates X (m)	Coordinates Y (m)
1	2	3	4	5	6	7	8	9	10	11	12
A											
				127 20 30							
B	128 57 32	+6	128 57 38							509.580	675.890
				178 22 52	40.510	+7 −40.494	+7 +1.144	−40.487	+1.151		
1	295 08 00	+6	295 08 06							469.093	677.041
				63 14 46	79.040	+14 +35.581	+15 +70.579	+35.595	+70.594		
2	177 30 58	+6	177 31 04							504.688	747.635
				65 43 42	59.120	+10 +24.302	+11 +53.894	+24.312	+53.905		
C	211 17 36	+6	211 17 42							529.000	801.540
D											

$f_\beta = +24''$

$\Sigma D = 178.670$ $f_x = -0.031$ $f_y = -0.033$

$f = +0.045$ $K = 1/3970$

单词与词组

minimize [ˈminimaiz] *vt*. 把……减至最低数量（程度）

diagram [ˈdaiəgræm] *n*. 图样，图表，图像，示意图，曲线图，电路图

misclosure [misˈkləuʒə] *n*. 闭合差

indicate [ˈindikeit] *vt*. 指示，指出，表明，显示，象征

· allowable angle misclosure 容许角度闭合差

· allowable relative length misclosure 容许长度相对闭合差

round [raund] *v*. 使成圆形，四舍五入成整数

list [list] *v*. 列出，列入

· in the light of 鉴于，由于，按照

· latitude misclosure 纵坐标增量闭合差

· departure misclosure 横坐标增量闭合差

· total length misclosure 全长闭合差

universal [ˌjuːniˈvəːsəl] *a*. 一般的，普遍的

symbolize [ˈsimbəˌlaiz] *vt*. 用符号表示，象征，代表

注释

[1] An equal distribution is acceptable method since each angle is measured in the same way and there is an equal chance of the misclosure having occurred in any of the angles. 此句由主句和 since 引导的原因从句组成。句中 having occurred in any of the angles 现在分词短语完成式修饰 misclosure，意为“具有发生在任一个角度的”全句译为：可以采用平均分配的方法，因为每个角度测量以相同的方式，每个角度产生闭合差的机率是相同的。

[2] “i^{th} side”表示“第 i 号边”。

关于序数词构词与缩写说明如下：

1 ~19 除了 first，second，third 有特殊形式外，其余均由基数词加后缀-th 构成。1 ~19 的序数词列如下：		十位（10 的倍数）序数词构词法：先将十位数的基数词尾 ty 中的 y 变为 i，然后加后缀-eth。例如：	十位序数词中包含 1 ~9 的个位数，十位数用基数词，个位数用序数词，中间用连字符“-”。缩写规则依序数词的词尾而定。例如：
first—1st	eleventh—11th	twentieth—20th	twenty-first—21st
second—2nd	twelfth—12th	thirtieth—30th	thirty-second—32nd
third—3rd	thirteenth—13th	fortieth—40th	forty-third—43rd
fourth—4th	forteenth—14th	fiftieth—50th	fifty-fourth—54th
fifth—5th	fifteenth—15th	sixtieth—60th	sixty-third—63rd
sixth—6th	sixteenth—16th	seventieth—70th	seventy-fifth—75th
seventh—7th	seventeenth—17th	eightieth—80th	eighty-eighth—88th
eighth—8th	eighteenth—18th	ninetieth—90th	ninety-ninth—99th
ninth—9th	nineteen—19th		
tenth—10th			

第 100 号：one hundredth—100th，第 1000 号：one thousandth—1000th。

第 i 号英文表示较为特殊：即 i^{th}（在 i 的右上角加 th）。

Chapter 8 Topographic Surveys

8.1 General

Topographic surveys are made to determine the configuration (relief) or three-dimensional characteristics of the Earth's surface, or part of it, and to locate the natural and artificial features thereon. The natural features include trees, streams, lakes and so on. The man-made features are highways, bridges, dams, buildings and so on. These characteristics of features and reliefs may be represented on topographic map, which is a graphic representation of an area by means of various lines and conventional symbols.

Topographic map is a kind of map that presents the horizontal and vertical positions of the features and reliefs. It is distinguished from a planimetric map by the addition of relief in measurable form. The distinctive characteristic of a topographic map is that the shape of the Earth's surface is described by contour lines.

Topographic maps are used by engineers in determining the most desirable and economical location of highways, railroads, bridges, buildings, canals, pipelines, transmission lines, reservoirs, and other facilities; by geologists in investigating mineral, oil, water, and other resources; by architects in housing and landscape design; and by agriculturists in soil conservation.

A topographic survey consists of (1) establishing, over the area to be mapped, a system of horizontal and vertical controls, which consists of key stations connected by measurements of high precision; and (2) locating the details, including selected ground points, by measurements of lower precision from the control stations.

Topographic surveys can be performed by aerial photogrammetric methods, ground survey methods, or digital mapping methods. The vast majority of topographic surveys are now performed using aerial surveying techniques, with the plans and digital elevation models (DEMs) constructed using modern computerized photogrammetric, or laser imaging. Smaller surveys are often performed using electronic equipment, such as total stations.

However, cround surveys are still frequently used, however, especially for preparing large-scale maps of smaller areas. Even when photogrammetry is used, ground surveys are necessary to establish control and field-check mapped features for accuracy. This chapter concentrates on ground methods, and several field procedures for locating topographic features will be discussed.

单词与词组

configuration [kənˌfigju ˈreiʃən] *n.* 结构，构造，外形，外貌，轮廓

relief [ri ˈli:f] *n.* 减轻，解除，地势的起伏

thereon [ðɛəˈɔn] *adv.* 在其上，在那上面，……之后立即
represent [ˌrepriˈzent] *vt.* 表现，描绘
representation [ˌreprizenˈteiʃən] *n.* 表示法，图像
graphic [ˈgræfik] a. 绘画的，图表的，图解的
· conventional symbol 惯用符号
planimetric [plæniˈmetrik] *a.* 平面的
· planimetric map 平面图
measurable [ˈmeʒərəbl] *a.* 可量的，可测量的，重要的
contour [ˈkɔnˌtuə] *n.* 外形，轮廓，略图（=contour line）等高线
railroad [ˈreilrəud] *n.*（=［英］railway）铁路，铁道
canal [kəˈnæl] *n.* 运河，渠（水）道
· transmission line 传输线，输电线
reservoir [ˈrezəvwɑː] *n.* 水库；蓄水池
facility [fəˈsiliti] *n.*（常用复）便利，设备，器材，工具，装置，机构
geologist [dʒiˈɔlədʒist] *n.* 地质学者，地质学家
mineral [ˈminərəl] *n.* 矿物；矿石
architect [ˈɑːkitekt] *n.* 建筑师，设计师
· landscape design 园林设计
agriculturist [ˈægrikʌltʃərist] *n.* 农学家
soil [sɔil] *n.* 土壤；土质

8.2 Contours

The most common method of representing the topography of a particular area is to use contour lines. A contour tine, or contour, may be defined as the line in which the surface of the ground is intersected by a level surface. It follows that every point on a contour has the same elevation—that of the assumed intersecting surface. Fig. 8-1 represents a hill which is shown intersected by a series of level surfaces at elevations of 320m, 330m, … 350m above datum. As a mater of fact, the edge of a still lake is a line of equal elevation or a contour line. If the water in that lake is lowered or raised, the edge of its new position would represent another contour line. These contours are shown on maps and plans to indicate the topography of the area.

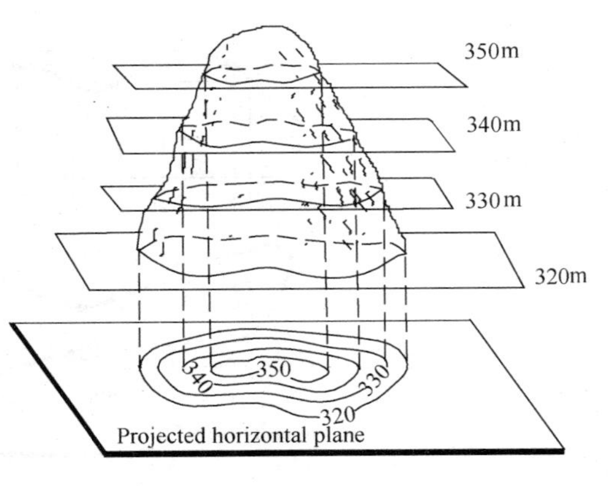

Fig. 8-1 The principle of contour lines

The contour interval of a map is the vertical distance between contour lines. The interval is determined by the purpose of the map and by the terrain being mapped (hilly or level). For normal maps, the interval varies from 2 to 20m, but it may be as small as 0. 5m for flat country and as large as 20 to 50m for mountainous country.

The selection of the contour interval is a very important topic. The contour interval selected depends on the scale of the map and the amount of relief in the area mapped. In areas with high relief the contour interval is usually larger to prevent the map from having too many contour lines, which would make the map difficult to read. Table 8-1 shows regulations of the contour interval of large scale topographic maps in China.

Table 8-1 Contour intervals of large scale topographic maps in China (m)

Terrain categories \ Scale	1:500	1:1000	1:2000
* Flat ground	0. 5	0. 5	0. 5 or 1
Foothills	0. 5	0. 5 or 1	1
Hilly countries	0. 5 or 1	1	2
Upland ground	1	1 or 2	2

There are four types of contour lines in all on topographic maps. They are basic interval contour, intermediate contour, auxiliary contour, and index contours.

Basic interval contours are those contours specified by regulations of large scale topographic maps (see Table 8-1), as example 100, 102, 104, 108, 10m and so on.

Intermediate contours are of a half of basic interval contours, these contour lines were drawn with dash lines on maps, as example 103m. This is also called one-half interval.

Auxiliary contours are of one-fourth of basic interval contours, these contour lines were drawn with dash lines on maps, as example 108. 5m. This is also called one-fourth interval contour.

Index contour are those wider contours, in order to read map, each fifth contour line is numbered and shown as a wider or heavier line. Such a line is called an index contour (see Fig. 8-2), as example 100m, 110m. The elevation value shown for index contour lines also helps the user estimate the elevations of adjacent contour lines.

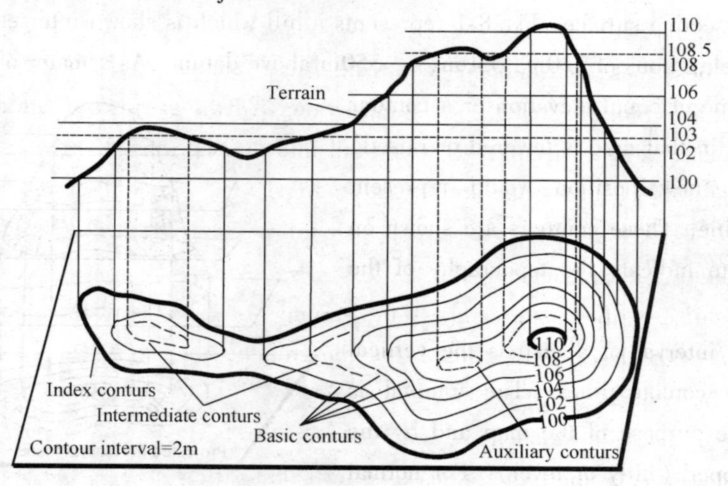

Fig. 8-2 Types of contour lines

单词与词组

topography [tə'pɔgrəfi:] *n.* 地形，地势，地貌，地形学，形貌学

edge [edʒ] *n.* 刀口，刀刃，刀锋，锐利，边，棱，边缘

hilly ['hili] *a.* 多小山的；多丘陵的

flat [flæt] *a.* 平的，平坦的，扁平的，平伸的，平展的

foothill ['futˌhil] *n.* 山麓，丘陵

upland ['ʌplənd] *n.* 高地，山地

· contour interval 等高线间隔，等高距

· basic interval contour 基本等高线，首曲线

· intermediate contour 中间等高线，间曲线

· auxiliary contour 辅助等高线，助曲线

· index contours 计曲线

注释

* 按照我国城市测量规范第4章有关地形的分类说明如下：

(1) 平地（flat ground）——地面倾角在2°以下地区。

(2) 丘陵地（foothills）——地面倾斜角在2° ~ 6°的地区。

(3) 山地（hilly countries）——地面倾斜角在6° ~ 25°的地区。

(4) 高山地（ulpine ground）——地面倾斜角在25°以上的地区。

8.3 Methods for detail surveys

8.3.1 Definition of details and selection of detail points

Once the topographic control has been established, the next step in a topographic survey is to locate the positions of topographic details from control point. These details consist of relief characteristics and all natural and artificial features. Attention must be paid that enough ground points and spot elevations to make the drawing of contour lines possible.

Detail can be subdivided into the following four main categories:

Hard detail describes well-defined features. These tend to be features which have been constructed, such as buildings, roads, walls and so on. Anything with a definite edge or having a precise position that can be easily located would fall into this category.

Soft detail describes features that are not well defined. These tend to be natural features such as river banks, bushes, trees and other vegetation. Anything where its edge or its precise position is in doubt would fall into this category.

Overhead detail describes features above the ground, such as power lines and telephone lines.

Underground detail describes features below the ground, such as water pipes and sewer runs.

Many types of symbols are used for representing detail and a standard format has yet to be universally agreed. Those symbols and abbreviations are fairly comprehensive and their use is recommended. However, at the large scales used for engineering surveys, the actual shapes of many features can be plotted to scale and, therefore, do not need to be represented by a symbol.

Selection of detail points is very important. For natural and man-made features, such as rivers, roads and buildings, the detail points should be at the deviation place of feature outlines, such as the corner points of buildings, the points of intersection and deviation of roads, etc. For topography, the detail points should be capable of representing ridges, valleys, peaks, depressions and neks.

The methods and the instruments used in topographic surveys depend upon the purpose of the survey, the order of precision needed, the nature of the terrain to be covered, the map scale, and the contour interval. For a high degree of accuracy, azimuths may be located with a gyro theodolite; horizontal distances may be measured with the EDM devices or total stations; elevations may be determined with a level.

8.3.2 Basic methods of locating points in the field

Six methods used to locate a point P in the field are illustrated in Fig. 8-3. All are based on horizontal control. One line (distance AB) must be known in each of the first four methods. The positions of three points must be known or identifiable on a chart to apply the sixth method, which is called the three-point problem. Known distances are shown in Fig. 8-3 as heavy lines. The quantities to be measured in the respective diagrams are:

(1) Two distances.

(2) Two angles.

(3) One angle and the adjacent distance.

(4) One angle and the opposite distance.

(5) One distance and a right-angle offset.

(6) Two angles at the point to be located.

Method 3 is used most often, but an experienced party chief employs whichever method is appropriate in a given situation; he must consider both field and office (computation and map) requirements.

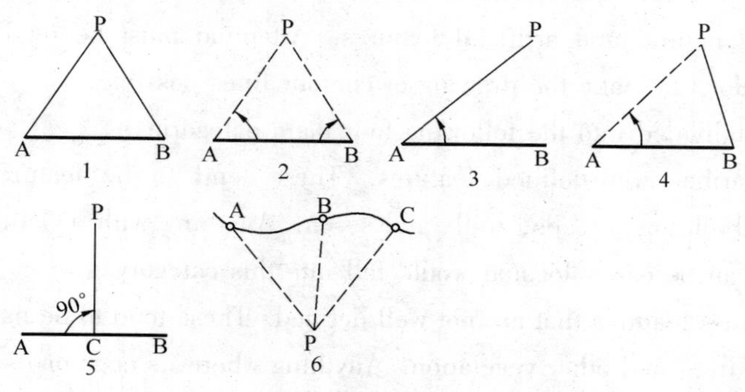

Fig. 8-3 Locating a point P

8.3.3 Conventional topographic survey

Conventional topographic survey is commonly used the theodolite and stadia method. When this method is used, horizontal distances and differences in elevation are directly determined by using subtended intervals and vertical angles observed with a theodolite on a leveling staff.

Stadia tacheometry is used to locate points of detail by the polar (or radiation) method, the basis of which is shown in Fig. 8-4, where d and β are measured in the field to locate P.

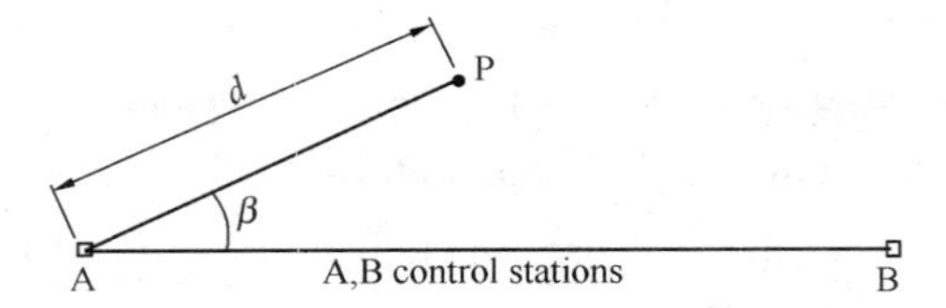

Fig. 8-4 Locating points of detail by the polar method

The component d is measured by stadia and β by reading the horizontal circle of the theodolite used. Tacheometry be used effectively in contouring, particularly in open areas where there are no points of clearly defined detail.

The tacheometry method is very suitable for detail collection. The steps of detail survey for one instrument station will be described as follows.

(1) Set the instrument up over control station A and centre and level it in the usual way. Measure and record the height (hi) of the trunnion axis above the station mark. For a detail survey it is standard practice to measure horizontal and vertical angles on one face only and hence the theodolite should be in good adjustment.

(2) Select a suitable control station as reference object (RO), e. g. station B in Fig. 8-5. Set the horizontal circle to zero along the direction to the RO. Measure the horizontal angle (β) between line AB and Al of from station A to detail point 1. The detail point 1 is the corner points of a building, the staff should be erectly held at the detail point 1.

Fig. 8-5 The principal of detail survey by stadia method

(3) Read stadia interval of the staff held on detail point l and the inclination angle (α) of collimation axis, and then the horizontal distance (d_{A1}) and the elevation difference (h_{A1}) between point A and point 1 can be calculated. The elevation of the detail point 1 (H_1) can be obtained by $H_1 = H_A + h_{A1}$.

(4) To plot the detail points, a protractor and a scale rule are required. Draw point 1 on the map sheet according to a given scale, its elevation is written next to its position.

Other detail points can also be surveyed and drawn on the map.

(5) Draw and interpolate contour lines according to surveyed points.

If the contours are to be drawn manually, they are usually obtained using eye estimating interpolation method. The procedure for this is as follows:

When drawing contour lines, the first use a pencil to draw the structure line, the ridge line is drawn with dashed lines, the valley line is drawn with solid lines, and then mark points through which the contour lines by interpolating. Owing to detail points are selected in the gradient change, the ground slopes between two adjacent spot heights can be uniform. That is the elevation difference between two points proportional to the horizontal distance between two points. On the joining line of two adjacent detail points, interpolate and mark with a pencil the positions of the required contour points, its method is so-called interpolation by eye estimating. In Fig. 8-6a, let two detail points A and C on the ground, the elevation values of A, C are 207. 4m, 202. 8m respectively, if the contour interval of the map is 1m, there must be five contour lines whose elevations are 203m, 204m, 205m, 206m, and 207m. According to the principle of the horizontal distance is proportional to the height difference, the first estimate position of the 203m elevation to mark point *m* and position of the 207m elevation to mark point *g* (see Fig. 8-6a), then *mq* distance is quartered to mark the elevations of 204m, 205m, 206m with three points i. e. *n*, *o*, *p*. The same way interpolate and mark those points through which contour lines between other two adjacent detail points must be. When all the contour positions have been located, those points having the same elevations are joined by smooth curves to form the contours as shown Fig. 8-6b.

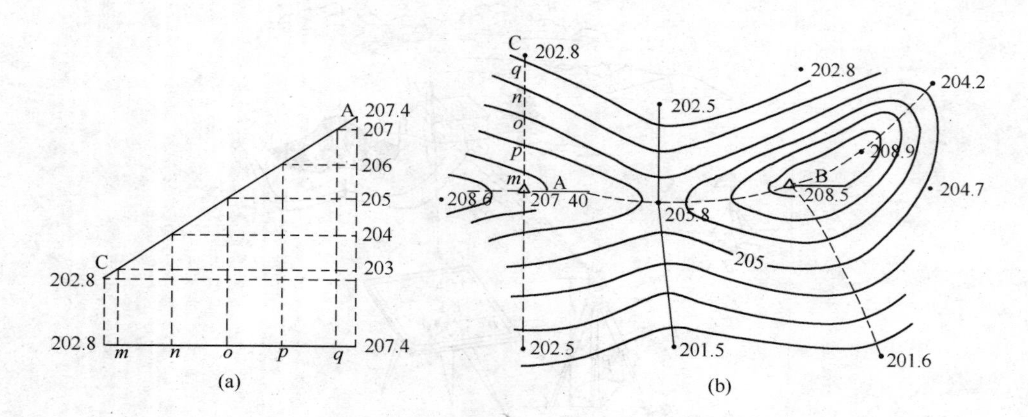

Fig. 8-6 Drawing contour lines

单词与词组

· spot elevation 独立高程点

gyro [ˈdʒaiərəu] *n.* (= gyroscope) 陀螺仪，回转仪

· gyro theodolite 陀螺经纬仪

abbreviation [əˌbriːvi ˈeiʃən] *n.* 省略，缩写，缩写词

comprehensive [ˌkɔmpri ˈhensiv] *a.* 广泛的，全面的，综合的

ridge [ridʒ] *n.* 山脊；岭，岗，分水岭，山脉

valley ['væli] *n.* 谷，峪，凹处，流域

· dashed line 虚线

· solid line 实线

peak [pi:k] n. 山峰，山顶

depression [di 'preʃən] *n.* 压低，降低，洼地，盆地

nek [nek] *n.* 鞍部，山峡

offset ['ɔ:fset] *n.* 分支，横距，支距；不均匀性；剩余，误差，残余偏差

subtend [səb 'tend] *vt.* 对着（向），（弦、边）对（弧、角）

· reference object（RO）参考目标，基准目标

e. g. = exempli gratia（拉） 例如

erectly [i 'rektli] *adv.* 直立地，挺直地，垂直地

protractor [prə 'træktə] *n.* 量角器

interpolate [in 'tə:pəuleit] *v.* 插入

gradient ['greidiənt] *n.* 坡度，梯度，陡度，斜率

smooth [smu:ð] *a.* 圆滑的，柔和的，安祥的

8.4 Detail surveying using total stations

8.4.1 Principle of radiation

A full description of total stations is given in Chapter 6, and their application to detail surveying is given here. A total station can be used in conjunction with a prism mounted on a detail pole to locate points of detail by the radiation method.

The principle of radiation is shown in Fig. 8-7, where a tree (P) is to be located from a survey line AB. The plan position of the tree can be fixed by measuring the horizontal distance, r, to it from A and also the horizontal angle, θ, subtended at A by the survey line AB and the line from point A to the tree. Hence the radiation method consists of measuring horizontal angles and horizontal distances. In addition, the reduced level of the point at which the detail pole is held can be obtained by measuring the vertical angle to the prism and using this in conjunction with the height of the prism and the height of the total station. Hence three-dimensional data is collected about the point being surveyed; that is, its plan position and its reduced level.

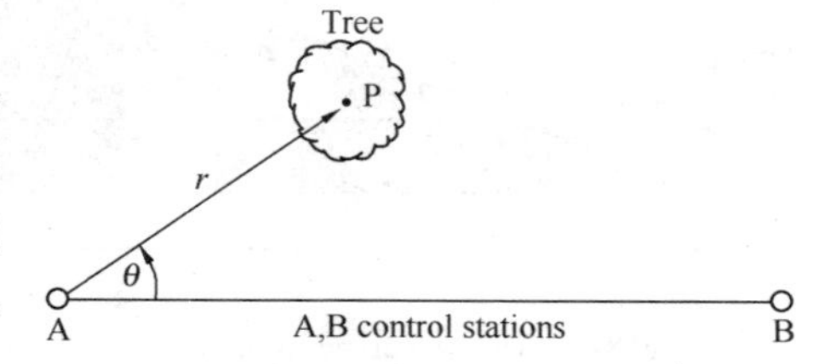

Fig. 8-7 Principle of radiation

Radiation using total stations can be used very effectively for contouring, particularly in open areas where there are no points of clearly defined detail. Also, because of its high accuracy capabilities, it can be used to pick up any type of detail, either hard or soft. Traditionally, two people have been required to locate detail using total stations: one operating the total station and the other holding the detail pole. However, in the past few years, total stations have been developed which are motorized and capable of locking onto and tracking prisms on detail poles using automatic target recognition. These are so-called robotic total stations which enable one-person operation. In this system, the instrument is set up as normal over a control point and left unattended with the operator working at the detail pole actually choosing and fixing the points of detail.

Although the use of such robotic instruments is on the increase, two-person total station detailing is still the norm. Hence the remainder of this section assumes that two people are involved: one using the instrument and the other holding the detail pole.

8.4.2 Fieldwork for detail surveying using total stations

During fieldwork, the total station is set up at each point in the control network in turn and used to locate detail from them as required. Because of its inherently high accuracy, it is possible to locate detail which lies a considerable distance from the control points using this method. However, if the distance becomes too great it can be very difficult for the two people carrying out the work to communicate effectively with each other. Hence if voice communication is being used, it is recommended that sighting distances should not exceed 50m or so. For distances greater than this the use of two-way radio communication is recommended. This, of course, is not a problem if a robotic system is being used.

Several methods of observation are possible and the following description is given as a general purpose approach only. It is assumed that the reduced levels of the control stations are known. Consider Fig. 8-8, which shows a detail surveying observation being taken from a control point A to a detail point P.

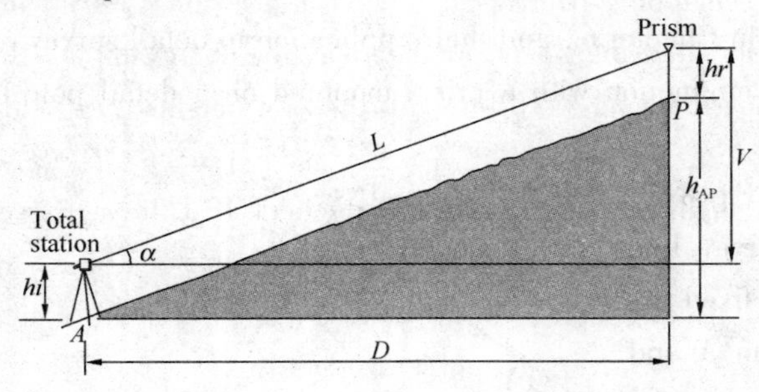

Fig. 8-8 Detail surveying using a total station

(1) Set instrument up over the station mark at control point A and centre and level it in the usual way. For a detail survey it is standard practice to measure horizontal and vertical angles on one face only, and hence the total station should be in good adjustment or calibration.

(2) Measure and record the height of the tilting axis (*hi*) above the station mark. Select a nearby control station as a reference object (RO), sight this point and record the horizontal circle reading. It may be necessary to erect a target at the RO if it is not well defined. All the detail in the radiation pattern will now be fixed in relation to this chosen direction. Some engineers prefer to set the horizontal circle to zero along the direction to the RO, although this is not essential.

(3) The actual observations taken at each pointing will depend on the type of instrument being used and whether they are being booked by hand or recorded automatically. In the simplest case, readings of the horizontal circle, the vertical circle and the slope distance *L* are booked or recorded. On instruments which give the horizontal distance D and the vertical component *V* of the slope distance directly, these two values can be booked or recorded together with the horizontal circle reading. If using an instrument which calculates and displays the coordinates (*X*, *Y*) and reduced level (*RL*) of the point on its screen, then these can be booked or recorded directly.

(4) If hand booking and drawing is being done, the type of detail should be noted on the booking form. If automatic booking and plotting is being done, a unique point number must be allocated to the detail point and a feature code must be assigned to it to so that it will be recognized by the software being used to process the data and plot the map. The height of the centre of the prism (reflector) above the bottom of the detail pole (*hr*) must be booked or recorded. Since detail poles are telescopic, this height can be set as required. If hand booking and calculations are being done, these are simplified if *hr* is set to the same value as the height of the instrument above the control station *hi*. In such a case, *hi* is cancelled out by *hr* and the vertical component *V* of the slope distance *L* is equal to the difference in height between ground points A and P. If the observations are being recorded in a data collector or on a memory card then *hr* can be set to any convenient value. However, once it has been set, it is recommended that the height of the prism is not altered unless absolutely necessary, since every time *hr* is changed its new value must be keyed into the data collector or instrument keyboard by hand. As well as being time-consuming, this can easily be forgotten, causing errors.

(5) Once point P has been located the detail pole is moved to the next point and the procedure is repeated for each detail point in turn until all the observations have been completed. As far as is practicable, each of the detail points should be selected in a clockwise order to keep the amount of walking done by the person holding the detail pole to a minimum. However, if the observations are being recorded for use with a computer software mapping package it is advisable to pick up all the points needed to define a particular feature before moving on to another feature; for example, the corners of the same building, all the points along the same kerb line, all the spot heights in a field, and so on.

(6) Before packing up, the final sighting should be back to the RO to check that the setting of the horizontal circle has not been altered during observations. If it has, all the readings are unreliable and should be remeasured. Hence it is advisable, during a long series of total station readings, to take a sighting back to the RO after, say, every 10 points of detail.

单词与词组

conjunction [kən 'dʒʌkʃən] *n*. 连合（接），联合
unattended [ˌʌnə'tendid] *a*. 无人监视的，自动（化）的，无人伴随的
integral ['intigrəl] *a*. 构成整体所必需的，组成的，主要的，必备的
norm [nɔ:m] *n*. 模范，典型，标准，规范
inherently [in 'hiərəntli] *adv*. 天性地，固有地
calibration [ˌkælə'breiʃən] *n*. 调准，校准
time-consuming ['taimkənˌsu:miŋ] *a*. 费时的，旷日持久的
package ['pækidʒ] *n*. 包裹，包；捆

8.5 Digital mapping techniques

In recent years, a number of advances have been made in computer technology such that many powerful desk top computers are now available with sophisticated peripherals. Using these, many survey organizations and large civil engineering contractors have developed their own in-house systems for the plotting of survey plans by computer. In addition, manufactured systems are also available for automated plotting of survey plans. The elements of an automated survey plotting system are shown in block form in Fig. 8-9. The various stages involved are described in the following sections.

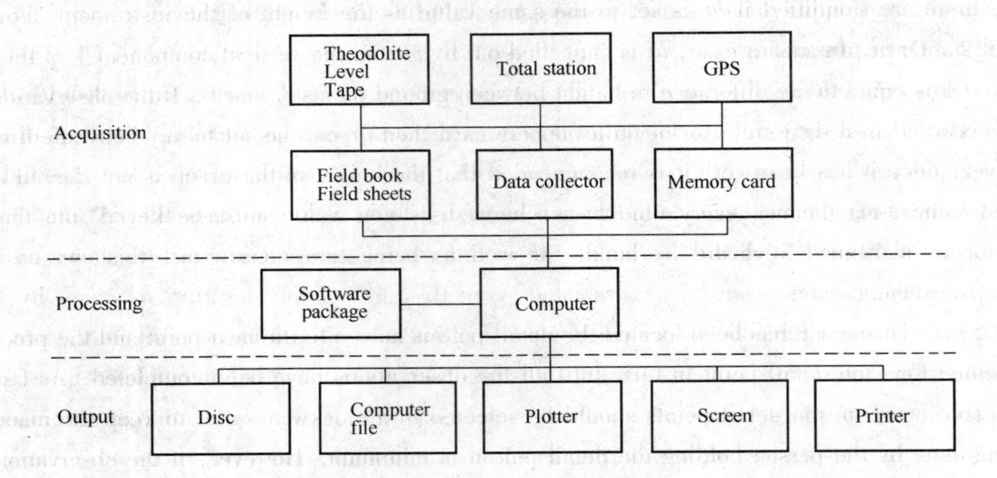

Fig. 8-9　Computer aided plotting system

8.5.1　Data acquisition

The raw data is normally acquired using a range of surveying equipment as indicated in Fig. 8-9. Traditionally, all observations were recorded by hand on field sheets or in field books. Nowadays, observations are recorded electronically either on data collectors or on memory cards plugged into total stations or using GPS receivers. The data from electronic sources can be very quickly transferred into the computer through a standard RS232 or USB interface. If the field sheets have been used, then the observations must be carefully keyed in by hand and this can be very time-consuming as well as a potential source of error.

For each point surveyed in the field, a code is used to define the type of detail being observed, the annotation (if any) to be shown on the plan at the point of detail and any other information such as running dimensions or names of features. In effect, this code replaces the field sketch but it is recommended that sketches be drawn as aids to the subsequent plotting.

The total station is capable of recording angles and distances with their appropriate codes directly into a data storage unit. The contents of the data storage unit can be transferred directly from the data storage unit into the computer in order to the subsequent processing. Suitable interfaces are required for this movement of data.

Combined theodolite and EDM systems can also be used in conjunction with a data storage unit but in many cases the data acquisition is semi-automatic in that the readings from the theodolite and EDM unit, with their appropriate codes, have to be entered into the data storage unit by hand.

Using the traditional approach, all field observations can be entered on to field sheets and, with the aid of accompanying sketches, the data is coded and transferred into the computer using the keyboard.

The acquisition stage is the most important part of the whole process. The quality of the field observations and feature coding will greatly influence the outcome of the survey. If the acquisition stage is carried out properly with all the correct codes being assigned to the points, the subsequent processing and output stages should proceed quickly without any problems.

8.5.2 Data processing

Once entered into the computer, all data are stored in a field observation file. With the aid of specially devised software, the computer operator checks all readings and then the three-dimensional coordinates are automatically computed for each point of detail surveyed in the field. This information, together with the code for each point, is stored in a coordinate file. At this stage, it is necessary to begin editing the coordinate file in order to ensure that information shown on the final plot is in the correct position, is annotated correctly and fits the descriptions given by the field surveyor. This editing process is carried out using software often known as an interactive graphics routine. The interactive graphics enable any small area of the coordinate file to be selected for display on the graphics screen. Information viewed can then be changed, moved or erased on the graphics screen as desired by using a light pen or an electronic cursor. As points are changed, new coordinates are computed and the point code altered accordingly. This information is displayed, via the coordinate file, on the alphanumeric monitor. Various layers of information or combination of layers can be presented to the screen for editing. These layers are made up by the coordinate file using the point codes and the layers can contain such data as characters, line work, symbols, buildings, spot levels, roads, water features, underground services, overhead lines and so on.

The coordinate file is also used to produce computer generated contours. Since contours should not be drawn across certain features such as embankments, cuttings, ditches, buildings and so on, the computer is instructed not to draw contours through these features. This is achieved by labeling, in the coordinate file, the edges of the embankments and so on with suitable codes. The contour information is usually contained in a separate information layer.

8.5.3 Data output

Once the plotting files have been established, data can be plotted in a variety of ways. Data can be plotted onto a high-resolution graphics screen. The plot can be checked for completeness, accuracy, and so forth. If interactive graphics are available, the plotted features can be deleted, enhanced, corrected, crosshatched, labeled, dimensioned, and so on. At this stage, a hard copy of the screen display can be printed either on a simple printer or on an ink-jet color printer. Plot files can be plotted directly on a digital plotter similar to that shown in Figure 8-10. The resultant plan can be plotted to any desired scale, limited only by the paper size. Some plotters have only one or two pens, although plotters are available with four to eight pens: a variety of pens permits colored plotting, or it permits using various line weights. Plans, and plan and profiles, drawn on digital plotters are becoming more common on construction sites.

Fig. 8-10 Hewlett 8-pen digital plotter

单词与词组

sophisticated [səˈfistikeitid] *a.* 老练的，复杂的，精致的
peripheral [pəˈrifərəl] *a.* 周界的，外围的，边缘的；*n.* 外围设备，周边设备
time-consuming [ˈtaimkənˌsu：miŋ] *a.* 费时的
potential [pəˈtenʃəl] *a.* 潜在的，有可能的
annotation [ˌænou ˈteiʃən] *n.* 注解，注释
annotate [ˈænəuteit] *v.* 注解，注释
accompanying [əˈkʌmpəniŋ] *a.* 陪伴的，附随的
outcome [ˈautkʌm] *n.* 结果，成果，后果
devise [di ˈvaiz] *vt.* 设计，想出，创造，发明
description [dis ˈkripʃən] *n.* 叙述，描写，说明
routine [ru：ˈti：n] *n.* 日常工作，常规，手续，程序
cursor [ˈkə：sə] *n.* 光标，游标，指针
alphanumeric [ˌælfənju：ˈmerik] *a.* 字母数字的
layer [ˈleiə] *n.* 层，阶层，地层
· line work 线划清绘，线划图，轮廓线图形
· spot level 独立高程点
· overhead line 架空管道，架空线
edge [edʒ] *n.* 边，棱，边缘，界线 *vt.* 给……形成边，使锐利
embankment [em ˈbæŋkmənt] *n.* 路堤，（河流的）岸堤
cutting [ˈkʌtiŋ] n. 切片，切断，切削，采掘，路堑，挖方
label [ˈleibl] *n.* 标签，签条，标记，符号；*vt.* 贴标签于，加标签，把……列为
· and so forth / and so on 等等
crosshatch [ˈkrɔshætʃ] *v.* 画出交叉阴影线
ink-jet [ˈiŋkˈdʒet] *a.* 喷墨的

8.6 Digital terrain models（DTMs）

The name often given to a mathematical representation of part of the Earth's natural surface is a digital terrain model or DTM, since the data is stored in digital form, that is, X, Y and RL. However, there are a number of other names which have been applied to such a representation, for example digital elevation model (DEM) and digital ground model (DGM). But the one most applicable to engineering surveying is digital terrain model because this is considered by most people to include both planimetric data (X, Y) and relief data (RL, geographical elements and natural features such as rivers, ridge lines and so on).

To form a DTM, a detail survey is carried out in the area for which the DTM is required. Since the shape of natural surfaces varies in a random way, the network of points surveyed to represent the shape of the ground will usually form a random pattern consisting of horizontal coordinates with associated heights.

In many cases, photogrammetric methods involving aerial surveys are used to provide surface information for DTMs. These methods are well suited to obtaining three-dimensional information over large areas where ground techniques would become laborious.

A DTM is usually formed from the field data using one of the following techniques.

1. Grid-based terrain modelling

In this technique, a regular square grid is established over the site and the RL values at each of the grid nodes are interpolated from the field data points.

A grid size is chosen such that it is small enough to give an accurate representation of the irregular surface on which it is based. One disadvantage of this method is that the grid nodes do not coincide with the actual field data points. This tends to smooth out any surface irregularities and causes any contours generated from the grid to be less representative of the true surface. *[A further drawback is that any ridges and other changes of slope which have been carefully surveyed in the form of a string feature on site cannot be accurately reproduced on the grid.] This will also affect the shape of any contours generated.

2. Triangulation-based terrain modelling

This method uses the actual field survey points as node points in the DTM. The software joins together all the data points as a series of non-overlapping contiguous triangles with a data point at each node to produce a Triangulation Irregular Network (TIN).

Such a technique has none of the disadvantages of the grid-based method outlined above. A much truer representation of the surface is obtained and features which have been carefully surveyed as strings, such as the tops and bottoms of embankments, are faithfully reproduced and can be taken into consideration if contours are generated. It is also possible to set up areas of the surface from which contours can be excluded; this is not possible with the grid-based system.

In addition to the generation of contours, DTMs have numerous applications in engineering surveying. They can be viewed from different ang wireframe views which can highlight areas of specific interest. Examples of grid-based and TIN-based wire frame views are shown for the same area inFig. 8-11a and Fig. 8-11b. Once a DTM has been created, volumes of features such as lakes, spoil heaps, stockpiles and quarries can easily be obtained and longitudinal and cross-sections quickly produced.

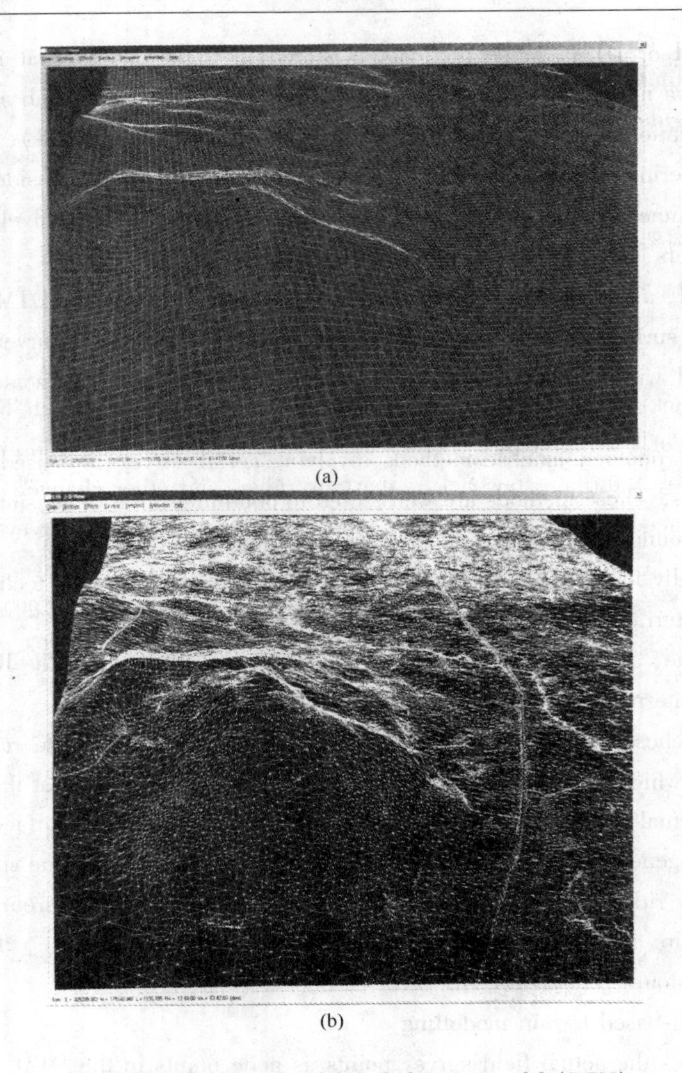

(a)

(b)

Fig. 8-11 The example of digital elevation model (DEM)

(a) Wireframe of the DTM showing a 50m grid. (b) Wireframe of the DTM showing the TIN.

单词与词组

· digital terrain model (DTM) 数字地形模型

· digital elevation model (DEM) 数字高程模型

· digital ground model (DGM) 数字地面模型

node [nəud] *n.* 节，结；瘤

disadvantage [ˈdisəd ˈvɑːntidʒ] *n.* 不便，不利，缺点，缺陷

tend [tend] *v.* 照料（看），看护，管理，势必会，倾向

smooth [smuːð] *a.* 滑溜的，圆滑的，平坦的，平静的，安稳的

representative [ˌrepri ˈzentətiv] *a.* 表现的，代表性的；*n.* 代表

drawback [ˈdrɔːbæk] *n.* 缺点，缺陷，瑕疵，障碍（物）

ridge [ridʒ] *n.* 脊，脊背，山脊，岭，岗，分水岭，山脉

overlap [əuvə 'læp] *v*. 重叠，搭接，超越，交错
contiguous [kən 'tigju:əs] *a*. 连接的，邻近的，接近的
· Triangulation Irregular Network (TIN) 不规则三角网
string [striŋ] *n*. 线，带，绳子，一连串，一系列，一串，(pl.)(附带)条件，限制
exclude [iks 'klu:d] *vt*. 排除，把……关在外面，不包括在内
frame [freim] *n*. 构架，骨架结构，框架，框子，机架
heap [hi:p] *n*. 堆，炼焦堆，(口)许多，大量
hedge [hedʒ] *n*. 灌木树篱，栅栏，隔板，障碍物，界限

注释

* A further drawback is that any ridges and other changes of slope which have been carefully surveyed in the form of a string feature on site cannot be accurately reproduced on the grid。此句为主从复合句，主句是 A further drawback is that any ridges and other changes of slope……cannot be accurately reproduced on the grid。从句为 which have been carefully surveyed in the form of a string feature on site。它是定语从句修饰主句中的 ridges（山脊）与 changes（变化），定语从句含义是：工地上一系列地貌已按一定形式仔细测量。全句译为：更多的缺点在于在工地上一系列地貌已按一定形式仔细测量的任何山脊以及斜坡上另外的变化都不能在方格网上精确地再现。

Chapter 9 Highway Curve Surveys

9.1 Circular curves

In the design of highways and railways, straight sections of roads are connected by curves of constant or varying radius as shown Fig. 9-1. The purpose of the curves is to deflect the road through the angle between the two straights, Δ. For this reason, Δ is known as the deflection angle.

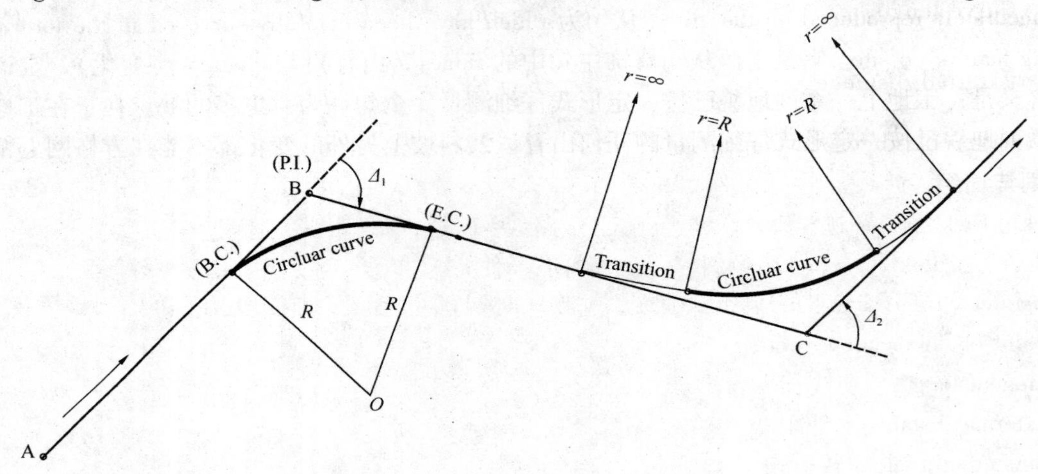

Fig. 9-1 Horizontal curves of highway

The curves shown Fig. 9-1 are horizontal curves of highway since all measurements in their design and construction are considered in the horizontal plane. The two main types of horizontal curve are:

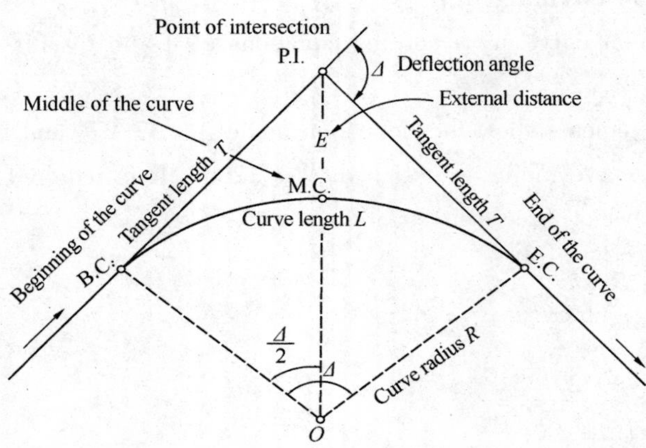

Fig. 9-2 Circular curve elements

(1) Circular curves, which are curves of constant radius as shown in the left part of Fig. 9-1

(2) Transition curves, which are curves of varying radius as shown in the right part of Fig. 9-1.

A simple circular curve consists of one arc of constant radius, as shown in Fig. 9-2.

Circular-curve elements are shown in Fig. 9-2. The *point of intersection of the tangents* (P. I.) is also called the vertex (V). The *beginning of the curve* (B. C.) and the *end of the curve* (E. C.) are also termed the *point of tangent to curve* (T. C.) and the *point of curve to tangent* (C. T.). The *middle of the curve* (M. C.) is the mid-point of the circular curve. The B. C., M. C. and E. C. are called *three main points of circular curve*.

The distance from the B. C. to the P. I., and from the P. I. to the E. C., is *the tangent length* (T). The *length of curve* (L) is the distance from the B. C. to the E. C. measured along the curve for the arc definition. The *external distance* (E) is the distance from the vertex to the curve on a radial line. The change in direction of the two tangents is the *deflection angle* Δ, which is equal to the central angle.

Circular-curve elements and their relationship are as shown bellows:

Tangent length $$T = R \cdot \tan\frac{\Delta}{2} \qquad (9\text{-}1)$$

Curve length $$L = R \cdot \frac{\Delta}{\rho} \qquad (9\text{-}2)$$

External distance $$E = R\left(\sec\frac{\Delta}{2} - 1\right) \qquad (9\text{-}3)$$

Difference between tangent and curve $$D = 2T - L \qquad (9\text{-}4)$$

单词与词组

track [træk] *n.* 踪迹，痕迹，足迹，路径，路线，轨道；*vt.* 跟踪，追踪

deflect [di'flekt] *v.* （使）偏斜，（使）转向，（使）弯曲

transition [træn 'ziʒn] *n.* 转变，演变，过渡（时期），过渡段，渐变段，缓和

- point of intersection 交点
- tangent length 切线长
- external distance 外距
- deflection angle 转折角

9.2 Setting out the main points of circular curve

9.2.1 Computations of the main points of circular curve

Computations of the main points of circular curve are according equations (9-1) to (9-4), to give examples as follows:

Example 9.1 It is known that field measurements show the deflection angle $\Delta = 39°27'$ and the mileage of P. I. = K5 + 178.64. The circular curve radius is designed, $R = 120$m. It is required to compute circular-curve elements and the number of three main points.

Solution:

$$T = 120 \cdot \tan\frac{39°27'}{2} = 43.03\text{m}$$

$$L = 120 \cdot \frac{39°27'}{3437.75'} = 82.62\text{m}$$

$$E = 120\left(\sec\frac{39°27'}{2} - 1\right) = 7.48\text{m}$$

$$D = 2 \times 43.025 - 82.624 = 3.44\text{m}$$

Mileage of intersection point is essentially the distance measured from the starting point of road, therefore

$$\left.\begin{aligned}&\text{Mileage of B. C.} = \text{mileage of P. I.} - T\\&\text{Mileage of M. C.} = \text{mileage of B. C.} + L/2\\&\text{Mileage of E. C.} = \text{mileage of M. C.} + L/2\end{aligned}\right\}\qquad(9\text{-}5)$$

To avoid mistake of the calculation, check the calculation using following formula:

$$\text{Mileage of E. C.} = \text{mileage of P. I.} + T - D \qquad (9\text{-}6)$$

This example, mileage of P. I. is K5 + 178.64, the calculation procedures are as bellow

Mileage of P. I.	=	K5 + 178.64
$-T$		43.03
Mileage of B. C.	=	K5 + 135.61
$+L/2$		41.31
Mileage of M. C.	=	K5 + 176.92
$+L/2$		41.31
Mileage of E. C.	=	K5 + 218.23

Check the calculation using equation (9-6):

$$\text{Mileage of E. C.} = \text{K5} + 178.64 + 43.03 - 3.44 = \text{K5} + 218.23$$

The twice calculations of mileage of E. C. are exactly the same, it can be sure that all the calculations above are correct.

9.2.2 Setting out procedure for the main points of circular curve

1. Setting out B. C. and E. C. of circular curve

(1) The theodolite is set up on the intersection point, and centred and levelled using the method previously described.

(2) Sight backward to the original point (A) of road as shown in Fig. 9-1, or to the back intersection point, and measure tangent length T from the intersection point along the sighting direction, and then drive a wooden stake with the B. C. point's number written ahead of time into the ground, this is point B. C..

(3) Similarly, sight forward to the next intersection point C (see Fig. 9-1) and measure tangent length T from the intersection point P. I. along the sighting direction, and then drive a wooden stake with the E. C. point's number written ahead of time into the ground, this is point E. C..

2. Setting out M. C. of circular curve

(1) The theodolite which has been set up at station B now don't is moved, but you must pay attention to check centring and levelling (see Fig. 9-1).

(2) Sight forward to the next intersection point C, and set the horizontal circle reading to 0°00′00″.

(3) Rotate the alidade of theodolite and set the horizontal circle to read calculated angle $\frac{180° - \Delta}{2}$ (refer to Fig. 9-1 and Fig. 9-2), at this time, the line of sight of telescope is in the angle bisector of the angle CBA. Measure external distance E from the intersection point P. I. along the angle bisector and then drive a wooden stake with the M. C. point's number written ahead of time into the ground, this is point M. C..

单词与词组

mileage ['mailidʒ] *n.* 里数，里程

stake [steik] *n.* 桩，竖管，支柱

bisector ['baiˌsektə] *n.* 二等分线，等分角线

9.3 Setting out the detail points of circular curve

The purpose of setting out the detail points of circular curves are for establishing the centre-line of the curve on the ground by means of pegs at some distance intervals. In common condition, when length of circular curve is less than 40 m, setting out the three main point of circular curve can settle for needs of construction in sites. However, for wavy terrain, or larger curve length, or circular curves of shorter radius, in the situation, the detail points of circular curves must be set out by driving pegs at 10m or 20m intervals.

In centerline survey, the number of centerline stake must be adopted integer stake, hence, the number of the first stake near the curve point B. C. becomes to an integral stake number. See the following practical example:

It is known that a highway curve with mileage of B. C is K8 + 720. 56 and that mileage of E. C is K8 + 767. 18. If the curve needs to be staked out at 10m arc length intervals, then the first mileage of P_1 stake near the curve point B. C. should be K8 + 730, the second stake P_2 should be K8 + 740, the third stake P_3 should be K8 + 750, the fourth stake P_4 should be K8 + 760 till to the terminal point E. C..

Setting out the detail points of circular curves there are mainly following three methods:

1. Tangent offset method (the rectangular coordinate method)

Tangent offset method is base on the rectangular coordinate system whose origin point is curve point B. C. or curve point E. C., the tangent of passed curve point is x-axis, the radius direction of origin point is y-axis. The coordinates x, y of detail points on curve should be calculated firstly before field location. The measuring principle is as shown in Fig. 9-3. Mainly computing formulas are as follows:

$$\left.\begin{aligned} \varphi_i &= \frac{l_i}{R}\left(\frac{180^\circ}{\pi}\right) \\ x_i &= R\sin\varphi_i \\ y_i &= R(1-\cos\varphi_i) \end{aligned}\right\} \tag{9-7}$$

Where l_i = arc length from detail point i to curve point B. C. (or E. C.), φ_i = the central angle subtended arc length l_i, x_i = ordinate along tangent (see Fig. 9-3b), y_i = offset length or departure perpendicular to x-axis (see Fig. 9-3b).

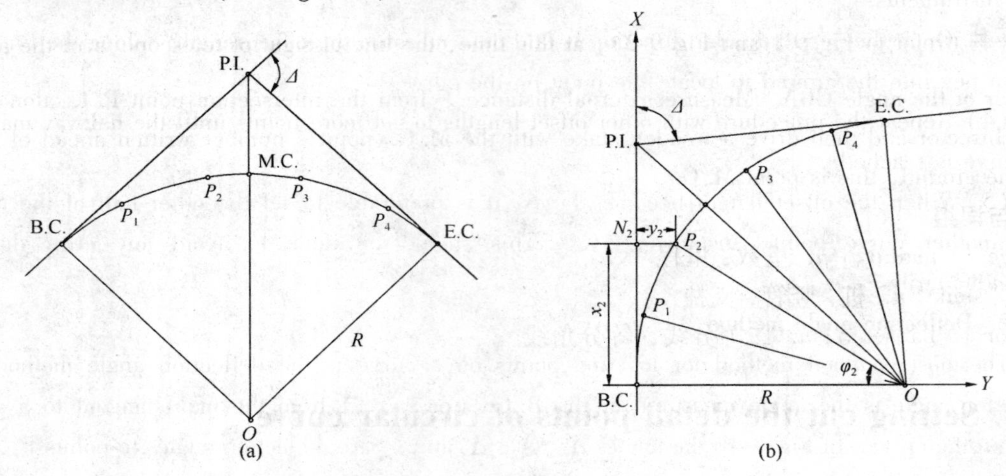

Fig. 9-3 Tangent offset method

Example 9.2 It is assumed that mileage of P. I. K8 + 745. 72 and that the deflection angle $\Delta = 53''\ 25'20''$. The curve radius is designed $R = 50$m. It is required to locate center line stakes every 10m interval. Try to calculate circular curve elements and all setting-out data.

Solution

From formulas (9-1) to (9-4) give:

$T = 25.16$m, $L = 46.62$m, $E = 5.97$m, $D = 3.70$m, and

mileage of B. C. = K8 + 720. 65, mileage of M. C. = K8 + 743. 87, mileage of E. C. = K8 + 767. 18.

Substituting the different arc lengths l_i into formula (9-7), thier results list in Table 9-1.

Table 9-1 Locating calculates by the tangent offset method

Main points	Detail points		Total arc length l_i	x (m)	y (m)	Chord (m)
B. C. K8 + 720. 56				0. 00	0. 00	9. 43
	P_1	+ 730	9. 44	9. 38	0. 89	9. 98
	P_2	+ 740	19. 44	18. 95	3. 73	3. 87
M. C. K8 + 743. 87			23. 31	22. 47	5. 34	6. 13
	P_3	+ 750	17. 18	16. 84	2. 92	9. 98
	P_4	+ 760	7. 18	7. 16	0. 51	7. 17
E. C. K8 + 767. 18				0. 00	0. 00	

The steps for setting out the curve are as follows:

(1) Measure ordinate x from the curve point B. C. along the tangent and insert a chaining pin

in the ground.

(2) Lay the offset perpendicular to the tangent. This can be done using an optical square or other instruments.

(3) Once the perpendicular direction is laid out, measure the offset length y along the line and drive a peg into the ground to locate the point on the curve.

(4) Repeat the procedure with other offset lengths to get more points until the halfway mark of the curve is reached.

(5) When the offset length becomes large, it is preferable to set the other half of the curve from another curve point, instance E. C.. This should be done to avoid any error due to perpendicularity.

2. Deflection angle method

The most common method for locating points on a curve is the deflection angle method. A deflection angle is the angle measured at the B. C. (or E. C.) from the main tangent to a given point on the curve. In Fig. 9-4b the angles Δ_1, Δ_2, Δ_3 and Δ_4 are deflection angles to points P_1, P_2, P_3 and P_4, respectively.

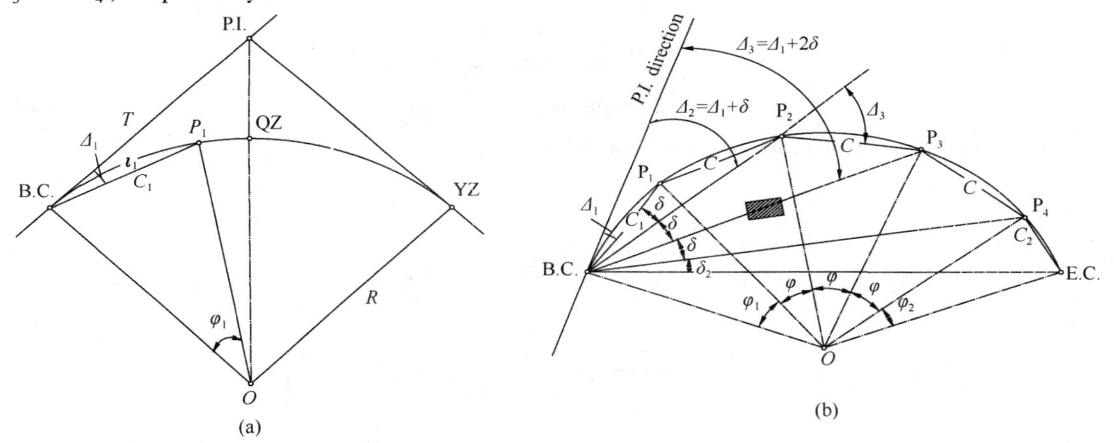

Fig. 9-4 Deflection angle method

Deflection angle method is generally set out integer stake laid on the curve. The mileage stakes spacing on the curve are more dense than straight line segments, according to the provisions of 5m, 10m, 20m, etc., in order to the mileage stakes on the curve are integer, then the curve head and tail two sections of arc lengths usually are not integer, the first section of sub-arc denotes l_1 and the tail section sub-arc denotes l_2, corresponding to the two chords denote C_1 and C_2, respectively. The entire arc adjacent two points in the middle part of curve denote l_0, their corresponding chords denote C.

In Fig. 9-4, the first section of sub-arc l_1 is from B. C. point to the P_1 point, setting out data of P_1 point from Fig. 9-4a can be obtained. The central angle φ_1 corresponding to arc length l_1 is calculated by the following formula:

$$\varphi_1 = \frac{l_1}{R}\left[\frac{180^\circ}{\pi}\right] \tag{9-8}$$

Deflection angle Δ_1 and its chord length C_1 corresponding to first section of sub-arc l_1 are calculated respectively by

$$\Delta_1 = \frac{\varphi_1}{2} = \frac{l_1}{R}\left[\frac{90^\circ}{\pi}\right] \tag{9-9}$$

$$C_1 = 2R\sin\Delta_1 \tag{9-10}$$

The central angle corresponding to tail section sub-arc l_2 is φ_2, the circumferential angle and chord length corresponding to tail section sub-arc are δ_2 and C_2 respectively. They are calculated by

$$\Delta_2 = \frac{\varphi_2}{2} = \frac{l_2}{R}\left[\frac{90^\circ}{\pi}\right] \tag{9-11}$$

$$C_2 = 2R\sin\delta_2 \tag{9-12}$$

In the middle part of circular curve, the lengths of entire arc adjacent two points are l_0, the central angles corresponding to entire arc are φ. The circumferential angles δ and chord lengths C corresponding to entire arc are calculated respectively by

$$\delta = \frac{\varphi}{2} = \frac{l_0}{R}\left[\frac{90^\circ}{\pi}\right] \tag{9-13}$$

$$C = 2R\sin\delta \tag{9-14}$$

Deflection angle Δ of every detail points list as follows:

point P_1 $\quad \Delta_1$

point P_2 $\quad \Delta_2 = \dfrac{\varphi_1 + \varphi}{2} = \Delta_1 + \delta$

point P_3 $\quad \Delta_3 = \dfrac{\varphi_1 + 2\varphi}{2} = \Delta_1 + 2\delta$

$\vdots \qquad \vdots$

point E. C. $\quad \Delta_{E.C.} = \dfrac{\varphi_1 + n\varphi + \varphi_2}{2} = \Delta_1 + n\delta + \delta_2$

$$= \frac{\Delta}{2}\text{(checked)}$$

Setting out circle curve using deflection angle method works successively, and the deflection angle of each point is calculated by the cumulative method called the "cumulative deflection angle". In order to make a checking of the calculation, the last cumulative deflection angle should equal to $\frac{\Delta}{2}$.

Example 9.3 It is known that the mileage of P. I. K5 + 135.22, and that the deflection angle $\Delta = 40°21'10''$. The circular curve radius is designed, $R = 100$m. It is now required to locate every detail point stakes at 20m interval, try to calculate locating data using deflection angle method.

From formulas (9-1) ~ (9-4) can be obtained: $T = 36.75\text{m}$, $L = 70.43\text{m}$, $E = 6.54\text{m}$, $D = 3.07\text{m}$. Setting out the detail points of circular curves by the deflection angle method, all data are calculated according equations (9-8) ~ (9-14), their results list in Table 9-2

Table 9-2 Locating calculates by the deflection angle method

Main points	Detail points	Total arc length (m)	Deflection angle Δ (° ′ ″)	Cumulative deflection angle Δ (° ′ ″)	Total chord length C (m)
B. C. K5 +098. 47				0 00 00	
		1. 53	0 26 18		1. 53
	P_1 +100			0 26 18	
		20. 00	5 43 46		19. 97
	P_2 +120			6 10 04	
		20. 00	5 43 46		19. 97
	P_3 +140			11 53 50	
		20. 00	5 43 46		19. 97
	P_4 +160			17 37 36	
		8. 90	2 32 59		8. 90
E. C. K5 +168. 90				20 10 35	

Setting out procedure is as follows:

(1) Set up a theodolite at B. C. (or E. C.), and then loose the alidade to sight the point of intersection P. I. and set the horizontal circle reading to 0°00′00″.

(2) Turn the alidade to set the horizontal circle to read desired deflection angle Δ_1 (0°26′18″), and then measure chord length C_1 (1.53m) from the instrument station along the sighting direction and drive a peg into the ground to locate first point on the curve, i. e. P_1 (K5 + 100).

(3) The theodolite don't move, turn the alidade to set the horizontal circle to read desired cumulative deflection angle Δ_2 (6°10′04″). A one person holds the fore end of the tape to make the tape reading C (i. e. a chord length 19.97m) at point P_1, and this time the other person holds the rear end of the tape (zero graduation) at round P_2 taut pulling and swing the rear end of the tape until it is intersected by the line of sight of the theodolite, this time insert immediately a chaining pin at zero graduation. This is the point P_2 (K5 + 120) on the curve.

This procedure is continued until the point E. C. is reached.

(4) Theoretically, the last point to be so located should coincide with the already staked E. C. , however, some misclosure may be encountered as a result of accumulated errors or other imperfections in measurement. If the misclosure is unacceptable, all measurements should be checked and adjusted until a satisfactory stake position is achieved.

Deflection angle method is a high accuracy, practicality, flexibility, common method of setting out the detail points of curves. The method can be set up station at any point on the curve or the intersection point P. I. . However, because the distance is the continuous measurement point by point, the position errors of having located points are bound to affect measuring accuracy of succeeding points, the position error is gradually accumulated. Therefore, for long curves it is considered better to run in the first half of the curve from the B. C, and the second half back from the E. C. the small errors occur can be adjusted at the middle of the curve.

3. The polar coordinate method (settring-out using total station)

When setting-out by total station, the deflection angle of each point (Δ_1, Δ_2, Δ_3, etc) and the total distance from B. C. to the each point (C_1, C_2, C_3, etc) should be calculated as shown in Fig. 9-5. Therefore, all the data required as shown Table 9-3 must firstly be calculated (see Table 9-3).

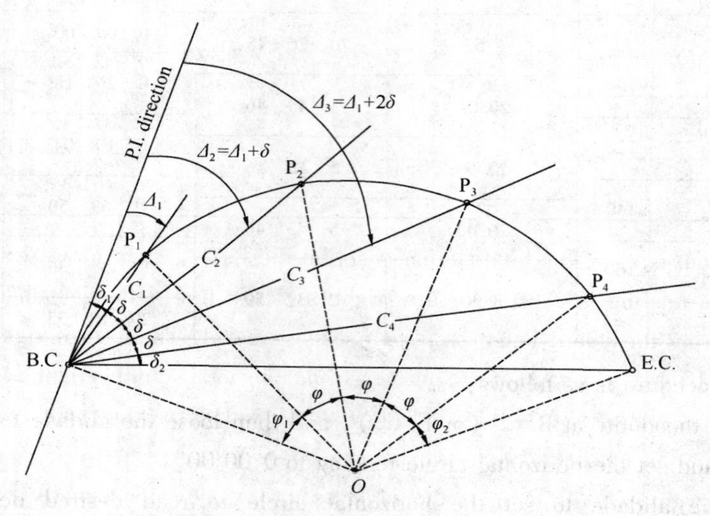

Fig. 9-5 The polar coordinate method

On the formula of deflection angle Δ_1 and chord length C_1 have been deduced in the previous section, that is equations (9-9) and (9-10):

$$\Delta_1 = \frac{\varphi_1}{2} = \frac{l_1}{R}\left[\frac{90°}{\pi}\right]$$

$$C_1 = 2R\sin\Delta_1$$

$$\delta = \frac{\varphi}{2} = \frac{l_0}{R}\left[\frac{90°}{\pi}\right]$$

From Fig. 9-5 can obtain:

$$\left.\begin{aligned} \Delta_2 &= \Delta_1 + \delta \quad & C_2 &= 2R\sin\Delta_2 \\ \Delta_3 &= \Delta_1 + 2\delta \quad & C_3 &= 2R\sin\Delta_3 \\ \Delta_4 &= \Delta_1 + 3\delta \quad & C_4 &= 2R\sin\Delta_4 \\ &\vdots & &\vdots \end{aligned}\right\}$$

Example 9.4 All the raw data are exactly the same with Example 9.3, computations of setting-out circular cirve by the polar coordinate method list in Table 9-3:

Table 9-3 Locating calculates by the polar coordinate method

Main points	Detail points	Arc length (m)	Deflection angleΔ (° ′ ″)	cumulative deflection angle Δ (° ′ ″)	Total chord length C (m)
B. C. K5 +098. 47				0 00 00	
		1. 53	0 26 18		
	P1 +100			0 26 18	1. 53
		20. 00	5 43 46		
	P2 +120			6 10 04	21. 48
		20. 00	5 43 46		
	P3 +140			11 53 50	41. 23
		20. 00	5 43 46		
	P4 +160			17 37 36	60. 56
		8. 90	2 32 59		
E. C. K5 +168. 90				20 10 35	68. 98

When setting out curves, the total station is set on the B. C., sight forward to the P. I. and set the horizontal circle reading to 0°00′00″, turn rightwards the first deflection angle 0°26′18″ to set point P_1 by measuring the chord length 1. 53m (see Fig. 9-5). Then, turn rightwards the second deflection angle 6°10′04″ to set point P_2 by measuring the total chord length 21. 48m. Similarly, each point is set successively by setting deflection angle and measuring total chord length. When complete to set out point P_4 in Fig. 9-5, we should measure distance between P_4 and point E. C. to see if the distance value equals to their theoretical chord length. The curve should be inspected visually for consistency and smoothness. Obviously, this procedure requires clear lines of sight from the B. C. to all stations.

单词与词组

peg [peg] *n*. 木（金属，竹）钉，栓，销，桩，棒

settle ['setl] *v*. 安排，安放；安家，定居；解决，决定，调停

wavy ['weivi:] *a*. 波状的，起伏的

· centerline survey（道路）中线测量

stake [steik] *n*. 桩，竖管，支柱

offset ['ɔ:fset] *n*. 分支，支脉，偏置，偏距，横距，支距，残余偏差

· tangent offset method 切线支距法

· central angle 圆心角

· circumferential angles 圆周角

ordinate ['ɔ:dinit] *n*. 纵坐标

departure [di 'pɑ:tʃə] *n*. 东西距离，横距

pin [pin] *n*. 针，别针，大头针，插头

· chaining pin 测钎

perpendicularity ['pə:ˌpəndikju 'læriti] *n*. 垂直度，直立，正交

· deflection angle method 偏角法

deflection [di 'flekʃən] *n*. 偏转，偏斜，偏差

accumulate [ə 'kju:mjuleit] *vt.* 积累，存储，蓄积
flexibility [ˌfleksi 'biliti] *n.* 柔（韧）性，可曲性，机动性，适应性，灵活性
inspect [in 'spekt] *vt.* 检阅，检查，检验
consistency [kən 'sistənsi:] *n.* 一致性，连贯性
smoothness ['smu:ðnis] *n.* 平滑，光滑，光顺性

9.4 Transition curves

9.4.1 General

The transition curve is a curve of constantly changing radius. If used to connect a straight to a curve of radius R, then the commencing radius of the transition will be the same as the straight (∞), and the final radius will be that of the curve R (see Fig. 9-6).

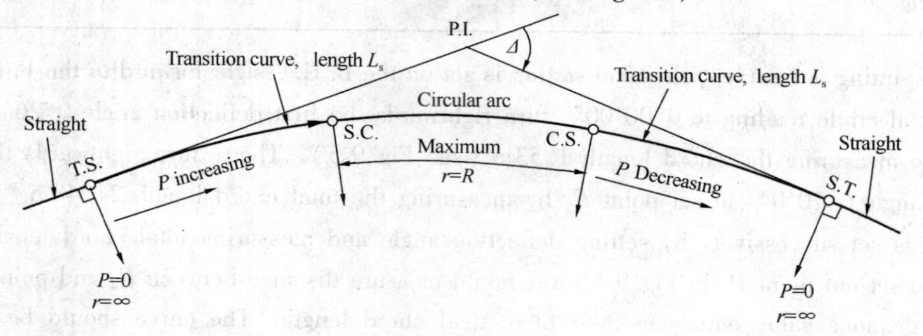

Fig. 9-6 Transition curve

Consider a vehicle travelling at speed (v) along a straight. The forces acting on the vehicle will be its weight w, acting vertically down, and equal an opposite force acting vertically up through the wheels. When the vehicle enters the curve of radius R at tangent point T. S., an additional centrifugal force (P) acts on the vehicle. It is known that $P = \dfrac{wv^2}{gr}$, where w is the weight of the vehicle, g is the gravity acceleration $g = 9.8\text{m/sec}^2$, and r is the radius of curvature at a point. If P is large the vehicle will be forced to the outside of the curve and may skid or overturn.

On a straight road, $r =$ infinity, therefore $P = 0$

On a circular curve of radius R, $r = R$, therefore $P = \dfrac{wv^2}{gR}$

9.4.2 The intrinsic equation of spiral

Usually the road curve consists of two transitions and a circular curve (see Fig. 9-6). For a vehicle travelling from T. S. to S. T., the force P gradually increases from zero to its maximum on the circular curve and then decreases to zero again. This greatly reduces the tendency to skid and reduces the discomfort experienced by passengers in the vehicles.

For a constant speed v, the force P acting on the vehicle is (mv^2/gr). Since any given curve is designed for a particular speed and the weight of a vehicle can be assumed constant, it follows that $P \propto \dfrac{1}{r}$.

However, if the force is allowed to increase uniformly along the curve, it also follows that P must be proportional to l, i. e. $p \propto l$, where l is the length along the curve from the entry tangent point to the point in question.

Combination of these two requirements gives $l \propto \frac{1}{r}$ or $rl = K$, where K is a constant. If L_S is the total length of each transition and R the radius of the circular curve, then $RL_S = K$.

The fundamental requirement of a transition curve is that its radius of curvature r must vary as the length l from the beginning of the transition curve. The product of length measured along the curve from the tangent point and the radius of curvature at that point has a constant value.

A type of curve satisfying the fundamental requirement of a transition curve is the clothoid or true spiral. This curve is somewhat modified in practice by reducing it to a cubic spiral or cubic parabola.

A spiral is shown in Fig. 9-7. T. S. = tangent to spiral. S. C. = spiral to curve. C is any point on the spiral at l from T. S.. The tangent to the curve at S. C. intersects the tangent AX at point M. r is the radius of the spiral at C, *1 [β the angle made by the tangent at C with the initial tangent AX, β_S the deflection angle or spiral angle, i. e. , angle between the tangents to the spiral at its ends,] R the radius of the circular curve, and L_S is the length of the transition curve. From the fundamental requirement of a transition curve, we have

$$l\,r = L_S R \tag{9-15}$$

The radius (r) of the spiral at any point can be obtained from formula above, that is

$$r = \frac{RL_S}{l} \tag{9-16}$$

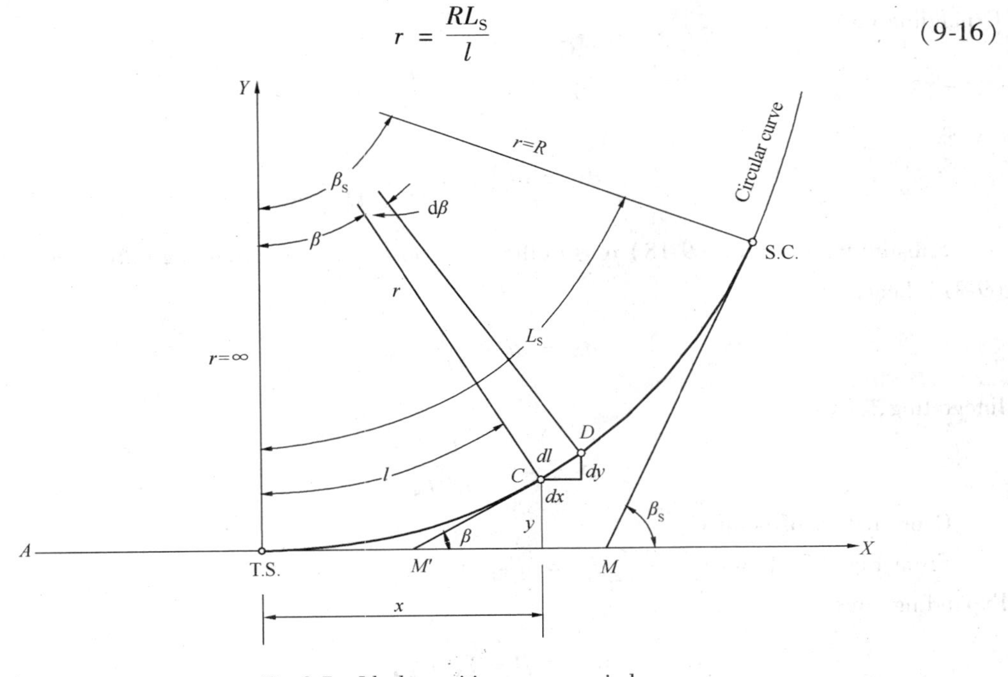

Fig. 9-7 Ideal transition curve—spiral

From Fig. 9-7 shows, consider to take a micro-segment dl on the spiral at any point, Central angle corresponding to micro-segment dl is $d\beta$, we have

$$r d\beta = dl \tag{9-17}$$

$$d\beta = \frac{dl}{r} = \frac{l}{R}\frac{dl}{L_S}$$

Integrating,

$$\beta = \frac{l^2}{2RL_S} \tag{9-18}$$

This is the intrinsic equation of an ideal transition curve.

When $l = L_S$, equation (9-18) becomes

$$\beta_S = \frac{L_S}{2R} \tag{9-19}$$

9.4.3 Rectangular coordinates for points on the curve

It will be easier to set out the curve from tangent point T. S.. To do this, rectangular coordinate system of a point on the curve are required. Referring to Fig. 9-7, *T. S.* is the origin of the coordinate system (x, y) are the coordinates of point C, D is any point having coordinates ($x + dx$) and ($y + dy$), dl is the distance between C and D along the curve, β is the angle made by the tangent at C with AX, and $\beta + d\beta$ is the angle made by the tangent at D with AX.

Coordinate x of point C:

From Fig. 9-7 shows $\quad dx = dl\cos\beta \quad$ (9-20)

Expanding $\cos\beta$

$$\cos\beta = 1 - \frac{\beta^2}{2!} + \frac{\beta^4}{4!} - \cdots$$

$$\therefore \quad dx = dl\left(1 - \frac{\beta^2}{2!} + \frac{\beta^4}{4!} - \cdots\right) \tag{9-21}$$

Substituting equation (9-18) of β in the equation above and neglecting third term, the equation (9-21) becomes

$$dx = dl - \frac{l^4}{8R^2L_S^2}dl$$

Integrating,

$$x = l - \frac{l^5}{40R^2L_S^2} \tag{9-22}$$

Coordinate y of point C:

From Fig. 9-7 shows $\quad dy = dl\sin\beta \quad$ (9-23)

Expanding $\sin\beta$,

$$\sin\beta = \beta - \frac{\beta^3}{3!} + \frac{\beta^5}{5!} - \cdots$$

$$\therefore \qquad dy = dl\left(\beta - \frac{\beta^3}{3!} + \frac{\beta^5}{5!} - \cdots\right) \tag{9-24}$$

Substituting equation (9-18) of β in the equation above and neglecting third term, the equation (9-24) becomes

$$dy = dl\left(\frac{l^2}{2RL_S^2} + \frac{l^6}{48R^2L_S^2}\right)$$

Integrating,

$$y = \frac{l^3}{6RL_S} - \frac{l^7}{336R^2L_S^2} \tag{9-25}$$

The coordinates x and y can be obtained from these expressions.

9.4.4 Computations and layouts of the transition curve

1. The computations of the transition curve

Fig. 9-8 illustrates how the circular curve is moved inward (toward the center of the curve), leaving room for the insertion of a spiral at either end of the shortened circular curve. *2[The amount that the circular curve is shifted in from the main tangent line is known as p.] This shift results in the curve center (O) being at the distance $(R+p)$ from the main tangent lines.

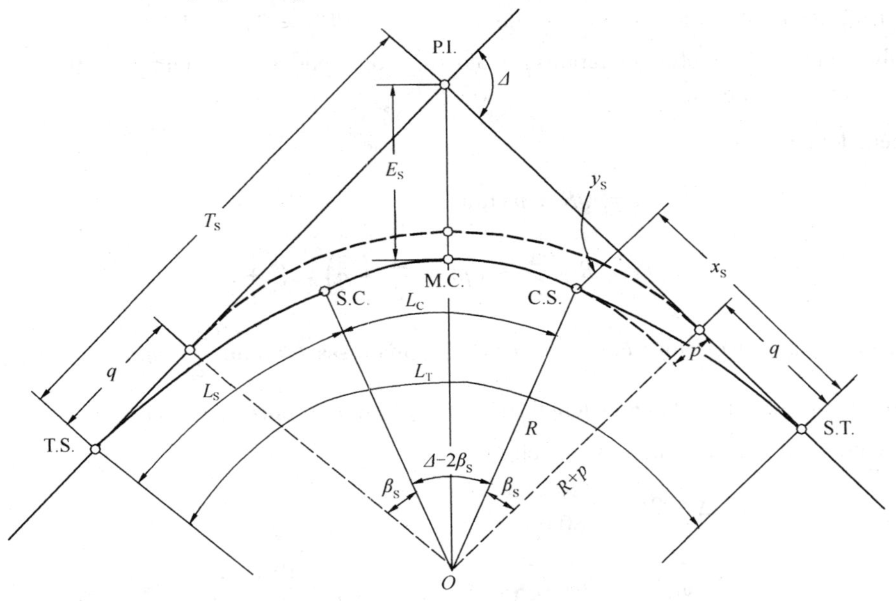

Fig. 9-8 Spiral geometry and spiral symbols

Fig. 9-8 can show, the inward shift value p

$$p = y_S - (R - R\cos\beta_S) \tag{9-26}$$

Expanding $\cos\beta$ and substituting equation (9-26), we have

$$p = \frac{L_S^2}{24R} \tag{9-27}$$

Fig. 9-8 can show, the tangent growth value q

$$q = x_S - R\sin\beta_S \tag{9-28}$$

Expanding $\sin\beta$ and substituting equation (9-28), we have

$$q = \frac{L_S}{2} - \frac{L_S^3}{240R^2} \tag{9-29}$$

From equation (9-29) can show, when a circular curve radius is large enough, the second of formula is minimal, it can be ignored. The growth of tangent value is a half of length of transition curve approximately, i. e. $q = \frac{L_S}{2}$.

Transition curve elements and their relationship are as shown bellow:

(1) Tangent length $T_S = (R + p)\tan\frac{\Delta}{2} + q$ (9-30)

(2) Master curve (circular curve arc) length $L_C = R(\Delta - 2\beta_S)\frac{\pi}{180°}$ (9-31)

(3) Total curve length $L_T = L_C + 2L_S$ (9-32)

(4) External distance

$$E_S = (R + p)\sec\frac{\Delta}{2} - R \tag{9-33}$$

(5) Difference between tangent and curve $D_S = 2T_S - L_T$ (9-34)

For the convenience of calculations, make a little change to equations (9-30), (9-31), (9-32), (9-33) and (9-34):

Tangent length T_S:

$$T_S = (R + p)\tan\frac{\Delta}{2} + q$$

$$T_S = R\text{tg}\frac{\Delta}{2} + \left(p\tan\frac{\Delta}{2} + q\right) = T + t \tag{9-35}$$

Where tangent length of circular curve $T = R\tan\frac{\Delta}{2}$, mantissa $t = p\tan\frac{\Delta}{2} + q$. i. e. , Tangent length of transition curve T_S equals tangent length of circular curve T plus mantissa t.

Master curve (circular curve arc) length L_C,

$$L_C = R(\Delta - 2\beta_S)\frac{\pi}{180°}$$

$$= R\Delta\frac{\pi}{180°} - 2R\beta_S\frac{\pi}{180°} = L - 2R\left(\frac{L_S}{2R}\frac{180°}{\pi}\right)\frac{\pi}{180°} = L - L_S \tag{9-36}$$

i. e. , Master curve (circular curve arc) length L_C equals circular curve length L minus spiral length L_S.

Total curve length L_T

$$L_T = L_C + 2L_S = (L - L_S) + 2L_S = L + L_S \tag{9-37}$$

i. e. , Total curve length L_T equals circular curve length L plus spiral length L_S.

External distance E_S

$$E_S = (R + p)\sec\frac{\Delta}{2} - R$$

$$= \left(R\sec\frac{\Delta}{2} - R\right) + p\sec\frac{\Delta}{2} = E + e \tag{9-38}$$

where external distance of circular curve $E = R\sec\frac{\Delta}{2} - \mathrm{R}$, mantissa $e = p\sec\frac{\Delta}{2}$.

That is, External distance of transition curve E_S equals External distance of circular curve E plus mantissa e.

Difference between twice the tangent length and the curve length denotes D_S, i. e.

$$D_S = 2T_S - L_T \tag{9-39}$$

Circular curve radius R and the length of transition curve L_S are determined based on the level of road and terrain conditions; the deflection angleΔ has been measured, and thus these data required about survey curves can obtained according to the formula listed above.

The mileages of main points of the transition curve are as follows:

$$\left.\begin{aligned}
&\text{Mileage of T. S.} = \text{Mileage of P. I.} - \text{Tangent length } T_S\\
&\text{Mileage of S. C.} = \text{Mileage of T. S.} + \text{Spiral length } L_S\\
&\text{Mileage of M. C.} = \text{Mileage of S. C.} + \text{Master curve length } L_C/2\\
&\text{Mileage of C. S.} = \text{Mileage of M. C.} + \text{Master curve length } L_C/2\\
&\text{Mileage of S. T.} = \text{Mileage of C. S.} + \text{Spiral length } L_S
\end{aligned}\right\} \tag{9-40}$$

For avoid mistake of calculation, check the calculation by following formula:

$$\text{Mileage of S. T.} = \text{Mileage of P. I.} + T_S - D \tag{9-41}$$

Example 9. 5 It is known that a highway design speed is 120km/h, that the mileage of P. I. is K9 +658. 86, that the deflection angle$\Delta = 20°18'26''$ (right deviation), that designed circular circle radius $R = 600$m, and that designed length of transition curve $L_S = 100$m. It is required that calculating curve elements and that caculating the mileage of the main points of transition curve.

All of calculations are listed in Table 9-4. It is preferable that programming calculation with the notebook computer, the efficiency of the calculations will greatly be improved.

Table 9-4 The calculations of setting out the transition curve

Engineering project: ××××× Address: ××××××

Known data	Intersection: P. I. 8 Number: K9 +658. 86 Deflection angle $\Delta = 20°18'26''$ (right deviation) Circular circle radius $R = 600$m Length of transition curve $L_S = 100$m
Calculations of intrinsic parameters of the transition curves	The tangent growth value $q = \frac{l_h}{2} - \frac{l_h^3}{240R^2} = 50.00$m
	The inward shift value $p = \frac{l_h^2}{24R} = 0.69$m
	The spiral angle $\beta_h = \frac{l_h}{2R} \times \frac{180°}{\pi} = 4°46'29''$

Continued

Calculations of transition curve elements	T_S	The tangent length of circle curve	$T = R\tan\frac{\Delta}{2} = 107.46\text{m}$
		The mantissa of tangent length	$t = p\tan\frac{\alpha}{2} + q = 50.12\text{m}$
		The tangent length of transition curve	$T_S = T + t = 107.46 + 50.12 = 157.58\text{m}$
	L_T	The length of circle curve	$L = R\Delta\frac{\pi}{180°} = 212.66\text{m}$
		The length of transition curve	$L_S = 100\text{m}$
		Total curve length	$L_T = L + L_S = 212.66 + 100 = 312.66\text{m}$
	L_C	The master curve length	$L_C = L - L_S = 112.66\text{m}$
	E_S	The external distance of circle curve	$E = R\left(\sec\frac{\Delta}{2} - 1\right) = 9.55\text{m}$
		The mantissa of external distance	$e = p\sec\frac{\Delta}{2} = 0.70\text{m}$
		The external distance of transition curve	$E_S = E + e = 9.55 + 0.70 = 10.25\text{m}$
	D_S	Difference between tangent and curve	$D_S = 2T_S - L_T = 2.26\text{m}$

Calculations of main points of the transition curve	Tangent to Spiral $TS = PI - T_S = K9 + 501.28$	Sketch map 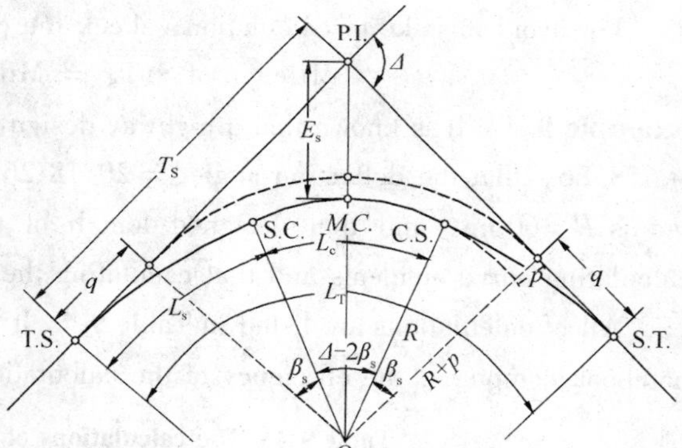
	Spiral to circle $SC = TS + L_S = K9 + 601.28$	
	Middle of circle $MC = SC + \frac{L_C}{2} = K9 + 657.61$	
	Circle to Spiral $CS = MC + \frac{L_C}{2} = K9 + 713.94$	
	Spiral to Tangent $ST = CS + L_S = K9 + 813.94$	
Checked	$ST = PI + T_S - D_S$ $= K9 + 813.94$	

(1) Computations of transition curve parameter:

Compute spiral angle by equation (9-19)

$$\beta_S = \frac{L_S}{2R} \times \frac{180°}{\pi} = \frac{100 \times 180}{2 \times 600 \times \pi} = 4°46'29''$$

Compute the inward shift value p by equation (9-27)

$$p = \frac{L_S^2}{24R} = \frac{100^2}{24 \times 600} = 0.69\text{m}$$

Compute the tangent growth value q by equation (9-29)

$$q = \frac{L_S}{2} - \frac{L_S^3}{240R^2} = \frac{100}{2} - \frac{100^3}{240 \times 600^2} = 50\text{m}$$

Compute the coordinates x and y of any point on the transition curve by equation (9-22) and (9-25) respectively, i. e.

$$x = l - \frac{l^5}{40R^2L_S^2} \quad \text{and} \quad y = \frac{l^3}{6RL_S} - \frac{l^7}{336R^2L_S^2}$$

When $l = L_S$, substituting L_S formulas above listed, the coordinates x_S and y_S of the point S. C. are obtained, i. e.

$$x_S = L_S - \frac{L_S^3}{40R^2} = 100 - \frac{100^3}{40 \times 600^2} = 99.93\text{m}$$

$$y_S = \frac{l^2}{6R} = \frac{100^2}{6 \times 600} = 2.78\text{m}$$

(2) Computations of transition curve main elements

Tangent: $T_S = T + t$

$$T = R\tan\frac{\Delta}{2} = 600 \times \tan\frac{20°18'26''}{2} = 107.46\text{m}$$

$$t = p\tan\frac{\Delta}{2} + q = 0.69 \times \tan\frac{20°18'26''}{2} + 50.00 = 50.12\text{m}$$

$$\therefore \quad T_S = T + t = 107.46 + 50.12 = 157.58\text{m}$$

Total curve length L_T:

$$L = R\Delta\frac{\pi}{180°} = 600 \times 2°18'26'' \times \frac{\pi}{180°} = 212.66\text{m}$$

$$\therefore \quad L_T = L + L_S = 212.66 + 100 = 312.66\text{m}$$

Master curve length L_C: $\quad L_C = L - L_S = 212.66 - 100 = 112.66\text{m}$

External distance: $E_S = E + e$

$$E = R\left(\sec\frac{\Delta}{2} - 1\right) = 600\left(\sec\frac{20°18'26''}{2} - 1\right) = 9.55\text{m}$$

$$e = p\sec\frac{\Delta}{2} = 0.69 \times \sec\frac{20°18'26''}{2} = 0.70\text{m}$$

$$\therefore \quad E_S = E + e = 9.55 + 0.70 = 10.25\text{m}$$

Difference between tangent and curve D_S:

$$D_S = 2T_S - L_T = 2 \times 157.58 - 312.66 = 2.50\text{m}$$

(3) Computations for the mileage of main points of transition curve

Mileage of P. I.	K9 + 658.86
$-T_S$	157.58
Mileage of T. S.	K9 + 501.28
$+L_S$	100.00
Mileage of S. C.	K9 + 601.28

$+L_C/2$	56. 33
Mileage of M. C.	K9 +657. 61
$+L_C/2$	56. 33
Mileage of C. S.	K9 +713. 94
$+L_S$	100. 00
Mileage of S. T.	K9 +813. 94

Checked: Mileage of S. T. = Mileage of P. I. + $T_S - D$ = K9 +658. 86 +157. 58 −2. 5 = K9 +813. 94

The checking has been passed, so the procedure of calculation above is correct.

2. Setting out procedure for the main points of transition curve

(1) The theodolite set up over the intersection point P. I. and then centering and levelling. Sight backward to the central line of road and measure tangent length T_S = 157. 58m from the intersection point P. I. along the central line and drive a wooden peg with the number (K9 + 501. 28) written ahead of time into the ground.

Similarly, sight forward to the central line of road and measure tangent length T_S from the intersection point P. I. along the central line and drive a wooden peg with the number (K9 + 813. 83) written ahead of time into the ground.

(2) The instrument does not move, sight forward to point S. T. and set horizontal circle reading 0°00′00″, and then turn theodolite alidate and set horizontal circle reading $\frac{180° - \Delta}{2}$, at this time the telescope sighting line is in the angle bisector. Measure external distance E_S from the intersection point P. I. along the angle bisector and drive a wooden peg with the number (K9 + 657. 55) written ahead of time into the ground.

(3) It is assumed that T. S. point is the origin of coordinate, the tangent direction established from T. S. to P. I. defines as the x-axis of the coordinate system, perpendicular to the tangent direction defines as the y-axis. Use the tangent offset method to measure length x_S = 99. 93m, y_S =2. 78m get a point of S. C., its mileage number is K9 +601. 28.

(4) Similarly, assumed S. T. points as the coordinate origin, the tangent direction established from S. T. to P. I. defines as the x-axis of the coordinate system, the perpendicular to the tangent direction defines as the y-axis. Use tangent offset method to measure length x_S =99. 93m, y_S = 2. 78m get a point of C. S., its mileage number is K9 +713. 83.

(5) It is noted that when driving pegs, the exact point position should be marked by nails in the tops of the pegs.

单词与词组

centrifugal [sen ′trifjəgəl] *a*. 离心的，利用离心力的

skid [skid] *v*. 滑向一侧，打滑

overturn [ˌəuvə ′təːn] *v*. 打翻，（使）翻过来，（使）倒转

intrinsic [in ′trinsik] *a*. 固有的，内在的，本质的，特征的

spiral [′spaiərəl] *a*. 螺旋形的；*n*. 螺旋线

clothoid [ˈkləuθɔid] *n*. 回旋曲线，回旋螺线
parabola [pəˈræbələ] *n*. 抛物线，抛物面，反射器
· spiral angle 螺旋线角，缓和曲线角
micro-segment [ˈmaikrə-segmənt] *n*. 微分段
inward [ˈinwəd] *a*. 内部的，内在的；精神上的，向内的；输入的，进口的
shorten [ˈʃɔːtn] *vt*. 弄（缩）短，减少
mantissa [mænˈtisə] *n*. 尾数，尾加数

注释

[1] β the angle made by the tangent at C with the initial tangent AX, β_s the deflection angle or spiral angle, i. e., angle between the tangents to the spiral at its ends，第1句难在对 initial 的理解，该词有“最初的，开始的，开头的，固有的，原始的”等含义，从专业知识了解到每个转折点的切线有两条，按路线前进方向分前后两条切线，initial tangent AX 指的是后面的一条切线。第1句译为：β 是在 C 点切线同后一边的切线 AX 的夹角。
第2句在句末省略与前句相同的部分，即 with the initial tangent AX，这是英语的语言现象，省略后不影响对语言的理解，使语言简洁。补充为完整的句子应是：β_s the deflection angle or spiral angle, i. e., angle between the tangents to the spiral at its ends with the initial tangent AX。全句译为：β_s 为偏转角或螺旋线角，即螺旋线终点的切线与后一边的切线 AX 组成的夹角。

[2] The amount that the circular curve is shifted in from the main tangent line is known as p. 这是复合句，句中下划线的部分是定语从句修饰 amount，is shifted in 意为向内移动，而 amount 是主句的主语，谓语是 is known as（被称为），from the main tangent line 是方式状语，所以全句译为：圆曲线从主切线向内移动量称为 p。

9.5 Vertical curves

The curves used in a vertical plane to provide a smooth transition between the grade lines of highways are called vertical curves, whose functions are to ensure that passengers in vehicles travelling along the curves are transported safely and comfortably, and to ensure that *[there is adequate sighting distance to enable oncoming vehicles to be able either to stop safely or to overtake safely].

When the two gradients form a hill, the curve is called a summit or crest curve (as shown in Fig. 9-9a). When the gradients form a valley, a sag or valley curve is produced (as shown in

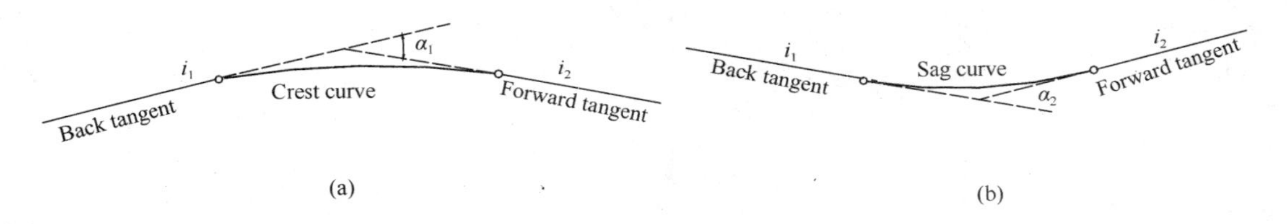

Fig. 9-9 Two types of vertical curve

Fig. 9-9b). Assume that a gradient of the back tangent is termed i_1 and a gradient of the forward tangent is termed i_2, usually expressed in percent. The algebraic difference of the gradients of two adjacent grade lines is denoted by α, that is

$$\alpha = i_1 - i_2 \tag{9-42}$$

In above equation, α is also termed the total change in grade. A positive α indicates a vertical curve to be a crest curves, while a negative α denotes a vertical curve to be a sag curves.

Fig. 9-10 shows the nomenclature used for vertical curves. When moving along the road, the first of the grade lines that will be encountered is called the back tangent. The other one is called the forward tangent. The change of gradient from the back tangent to the forward tangent is required to be smooth and gradual, so it is necessary that a sort of curve is chosen. A circular curve is practicable. The intersection point of two grade lines is abbreviated as V, the beginning of a vertical curve as BVC, the middle of the vertical curve as MVC, the end of vertical curve as EVC.

As with horizontal curves, the vertical curve elements and their relationship are express following formulas:

From Fig. 9-10 can obtain the tangent length:

$$T = R\text{tna}\frac{\alpha}{2}$$

Since α is very small, so $\tan\frac{\alpha}{2} = \frac{\alpha}{2}$.

$$\therefore \qquad T = \frac{1}{2}R(i_1 - i_2) \tag{9-43}$$

From Fig. 9-10 can obtain the curve length:

$$L = R(i_1 - i_2) = 2T \tag{9-44}$$

In Fig. 9-10, triangle ΔVBO is a right-angled triangle, thus

$$T^2 + R^2 = (E + R)^2 = E^2 + 2ER + R^2$$

$$T^2 = E(E + 2R)$$

In above equation, E is very small as compared with the designed radius of a vertical curve R,

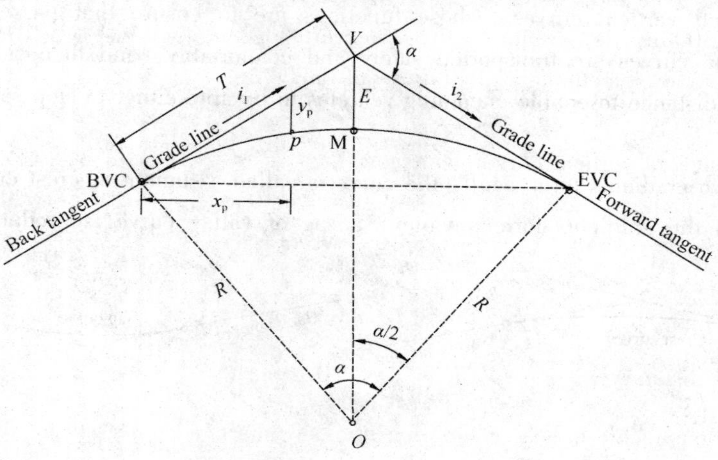

Fig. 9-10 Setting out a vertical curve

the error occurred using $2R$ instead of $(E+2R)$ is very small, so the external distance (E) can be written

$$E=\frac{T^2}{2R} \tag{9-45}$$

In order to meet the needs of the construction and earthwork calculation, the elevation corrections of various points on the vertical curve must be calculated, that is, the distance from the grade line in a vertical direction to the curve must be calculated. From Fig. 9-10 can see that the elevation correction of the middle (M) of the vertical curve is the external distance E. In order to calculate elevation corrections of other points on the grade line, it is necessary to establish the plane rectangular coordinate system. A starting point (BVC) or end point (EVC) of the vertical curve defines as origin of coordinates, the horizontal direction pass origin of coordinate defines as the x-axis, the vertical direction as the y-axis. Therefore, the elevation correction at any point p on the vertical curve can be calculated on following formula:

$$y_p=\frac{x_p^2}{2R} \tag{9-46}$$

Where x_p = the horizontal distance from any point p on the vertical curve to the BVC or the EVC,

y_p = the elevation correction of any point p on the vertical curve, it is also called tangent offset that is distance from the tangents in a vertical direction to the curve.

Therefore, the designed elevation of any point p on the vertical curve can be written.

$$H_p=H'_p\pm y_p \tag{9-47}$$

Where H_p = the designed elevation of any point on the vertical curve;

H'_p = the practical elevation on the grade line corresponding any point p.

Using equation (9-47), the second term should take a negative sign for crest curve, whereas take a positive sign for sag curve.

Example 9.6 Assume that the mileage of the intersection point for a highway is K6 + 144 and that the elevation of the intersection point is 44.50m. The gradients of two adjacent grade lines are designed as $i_1=+0.6\%$ and $i_2=-2.2\%$, respectively. The designed radius of a vertical curve for crest curve is $R=3000$m. It is required to locate vertical curve stakes every 10m interval. Try to calculate vertical curve elements and all setting-out data.

Solution

(1) Calculations of vertical curve elements

The total change in grade $\alpha=i_1-i_2=0.006-(-0.022)=0.028$

Tangent length: $T=\frac{1}{2}R(i_1-i_2)=\frac{1}{2}\times 3000\times 0.028=42\text{m}$

Curve length: $L=2T=84\text{m}$

External distance: $E=\frac{T^2}{2R}=0.29\text{m}$

(2) Calculations of the mileages and elevations of the main points for the vertical curve

Mileage of the BVC: K6 + 144 − T = K6 + 102

Elevation: 44.50 − 0.6% × 42 = 44.25m

Mileage of the MVC: K6 + 102 + L/2 = K6 + 144

Elevation: 44.50 − 0.29 = 44.21m

Mileage of the EVC: K6 + 102 + L = K6 + 186

Elevation: 44.50 − 2.2% × 42 = 43.58m

(3) Calculations of the detail points on the vertical curve

All the calculations are listed in Table 9-5.

Table 9-5 The calculations of the vertical curve

Main points	Station number	Distances from a point to the BVC or the EVC x (m)	Elevation corrections y (m)	Practical elevations on the grade line H' (m)	Designed elevations on the vertical curve H (m)
BVC	K6 + 102	0	0.00	44.25	44.25
	+ 112	10	0.02	44.31	44.29
	+ 122	20	0.07	44.37	44.30
	+ 132	30	0.15	44.43	44.28
V	K6 + 144	42	0.29	44.50	44.21
	+ 156	30	0.15	44.24	44.09
	+ 166	20	0.07	44.02	43.95
	+ 176	10	0.02	43.80	43.78
EVC	K6 + 186	0	0.00	43.58	43.58

单词与词组

· grade line 纵坡线，坡度线

transport [træns ˈpɔːt] *vt.* 运输，输送，搬运

enable [i ˈneibl] *vt.* 使能够，使成为可能

oncoming [ˈɔnkʌmiŋ] *a.* 迎面而来的，接近的

gradient [ˈgreidiənt] *n.* 斜坡，坡道

summit [ˈsʌmit] *n.* 峰顶，顶点 *a.* 峰顶的，最高级的

crest [krest] *n.* 浪尖，冠，顶部，凸形

sag [sæg] *vi.* 下陷，压弯，下垂，凹形 *n.* 下垂，下陷

abbreviate [ə ˈbriːvieit] *vt.* 节略，省略，缩写

nomenclature [ˈnomənˌkletʃə] *n.* 术语，命名（过程）

gradual [ˈgrædjuəl] *a.* 逐渐的，逐步的

注释

* there is adequate sighting distance to enable oncoming vehicles to be able either to stop safely or to overtake safely. 不定式短语 to enable oncoming vehicles 作 sighting distance 后置定语，所以 sighting distance to enable oncoming vehicles，译为“相会的车辆能有的视线距离”，后面一个不定式短语 to be able either to stop safely or to overtake safely 作为目的状语，这里 to be able = can，译为“能够安全停靠或安全通过”，所以全句译为“使相会的车辆能有足够视线距离以便于车辆能够安全停靠或安全通过。”

Chapter 10 Calculation of Areas and Volumes

10.1 Calculation of areas

The computation of areas may be based on data scaled from plans or drawings, or data gained directly from survey field data. The computation of areas in common use will be discussed in following sections.

10.1.1 Transparency method

To measure the areas enclosed by irregular lines, a transparency with squared grids may be used. The transparency is laid over the plan. Within the boundary of the area, the number of full squares is counted. Along the boundary of the area it will be necessary to use judgment to find the area. Generally, the exactly half squares are counted, more than half squares are counted as full squares and less than half squares are discarded. This is called balancing or equalizing (see Fig. 10-1).

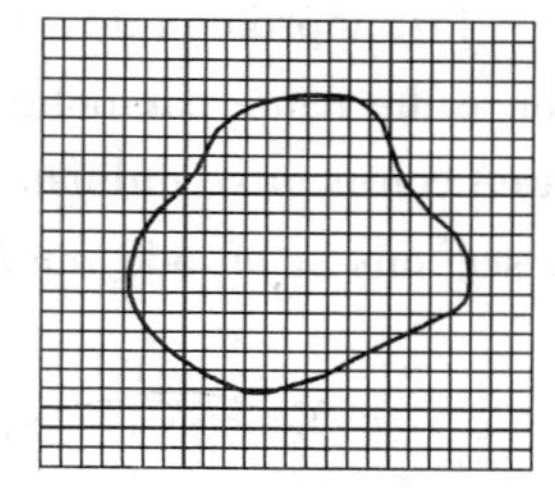

Fig. 10-1 Transparency method

$$\text{Total area} = (\text{number of grids}) \times (\text{area of each grid}) \times M \quad (10\text{-}1)$$

Where M = the denominator of map scales.

10.1.2 Triangulation method

The straight-sided figure can be divided into well-conditioned triangles (see Fig. 10-2), the areas of which can be calculated using one of the following formulae.

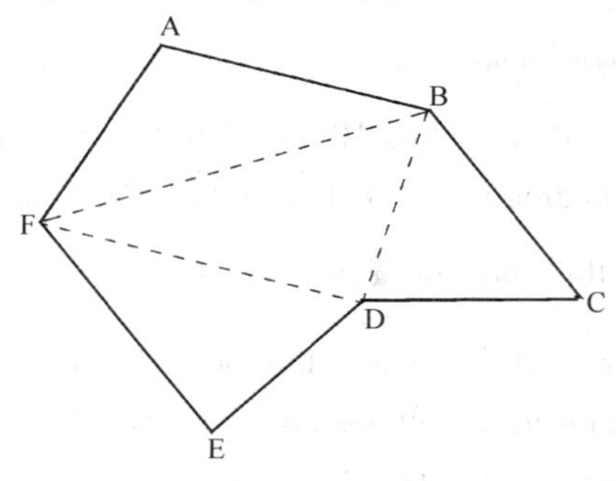

Fig. 10-2 Triangulation method

(1) $A = \sqrt{s(s-a)(s-b)(s-c)}$

where a, b and c are the lengths of the sides of the triangle and

$$s = \frac{1}{2}(a+b+c)$$

(2) $A = \frac{1}{2}$ (base of triangle × height of triangle).

(3) $A = \frac{1}{2}ab\sin C$

where C is the angle contained between side lengths a and b.

The area of any straight-sided figure can be calculated by splitting it into triangles and summing the individual areas.

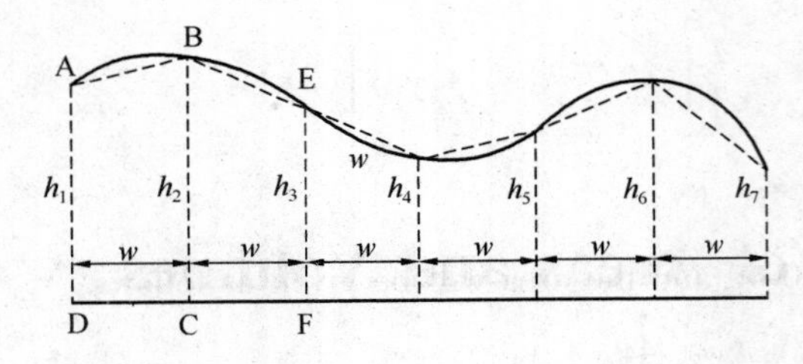

Fig. 10-3 Trapezoidal rule

10. 1. 3 Trapezoidal rule

In Figure 10-3:

$$\text{Area of 1st trapezoid ABCD} = \frac{h_1 + h_2}{2} \times w$$

$$\text{Area of 2nd trapezoid BEFC} = \frac{h_2 + h_3}{2} \times w$$

and so on.

Total area = sum of trapezoids

$$= A = w\left(\frac{h_1 + h_7}{2} + h_2 + h_3 + h_4 + h_5 + h_6\right) \tag{10-2}$$

N. B. (1) If the first or last ordinate is zero, it must still be included in the equation. (2) The formula represents the area bounded by the broken line under the curving boundary; thus, if the boundary curves outside then the computed area is too small, and vice versa.

10. 1. 4 Simpson's rule

In Simpson's rule, the boundary line segment between the ordinates is assumed to be a parabola hence the name parabolic rule. The area is shown in Fig. 10-4a. The formula finds the area using three consecutive ordinates and two sections at a time. Hence, the formula requires an even number of sections or an odd number of ordinates.

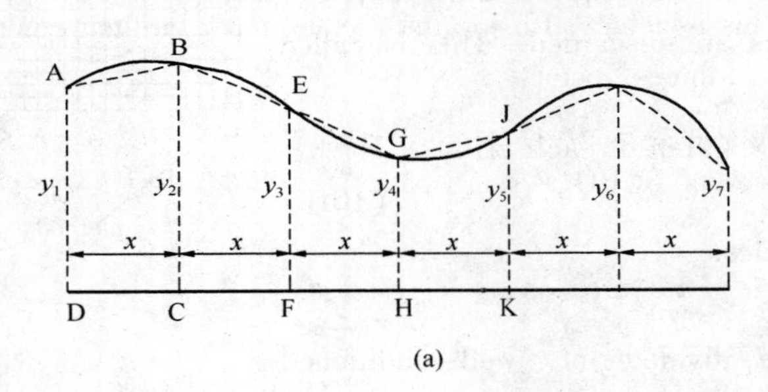

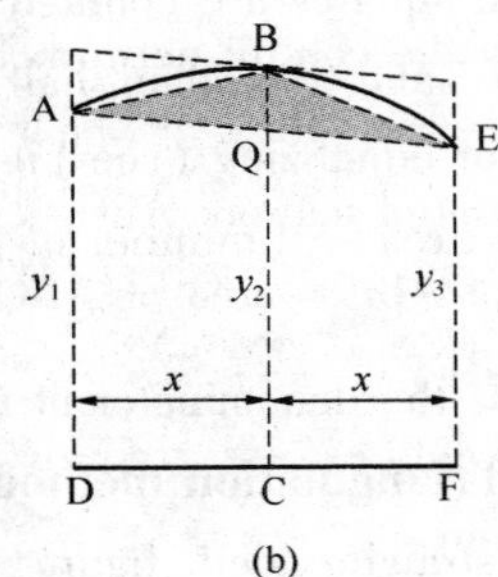

Fig. 10-4 Simpson's rule

(a) Boundary segment as parabola; (b) Area between two sections

Let us consider the two consecutive sections of the area shown in Fig. 10-4b. Join the first and last ordinate points A and E. DAQEFCD is a trapezium and its area can be calculated by the formula $\left[\frac{1}{2}(y_1 + y_3)\right] \times 2x$, where y_1 and y_3 are the ordinates at the end and x is the constant interval between the ordinates. The total area consists of this trapezium and the area of the parabolic segment shown shaded. This area is two-thirds of the area of the enclosing parallelogram. The base of this parallelogram can be taken as $2x$ and the height as BQ. The height BQ = BC − QC. BC = y_2 and QC = $\frac{1}{2}(y_1 + y_3)$. Thus the shaded area is the following:

$$\text{Shaded area} = \frac{2}{3}\left[y_2 - \frac{1}{2}(y_1 + y_3)\right](2x)$$

The total area between the first and the third ordinates is the sum of these two areas:

$$\text{Area ABEFCDA} = \frac{1}{2}(y_1 + y_3)(2x) + \frac{2}{3}\left[y_2 - \frac{1}{2}(y_1 + y_3)\right](2x)$$

$$= \frac{1}{3}(x)[y_1 + 4y_2 + y_3]$$

The next two sections will be as shown in Fig. 10-4a and the area will be given as follows:

$$\text{Area EGJKHFE} = \frac{1}{3}(x)[y_3 + 4y_4 + y_5]$$

When we sum up these partial areas,

$$\text{Total area} = \frac{1}{3}(x)[y_1 + y_7 + 2(y_3 + y_5) + 4(y_2 + y_4 + y_6)] \quad (10\text{-}3)$$

Simpson's rule can be stated as follows: To the sum of the first and last ordinates, add twice the sum of the third, fifth, seventh, etc. ordinates and four times the sum of the second, fourth, sixth, etc. ordinates and multiply this sum by $x/3$, where x is the distance between the ordinates.

N. B. (1) This rule assumes a boundary modeled as a parabola across pairs of sections and is therefore more accurate than the trapezoidal rule. (2) The equation requires an odd number of ordinates and consequently an even number of sections.

10.1.5 Areas from coordinates

In traverse, triangulation and trilateration calculations, the coordinates of the junctions of the sides of a straight-sided figure are calculated and it is possible to use them to calculate the area enclosed by the control network lines. This is achieved using the rectanqular coordinate method.

Consider Fig. 10-5, which shows three points clockwise control network ABC. The required area = ABC.

Area of ABC = area of ABQP + area of BCRQ − area of ACRP

These figures are trapezia for which the area is obtained from

$$\text{area of trapezium} = \frac{1}{2}(\text{height} \times \text{width})$$

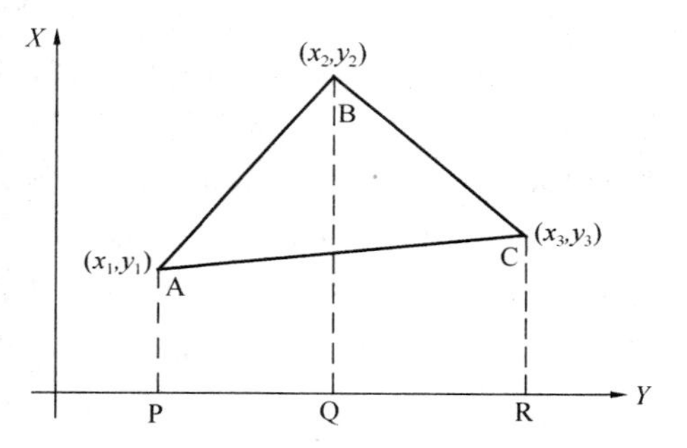

Fig. 10-5 Areas from coordinates

Therefore

$$\text{area of ABQP} = \frac{1}{2}(x_1 + x_2)(y_2 - y_1)$$

Hence area of ABC is calculated by

$$\text{area ABC} = \frac{1}{2}(x_1 + x_2)(y_2 - y_1) + \frac{1}{2}(x_2 + x_3)(y_3 - y_2) - \frac{1}{2}(x_1 + x_3)(y_3 - y_1)$$

Therefore

$$2 \times \text{areaABC} = x_1y_2 - x_1y_1 + x_2y_2 - x_2y_1 + x_2y_3 - x_2y_2 + x_3y_3 - x_3y_2 - x_1y_3 + x_1y_1 - x_3y_3 + x_3y_1$$

Rearranging, this gives

$$2 \times \text{area ABC} = x_1y_2 + x_2y_3 + x_3y_1 - (y_1x_2 + y_2x_3 + y_3x_1)$$

Although the example given is only for a three-sided figure, the formula can be applied to a figure containing n sides and the general formula for such a case is given by

$$2 \times \text{area} = 2 \times A = x_1y_2 + x_2y_3 + \cdots + x_{n-1}y_n + x_ny_1 - (y_1x_2 + y_2x_3 + \cdots + y_{n-1}x_n + y_nx_1)$$

Using a symbol "Σ", the equation above becomes:

$$A = \frac{1}{2}\left(\sum_{i=1}^{n} x_iy_{i+1} - \sum_{i=1}^{n} y_ix_{i+1}\right) \tag{10-4}$$

When applying this formula, it must pay attention to the subscripts of x and y. For a closed polygon, if the number in the equation (10-4) $i = n$, then $y_{i+1} = y_{n+1} = y_1$ and $x_{i+1} = x_{n+1} = x_1$. It is possible to easy calculate the area within closed polygon by the formula, the procedures are as follows:

(1) For convenient calculation, the coordinates of every point are listed in the following form as shown in Fig. 10-6, note that the first point is repeated at the end.

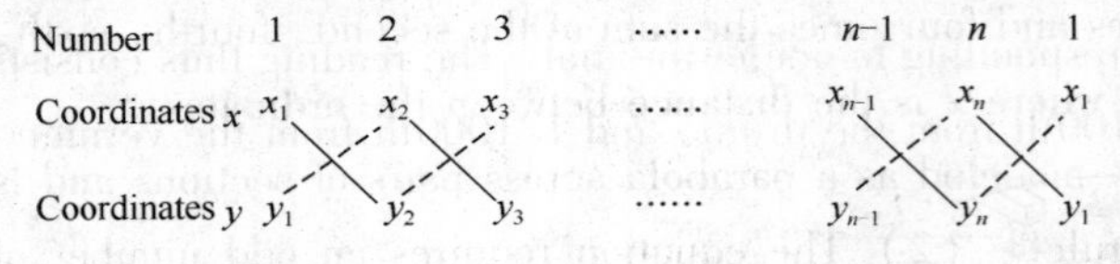

Fig. 10-6 The diagram of the coordinates of every point are arranged

(2) A solid diagonal line is drawn from x_1 to y_2, from x_2 to y_3, ……, and so on. Then a dashed diagonal line is drawn from y_1 to x_2, from y_2 to x_3, ……, and so on.

(3) The summation of the products of the coordinates joined by the solid lines minus the summation of the products of the coordinates joined by the dashed lines equals twice the area within the closed polygon, i. e.

2A = summation of solid-line coordinates products minus the summation of the dashed-line coordinates products.

In applying the coordinate method, if the computed area is negative, the negative sign should be ignored.

Example 10. 1 It is known that the coordinates of the four corners of a quadrangle are noted in Fig. 10-7, try to calculate its area.

Solution

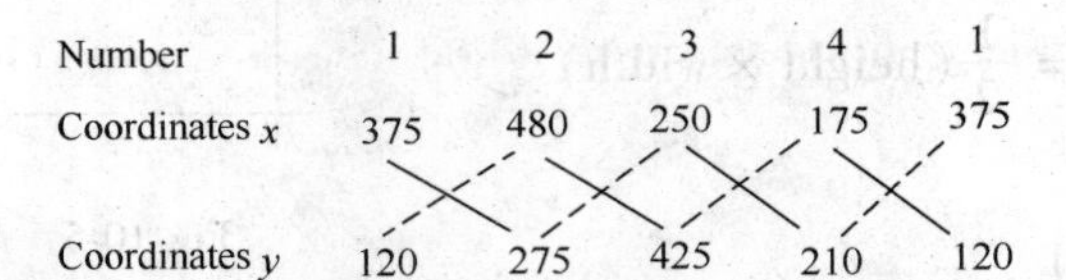

$$\begin{aligned} 2A &= 375 \times 275 + 480 \times 425 + 250 \times 210 \\ &\quad + 175 \times 120 - (120 \times 480 + 275 \\ &\quad \times 250 + 425 \times 175 + 210 \times 375) \\ &= 380625 - 279475 \\ &= 101150\text{m}^2 \end{aligned}$$

$$A = 50575\text{m}^2$$

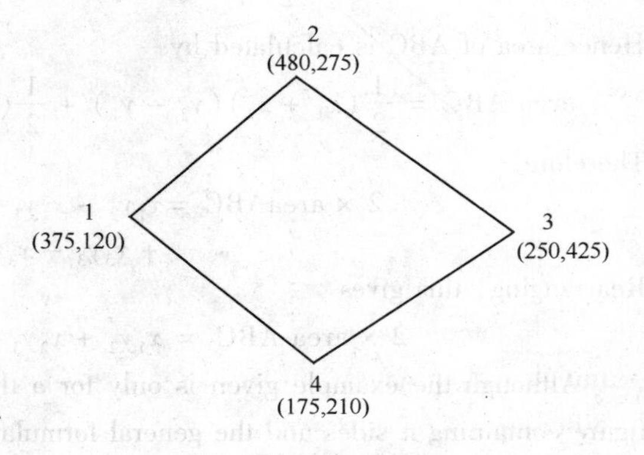

Fig. 10-7 Quadrangle

10. 1. 6 The planimeter

1. A mechanical polar planimeter

The planimeter is a mechanical device for determining the area of any irregular-sided plane figure. A high degree of accuracy can be achieved. Fig. 10-8 shows the main features of a polar planimeter.

The instruments consist of two arms, polar arm JP and tracing arm (JT), which are free to move relative to each other through the hinged point at J but fixed to the plan by an anchor pin of the pole block at P. M is the graduated measuring wheel and T the tracing point. As T is moved around the perimeter of the area, the measuring wheel partly rotates and partly slides over the plan with the varying movement of the tracing point (Fig. 10-8). The measuring wheel is divided into ten units each of which is subdivided into ten parts. The drum therefore reads directly to hundredths of a revolution and a vernier reading against the drum allows thousandths of a revolution to be measured. The wheel is geared to a horizontal counting dial, which shows the number of revolutions made by the wheel, one revolutions of the wheel corresponding to one of the dial. The reading thus consists of four digits, index reading, 1/10th and 1/100th from the drum, and 1/1000th from the vernier

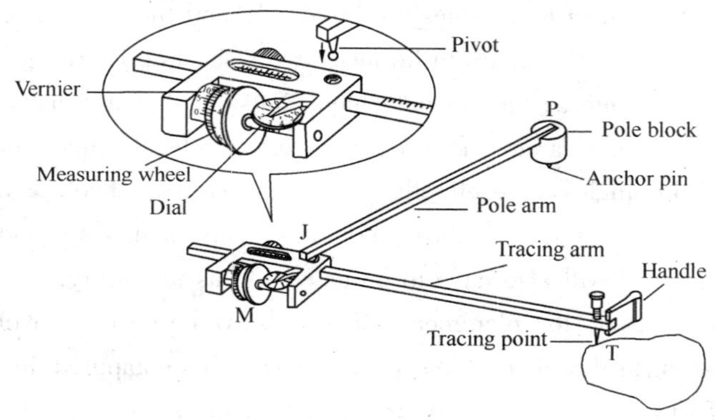

Fig. 10-8 A mechanical polar planimeter (traditional type)

A planimeter can be used in two ways:

(1) With the pole block outside the figure to be measured (see Fig. 10-9).

(2) With the pole block inside the figure to be measured (see Fig. 10-10).

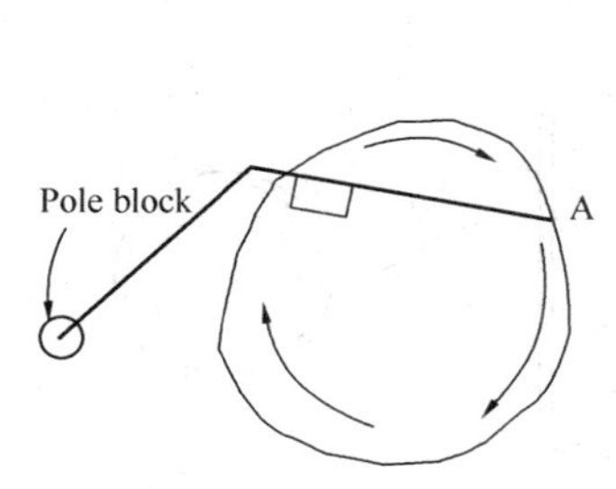

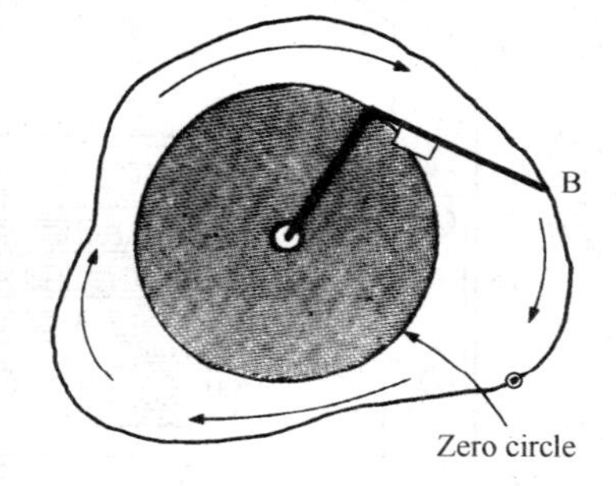

Fig. 10-9 The pole block outside the figure

Fig. 10-10 The pole block inside the figure

In the normal way, the pole block (anchor pin) P is fixed outside the area to be measured, the initial reading noted, the tracing point carefully around the area and the final reading noted. The difference of the two readings gives the number of revolutions of the measuring wheel. On a fixed tracing arm instrument the readings are obtained directly in mm^2 and then have to be converted according to the plan scale to obtain the ground area. On a movable arm instrument the tracing arm length can be set to particular values depending on the plan scale such that the readings obtained give the ground area directly.

With the pole outside the figure the procedure for measuring any area, the plan being on a flat horizontal surface, is as follows.

(1) Place the pole outside the area in such a position that the tracing point can reach any part of the outline.

(2) Keep the tracing point on the boundary of the area at the initial point marked, read the vernier.

(3) Move the tracing point clockwise around the outline, back to the initial point, and read the vernier again.

(4) The difference between the two readings, multiplied by the scale factor, gives the area.

(5) Repeat until three consistent values are obtained, and the mean of these is taken.

With the pole block inside the figure to be measured (see Fig. 10-10):

This is essentially the same as for the pole block outside, but a constant must be added, as shown in Fig. 10-10. The shaded area is known as the zero circle. In this situation the tracing arms and pole arm are at right angles to each other. The measuring wheel is normal to its path of movement and so slides without rotation, thus producing a zero change in reading. This constant is given with the planimeter and will also have to be converted as necessary.

Whichever method is used, the planimeter should always be checked over a known area and if a discrepancy is found a further correction factor should be computed and applied to all the planimeter readings. Testing bars are usually provided with the planimeter for this purpose.

2. A digital planimeter

Fig. 10-11 illustrates Sokkia KP90N electronic planimeter, which incorporates integrated circuit

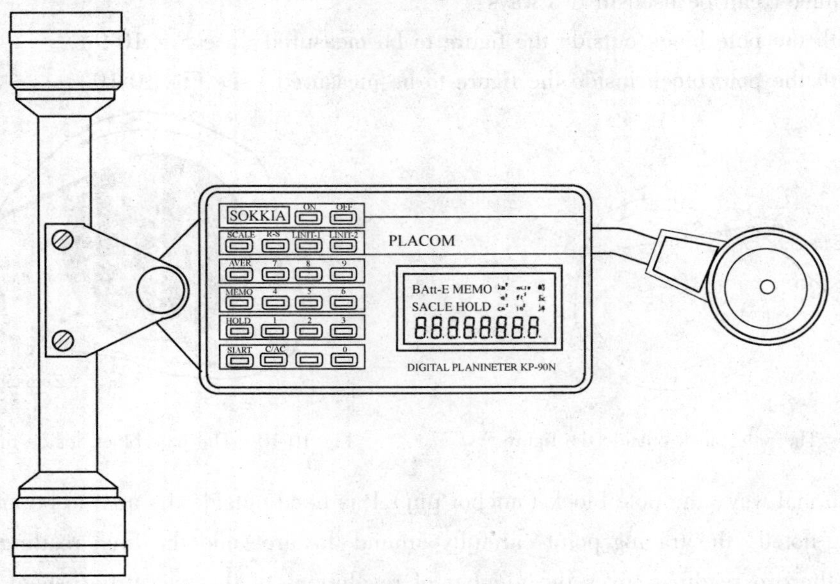

Fig. 10-11 Digital planimete (Sokkia KP-90N electronic planimeter)

technology. It has a tracer arm with a tracer lens and tracer point, but the pole block and pole arm have been supplanted by a roller.

Initially, the tracer arm is set on the approximate centre line of the area to be measured. The power, supplied by a nickel-cadmium battery, is switched on; the unit of measurement, e. g. m^2, is selected, and the scale is fed in via the relevant keys on the keyboard. A reference start point is selected as above, or marked, on the perimeter, and the tracer point is positioned thereon. The "start" key is activated, causing the display to register zero, and the tracer point is moved clockwise along the perimeter to return to the reference point. Motion of the system is sensed by an electro shaft-encoder, which generates pulses that are processed electronically so that the measured area is displayed digitally. This instrument can cater for dual scales, i. e. different horizontal and vertical scales, and has various other facilities also.

单词与词组

transparency [træns 'pεərənsi] *n.* 透明度；透明物体，透明图片，幻灯片
judgment ['dʒʌdʒmənt] *n.* 判断
discard [dis 'kɑ:d] *vt.* 放弃，丢弃，抛弃
trapezoidal [ˌtræpi 'zɔidəl] *a.* 梯形的
· N. B. /n. b. nota bene（拉）注意
· Simpson's rule 辛普生规则
parabola [pə'ræbələ] *n.* 抛物线，抛物面，反射器
odd [ɔd] *a.* 奇数的，单只的，零散的
trapezium [trə 'pi:zi:əm] *n.* 梯形，（美）不规则四边形
parallelogram [ˌpærə 'leləgræm] *n.* 平行四边形
trapezia [trə'pi:ziə] *n.* 梯形
polygon ['pɔligən] *n.* 多边（角）形
summation [sə 'meiʃən] *n.* 总结，总数，总和，加法，求和
planimeter [plæ 'nimitə] *n.* 求积仪
· polar arm 极臂
· tracing arm 描迹臂
hinge [hindʒ] *n.* 铰链，折叶，活页，转轴
anchor ['æŋkə] *n.* 锚，固定器，支撑物，铰钉；*vt.* 抛锚，使稳定，使固定，扣牢
vernier ['və:niə] *n.* 游标（尺），微分尺
· zero circle 基圆，零圆
discrepancy [dis 'krepənsi] *n.* 差异，不一致，矛盾，偏差，误差
· digital planimeter 数字求积仪，电子求积仪
supplant [sə 'plɑ:nt] *vt.* 代替，取代
roller ['rəulə] *n.* 滚转物，滚筒

10. 2 Calculation of volumes

10. 2. 1 Calculation formula of volumes

1. The prismoidal formula

A prisrnoid (see Fig. 10-12) is defined as any solid bounded by two plane parallel figures and a surface generated by a straight line moving in contact with the perimeters of the two figures. The prismoidal formula can be found in any text on solid geometry. It is stated as follows:

$$V = \frac{h}{6}(A_1 + 4M + A_2) \tag{10-5}$$

Where V = volume of the solid defined, h = perpendicular distance between the two bounding planes or bases, A_1 = area of one base, A_2 = area of the other base, M = area of the mid-scction, a section of the solid cut by a plane parallel to and halfway between the bases.

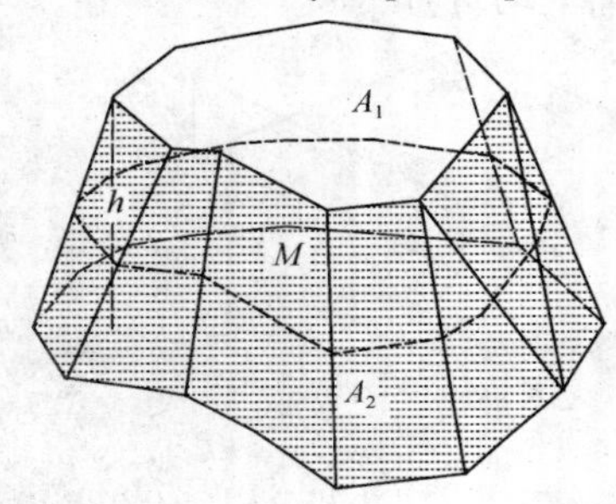

Fig. 10-12 Prismoid

Application of the prismoidal formula:

The areas A_1 and A_2 can be found for each layer by tracing the contour with a planimeter. The value h is the contour interval. But the area of the midsection cannot be found. It can be estimated, however, if it is assumed that areas bounded by adjacent contours are similar figures and that the midsection is similar to both. The linear dimensions of the midsection would then be equal to the average of those of the bases. Then, since linear dimensions of similar figures are to each other as the square roots of the areas of the figures,

$$\sqrt{M} = \frac{\sqrt{A_1} + \sqrt{A_2}}{2}$$

$$M = \frac{A_1 + 2\sqrt{A_1 A_2} + A_2}{4}$$

Substituting in the prismoidal formula,

$$V = \frac{h}{6}\left(A_1 + 4\frac{A_1 + 2\sqrt{A_1 A_2} + A_2}{4} + A_2\right)$$

$$= \frac{h}{3}(A_1 + \sqrt{A_1 A_2} + A_2) \tag{10-6}$$

2. The end-area formula

The end-area formula is an approximate formula often used for simplicity:

$$V = \frac{A_1 + A_2}{2} \times h \tag{10-7}$$

Where the symbols are the same as in equation (10-5). When several volumes are added, this becomes

$$V = h\left(\frac{A_1 + A_2}{2} + A_2 + A_3 + \cdots + A_{n-1}\right) \tag{10-8}$$

10.2.2 Volumes from spot levels (grid method)

This method is used to obtain the volume of large deep excavations such as basements, underground tanks and so on.

A square, rectangular or triangular grid is established on the ground and spot levels are taken at each grid intersection. The smaller the grid the greater will be the accuracy of the volume calculated but the amount of fieldwork increases so a compromise is usually reached.

The formation level at each grid point must be known and hence the depth of cut from the existing to the proposed level at each grid intersection can be calculated.

Fig. 10-13 shows a 10m square grid with the depths of cut marked at each grid intersection. Consider the volume contained in grid square $h_1h_2h_6h_5$; this is shown in Fig. 10-14.

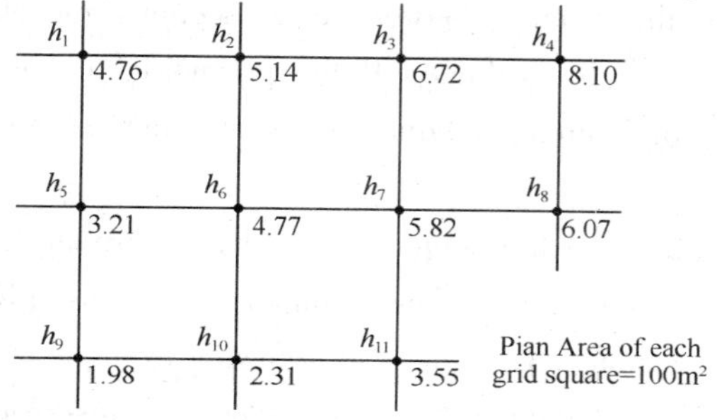

Fig. 10-13 Square grid

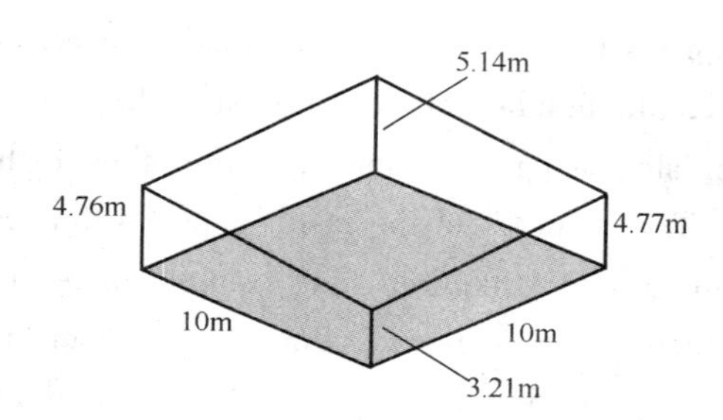

Fig. 10-14 An individual grid square

It is assumed that the surface slope is constant between grid intersections, hence the volume is given by

$$\text{volume} = \text{mean height} \times \text{plan area}$$

$$= \frac{1}{4}(4.76 + 5.14 + 4.77 + 3.21) \times 100 = 447\text{m}^3$$

A similar method can be applied to each individual grid square and this leads to a general formula for square or rectangular grids

$$\text{total volume} = \frac{A}{4}(\sum h_S + 2\sum h_D + 3\sum h_T + 4\sum h_Q) + \delta V \qquad (10\text{-}9)$$

where A = plan area of each grid square, h_S = single depths, such as h_1, h_4, h_8, h_9 and h_{11} which are used once, h_D = double depths, such as h_2, h_3, h_5 and h_{10} which are used twice, h_T = triple depths, such as h_7 which are used three times, h_Q = quadruple depths, such as h_6 which are used four times, δV = the all volume outside the grid which is calculated separately.

Hence, in the example shown in Fig. 10-13

$$\text{Volume contained within the grid area} = \frac{100}{4}[4.76 + 8.10 + 6.07 + 1.98 + 3.55$$
$$+ 2(5.14 + 6.72 + 3.21 + 2.31) + 3 \times 5.82$$
$$+ 4 \times 4.77]$$
$$= 25\ (24.46 + 34.76 + 17.46 + 19.08)$$
$$= 2394\text{m}^3$$

The result is only an approximation since it has been assumed that the surface slope is constant between spot heights.

If a triangular grid is used, the general formula must be modified as follows

(1) $\frac{A'}{3}$ must replace $\frac{A}{4}$ where A′ = plan area of each triangle and

(2) depths appearing in five and six triangles must be included.

10.2.3 Volumes from contours

This method is particularly suitable for calculating very large volumes such as those of reservoirs, earth dams, spoil heaps and so on.

The system adopted is to calculate the plan area enclosed by each contour and then treat this as a cross-sectional area. The contour interval provides the distance between cross sections and either the prismoidal or end areas method is used to calculate the volume. If the prismoidal method is used, the number of contours must be odd. The plan area contained by each contour can be calculated using a planimeter or the other methods.

The accuracy of the result depends to a large extent on the contour interval but normally great accuracy is not required, for example in reservoir capacity calculations, volumes would usually be rounded to the nearest $1000m^3$, that is more than adequate. Consider the following example.

Fig. 10-15 shows a plan of a proposed reservoir and dam wall, which shows a cross section through the reservoir and the plan areas enclosed by each contour and the dam wall.

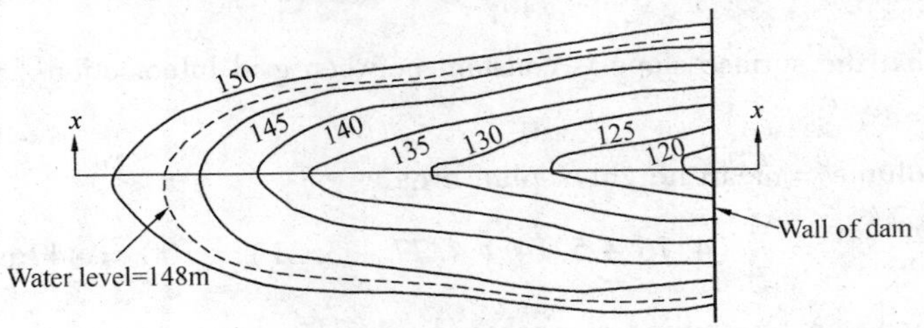

Fig. 10-15 A map of a proposed reservoir

The volume of water that can be stored between the contours can be found by reference to Fig. 10-16. The vertical interval is 5m and the water level of the reservoir is to be 148m. The capacity of the reservoir is required.

total volume = volume between 148m and 145m contours

+ volume between 145m and 120m contours + small volume below 120m contour.

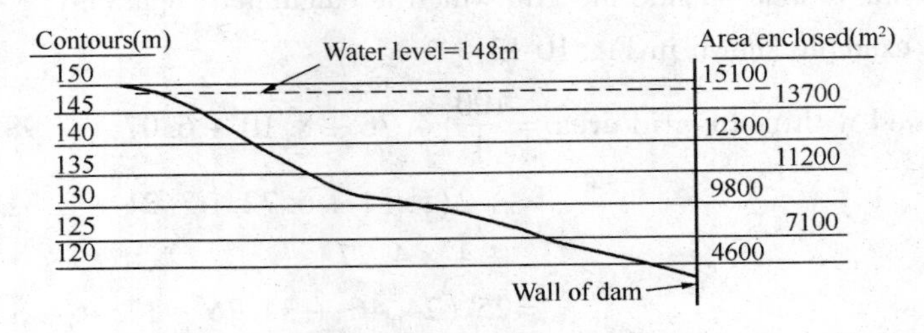

Fig. 10-16 The vertical section of reservoir

Volume between 148m and 145m contours is found by the formula (10-7), i. e.

$$=\frac{15100+13700}{2}\times 3 = 43200\text{m}^3$$

Volume between 145m and 120m contours can be found by the formula (10-8), i. e.

$$=\frac{5}{2}(13700+4600+2\times(12300+11200+9800+7100)) = 247750\text{m}^3$$

The small volume below the 120m contour can be found by decreasing the contour interval to say, 1m and using the end areas method or the prismoidal formula. Alternatively, if it is very small, it may be neglected. Let this volume = δV. Therefore

$$\text{total volume} = 43200 + 247750 + \delta V \qquad (10\text{-}10)$$
$$= (2909504 + \delta V)\ \text{m}^3$$

(this would usually be rounded to the nearest 1000m^3). The second term in equation (10-10) was obtained by the end areas method applied between contours 145m and 120m. Alternatively, the prismoidal formula could have been used between the 145m and 125m contours (to keep the number of contours ODD) and the end areas method between the 125m and 120m contours. If this is done, the volume between the 145m and 120m contours is calculated to be 248 583m^3.

单词与词组

prismoid [′prizmɔid] *n.* 平截头棱锥体，棱柱体
prismoidal [′prizmɔidəl] *a.* 似棱形的，棱柱体的
bounded [′baundid] *a.* 有界限的，有限制的
solid [′sɔlid] *a.* 固体的，结实的，稳定的，立方体的，立方的，三维的
· solid geometry 立体几何
perimeter [pə ′rimitə] *n.* 周，周长，周边，周围，周界线
midsection [mid ′sekʃən] *n.* 中部，中间部分
depth [depθ] *n.* 深；深度，浓度，纵深
quadruple [′kwɔdru：pəl] *a.* 四倍的
· spoil heap 矸子山
round [raund] *a.* 圆形的，球形的，用整数表示的，取整数的
compromise [′kɔmprəmaiz] *n.* 妥协，和解，折中，折中方案
formation [fɔ:′meiʃəm] *n.* 组织，构造，形成物，结构岩层

Chapter 11 Construction Surveys

11.1 General

Construction survey, which is known as construction layout or setting-out, is the process of using the surveying instruments and techniques to transfer constructions from a plan to the ground. As such it is the opposite of surveying. In other words, construction surveying involves transfer of the dimensions on the drawing to the ground so that the work is done in its correct position. *1 [This type of surveying is sometimes called setting lines and grades.] The work of the surveyor for construction projects is often referred to as layout work.

Construction surveys provide line and grade for a wide variety of construction projects for example, highways, streets, pipelines, bridges, and buildings. The construction layout marks the horizontal location (line) as well as the vertical location or elevation (grade) for the proposed work. The contractor can measure from the surveyor's markers to the exact location of each component of the facility to be constructed.

*2 [To begin with, a topographic survey and map showing the location of whatever is to be constructed are required before construction begins.] During the construction process, surveyors lay out the structure and perform other tasks as needed. As-built surveys are made after a construction project is complete, to provide the positions and dimensions of the features of the projects as they were actually constructed.

Construction surveying, as the basic for all construction and part of it, has become increasingly important. It is estimated that 60 percent of all surveying man-hours are spent on location-type surveys giving line and grade.

The various parts of the structure should be placed at the corrected positions. To accomplish this goal the construction surveyor will establish the reference lines or base lines before the actual layout measurements, which will be discussed in section 11.2.

In construction layout measurement, the data necessary to establish the direction and distance from a control point to locate a construction point can be entered into the instrument via the keyboard or directly from an office computer. Then the surveyor guides the person holding the prism along the line of computed direction until the distance to the point to be located agrees with the computed distance. Various techniques may be applied to accomplish this goal. A very popular technique called free station permits the surveyor to set up the total station at any convenient position and then to determine the coordinates and elevation of that instrument position by sighting previously coordinated reference stations. After the instrument has been set up over this instrument position (a control point) and properly oriented, angles or azimuths and distances from the control point to each layout point may be indicated. Now many total stations have such functions that the coordinates and elevations of the layout points may be uploaded into the total station, the instrument's display shows the left/right, forward/back, and up/down movements needed to place the prism in each of the desired positions.

11.1.1 Tasks of construction survey

The tasks of construction survey include at least the following four aspects:

(1) Survey existing conditions of the work site, including topography, existing buildings and infrastructure, and underground infrastructure whenever possible (for example, measuring invert elevations and diameters of sewers at manholes).

(2) Stake out reference points and markers that will guide the construction of new structures.

(3) Verify the location of structures during construction.

(4) As-built surveys are made after a construction project is complete, to provide the positions and dimensions of the features of the projects as they were actually constructed. These surveys not only provide a record of what was constructed but also provide a check to see if the work was done according to the design plans.

Setting-out is a three-part problem, which is to say that whatever is being set out must be

(1) In the correct position,

(2) At the correct level, and

(3) Ensure verticality of the structure.

11.1.2 Staking out buildings

In staking out buildings, wooden stakes with a tack can be driven into the ground to mark a structure's corners. Such stakes will be temporary, as they are of necessity removed during digging and construction of the foundation. To provide more permanent marking of corners during foundation construction, "batter boards" are placed near each corner.

Batter boards consist of several stakes driven into the ground with boards nailed to the stakes horizontally at some particular elevation. Batter boards are set 1 ~ 2m off line in each direction at each corner. Nails are driven (or notches cut) at proper locations on the horizontal boards so that strings can be attached at the nails (or notches) and stretched from one corner to another. A corner of the building is defined where two strings cross. Strings also provide a reference elevation from which vertical measurements can be made to maintain levelness. Batter boards may be better understood by referring to Fig. 11-1 and 11-2. Fig. 11-1 shows them (in perspective) at one corner; Fig. 11-2 shows them (in plan view) as they define the four corners (C, D, J, and I) of a simple building.

Batter boards should be established firmly in the ground and braced, if necessary, to prevent movement. Strings may be removed and later put back as necessary during construction. For example, strings might be needed to determine (by measuring downward) how much cut is needed for the foundation trench. They could then be removed to allow a person or machine to dig in the trench. Later they would be replaced for reference in doing the foundation masonry.

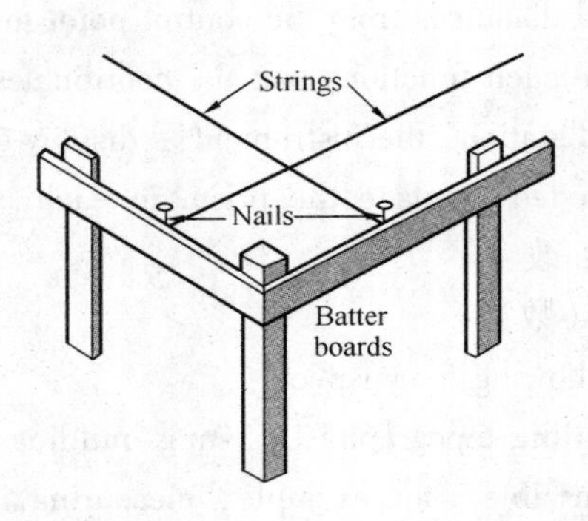

Fig. 11-1 Batter boards

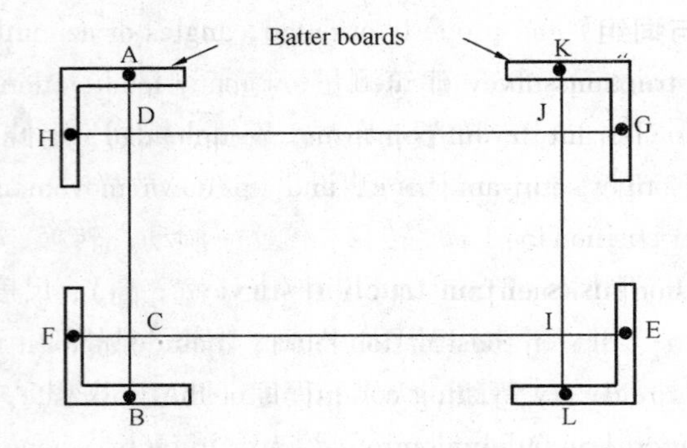

Fig. 11-2 Batter boards in plan view

A procedure for setting batter boards is as follows:

(1) Locate each corner of the building on the ground by temporary stakes with a tack.

(2) Drive batter board stakes near each corner and nail boards horizontally to the stakes at the desired elevation. They can be set at the desired elevation using a level and rod.

(3) Place nails for one string (e. g., points A and B in Fig. 11-2) defining one side of the building, so that the string passes directly over the desired corners (points C and D in Fig. 11-2). Stretch the string tightly between A and B.

(4) Set a theodolite over corner C. sight on A, turn a right angle, and place a nail at E so that ACE is a right angle.

(5) Stretch a string tightly from E directly under the theodolite (i. e., over corner C) and locate point F on the batter board. Drive a nail at F and attach the string at F. The intersection of strings EF and AB defines corner C.

(6) Lay off building dimension CD to establish corner D exactly.

(7) Move the theodolite to D, sight on B, turn a right angle, and place a nail at G so that GDB is a right angle.

(8) Locate point H by stretching a string from G through D (same procedure as step 5), and place a nail at H. Attach string from G to H. The intersection of strings AB and GH defines corner D.

(9) Lay off building dimensions CI and DJ to establish corners I and J.

(10) Locate two points K and L so that a string stretched between them crosses corners I and J exactly. Place nails at K and L and stretch string. The intersection of strings KL and FE defines corner I, and that of KL and HG locates corner J.

If these steps are done perfectly, the building should be laid out correctly. One should certainly measure IJ to make sure it is right, and it is also good practice to measure the diagonals to see if they are the correct length.

The preceding discussion covered a very simple building with right-angle corners and the same elevation throughout. Certainly many buildings will be more complicated some of them very much so. Nevertheless, the general procedure described above gives the fundamental steps required.

单词与词组

· construction survey 建筑测量，施工测量
layout['lei,aut] *n.* 布置，布局，安排，规划，定位，定线
setting-out['setiŋ-aut] *n.* 测设，定线，放样
transfer[træns 'fə:] *vt.* 转移，传送，调动，转换，变换，改变
dimension[di 'menʃən] *n.* 尺寸(长，宽，高)，尺度，维(数)
drawing['drɔ:iŋ] *n.* 绘图，图画，图样，制图
grade[greid] *n.* 等级，级别，程度，坡度，高程（更多含义，见注释[1]）
contractor[kən 'træktə] *n.* 立约人，承包人
marker['mɑ:kə] *n.* 标志，标记，指标，标杆，旗标；书签
· to begin with 首先
· as-built survey 竣工测量
orient['ɔ:riənt] *vt.* 使定向，为定……方位
infrastructure['infrə,strʌktʃə] *n.* 基础，基础结构，基础设施
sewer['su:ə] *n.* 阴沟，污水管，下水道
verify['verifai] *vt.* 证实，查证，鉴定，验证
verticality[,və:ti 'kæləti] *n.* 垂直性，垂直状态
prefabricate[pri:'fæbri,keit] *vt.* 预制，预加工
property['prɔpəti] *n.* 财产，地产，房地产，所有权
· property line 建筑红线
ordinance['ɔ:dinəns] *n.* 法令，条例
regulation[,regju 'leiʃən] *n.* 管理，控制，规则，规定，条例
adhere[əd 'hiə] *vi.* 坚持，遵守(to)
tack[tæk] *n.* 大头钉，平头钉
· batter board 龙门板
notch[nɔtʃ] *vt.* 在(某物)上刻 V 形痕
perspective[pə 'spektiv] *n.* 透视(画，画法)，远景，景色，前景，眼界
brace[breis] *n.* 支撑物，支架 *vt*，支撑，加固

注释

[1] This type of surveying is sometimes called setting lines and grades。这句仅是简单句，难点在于如何理解"grade"一词此处摘录一段国外教材的解释(详见参考文献[11]的 14. 1. 1 节)，原文如下：

The word grade has several different meaning. In construction work alone, it is often used in three distinctly different ways to refer to:

1. A proposed elevation.（拟用高程）
2. The slope of profile line (i. e. , gradient).（纵断面线坡度）
3. Cuts and fills (vertical distances below or above reference marks on grade stakes).（挖深与填高）

The surveyor should be aware of these difference meanings and always notes the context in which the word grade is being used.

因此，课文中这一句的词 grades 包含第一、二两种含义，故全句译为：这种类型测量有时称为定线和测设高程与坡度。

［2］ 此句中 showing the location of whatever is to be constructed 分词短语作 map 的后是定语，意为“展示拟建区域中各种物体位置的”，全句译为：在工程开始之前，首先必须进行地形测量与绘制以表示拟建区域的一切物体的位置。

11.2 Base lines

Before the actual layout measurements can begin, it is necessary for reference lines or base lines to be carefully established. A baseline is a line running between two points of known position. The base line is also a continuation of the building line of the existing adjacent buildings. Any baselines required to set out a project should be designed on the plan by surveyors.

Primary site control points, such as traverse stations E and F in Fig. 11-3, can be used to establish a baseline AB by angle α and distance l values as shown; Subsidiary offset lines can then be set off at right angles from each end of the baseline to fix two corners R and S of building Z as shown. Once R and S have been pegged out, the horizontal length of RS is measured and checked against its designed value. If it is within the required tolerance, points R and S can be used as a baseline to set out the corners T and U.

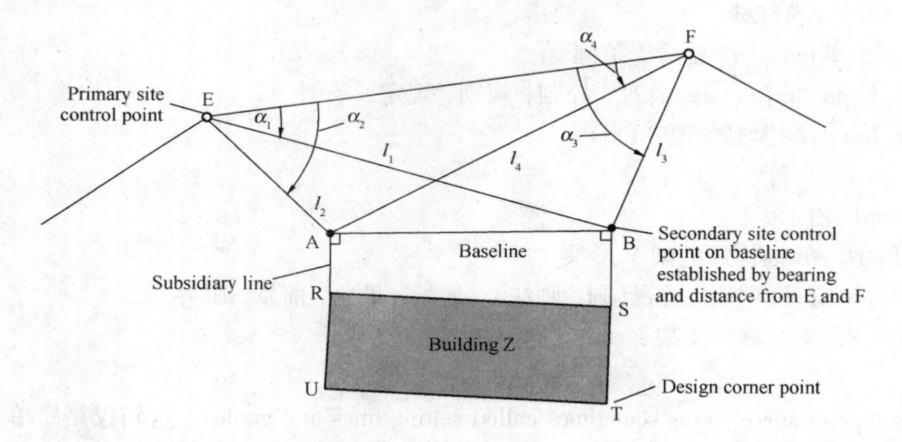

Fig. 11-3 A baseline

In some cases, the designer may specify a baseline that runs between points on two existing buildings. Design points are then set out from this line either by offsetting at right angles or by measuring distances from points on the line. The accuracy of this method depends greatly on how well the baseline can be established.

单词与词组

continuation [kənˌtinjuːˈeiʃən] *n.* 继续，连续，持续，延续，续篇，（线路等的）延长，续刊，增刊

subsidiary[səb 'sidiəri] *a.* 辅助的，附属的，次要的
fix[fiks] *vt.* 使固定，钉牢，安装，安排，确定

11.3　Use of grids

Many structures in civil engineering consist of steel or reinforced concrete columns supporting floor slabs.

As the disposition of these columns is inevitably that they are at right angles to each other, the use of a grid, where the grid intersections define the position of the columns, greatly facilitates setting out. Several different grids can be used in setting out.

1. *Survey grid*: the rectangular coordinate system on which the original topographic survey is carried out and plotted (Fig. 11-4).

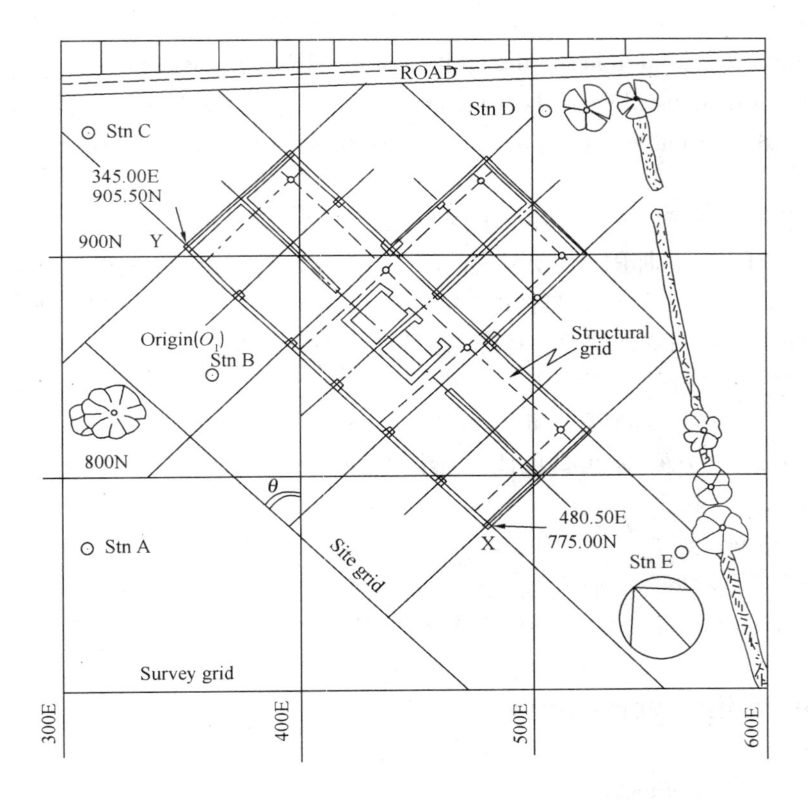

Fig. 11-4　Survey, site and structural grids

2. *Site grid*: the *site grid* is used by the designer.

The site grid defines the position and direction of the main building lines of the project, as shown in Fig. 11-4. The best position for such a grid can be determined by simply moving a tracing of the site grid over the original plan so that its best position can be located in relation to the orientation of the major units designed thereon.

In order to set out the site grid, it may be convenient to translate the coordinates of the site grid to those of the survey grid using the well-known transformation formula:

$$E = \Delta E + E_1 \cos\theta - N_1 \sin\theta$$
$$N = \Delta N + N_1 \cos\theta + E_1 \sin\theta$$

where ΔE, ΔN = difference in easting and northing of the respective grid origins,

E_1, N_1 = the coordinates of the point on the site grid,

θ = relative rotation of the two grids,

E, N = the coordinates of the point transformed to the survey grid.

Thus, selected points, say X and Y (Fig. 11-4) may have their site-grid coordinate values trans-formed to that of the survey grid and so set-out by polars or intersection from the survey control. Now, using XY as a baseline, the site grid may be set out using theodolite and steel tape, all angles being turned off on both faces and grid intervals carefully fixed using the steel tape under standard tension.

3. *Structural grid*: used to locate the position of the structural elements within the structure and is physically established usually on the concrete floor slab (seeFig. 11-4). The structural grid is usually established from the site grid points and uses the same cocrdinate system.

单词与词组

grid[grid] *n.* 格栅，格子，坐标方格，电网
reinforce[ˌriːin'fɔː] *v.* 加固，使更结实，加强，充实；求援，给予更多的支持
concrete['kɔnkriːt] *a.* 实际的，具体的，特定的，固结成的；混凝土制的，水泥的
column['kɔləm] *n.* 圆柱，柱状物，专栏，纵队
slab[slæb] *n.* 厚板，平板，厚片
disposition[dispə'ziʃən] *n.* 布置，配置，安排，部署
inevitably[in'evitəbli] *adv.* 不可避免地，必然地 *a.* 不可避免的，照常的
· survey grid 测量格网
· site grid 施工格网
· structural grid 结构柱网
transformation[ˌtrænsfə'meiʃən] *n.* 转变，转化，变形

11.4 Controlling verticality

There are a number of methods in controlling verticality. Some common methods are as follows: (1) plumb-bob, (2) theodolite, (3) optical plumbing.

11.4.1 Using a plumb-bob

In low-rise construction a heavy plumb-bob (5 ~ 10kg) may be used as shown in Fig. 11-5. If the external wall were perfectly vertical then, when the plumb-bob coincides with the centre of the peg, distance *d* at the top level would equal the offset distance of the peg at the base. This concept can be used internally as well as externally, provided that holes and openings are available. The plumb-bob should be large, say 5kg, and both plumb-bob and wire need to be protected from wind.

The motion of the plumb-bob may need to be damped by immersing the plumb-bob in a drum of water. The considerations are similar to those of determining verticality in a mine shaft but less critical. To ensure a direct transfer of position from the bottom to the top floor, holes of about 0.2m diameter will need to be left in all intermediate floors. This may need the agreement of the building's designer.

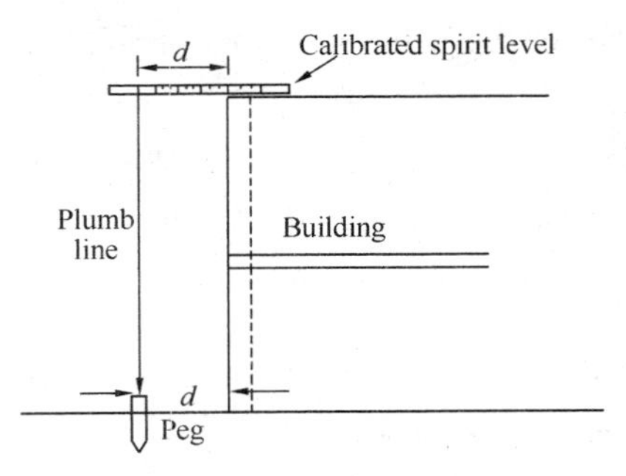

Fig. 11-5 Plumb-bob for verticality

11.4.2 Using a theodolite

This method assumes that the theodolite is in perfect adjustment so that its line of sight will describe a vertical plane when rotated about its tilting axis. The method is discussed below.

The theodolite is set up on extensions of each reference line marked on the ground floor slab in turn and the telescope is sighted on to the particular line being transferred. The telescope is elevated to the required floor and the point at which the line of sight meets the floor is marked. This is repeated at all four corners, and eight points in all are transferred, as shown in Fig. 11-6. Once the eight marks have been transferred, they are joined and the distances between them and their diagonal lengths are measured as checks.

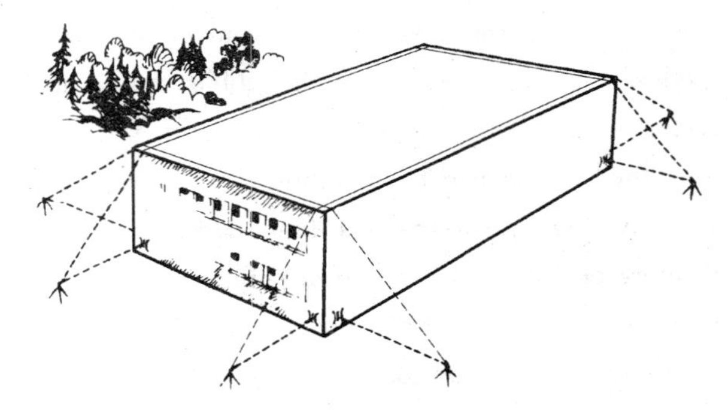

Fig. 11-6 Transfer of control in a multi-storey structure

If the centre lines of a building have been established, a variation of this method is to set up a theodolite on each in turn and transfer four points instead of eight, as shown in Fig. 11-7. This establishes two lines at right angles on each floor from which measurements can be taken.

If the theodolite is not in perfect adjustment, the points must be transferred using both faces and the mean position used. In addition, because of the large angles of elevation involved, the

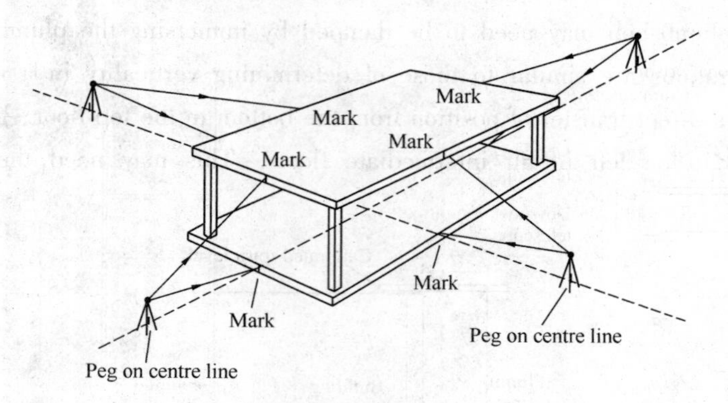

Fig. 11-7 Transfer of centre lines

theodolite must be carefully levelled and a diagonal eyepiece attachment may be required to enable the operator to look target through the telescope.

11.4.3 Using optical plumbing

For high-rise building the instrument most commonly used is an auto plumb. Various types of plummet are available for upwards and downwards sighting to allow the establishment of a vertical line, and these are normally manufactured so as to be interchangeable with theodolites on their tripods. Fig. 11-8 shows one such instrument, which has two telescopes, one low power for sighting downwards and locating the instrument over the ground mark, and the other of high power for sighting upwards onto a target. In this instrument the upwards-sighting telescope is fitted with the same type of compensator which was formerly used in the automatic levels and this automatically ensures a vertical line of sight, even if the instrument is tilted by several minutes of arc. The instrument is attached to a three-screw levelling base which can be mounted on a tripod, and a centring motion is available for positioning over ground points. Since the downward-sighting telescope is not compensated, it is essential that the instrument be levelled using a plate bubble, positioned parallel to the telescope axes, and a circular bubble, found at the top of the case.

This instrument provides a vertical line of sight to an accuracy of ±1 second of arc (1mm in 200m). Any deviation from the vertical can be quantified and corrected by rotating the instrument through 90° and observing in all four quadrants; the four marks obtained would give a square, the diagonals of which would intersect at the correct centre point.

A base figure is established at ground level from which fixing measurements may be taken. If this figure is carried vertically up the structure as work proceeds, then identical fixing measurements from the figure at all levels will ensure verticality of the structure (Fig. 11-9).

To fix any point of the base figure on an upper floor, a perspex target is placed over the hole left in the upper floor (Fig. 11-9b) and the centre point fixed as above.

The shape of the base figure will depend upon the plan shape of the building. In the case of a long rectangular structure a simple base line may suffice but T shapes and Y shapes are also used.

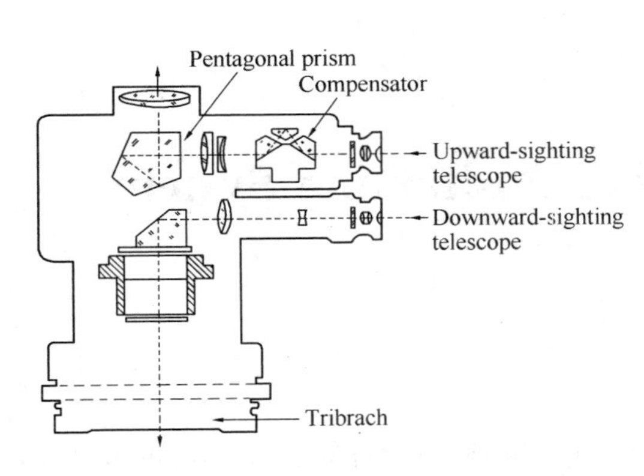

Fig. 11-8 The optical system of the auto plumb

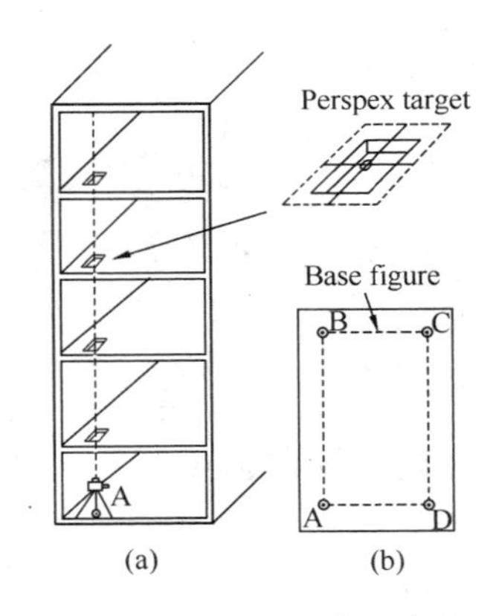

Fig. 11-9 Checking verticality of the structure
(a) Elevation; (b) Plan

单词与词组

low-rise[ˈləuˌraiz] *a.* 层数少而没有电梯的，不高的，低层的
external[eksˈtəːnl] *a.* 外部的，外面的，外界的，客观的；物质的
hole[həul] *n.* 洞，穴，孔眼，破洞，缺陷，缺点
opening[ˈəupniŋ] *n.* 开；开放，口，孔，槽，空隙，通道
immerse [iˈməːs] *vt.* 沉浸，浸入
mine[main] *n.* 矿，矿井
shaft[ʃɑːft] *n.* 升降机井，矿井，竖井；通风管道，烟囱
critical[ˈkritikəl] *a.* 紧要的，关键性的，严重的，危险的，危急的，急需的
plumb[plʌm] *n.* 铅锤，测锤，垂直；*a.* 垂直的 *vt.* 用铅垂线校正
plummet[ˈplʌmit] *n.* 铅锤，测锤；*vi.* 垂直落下，骤然跌落
· multi-storey structure 多层结构，多层建筑
· high-rise building 高层建筑
slab[slæb] *n.* 厚板，平板，厚片
extension [iksˈtenʃən] *n.* 扩大，伸展，广度，延长，延期
diagonal[daiˈægənl] *a.* 斜的，对角线的
· diagonal eyepiece 折轴目镜
quadrant[ˈkwɔdrənt] *n.* 象限，圆周的四分之一
perspex[ˈpəːspeks] *n.* 塑胶(有机)玻璃，透明塑胶

Chapter 12 The Global Positioning System

12.1 Introduction

The Global Positioning System (GPS) is a satellite-based system that can be used to locate positions anywhere on the earth, which was developed by the US Department of Defense (DoD) in the early 1970s. GPS provides continuous (24 hours/day), real-time, 3-dimensional positioning, navigation and timing worldwide. Any person with a GPS receiver can access the system, and it can be used for any application that requires location coordinates.

The concept of satellite position fixing commenced with the launch of the first Sputnik satellite by the USSR in October 1957. This was rapidly followed by the development of the Navy Navigation Satellite System (NNSS) by the US Navy. This system commonly referred to as the Transit system. However, as the determination of position required very long observation periods and relative positions determined over short distances were of low accuracy, its application was limited to geodetic and low dynamic navigation uses. In 1973, the US Department of Defense (DoD) commenced the development of NAVSTAR (Navigation System Timing and Ranging) Global Positioning System (GPS), and the first satellites were launched in 1978.

GPS satellites circle the earth twice a day in a very precise orbit and transmit signal information to earth. GPS receivers take this information and use triangulation to calculate the user's exact location. Essentially, the GPS receiver compares the time a signal was transmitted by a satellite with the time it was received. The time difference tells the GPS receiver how far away the satellite is. Now, with distance measurements from a few more satellites, the receiver can determine the user's position and display it on the unit's electronic map.

The original purpose of the satellite system was to enable planes, ships, and other military groups to quickly determine their geodetic positions. Although the system was developed for military purposes, it is of tremendous benefit to other groups, such as the National Geodetic Survey, the private surveying profession, and much of the general public. The GPS system can be used to accomplish anything that can be done with conventional surveying techniques.

Now that GPS is fully operational, relative positioning to several millimeters, with short observation periods of a few minutes, have been achieved. For distances in excess of 5km GPS is generally more accurate than EDM traversing. Therefore GPS has a wide application in engineering surveying. The introduction of GPS has had an even greater impact on practice in engineering surveying than that of EDM. Apart from the high accuracies attainable, GPS offers the following significant advantages:

(1) The results from the measurement of a single line, usually referred to as a baseline, will yield not only the distance between the stations at the end of the line but their component parts in the *X/Y/Z* or Eastings/Northings/Height or latitude/longitude/height directions.

(2) No line of sight is required. Unlike all other conventional surveying systems a line of sight between the stations in the survey is not required. Each station, however, must have a clear view of the sky so that it can "see" the relevant satellites.

(3) Most satellite surveying equipment is suitably weatherproof and so observations, with current systems, may be taken in any weather, by day or by night. A thick fog will not hamper survey operations.

(4) Satellite surveying can be a one-person operation with significant savings in time and labour.

(5) Position may be fixed on land, at sea or in the air.

(6) Continuous measurement may be carried out resulting in greatly improved deformation monitoring.

Because satellites orbit the whole Earth, the coordinate systems that describe the positions of satellites are global rather than local. Thus, if coordinates are required in a local datum or on a projection, then the relationship between the local projection and datum, and the coordinate system of the satellite, must also be known.

单词与词组

· Global Positioning System (GPS) 全球定位系统

· US Department of Defense (DoD) 美国国防部

navigation['nævi 'geiʃən] *n.* 航行，航海，航空航行学，航海(航空)术导航，领航

· NAVSTAR (Navigation Satellite Timing and Ranging) 卫星导航授时测距

sputnik['spʌtnik] *n.* (苏联)人造卫星

navy['neivi] *n.* 海军，(集合名词)海军人员，舰队，(英)海军部

· Navy Navigation Satellite System (NNSS) 海军卫星导航系统

transmit[trænz 'mit] *vt.* 传送(达，播)，使(光、热、声等)透射，透过，传导

impact['impækt] *n.* 冲击(力)，碰撞，冲突，效果，影响

suitable['sju:təbl] *a.* 适合的，适当的，相配的(to, for)

waterproof ['weðəpru:f] *a.* 防风雨的，不受天气影响的，抗风化的；*vt.* 防风雨

12.2 The GPS system

The GPS system consists of three segments: (1) The space segment: the GPS satellites themselves; (2) The control system, operated by the U. S. military; and (3) The user segment, which includes both military and civilian users and their GPS equipment.

12.2.1 The space segment

The space segment is composed of satellites (See Fig. 12-1). There are now at least 24 satellites orbiting the earth. The satellites are in almost circular orbits, at a height of 20200 km. The six orbital planes are equally spaced (See Fig. 12-2), there are four satellites in each of 6 orbital planes. Each plan is inclined at 55° to the equator. The orbital period is 12 hours, meaning that each satellite completes two full orbits each 24-hour day. These 24 satellites make up a full GPS constellation. Individual satellites may appear for up to five hours above the horizon. The system has been designed so that at least four satellites will always be in view at least 15° above the horizon.

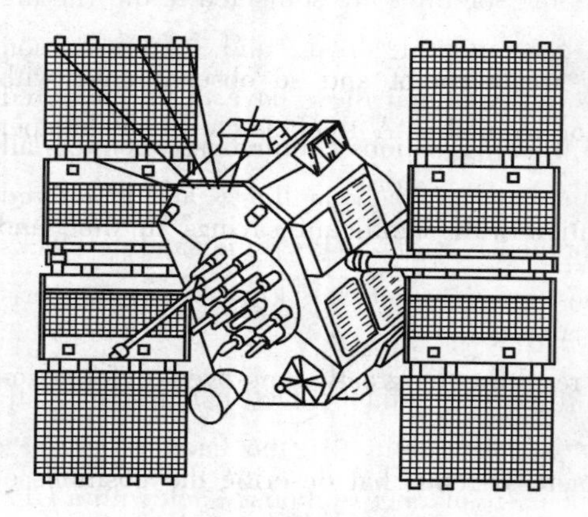

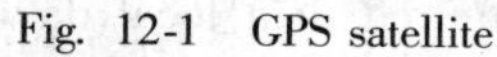

Fig. 12-1 GPS satellite

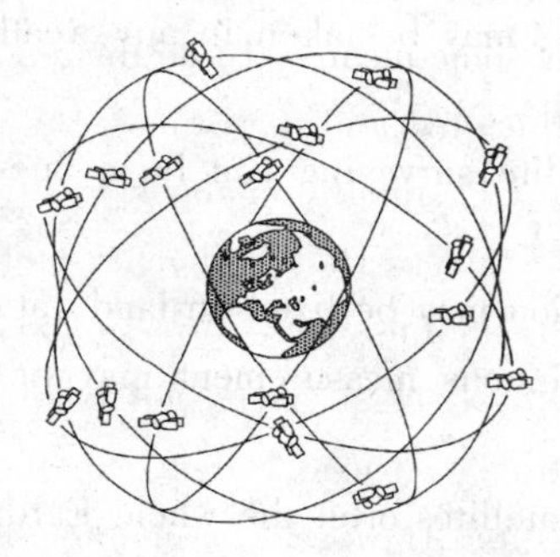

Fig. 12-2 The original planned GPS constellation

The design life of the satellites is 7. 5 years. The single-degree solar arrays cover a surface area of 7. 2 m^2 each when deployed. The surface of the solar panels is kept perpendicular to the direction of the sun. Power is retained during eclipse periods by three nickel-cadmium batteries. Reaction wheels control the orientation and position of the satellite in space. Antennae transmit the satellite's signals to the user. Each satellite carries two rubidium and two caesium atomic clocks to ensure precise timing.

12. 2. 2 The control segment

The control segment of the GPS system consists of one master control station, 3 Upload stations and 5 monitor stations (See Fig. 12-3).

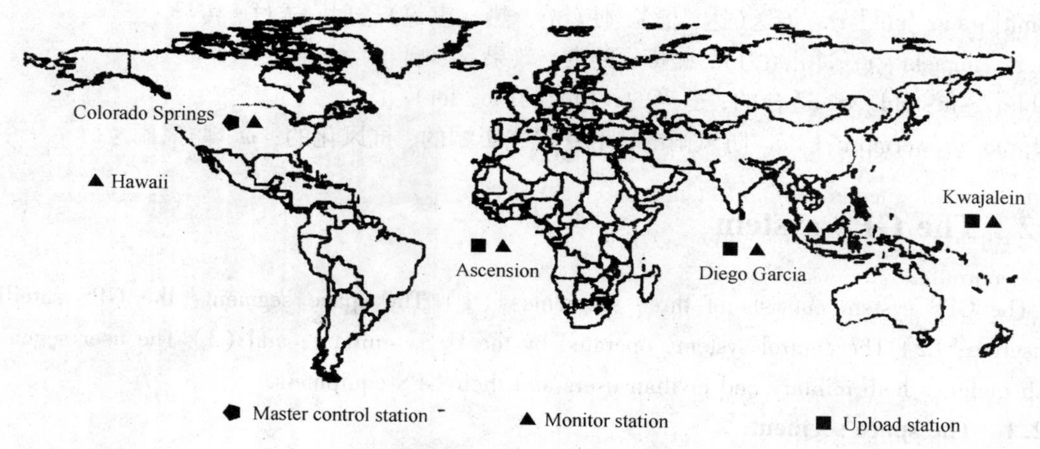

Fig. 12-3 The control segment of the GPS system

The *Master Control Station* (MCS) is located in the Consolidated Space Operations Center (CSOC) at Schriever Air Force Base near Colorado Springs in the USA. To be able to position accurately with GPS, the exact position of each satellite has to be known at all times and, despite their phenomenal accuracy, the satellite clocks do drift and they must be kept synchronized with GPS time as defined at the CSOC. As they orbit the Earth, the satellites are subjected to the varying gravitational attraction of the Earth, the attractions of the Sun and Moon, and solar radiation pressure. All of these cause the satellite orbits to change with time and these have to be measured and predicted by some means. To do this, a network of 13 tracking stations continuously monitors all of the GPS satellites in view at all times. As well as this, the clock in each satellite is also monitored and compared to GPS time to enable corrections to be computed in order to keep the satellite clocks in step with GPS time. The predicted satellite orbital positions (which are known as ephemeris predictions) and satellite clock corrections computed at the MCS .

Monitor stations set at Ascension Island (in the middle of the Atlantic Ocean between South America and Africa), Diego Garcia (in the Indian Ocean), Kwajalein (in the Pacific Ocean), Colorado Springs and Hawaii. The monitor stations are remote, unmanned stations, each with a GPS receiver, a clock, meteorological sensors, data processor and communications. Their functions are to observe the broadcast satellite navigation message and the satellite clock errors and drifts.

The data is automatically gathered and processed by each monitor station and is transmitted to the master control station. By comparing the data from the various monitor stations the master control station can compute the errors in the current navigation messages and satellite clocks, and so can compute updated navigation messages for future satellite transmission.

Upload stations set at Ascension, Diego Garcia and Kwajalein, USA. The Master Control Station periodically sends the corrected position and clock-timing data to the upload stations which then upload those data to each of the satellites. Finally, the satellites use that corrected information in their data transmissions down to the end user.

12.2.3 The user segment

The GPS receiver system consists of the antenna, the receiver itself, a command entry and display unit, and power supply. Fig. 12-4 illustrates a GPS system in which the receiver and antenna are contained in a compact unit mounted on the tripod. The command entry and display unit or controller is shown attached to one leg of the tripod, but maybe remounted and held by the operator for more convenient use. The power-supply pack is clamped in another leg of the tripod. The antenna receives the signal from the satellites and converse it into electrical energy usable in the receiver. The receiver, under

Fig. 12-4 GPS receiver system

control of a microprocessor, processes the signal converts it to a pseudorange, and computes approximate coordinates for the receiver. A data storage unit that is internal or has an output connection that allows interface with another computer also is found on most systems.

A GPS receiver usually has one or more channels, where a channel consists of hardware and software necessary to track a satellite's code and/or carrier phase measurement continuously. A receiver may have four or more dedicated channels, each tracking a different satellite simultaneously. In such a receiver, ranges and phase data from four more satellites can be obtained, allowing instantaneous determination of the receiver's position and receiver clock error.

单词与词组

constellation[kɔnstə 'leiʃən] *n.* 星座，星群
array[ə'rei] *n.* 陈列，一系列，数组，阵列
eclipse[i 'klips] *n.* 漆黑，晦暗，遮蔽
· reaction wheel 反冲式叶轮
antenna[æn 'tenə] *n.* 天线
rubidium[ru:'bidiəm] *n.* 铷（37号元素，符号Rb）
caesium['si:ziəm] *n.* 铯(55号元素，符号Cs)
· master control station 主控站
· upload station 注入站
· monitor station 监测站
consolidated[kən 'sɔlideitid] *a.* 加固的，整理过的，统一的
Schriever（地名）施里佛
Colorado Springs（地名）科罗拉多斯普林斯
synchronize['siŋkrə 'naiz] *vt.* 把(钟表)拨至相同的时间，校准，同步
drift[drift] *n.* 漂移，漂流
gravitation[ˌgrævi 'teiʃən] *n.* 重力，引力作用，万有引力，地心吸力
attraction [ə'trækʃən] *n.* 吸引，吸引力
ephemeris[i 'feməris] *n.* 星历表，历书
Diego Garcia（地名）迪戈加西亚
Kwajalein（地名）卡瓦加兰
Hawaii（地名）夏威夷
unmanned[ʌn 'mænd] *adj.* 无人操纵的，不载人的，遥控的
meteorological[ˌmi:tiərə 'lɔdʒikəl] *a.* 与气象学有关的，气象的
Ascension（地名）阿森松
periodical[ˌpiəri 'ɔdikəl] *a.* 周期的；定时的
pseudorange[psju:dəureindʒ] *n.* 伪距
channel['tʃænl] *n.* 河床，航道，磁道，频道，通道；管路，电路
dedicate['dedikeit] *vt.* 献给，献身；致力，专门用于

12.3 GPS signal structure

The relationship between the various GPS signals and the fundamental frequency is shown in Fig. 12-5.

Each satellite in the GPS constellation continuously broadcasts two electromagnetic signals that are in the L-band used for radio communication. The precise atomic clocks in the satellites have a fundamental clock rate or frequency f_0 of 10. 23 MHz and the two L-band frequencies assigned to GPS are derived from the atomic clocks as L1 = $154 \times f_0$ = 1575. 42 MHz and L2 = $120 \times f_0$ = 1227. 60 MHz. Both the L1 and L2 signals act as carrier waves。

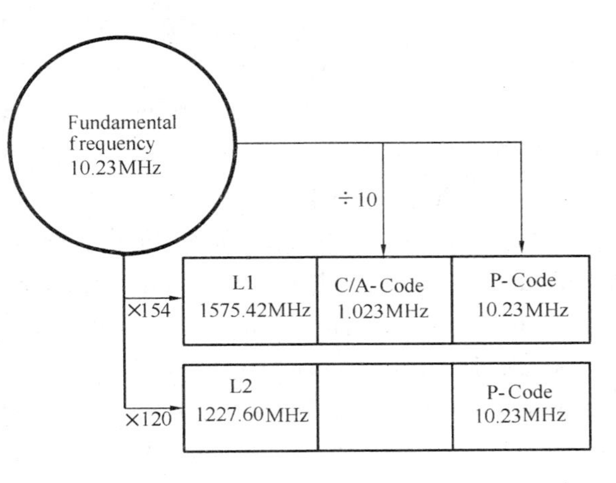

Fig. 12-5 GPS signal structure

Each satellite transmits a navigation message, which is a low rate of 50 bps data stream and modulated onto both the L1 and the L2 carriers. A complete navigation message consists of 25 frames of 1, 500 bits each, or 37, 500 bits in total (i. e. , a complete navigation message takes 750 seconds, or 12. 5 minutes, to be transmitted). Each of the frames is divided into 5 sub – frames of 300 bits each. The navigation message contains, along with other information, the coordinates of the GPS satellites as a function of time (broadcast ephemeris), the satellite clock correction parameters, the satellite health status, the satellite almanac, and atmospheric correction parameters.

As well as the navigation message, two binary codes are continuously modulated onto the L-band signals; these are called the coarse/acquisition or C/A code and the precise or P code. But, these are not the same as the data contained in the navigation message and are not uploaded to the satellites by the control segment, whereas they are continually broadcast by each satellite。

C/A code and P code comprised of zeros and ones, and each satellite transmits codes unique to that satellite. They are carefully structured, but the codes sound like random electronic "noise", they have been given the name psudo random noise (PRN). Although the two codes appear to be random, but they are generated by a mathematical algorithm, and they have some sequences that can be replicated. The C/A-code is modulated onto the L1 carrier, whereas the P-code is modulated on both L1 and L2. The C/A code on the L1 carrier at a rate of 1. 023 million bits per second (Mbps or a frequency of $f_0 \div 10 = 1.023$MHz), and this is repeated every millisecond. Because this gives a chip length (equivalent to 1 bit of information) of 293m, this means that measurement of distance on the C/A code is carried out using a wavelength of 293m.

The precision P-code is the principal code used for navigation. The chipping rate of the P-code is that of the fundamental frequency, that is, 10.23 MHz (frequency of $f_0 = 10.23$ MHz and chip length 29.3m). The P－code is a very long sequence-of binary digits that repeats itself after 266 days. It is also 10 times faster than the C/A-code. Multiplying the time it takes the P-code to repeat itself, 266 days, by its rate, 10.23 Mbps, tells us that the P-code is a stream of about 2.35×10^{14} chips! The 266-day-long code is divided into 38 segments; each is one week long. Of these, 32 segments are assigned to the various GPS satellites. That is, each satellite transmits a unique one-week segment of the P-code, which is initialized every Saturday/Sunday midnight crossing.

C/A-code is provided in the GPS Standard positioning Service (SPS), the level of service authorized for civilian users. The Precise Position Service (PPS), which provides access to both the C/A-code and the P-code, is designed primarily for military uses.

In the 1980s, the US Department of Defense made the system available for civilian use and implemented restrictive policies of Selective Availability (SA) and Anti-Spoofing (AS). SA was designed to degrade signal accuracy, it resulted in horizontal position accuracy of 100m. SA was permanently deactivated since May 2000. This policy change improved GPS accuracies in basic point positioning by a factor of 5 (e. g., down from 100m to about 10m to 20m). Another method designed to deny civilian users is called Anti-Spoofing (AS), in which the P code is encrypted and changed to a secret Y code. Differential positioning is used, AS and most natural and other errors can be eliminated.

单词与词组

· carrier wave 载波

· bps 即 bit per second 缩写，每秒传播比特数

almanac[ˈɔːlmənæk; ˈɔlmə, næk] *n.* 历书，年历，年鉴

malfunction[mælˈfʌŋkʃən] *n.* 失灵，机能失常，故障

· pseudo-random noise (PRN) 伪随机噪声

chip [tʃip] *n.* (集成)电路片，基片，芯片，码元(即 1bit)

exclusive[iksˈkluːsiv] *a.* 排他的，排外的，独家享有的，专有的，唯一的

encrypt [inˈkript] *vt.* 译成密码，编码

· anti-spoofing (AS) 反电子欺骗

12.4 Basic principle of position fixing

12.4.1 The basic idea

The idea behind GPS is rather simple. If the distances from a point of the Earth (a GPS receiver) to three GPS satellites are known along with the satellite locations, then the location of the point (or receiver) can be determined by simply applying the well-known concept of resection. How can we get the distances to the satellites as well as the satellite locations?

As mentioned before, each GPS satellite continuously transmits a microwave radio signal composed of two carriers, two codes, and a navigation message. When a GPS receiver is switched on, it will pick up the GPS signal through the receiver antenna. Once the receiver acquires the GPS signal, it will process it using its built-in processing software. The partial outcome of the signal processing consists of the distances to the GPS satellites through the digital codes (known as the pseudoranges) and the satellite coordinates through the navigation message.

Theoretically, only three distances to three simultaneously tracked satellites are needed. In this case, the receiver would be located at the intersection of three spheres; each has a radius of one receiver-satellite distance and is centered on that particular satellite (Fig. 12-6). From a practical point of view, however, a fourth satellite is needed to account for the receiver clock offset.

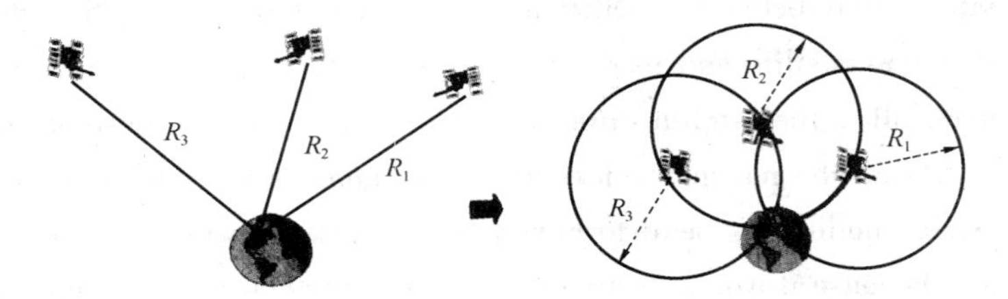

Fig. 12-6 The basic of GPS positioning

Position measurements generally fall into one of two categories: code measurement and carrier phase measurement. Civilian code measurement is restricted to the C/A code, which can provide accuracies only in the range of 10 ~ 15m when used in point positioning, and accuracies in the sub-meter to 5m range when used in various differential positioning techniques. P code measurements can apparently provide the military with much better accuracies. Point positioning is the technique that employs one GPS receiver to track satellite code signals so that it can directly determine the coordinates of the receiver station.

12.4.2 Code measurements

To locate a position, a GPS receiver determines distances to a number of different GPS satellites and then uses these to compute the coordinates of the receiver antenna. In GPS, there are two observables that can be used to determine distance pseudo-ranges and the carrier phase. Most GPS receivers are capable of determining pseudo-ranges by code ranging, and this provides low-accuracy, generally instantaneous, point positions with handheld and mapping-grade receivers. Carrier phase measurements are used in surveying for high-precision work using geodetic receivers.

Code ranging is the simplest form of GPS positioning and is carried out with a single receiver. To determine the distance to each satellite by code ranging, a receiver measures the time taken by the C/A code to travel from a satellite to the receiver and then calculates the distance or range R between the two as $R = c \times \Delta t$, where c is the signal velocity (speed of light) and Δt is the time taken for the codes to travel from satellite to receiver (called the propagation delay). There are some difficulties with these measurements because the process is one-way and the time at which the signal leaves the satellite must be determined as well as the time of arrival of the PRN codes at the receiver—both of these measurements are synchronized with GPS time, otherwise the transit time will be wrong and the distances will be incorrect.

To overcome these problems, the following technique has been developed for GPS code measurements. All the satellites use their on board atomic clocks to generate individual C/A codes and the relationship of these to GPS time is known. When a receiver locks onto a satellite, the incoming signal triggers the receiver to generate a C/A code identical to that transmitted by the satellite. The replica code generated by the receiver is then compared with the satellite code in a process known as autocorrelation, in which the receiver code is shifted in time until it is in phase, or correlated, with the satellite code. The amount by which the receiver-generated code is shifted is equal to the propagation delay between satellite and receiver. Multiplied by the speed of light, this gives the distance between satellite and receiver.

As already stated, all of the satellites are fitted with atomic clocks whose offsets from GPS time are precisely known through the navigation message. This enables the time at which the PRN codes were transmitted from a satellite to be determined almost exactly with respect to GPS time. For practical reasons, and because it would cost too much, it is not feasible to install atomic clocks in a GPS receiver. Instead, they have electronic clocks (oscillators) installed with accuracies much less than those of atomic clocks but still far better than everyday clocks. As a result, receiver clocks are not usually synchronized with GPS time and the receiver-generated replica codes will contain a chock error. Because of this, all distances (or ranges) determined by a receiver will be biased and are called pseudo-ranges—it is essential that the clock errors that cause these biases are removed from measurements.

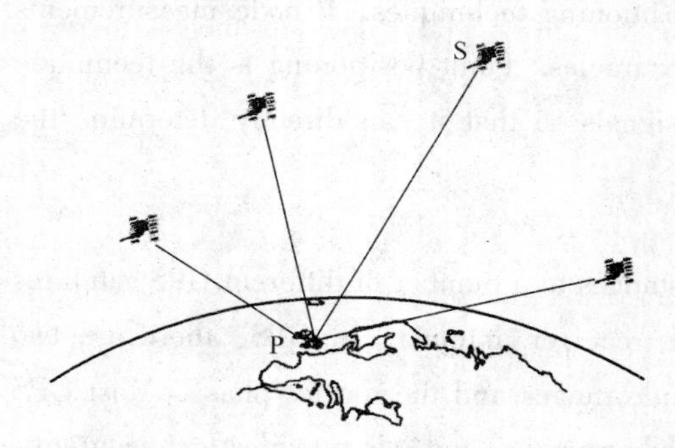

Fig. 12-7 Point position with GPS

As well as pseudo-range measurements, a receiver will also process the navigation code, and from this it computes the position of each satellite it tracks at the time the measurements are taken.

To compute position, the satellites are treated as control points of known coordinates to which distances are determined using a receiver located at a point whose position is unknown. In Fig. 12-7, the range front point P to any satellite S can be defined as

$$R_{SP} = \rho + c(dt - dT) \tag{12-1}$$

where

R_{SP} = the geometric range from satellite S to P,

ρ = the pseudorange measured from satellite S to P,

dt = the satellite clock offset from GPS time,

dT = the receiver clock offset from GPS time.

Since the satellite clock offset dt can be modeled using data contained in the navigation code, this equation reduces to $R_{SP} = \rho -$ cdT.

The geometric range R_{SP} is given by

$$R_{SP} = \sqrt{(X_S - X_P)^2 + (Y_S - Y_P)^2 + (Z_S - Z_P)^2} \tag{12-2}$$

in which X_S, Y_S and Z_S are the known coordinates of the satellite and X_P, Y_P and Z_P are the unknown coordinates of point P.

If observations are taken to four satellites simultaneously, four point positioning equations will be obtained as follows

$$\left.\begin{array}{ll}\text{For satellite 1} & \sqrt{(X_{S1}-X_P)^2+(Y_{S1}-Y_P)^2+(Z_{S1}-Z_P)^2}=\rho_1-cdT\\ \text{For satellite 2} & \sqrt{(X_{S2}-X_P)^2+(Y_{S2}-Y_P)^2+(Z_{S2}-Z_P)^2}=\rho_2-cdT\\ \text{For satellite 3} & \sqrt{(X_{S3}-X_P)^2+(Y_{S3}-Y_P)^2+(Z_{S3}-Z_P)^2}=\rho_3-cdT\\ \text{For satellite 4} & \sqrt{(X_{S4}-X_P)^2+(Y_{S4}-Y_P)^2+(Z_{S4}-Z_P)^2}=\rho_4-cdT\end{array}\right\} \quad (12\text{-}3)$$

These four equations can be solved to give the four unknowns: the receiver coordinates X_P, Y_P and Z_P and its clock offset dT.

As can be seen, it is necessary to track and measure to at least four satellites in any GPS survey. This does not, however, guarantee that it will be successful, and sometimes more are needed.

Code ranging is also known as the navigation solution and this fulfils the original aims of GPS. Because it is possible to determine the receiver clock error quite easily in this process, the clocks in GPS receivers need not be as accurate as atomic clocks and this makes the receivers much less expensive and suitable for field use. In any given navigation solution, the equations are solved many times using a computer and software installed in the receiver, the operator does not have to perform any calculations and merely reads the position displayed by the receiver in real time or records this electronically.

12.4.3 Carrier phase measurements

Another way of measuring the ranges to the satellites can be obtained through the carrier phases. GPS receivers measure the phase delay by comparing the phase of an incoming satellite signal with a similar signal generated by the receiver. These phase measurements are taken on the L1 and L2 carrier waves (not the modulated C/A and P codes). That is, the receiver removes the modulated PRN codes from the L1 and L2 signals using some technique. Therefore, the range to the satellites would simply be the sum of the total number of full carrier cycles plus fractional cycles at the receiver and the satellite, multiplied by the carrier wavelength (see Fig. 12-8). The ranges determined with the carrier phases are far more accurate than those obtained with the codes (i. e., the pseudoranges). Since the signal processing techniques are able to refine the observation resolution to approximately one percent of the carrier wavelength, the resulting potential precision with the carrier phases measurement is up to 1.9 mm (L1 signal has wavelength of 19cm) and 2.4mm (L2 signal of 24cm).

There is, however, one problem. The carriers are just pure sinusoidal waves, which means that all cycles look the same. Therefore, a GPS receiver has no means to differentiate one circle from another. In other words, the receiver, when it is switched on, cannot determine the total number of the complete cycles between the satellite and the receiver. It can only measure a fraction of a cycle

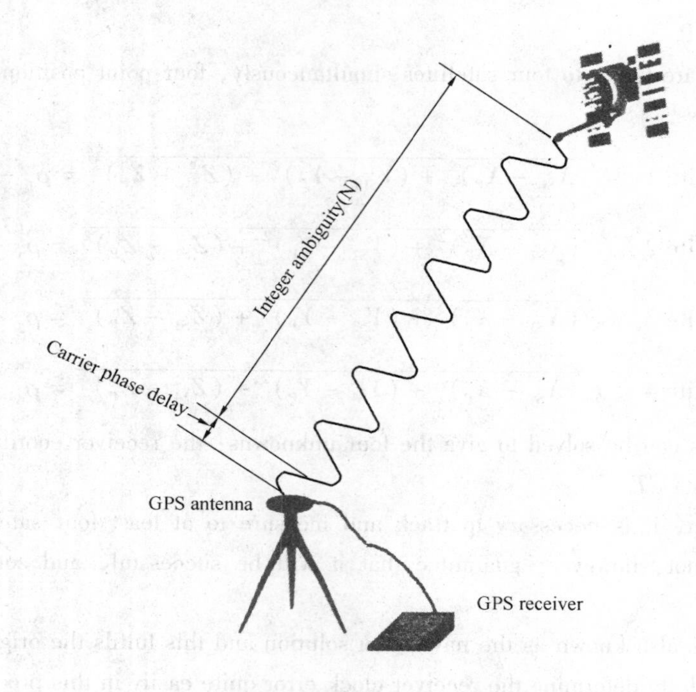

Fig. 12-8 Carrier-phase measurements

very accurately (less than 2mm), while the initial number of complete cycles remains unknown, or ambiguous. Fortunately, the receiver has the capability to keep track of the phase changes after being switched on. This means that the initial cycle ambiguity remains unchanged over time, as long as no signal loss (or cycle slips) occurs.

It is clear that if the initial cycle ambiguity parameters are resolved, accurate range measurements can be obtained, which lead to accurate position determination. This high-accuracy positioning can be routinely through the so-called relative (or differential) positioning techniques, either in real time or in the post processing mode. Unfortunately, this requires two GPS receivers simultaneously tracking the same satellites in view.

Many techniques have been developed for solving the carrier phase ambiguity both rapidly and reliably. One method is the receiver firstly to use code ranging to obtain an approximate position of the receiver, and then take carrier phase measurements. More about the various positioning techniques and the ways of resolving the ambiguity parameters are beyond the scope of this textbook.

单词与词组

· code measurement 代码测量

pseudorange[′su：dəurəndʒ] *n.* 伪距

· carrier phase measurement 载波相位测量

delay[di lei] *n.* 延迟，耽误，*vt.* 延期，推迟

sinusoidal [ˌsainə′sɔidəl] *a.* 正弦曲线的，正弦波的

ambiguous[ˌæm ′bigjuəs] *a.* 模棱两可的，多重性的，不确定性的

12.5 Relative positioning

GPS relative positioning, also called differential positioning, employs two (or more) GPS receivers simultaneously tracking the same satellites to determine their relative coordinates (Fig. 12-9). Of the two receivers, one is selected as a reference, or base, which remains stationary at a site with precisely known coordinates (i. e., pre-surveyed). The coordinates of the other receiver, known as the rover or remote receiver, are unknown. They are determined relative to the reference using measurements recorded simultaneously at the two receivers. The rover receiver may or may not be stationary, depending on the type of the GPS operation.

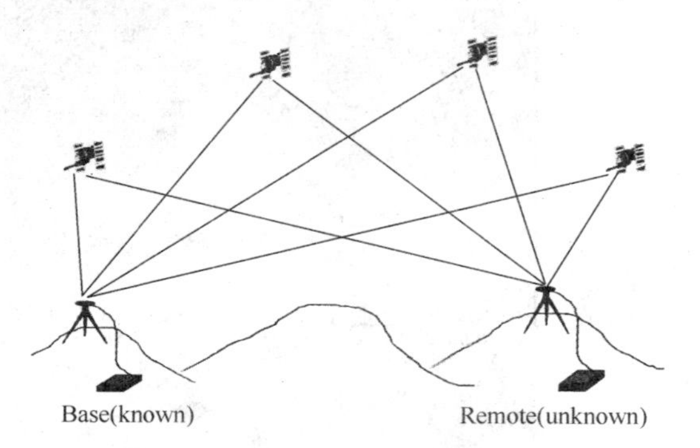

Fig. 12-9 Principle of relative positioning

A minimum of four common satellites is required for relative positioning. However, tracking more than four common satellites simultaneously would improve the precision of the GPS position solution. Carrier-phase and/or pseudorange measurements can be used in relative positioning. A variety of positioning techniques are used to provide positioning information in real time or at a later time (i. e. post-processing). Generally, GPS relative positioning provides a higher accuracy than that of autonomous positioning. Depending on whether the pseudorange or carrier-phase measurements are used in relative positioning, an accuracy level of a few meters to millimeters, respectively, can be obtained. This is mainly because the measurements of two (or more) receivers simultaneously tracking a particular satellite contain more or less the same errors and biases. The shorter the distance between the two receivers is, the more similar the errors. Therefore, if we take the difference between the measurements of the two receivers (hence the name differential positioning), common errors will be removed and those that are spatially correlated will be reduced depending on the distance between the reference receiver and the rover.

单词与词组

· differential positioning 差分定位

rover['rəuvə] *n.* 流浪者，徘徊者，流动站

12.6 Real-time kinematic GPS surveying (RTK GPS)

RTK is the abbreviation of Real Time Kinematic. RTK GPS surveying is a carrier phase-based relative positioning technique that employs two (or more) receivers simultaneously tracking the same satellites (Fig. 12-10).

In this method, the base receiver remains stationary over the known point and is attached to a

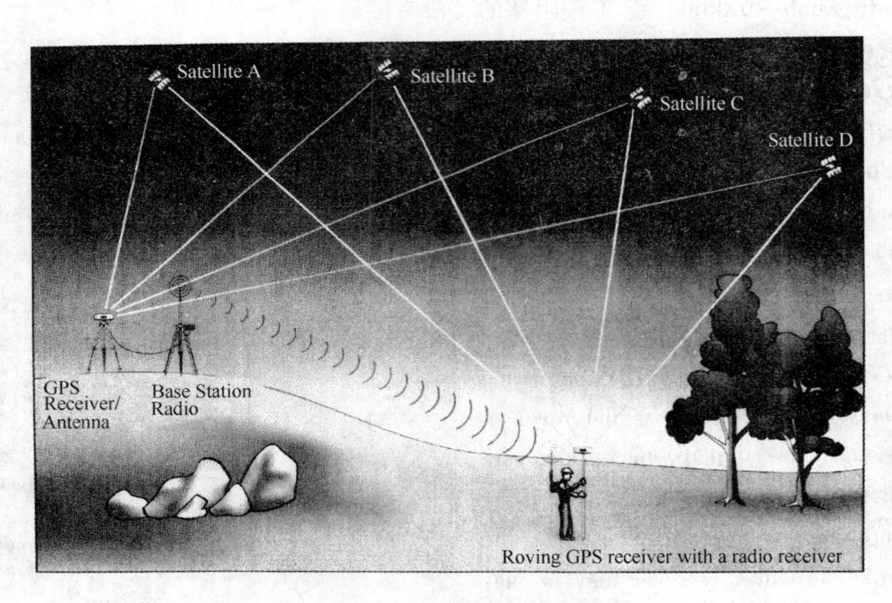

Fig. 12-10 RTK GPS surveying

radio transmitter (Fig. 12-10). The rover receiver is normally carried in a backpack (or mounted on a moving object) and is attached to a radio receiver. RTK requires a base station to measure the satellite signals, to process baseline corrections, and then to broadcast the corrections to any number of roving receivers via radio transmission.

The base station transmits code and carrier phase data to the roving receiver, which can use this data to help resolve ambiguities and to solve for change in coordinate differences between the base and the roving receivers.

The range of the radio transmission of the carrier phase data from the base to the roving receiver can be extended by booster radios to a distance of about 10km. Longer ranges require commercially licensed radios, as do even shorter ranges in some countries.

This method is suitable when (1) the survey involves a large number of unknown points located in the vicinity (i. e., within up to about 15 ~ 20km) of a known point; (2) the coordinates of the unknown points are required in real time; and (3) the propagation path between the two receivers is relatively unobstructed.

单词与词组

abbreviation[ə͵bri:vi ˈeiʃən] *n.* 省略，缩写，缩写词

kinematic[͵kaini ˈmætik] *a.* 运动学的

stationary[ˈsteiʃ(ə)nəri] *a.* 静止的，不动的，不变的，固定的

backpack [ˈbækpæk] *n.* （指登山者、步行者使用）背包

booster[ˈbu:stə] *n.* 增强器，放大器

licensed[ˈlaisnst] *a.* 得到许可的

vicinity[viˈsiniti] *n.* 附近，邻近，附近地区

unobstructed[ˌʌnəb ˈstrʌktid] *a.* 不被阻塞的，没有障碍的，畅通无阻的

12.7 Geodetic coordinate system

12.7.1 Concept of geodetic coordinate system

A coordinate system is defined as a set of rules for specifying the locations (also called coordinates) of points. This usually involves specifying an origin of the coordinates as well as a set of reference lines (called axes) with known orientation. Fig. 12-11 shows the case of a three-dimensional coordinate system that uses three reference axes (X, Y, and z) that intersect at the origin (C) of the coordinate system.

Coordinate systems may be classified as one-dimensional, two-dimensional, or three-dimensional coordinate systems, according to the number of coordinates required to identify the location of a point. For example, a one-dimensional coordinate system is needed to identify the height of a point above the sea surface.

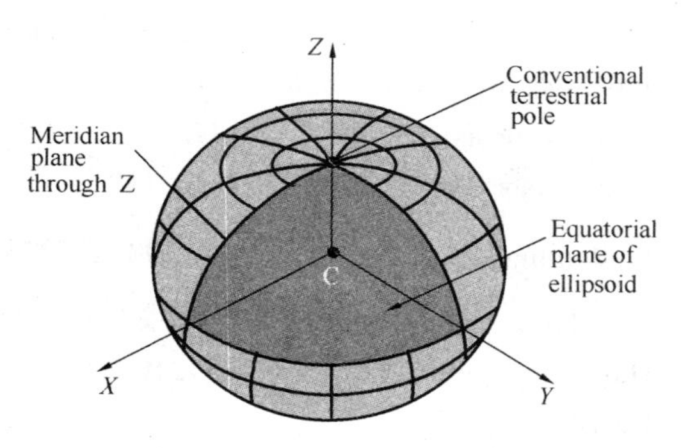

Fig. 12-11 Three dimensional coordinate system

Coordinate systems may also be classified according to the reference surface, the orientation of the axes, and the origin. In the case of a three-dimensional geodetic (also known as geographic) coordinate system, the reference surface is selected to be the ellipsoid. The orientation of the axes and the origin are specified by two planes: the meridian plane through the polar or z-axis (a meridian is a plane that passes through the north and south poles) and the equatorial plane of the ellipsoid (see Fig. 12-11 for details).

Of particular importance to GPS users is the three-dimensional geodetic coordinate system. In this system, the coordinates of a point are identified by the geodetic latitude (φ), the geodetic longitude (λ), and the height above the reference surface (h). Fig. 12-12 shows these parameters. Geodetic coordinates (φ, λ, and h) can be easily transformed to Cartesian coordinates (x, y, and z). To do this, the ellipsoidal parameters (a and f) must be known.

12.7.2 The WGS-84

The World Geodesy System of 1984 (WGS-84) is a three-dimensional, Earth-centered reference system. The origin of the WGS 84 system is the earth's center of mass. The z-axis pointing to the Conventional Terrestrial Pole (CTP) defined by the BIH 1984.0, the x-axis points to the intersection of the zero meridian plane of BIH 1984.0 and the CTP's equator, the y-axis and z, x axis form a right-handed coordinate system. Correspond to the WGS-84 geodetic coordinate system with a WGS-84 ellipsoid. The WGS-84 ellipsoid and related constant using recommended values at the 17th Congress of the International Association of Geodesy (IAG) and International Union of Geodesy and Geophysics(IUGG). This system is illustrated in Fig. 12-12.

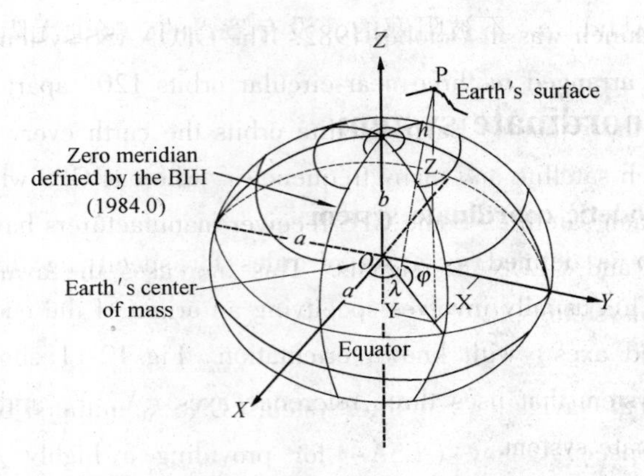

Fig. 12-12 The World Geodesy System of 1984

WGS-84 ellipsoid defining parameters:

Semi-major axis $a = 6378137.0\text{m}$

Semi-minor axis $b = 6356752.3142\text{m}$

Flattening $f = \frac{a-b}{a} = 1/298.257223563$

Angular velocity $\omega = 7.292115 \times 10^{-5}\ \text{rad/sec}$

Harmonic coefficient $C_{2,0} = -484.16685 \times 10^{-16}$

Gravity constant

$G_M = 398\ 600.5\ \text{m}^3/\text{sec}$

单词与词组

ellipsoid [iˈlipsɔid] *n.* 椭圆形，椭圆面，椭圆球

equatorial[ˌekwəˈtɔːriəl] *a.* 赤道的，赤道附近的

· World Geodesy System of 1984 (WGS-84) 1984 世界大地坐标系

· Conventional Terrestrial Pole (CTP) 协议地极

· Bureau International Heure (BIH) 国际时间局

· International Association of Geodesy (IAG) 国际大地测量协会

· International Union of Geodesy and Geophysics (IUGG) 国际大地测量与地球物理联合会

· semi-major axis 长半径轴

· semi-minor axis 短半径轴

flattening[ˈflætniŋ] *n.* 整平，扁率

· angular velocity 角速度

· harmonic coefficient 谐波系数

· gravity constant 重力常数

12.8 Other satellite navigation systems

12.8.1 GLONASS

The Russian government has developed a system, similar to GPS, called GLONASS. The first

GLONASS satellite launch was in October 1982. The GLONASS system is consists of 24 operational satellites. These are arranged in three near-circular orbits 120° apart at an altitude of 19 100 km inclined at 64.8° to the equator. Each satellite orbits the earth every 11h 15min. GLONASS uses the same code for each satellite and many frequencies, whereas GPS which uses two frequencies and a different code for each satellite. Some GPS receiver manufacturers have incorporated the capability to receive both GPS and GLONASS signals. This increases the availability of satellites and the integrity of combined system.

12.8.2 GALILEO

Galileo is a Global Navigation Satellite System (GNSS) initiated by the European Union (EU) and the European Space Agency (ESA) for providing a highly accurate, guaranteed global positioning service under civilian control. As an independent navigation system, Galileo will meanwhile be interoperable with the two other global satellite navigation systems, GPS and GLONASS. A user will be able to position with the same receiver from any of the satellites in any combination. Galileo will guarantee availability of service with higher accuracy.

The first Galileo satellite was launched in December 2005. The Galileo constellation consists of 30 Medium Earth Orbit (MEO) satellites in three orbital planes with nine equally spaced operational satellites in each plane plus one inactive spare satellite. The orbital planes are inclined 56°. Each Galileo satellite is in a nearly circular orbit with semi-major axis of 29600 km and a period of about 14 hours.

12.8.3 Beidou navigation satellite system (BNSS)

The Beidou navigation satellite system is independently developed by China, which is regional at present and will gradually extend to independent operation of the global satellite positioning system. The Beidou navigation satellite system is also known as BNSS. The first generation of the Beidou system consists of three satellites (two operational and one backup) placed in geosynchronous orbits as the BeiDou navigation test system. The two operational satellites were launched in October and December 2000, respectively, while the backup satellite was launched in May 2003.

Since 2007 April, China has launched a series of Beidou-2 navigation satellite. As of 2012 October, has launched 16 Beidou-2 navigation satellites and built a second generation of the Beidou navigation system. The Beidou navigation system provides navigation and communication services to both China and its surrounding areas. The Beidou system has a ground segment consisting of a central control station and ground correction stations. The user segment consists of a receiving/transmitting terminal.

China is planning to build the global satellite positioning and navigarion system around 2020, which contains five geosynchronous satellites and 30 non-geosynchronous orbit satellites .

单词与词组

Russian['rʌʃən] *n.* 俄国，俄罗斯人，俄语

GLONASS 格罗尼斯卫星系统

incorporate['inkɔpərit] *vt.* (使)合并，并入，合编

Galileo 伽利略卫星系统

· European Union（EU）欧盟

· European Space Agency（ESA）欧洲宇航局

meanwhile['miːnˌhwail] *adv.* 同时，其间

interoperable[ˌintər 'ɔpərəbl] *a.* 能共同操作的，能共同使用的

inactive[in 'æktiv] *a.* 不活动的，停业的，非现役的，备用的

· Medium Earth Orbit（MEO）中圆地球轨道

Beidou Navigation System 北斗卫星导航系统（简称北斗系统）

backup ['bækˌʌp] *n.* 备份，备份文件，后备

geosynchronous[ˌdʒiːəu 'siŋkrənəs] *a.* 与地球的相对位置不变的，与地球同步的

第二部分　课文参考译文

第1章　绪　　论

1.1　有关的定义

1.1.1　测量学与工程测量

测量学可定义为研究测定地上或地下天然的和人工的地物、地貌特征在三维空间位置的科学。这些地物、地貌可以模拟形式表示为地形图、平面图或曲线图，或以数字形式表示，如数字地面模型(DGM)。

工程测量定义为土木建筑和其他工程项目的定位、设计、施工、管理和经营所做的测量规划与实施的各种工作。这些工作包括各项工程的开始之前、施工过程和完工之后所实施的一切测量业务。

(1)在任何工程开始之前，都需要大比例的地形图或平面图作为设计的基础。因此要生产工程项目所在区域的最新的地形图。这些图的比例尺通常较其他方式的土地测量制作的地形图的比例尺要大。土建工程项目一般要用比例尺1∶500，1∶1000，1∶2000与1∶5000。城镇规划一般要用比例尺1∶5000和1∶10000。线路初测图一般要用比例1∶1000至1∶10000。绘制横断面图和纵断面图时放大垂直比例尺。

(2)新建筑物拟建的位置必须标定在实地，包括其平面位置和高程，通常把这种操作称为放样。测量员必须确保建筑物在实地相对位置和绝对位置的正确性。在土地测量和地籍测量中，经常需要计算土地的面积与体积。

(3)建筑工程完成之后要进行竣工测量，以便提供建成后实际工程容貌的位置与尺寸。竣工测量不仅提供工程建成后的记录，而且也成为今后保养管理、扩建与新建必须保存的非常重要文件。

(4)建立永久性控制点，以此监测建筑物，例如大坝与桥梁日后的位移。

工程测量是测绘学中最重要专业领域之一。

1.1.2　测绘学

现代测绘学英文术语GEOMATICS，它是如何构成的？它是由GEODESY(测地学)+GEOINFORMATICS(地球空间信息学)=GEOMATICS，或者说由GEO-(GEOSCIENCE地球科学)和-MATICS(INFORMATICS信息学)合成的。GEOMATICS这个术语第一次是在加拿大作为学术的学科出现的。近几年该术语已由许多高校在全世界广泛推介，其做法多半是通过对以前称为大地测量学或测量学进行改名以及通过增加许多计算机科学和GIS方向的课程。现在，这个术语涵盖了传统定义的测量学。随着新技术的发展，对不同类型有关空间信息不断增长的需求，尤其在测量和环境的监测方面，测量学重要性在逐步提高。

测绘学（GEOMATICS）是有关地理信息（地球空间数据）的收集、存储、处理、管理、分析和显示的一门科学与技术。它集成下列具体学科与技术：包括测量与制图、大地测量学、卫星定位测量、摄影测量、遥感、地理信息系统（GIS）、土地管理、计算机系统、环境可视化和计算机图形。

图 1-1　GIS 分层

测绘学不仅涵盖了传统测量员的工作，而且也反映了测量员在数据管理中角色的转变。这种情况的出现是由于测量学的发展，采用数字技术相对更加容易对大量的空间数据进行采集、处理和显示。这反过来又对各种数据源如地质、地理、水文、林业、交 通、政府和人力资源产生了一个巨大的需求。所有这些数据的收集和处理由计算机的地理信息系统（GIS）来进行 。这些是集成测量员提供空间数据的数据库，具有环境、地理和社会等不同的信息层（见图 1-1）。根据最终用户的需求，对这些图层进行组合，处理和以任意格式进行显示。毫无疑问，GIS 的最重要的部分是空间数据，它是其他所有的信息的基础，并且已为测量学提供了巨大的发展领域。

1.2　测量的分类

测量可以根据是否要考虑地球的曲率而分为两大类：

平面测量　在平面测量中，地球表面的弯曲特性不加考虑。地球表面上的线段被认为是直线。位于地球面上的三个点相连结形成平面三角形。因此，所有的铅垂线被看做是相互平行的。平面测量的方法适用于小范围测量。

大地测量　这种测量是适合于大区域和长距离。测量中必须考虑地球曲率。它用在测定精密的基本点的点位，以便建立对其他测量的控制。在大地测量中，站点通常相隔较远，大地测量较平面测量要求采用更精密的测量仪器与测量方法。

根据测量的不同目的，测量可分为以下几种类型：

控制测量　是建立水平和高程埋石点的控制网，这个控制网可作为其他测量的基准框架。小地区的控制测量可以用平面测量的方法，但通常采用大地测量的方法。

地形测量　是用于制作表示地面上天然和人工地物平面位置与地面点高程的地图。

施工测量　是为土木建筑工程项目提供建筑物的平面位置及高程，通常称为工程测量。

房地产测量　是确定地块轮廓角点、边界和地块面积。所以也称为土地测量，地籍测量和边界测量。

路线测量　是关于公路、铁路、管道、电力线、隧道以及不闭合至起始点的其他工程项目的测量。

水文测量　是制作岸线、湖底、河流、水库以及其他较大水体图。有时候把地形测量与水文测量合称地形制图测量。

航空测量　是从飞行器上进行的测量，它对地面以带状重叠方式进行摄影。这种测量覆盖广大地区，也称为航空摄影测量。这种方法成本很高。但对大的工程，地面测量可能很困难或不可能时建议采用此法。

1.3　测量的参考面

测量学是论述点位的测定，无论是控制点还是地形碎部点，因而需要某种格式的参考系。实际进行测量的地球物理表面，并非数学上可定义的表面，因此不能作为计算点位的参考基准面。我们发现水准面是合适的，因为多数仪器的竖直轴是借助于水准气泡以重力方向来安置的。这样的水准面都垂直于每一点的重力方向。在地球上，这种类型的自然面就是海洋面。如果没有外界的影响，例如潮汐、海浪、风力等等，则海平面处于平均的位置，这样的海面对整个地球而言是唯一的，可以认为地球的数学形状，并称为*大地水准面*(geoid 依据希腊“地球”一词)。

实际上，在地球物理表面上测量的点，开始计算时经常要归算到大地水准面上求得相应点位，即沿其重力方向投影到大地水准面。

根据重力而形成海平面。由于地球内部物质分布不均造成重力不规则，致使大地水准面不规则，因此不能用数学方法确定点位。适合于大地水准面的最简单的可定义的数学图形是*椭球体*，它是椭圆绕其短轴旋转而成的。根据 1967 年国际协议，赤道长半径设为 6378160m，旋转轴相对于赤道轴缩短系数为 1: 298. 25。这意味着旋转轴短于赤道轴 3%。旋转椭球面被用于某一国家作为测图系统的面，该旋转椭球面被称为*参考椭球面*。

图 1-2 表示这几个面之间的关系。

大多数工程测量是在有限范围内进行的，在这种情况下，基准面可以采取大地水准面的切平面，并可应用平面测量的原则。换句话说，即地球曲率可以不考虑，所以地球物理表面上各点可以正射投影到平面上。如果地面面积小于 $10km^2$，把地面假设为平的而无弯曲是允许的。当球面上大约 20km 弧长与相应的弦长之差只不过 8mm。

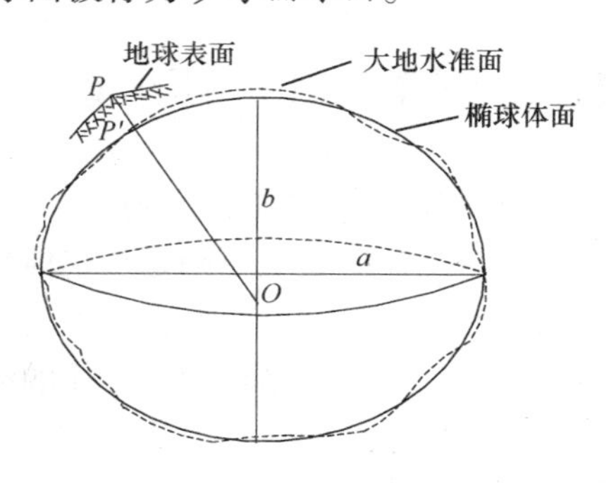

图 1-2　大地水准面、椭球面及物理表面

上述把地球表面假设为平的，对某些定位应用可以接受，但对求点的高程是不允许的。因为距切点 1km 的切平面与大地水准面的偏差约 80mm，10km 时偏差达 8m。因此，理论上无论如何对求高程都要考虑地球曲率，实际上通常使用平均海水面(MSL)。

1.4　基本的观测量

图 1-3 明显地表示定位 A、B、C 必要的基本观测量，它们的正射投影是 A′、B′、C′。假设已知 AB 的方向，那么测量斜距 AB 及 A 至 B 的垂直角，就可确定 B 相对于 A 的点位。观测 A 至 B 的垂直角，即可把斜距 AB 换算为相应的水平距 A′B′以便绘图。用类似的方法可

以确定 C 点，但此时还须在 A 点测量 B 至 C 的水平角 B′A′C′。定义 3 个点的相对高程的垂直距离也可以由斜距与垂直角求得，或者用针对于具体的参考基准面的直接水准测量求得。

基本的平面测量包含上述 5 个测量值，如图 1-4 所示，即 AB 斜距，AA′平距，A′B 垂直距离，BAA′垂直角 α，A′AC 水平角 θ。

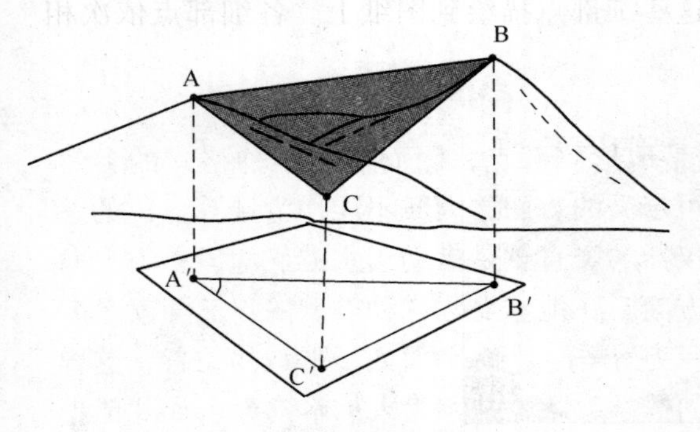

图 1-3　地面点正射投到水平面

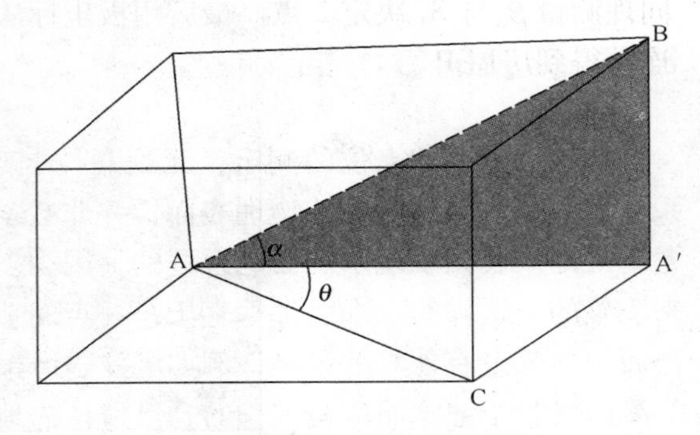

图 1-4　基本的观测量

从上述明显看出，在平面测量中所需量测的量是角度与距离。然而，现代技术已完全实现了这种简单数据的采集与处理。目前，用光学经纬仪或电子经纬仪测角精度达到秒级，电磁波测距仪(EDM)测距达到几千米，精度几毫米取决于测量的距离。激光和指北的陀螺仪成为隧道测量中实际的标准设备。轨道运行的卫星也已用于海岸测量。航测及地面摄影测量技术不断改进以及扫描设备能采集海量数据，因而成为重要的测量工具。最后，数据记录器以及电脑在处理数据与野外数据自动绘图中能采用最完善的程序。

1.5　测量工作的两个重要原则

1.5.1　测量工作的组织原则

测量工作的组织原则是从整体到局部，从控制测量到碎部测量，从高级到低级。第一层含义是对测量整体布局而言，首先确定对整个测区要采用什么方案，然后安排局部地区该怎么做。第二层含义是对测量工作的程序而言，先做控制测量，后做碎部测量。第三层含义是对测量精度来说的，首先执行高精度测量，然后执行低精度测量，即首先建立高等级的控制网，然后建立低等级的控制网。

遵循上述这个原则，可使测量结果避免误差的积累，确保测区内观测精度的一致性。由于测区可分几个部分同时测量，野外工作效率能增加好几倍。

1. 控制测量

当测量工作着手开始时，测量员应在测区内选择有限数量关系到全局意义的点，使用高精度的设备和方法建立的这些点称为控制点。控制点组成的几何图形称为控制网。建立控制点以及必要的观测与计算的专业工作称为控制测量。例如，图 1-5 中，选 A、B、C、D、E、F 各点作为控制点。使用高精度的仪器，测量控制点之间的距离以及相邻两条边之间水平夹角，最后计算出各控制点的坐标。为了求点的高程还必须测量各控制点彼此之间的高差。如果 A 点的高程为已知，就可求出其他控制点的高程。

2. 细部测量

细部测量就是测量所有地物与地貌特征点的位置，例如公路、铁路、桥梁、建筑物、山脉与河流等等。然后把这些地物地貌按比例画在图上。在图1-5的例子中，测量房屋P，就必须测定房屋的细部点1、2等点。为此，在A点测量水平夹角β_1与边长S_1即可决定1点；同理测量β_2与S_2决定2点。最后用极坐标法把这些细部点描绘到图纸上。各细部点依次相连结得到房屋P。

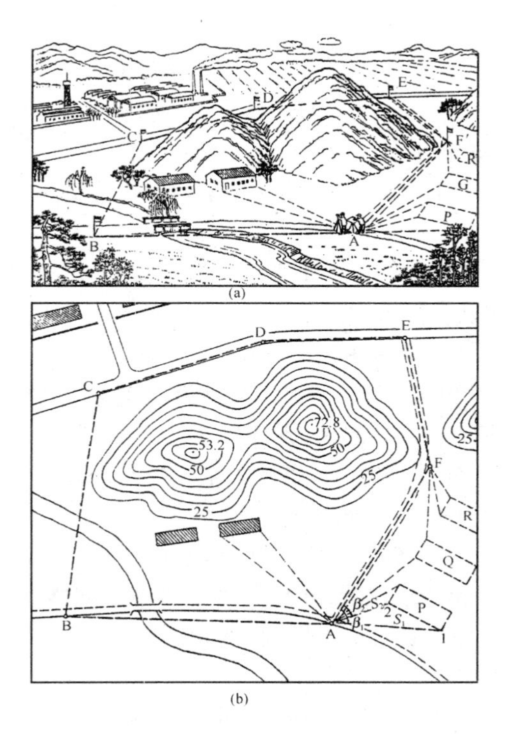

图1-5　某区域控制测量与细部测量的示意图

1.5.2　测量工作的实施原则

测量工作中不可避免地会出现各种误差或错误，为了避免错误与减少误差必须采取下列的实施原则：

(1)重复观测　许多未知量必须观测几次以避免错误的发生并且提高精度。例如测量角度或测量距离若干次，然后计算平均值。

(2)多余观测　多余观测就是超过绝对必要的观测，也可称为额外的观测。例如，假设我们关注于平面三角形的形状，如果测量三角形的2个角，则三角形的形状就能唯一地确定。然而现在测量三角形的所有3个角，那么就会产生一个多余观测。在数据处理时，应用最小二乘平差法去解算未知的参数。实施多余观测即可检查观测，又可提高最后成果的精度。

(3)步步检核　众所周知，测量工作的每个项目包含若干个步骤，只有在第一步结束之后才允许做第二步。每次测量必须立即检查。第一步工作成果未作检查或检查未达到全部要求，绝对不允许做第二步工作，这就是所谓步步检核。

测量工作包含有大量野外工作，遵循上述的实施原则非常重要。

第2章 误差理论

2.1 测量误差

2.1.1 误差的定义、来源与分类

所有的观测无论你如何仔细去做，都将包含误差。误差是某量的观测值与其真值之差，所以误差可定义为

$$误差 = 观测值 - 真值 \tag{2-1}$$

而误差的反号就是改正数，可以定义为

$$改正数 = 真值 - 观测值 \tag{2-2}$$

因为观测值的真值无法得知，因而误差的真实大小总是不可知的。即使你使用最尖端的设备，其观测值也仅是某量的估计值。这是因为仪器，以及使用仪器的人并非完美无缺，周围环境条件的变化不可能完全预测。然而，随着仪器的发展，外界环境条件的改善，观测者能力的提高，观测值将趋近于真值，但不可能绝对精确。

在测量与其他量测中都会产生各种误差，其误差分类如下：

(1) 仪器误差　这种误差是由于仪器结构不完善，仪器的调整或个别部件的位移而引起的。

(2) 环境误差　这种误差是由于周围环境变化而引起的。例如大气压、温度、风、折光、重力场和磁场等等。

(3) 人为误差　这种误差是由于人的视力、触觉和听力的局限性引起的。例如显微镜读数能力、气泡居中鉴别力等等。

在测量与其他量测中都能产生各种误差，其误差分类如下：

粗差　通常称为错误或过失，它一般比其他误差要大得多。在建筑和其他施工场地，错误通常是由于工程人员缺乏经验或测量员不熟悉仪器或使用方法不当造成的。因此可以说，粗差是由于作业员缺乏经验和粗心大意造成的，很多实例说明了这一点。一般的错误包括：读错水准标尺，带尺刻度不准确，在野外记录手簿写错，颠倒数字。例如 28. 342 写成 28. 432。在测量或放样过程中，如果没有检测到粗差，那就会导致工地上出现严重的问题。

系统误差　是遵循某种数学规律的误差。在相同条件下，一系列重复测量值具有相同大小与符号。换言之，如果仪器、观测者、外界条件没有改变，观测值存在的系统误差也不会改变。如果测量的条件改变，系统误差大小也随之改变。如果对系统误差导出适当的数学模型，那么，通过使用加改正数的方法可以消除系统误差。例如，在钢尺丈量中，不同因素的影响，如温度和拉力的影响可以通过应用简单的数学公式加以消除(每个公式对应某个因素的数学模型)。

消除系统误差的另一种方法是，检校测量仪器并确定误差大小，允许对随后的观测施加改正值。有必要经常对全站仪进行电子检测，以便测得仪器系统误差的大小。全站仪自动施

加了检测值，在随后的测量中加以改正。用全站仪进行距离测量中加棱镜常数改正就是仪器加系统误差改正的一个很好实例。如果忽略了棱镜常数或施加改正不正确，那么在所有的读数中都会存在某个误差。

选择观测的方法也可以消除系统误差的影响。例如，在角度测量中，采用盘左与盘右观测取平均数可以消除水平角与垂直角中的视准轴误差。在水准测量中，采用前后视距相等可以从高差中消除水准仪的视准轴误差。

随机误差(也称偶然误差)是消除了粗差和系统误差之后剩下的那些误差。它们是由于人的感官的局限、测量仪器的不完善以及外界环境变化的不可控制引起的。随机误差一般很小，但在测量中不可避免，必须按或然率法则进行处理。

2.1.2 偶然误差的特性

图2-1代表真误差Δ_i分布情况，它取自于同一个角度的160个观测值。误差按大小分组，即Δ_i按1″分组，将误差显示在横坐标轴上。每组画出矩形，其高与各组内误差数量成正比。可以看出：形成的直方图显示误差Δ出现的概率是误差大小的函数。

从上述可以看出：对某个量大量重复观测值的偶然误差(Δ)符合正态或高斯分布。误差大小对应概率的关系图是趋于光滑的钟形特征的曲线(图2-1)，这种曲线就是所谓正态误差分布曲线。正态误差分布的方程是

$$y = \frac{1}{\sigma\sqrt{2\pi}} e^{-\frac{\Delta^2}{2\sigma^2}} \tag{2-3}$$

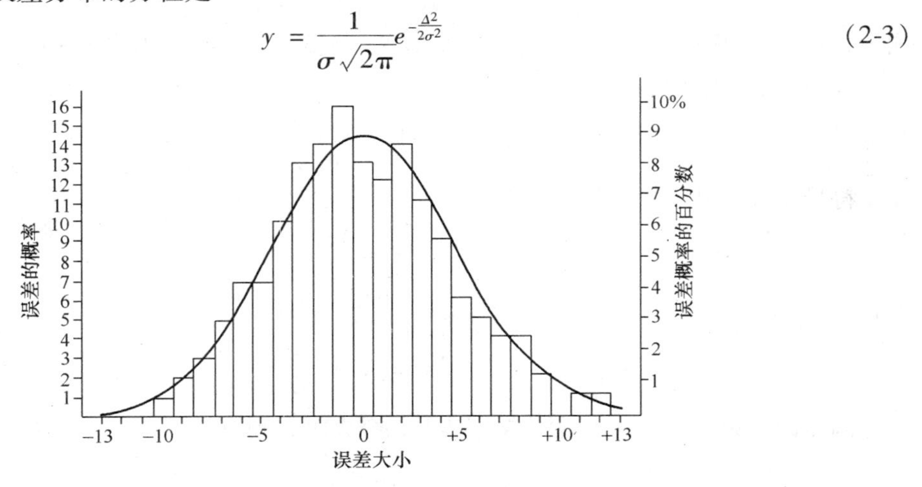

图2-1 正态分布曲线图

式中，y为曲线上某点的纵坐标，即偶然误差Δ发生的概率；σ称为某一组观测值的标准偏差或标准差；e为自然对数的底(2.718)。标准差的平方称为方差(σ^2)。

偶然误差或然率的一般规律如下：

(1)偶然误差不会超过某一数值。

(2)正负偶然误差出现概率相等。因为正态分布曲线是对称的，同样大小的正负偶然误差出现的概率是相等的。

(3)数量级小的误差出现机率大于数量级大的误差。

(4)当样本数趋于无穷大时，随机变量的算术平均值趋于零。

在测量中，真差Δ可以写为观测值(l_i)减真值(X)，即$\Delta_i = l_i - X$。消除系统误差之后的真误差就是偶然误差。当观测次数非常大，则有

$$\lim_{n\to\infty}\frac{\Delta_1+\Delta_2+\cdots+\Delta_n}{n}=\lim_{n\to\infty}\frac{[\Delta]}{n}=0$$

方差是偶然误差平方和的平均值，即

$$\sigma^2=\lim_{n\to\infty}\frac{\Delta_1^2+\Delta_2^2+\cdots+\Delta_n^2}{n}=\lim_{n\to\infty}\frac{[\Delta^2]}{n}=0$$

因此，标准差 σ 为

$$\sigma=\lim_{n\to\infty}\sqrt{\frac{[\Delta\Delta]}{n}} \tag{2-4}$$

2.1.3　最或然值(MPV)

最或然值(MPV)是从一系列观测值取得最接近于真值的值。如果某量等精度观测 n 次，其结果为 $l_1, l_2, \cdots, l_n$，相应的真误差为 $\Delta_1, \Delta_2, \cdots, \Delta_n$，设该量的真值为 X，则有

$$\left.\begin{aligned}\Delta_1&=l_1-X\\\Delta_2&=l_2-X\\&\vdots\\\Delta_n&=l_n-X\end{aligned}\right\} \tag{2-5}$$

方程(2-5)各式相加得

$$[\Delta]=[l]-nX \tag{2-6}$$

式中，符号[ ]表示“总和”，$[\Delta]=\Delta_1+\Delta_2+\cdots+\Delta_n$，$[l]=l_1+l_2+\cdots+l_n$，$n$ 为观测数。

(2-6)式两边除以 n 得

$$\frac{[\Delta]}{n}=\frac{[l]}{n}-X \tag{2-7}$$

使用符号 x 表示算术平均值，即

$$x=\frac{[l]}{n}$$

上式代入(2-7)式得

$$x=X+\frac{[\Delta]}{n}$$

根据偶然误差第 4 特性，当观测数 n 趋于无穷大时，则有

$$\lim_{n\to\infty}\frac{[\Delta]}{n}=0$$

$\therefore$

$$\lim_{n\to\infty}x=X \tag{2-8}$$

因此得结论，算术平均值就是同精度观测值的最或然值(MPV)，MPV 最接近于真值。

2.2　精度指标

2.2.1　标准差

标准差公式在 2.1.3 节的公式(2-4)已提到，即

$$\sigma=\lim_{n\to\infty}\sqrt{\frac{[\Delta\Delta]}{n}}$$

式中，σ 是特大样本的标准差，n 是特大样本的数量。但是，实际上观测的数量 n 总是有限

的数量。对于同精度的一组观测值，它们的标准差(也可称为均方差 m)可以写为

$$m = \pm\sqrt{\frac{[\Delta\Delta]}{n}} \tag{2-9}$$

式(2-5)中，用算术平均值 x 取代真值 X，则有

$$\left.\begin{aligned} v_1 &= l_1 - x \\ v_2 &= l_2 - x \\ &\vdots \\ v_n &= l_n - x \end{aligned}\right\} \tag{2-10}$$

式中 v_i 是每个观测值的残差，即每个观测值与算术平均值 x 的偏差。

(2-5)式减(2-10)式得

$$\left.\begin{aligned} \Delta_1 &= v_1 + (x - X) \\ \Delta_2 &= v_2 + (x - X) \\ &\vdots \\ \Delta_n &= v_n + (x - X) \end{aligned}\right\} \tag{2-11}$$

(2-11)式两边分别求和，并顾及$[v]=0$ 得

$$[\Delta] = n(x - X)$$

$$x - X = \frac{[\Delta]}{n}$$

(2-11)式两边平方，然后相加得

$$[\Delta\Delta] = [vv] + 2[v](x - X) + n(x - X)^2 \tag{2-12}$$

式中

$$(x - X)^2 = \frac{[\Delta]^2}{n^2} = \frac{\Delta_1^2 + \Delta_2^2 + \cdots + \Delta_n^2}{n^2} + \frac{2(\Delta_1\Delta_2 + \Delta_1\Delta_3 + \cdots + \Delta_{n-1}\Delta_n)}{n^2}$$

根据偶然误差的第 4 特性，上列方程的第 2 项趋于零，即

$$\lim_{n\to\infty}\frac{2(\Delta_1\Delta_2 + \Delta_1\Delta_3 + \cdots + \Delta_{n-1}\Delta_n)}{n^2} = 0$$

因此 $$(x - X)^2 = \frac{\Delta_1^2 + \Delta_2^2 + \cdots + \Delta_n^2}{n^2}$$

上列方程代入方程(2-12)得

$$[\Delta\Delta] = [vv] + n(x - X)^2 = [vv] + \frac{[\Delta\Delta]}{n}$$

$$\frac{[\Delta\Delta]}{n} = \frac{[vv]}{n - 1} \tag{2-13}$$

(2-13) 式代入 (2-9)式 得

$$m = \pm\sqrt{\frac{[vv]}{n - 1}} \tag{2-14}$$

公式(2-14)称为白塞尔公式。当真差不知道时，用此式计算一组同精度观测值的标准差。

2.2.2 相对精度

很多测量观测中经常要用到相对精度(也称相对误差)的概念。它就是用测量的精度与

测量值的本身的比值。例如，如果某段距离 D 测量值的误差为 m_D（即标准差），则相对精度表示为$1:\frac{D}{m_D}\left(例如\frac{1}{5000}\right)$。换言之，相对精度可表示为一个分子式，它的分子为 1，分母凑整到 100 单位。例如 $D=100\text{m}$，$m_D=\pm0.019\text{m}$，相对精度表示为：

$$K=\frac{|\pm m_D|}{D}=\frac{1}{\frac{D}{|\pm m_D|}}=\frac{1}{5263}\approx\frac{1}{5300}$$

这种方法用于定义卷尺丈量精度和定义导线测量精度（或准确度）。另一种方法是引用百万分之几，即 ppm$\left(即\frac{1}{1,000,000}\right)$，这也相当于每 1km 精度为 1mm。这两种表示法都用于全站仪的距离测量。在测量中，一旦已知观测值的精度，则观测值的相对精度就可计算，即开始测量之前预先规定观测值的精度，从而选择适当的仪器和方法，以便达到规定的相对精度。

2.2.3 置信度区间与容许误差

正态分布的一些有用特性：

（1）当 Δ 趋于 $\pm\infty$ 时，则 Δ 出现的 概率等于零。

（2）误差落在 Δ_1 与 Δ_2 区间的概率就是在误差 $\Delta=\Delta_1$ 与 $\Delta=\Delta_2$ 限定范围的曲线下面的面积。图 2-2 阴影的面积表示误差 Δ 落在平均值 $\mu\pm\sigma$（这里 μ 为 0）范围内的概率。偏离平均值头三个整数倍 σ 范围内的概率是：

$$P(-\sigma<\Delta<+\sigma)=0.6827$$
$$P(-2\sigma<\Delta<+2\sigma)=0.9546$$
$$P(-3\sigma<\Delta<+3\sigma)=0.9974$$

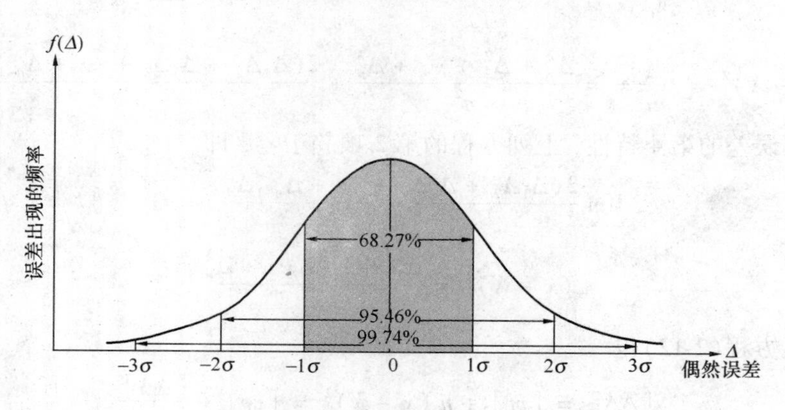

图 2-2 置信度区间

从上列式子可知，偶然误差 Δ 落在 -2σ 到 $+2\sigma$ 的概率为 95.5%，偶然误差 Δ 落在这个区间之外的概率仅为 5%；Δ 落在 -3σ 到 $+3\sigma$ 的概率为 99.7%，而 Δ 落在这个区间之外的概率仅为 0.3%。

因此，在实际测量中，容许误差规定为 2 倍或 3 倍 σ，即

$$\left.\begin{aligned}|\Delta_{容}|&=2|\sigma|\\|\Delta_{容}|&=3|\sigma|\end{aligned}\right\}\qquad(2\text{-}15)$$

如果实际误差大于 $|\Delta_{容}|$，测量数据应丢弃或重测。

2.3 精（密）度与准确度的定义

工程测量中，经常使用精（密）度与准确度的术语。在厂家的仪器说明书以及在工地测量员与工程师描述野外工作达到何种程度时都经常使用这两个术语。

精（密）度 是表示测量某个量的精密程度。换言之，就是一测量值对另一测量值的接近程度。如果对某个量测量几次，获得的数值彼此十分接近，则称该精（密）度是高的。

准确度 涉及到测量所获得数值的精准程度，它表示观测值接近于该量真值的程度。观测值离它的真值越远，它的准确度就越低。

精（密）度显示观测量的重复性且仅同随机误差有关。精度好的一系列观测值紧密地聚集在一起，它与样本的平均值 $\bar{x}$ 或最或然值 MPV 偏差小，并且具有正态分布，如图 2-3a 所示。另一方面，如果一组观测值很发散，则精度差，其正态分布如图 2-3b 所示。

相比之下，准确度被认为是全面估计测量中存在的误差，包含系统误差。某组观测值认为是准确的，其最或然值或样本平均值必定十分接近真值，如图 2-4a 所示。一组观测值可能很精密，但不准确，如图 2-4b 所示，此时真值与 MPV 之差很大，可能存在多个系统误差。如果所有的系统误差被消除，则准确度与精度相同。在这种情况，精（密）度可以作为准确度的指标。

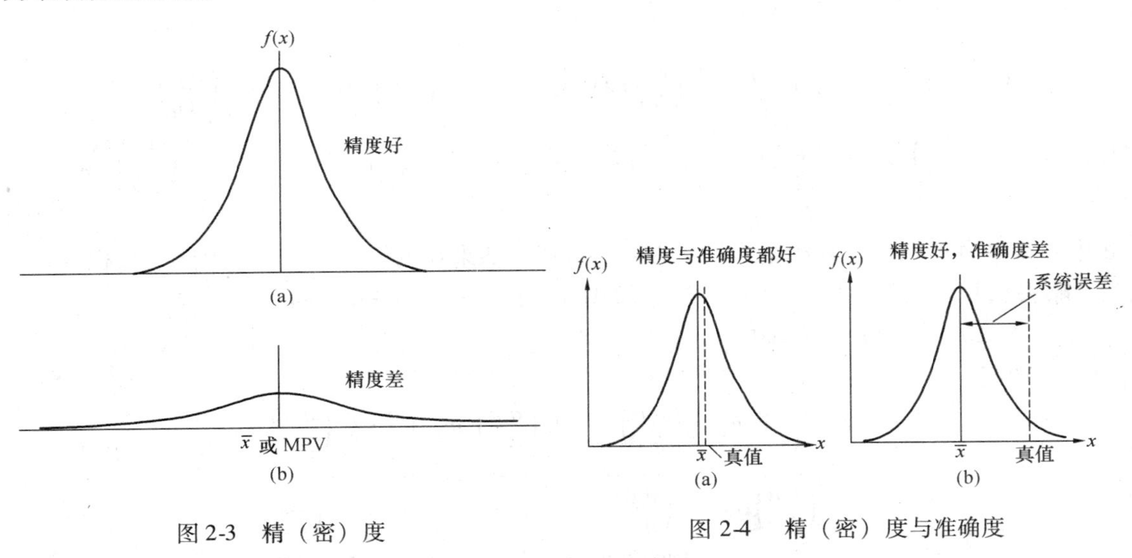

图 2-3 精（密）度

图 2-4 精（密）度与准确度

2.4 误差传播定律

测量中观测值，例如角度与距离常用数学的关系式推导出其他量。例如，水准测量高差由标尺读数相减求得，水平距是由斜距及垂直角通过计算而求得，坐标可通过联合水平角及距离推求。这些情况，原始观测值是随机分布并有误差，由此得出结论，由它们推导的任何量也必定存在误差。

设 u 是独立测量值 x，y，$\cdots$ q 的函数，即

$$u = f(x, y, z, \cdots q) \tag{2-16}$$

设每个独立变量（即互不相关）有微小变量 dx，dy，$\cdots$ dq，则 u 就有微小变量 du，由下面微分表达式给出：

$$du = \frac{\partial u}{\partial x}dx + \frac{\partial u}{\partial y}dy + \cdots + \frac{\partial u}{\partial q}dq$$

式中，$\frac{\partial u}{\partial x}$ 表示 u 对 x 的偏微分，而其他变量也是这样。应用这些到一组观测值，由于残差是微小量，设 $\Delta x_i = dx_i$，$\Delta y_i = -dy_i$，$\Delta z_i = dz_i \cdots \Delta q_i = dq_i$. 则 $\Delta u_i = du_i$，

$$\left.\begin{aligned}\Delta u_1 &= \frac{\partial u}{\partial x}\Delta x_1 + \frac{\partial u}{\partial y}\Delta y_1 + \frac{\partial u}{\partial z}\Delta z_1 + \cdots + \frac{\partial u}{\partial q}\Delta q_1 \\ \Delta u_2 &= \frac{\partial u}{\partial x}\Delta x_2 + \frac{\partial u}{\partial y}\Delta y_2 + \frac{\partial u}{\partial z}\Delta z_2 + \cdots + \frac{\partial u}{\partial q}\Delta q_2 \\ &\vdots \\ \Delta u_n &= \frac{\partial u}{\partial x}\Delta x_n + \frac{\partial u}{\partial y}\Delta y_n + \frac{\partial u}{\partial z}\Delta z_n + \cdots + \frac{\partial u}{\partial q}\Delta q_n\end{aligned}\right\} \tag{2-17}$$

方程（2-17）两边平方并且相加得，

$$\begin{aligned}\Delta u_1^2 &= \left(\frac{\partial u}{\partial x}\right)^2\Delta x_1^2 + 2\left(\frac{\partial u}{\partial x}\right)\left(\frac{\partial u}{\partial y}\right)\Delta x_1\Delta y_1 + \cdots + \left(\frac{\partial u}{\partial y}\right)^2\Delta y_1^2 + \cdots + \left(\frac{\partial u}{\partial q}\right)^2\Delta q_1^2 \\ \Delta u_2^2 &= \left(\frac{\partial u}{\partial x}\right)^2\Delta x_2^2 + 2\left(\frac{\partial u}{\partial x}\right)\left(\frac{\partial u}{\partial y}\right)\Delta x_2\Delta y_2 + \cdots + \left(\frac{\partial u}{\partial y}\right)^2\Delta y_2^2 + \cdots + \left(\frac{\partial u}{\partial q}\right)^2\Delta q_2^2 \\ &\vdots \\ \Delta u_n^2 &= \left(\frac{\partial u}{\partial x}\right)^2\Delta x_n^2 + 2\left(\frac{\partial u}{\partial x}\right)\left(\frac{\partial u}{\partial y}\right)\Delta x_n\Delta y_n + \cdots + \left(\frac{\partial u}{\partial y}\right)^2\Delta y_n^2 + \cdots + \left(\frac{\partial u}{\partial q}\right)^2\Delta q_n^2\end{aligned}$$

$$\sum \Delta u^2 = \left(\frac{\partial u}{\partial x}\right)^2\sum \Delta x^2 + 2\left(\frac{\partial u}{\partial x}\right)\left(\frac{\partial u}{\partial y}\right)\sum \Delta x\Delta y + \cdots + \left(\frac{\partial u}{\partial y}\right)^2\sum \Delta y^2 + \cdots + \left(\frac{\partial u}{\partial q}\right)^2\sum \Delta q^2 \tag{2-18}$$

式中一些平方项和交叉乘积项为简化方程而省略。如果观测值是独立的，则交叉乘积项趋于零，略去这些交叉乘积项，然后把方程（2-18）两边除以（$n-1$）得

$$\frac{\sum \Delta u^2}{n-1} = \left(\frac{\partial u}{\partial x}\right)^2\frac{\sum \Delta x^2}{n-1} + \left(\frac{\partial u}{\partial y}\right)^2\frac{\sum \Delta y^2}{n-1} + \left(\frac{\partial u}{\partial z}\right)^2\frac{\sum \Delta z^2}{n-1} + \cdots + \left(\frac{\partial u}{\partial q}\right)^2\frac{\sum \Delta q^2}{n-1}$$

$$m_u^2 = \left(\frac{\partial u}{\partial x}\right)^2 m_x^2 + \left(\frac{\partial u}{\partial y}\right)^2 m_y^2 + \left(\frac{\partial u}{\partial z}\right)^2 m_z^2 + \cdots + \left(\frac{\partial u}{\partial q}\right)^2 m_q^2$$

$$m_u = \sqrt{\left(\frac{\partial u}{\partial x}\right)^2 m_x^2 + \left(\frac{\partial u}{\partial y}\right)^2 m_y^2 + \left(\frac{\partial u}{\partial z}\right)^2 m_z^2 + \cdots + \left(\frac{\partial u}{\partial q}\right)^2 m_q^2} \tag{2-19}$$

公式（2-19）为任意函数误差传播的公式，可应用到任何具体函数。

（1）设 $u = x + y + z$，式中 x，y 和 z 为 3 个独立的观测量。从（2-19）式得

$$m_u = \pm\sqrt{m_x^2 + m_y^2 + m_z^2} \tag{2-20}$$

（2）设 $u = xy$，式中 x，y 为 2 个独立的观测量。从（2-19）式得

$$\because \quad \frac{\partial u}{\partial x} = y,\ \frac{\partial u}{\partial y} = x$$

$$\therefore \quad m_u = \pm\sqrt{y^2 m_x^2 + y^2 m_y^2} \tag{2-21}$$

（3）线性函数

$$u = k_1x_1 \pm k_2x_2 \pm \cdots \pm k_nx_n$$

式中，x_l，x_2，…，x_n 为独立变量，k_1，k_2，…，k_n 为常数，从公式（2-19）得

$$m_u = \pm\sqrt{k_1^2 m_1^2 + k_2^2 m_2^2 + \cdots + k_n^2 m_n^2} \tag{2-22}$$

（4）同精度观测值的算术平均值的标准差

算术平均值公式：

$$x = \frac{[l]}{n} = \frac{1}{n}l_1 + \frac{1}{n}l_2 + \cdots + \frac{1}{n}l_n$$

式中，$\frac{1}{n}$为常数，各独立观测值的标准差均为 m，把这些值代入，由公式（2-22）得方程：

$$m_x = \pm\sqrt{\left(\frac{1}{n}\right)^2 m^2 + \left(\frac{1}{n}\right)^2 m^2 + \cdots + \left(\frac{1}{n}\right)^2 m^2}$$

因此

$$m_x = \pm\frac{m}{\sqrt{n}} \tag{2-23}$$

从上式可以看出：增加公式（2-23）中观测次数 n 可以减少标准差 m_x。但是，当观测次数足够大后，标准差的减少很有限。

一般来说，应用误差传播定律归纳为下列 3 个步骤：

（1）列出函数方程式：$u = f(x, y, z, \cdots, q)$

例如，测量长方形地块面积，边长 $a = 800.00 \pm 0.12$m，边长 $b = 550.00 \pm 0.07$m，问地块面积及面积中误差为多少？

本例给出函数方程式：

$$A = a \cdot b$$

（2）对上式全微分：

$$dA = \frac{\partial A}{\partial a}da + \frac{\partial A}{\partial b}db = bda + adb$$

（3）转变上式微分表达式为中误差表达式，按公式（2-19）计算 m_A，即

$$m_A = \pm\sqrt{b^2 m_a^2 + a^2 m_b^2} = \sqrt{550^2 \times 0.12^2 + 800^2 \times 0.07^2} = \pm 86.6\text{m}^2$$

例 2.1 水准测量中，如果后视读数中误差 $m_a = \pm 0.002$m，前视读数中误差 m_b 也是 ± 0.002m，两点的高差中误差 m_h 为多少？

【解】

两点的高差等于后视读数减前视读数，即 $h_{AB} = a - b$.

两点的高差中误差

$$m_h = \pm\sqrt{m_a^2 + m_b^2}$$
$$= \pm\sqrt{0.002^2 + 0.002^2} = \pm 0.0028\text{m}$$

例 2.2 同一观测者用相同仪器观测某角度 10 次，其结果如下：47°56′38″，47°56′40″，47°56′35″，47°56′33″，47°56′40″，47°56′34″，47°56′42″，47°56′39″，47°56′32″，47°56′37″。计算 MPV 及其中误差。

【解】

MPV 是　$\alpha_x = 47°56' + \frac{(38 + 40 + 35 + 33 + 40 + 34 + 42 + 39 + 32 + 37)''}{10}$

$= 47°56'37.0''$

根据 MPV 计算残差 v 及其平方列于表 2-1：

表 2-1　例 2-2 列表计算

i	α_i	$v_i = (\alpha_i - \alpha_x)$	v_i^2
1	47°56′38″	+1	1
2	47°56′40″	+3	9
3	47°56′35″	−2	4
4	47°56′33″	−4	16
5	47°56′40″	+3	9
6	47°56′34″	−3	9
7	47°56′42″	+5	25
8	47°56′39″	+2	4
9	47°56′32″	−5	25
10	47°56′37″	0	0
	$\alpha_x = 47°56'37''$	$\sum v_i = 0$	$\sum v_i^2 = 102$

使用表 2-1 中的这些数值，由公式（2-14）得

$$m = \pm\sqrt{\frac{[vv]}{n-1}} = \pm\sqrt{\frac{102}{10-1}} = \pm 3.4''$$

角度 α 的 MPV 的标准差 m_x 计算如下：

$$m_x = \frac{m}{\sqrt{n}} = \frac{\pm 3.4''}{\sqrt{10}} = \pm 1.1''$$

角度 α 的 MPV 写为

$$\alpha_x = 47°56'37'' \pm 1.1''$$

例 2.3　测量建筑物一层两边：$x = 32.00 \pm 0.01\text{m}$ 与 $y = 18.00 \pm 0.02\text{m}$，计算建筑物一层面积及其中误差。

【解】

建筑物一层面积：$A = x \cdot y = 32.00 \times 18.00 = 576.00\ \text{m}^2$

用误差传播定律

$$\begin{aligned} m_A &= \sqrt{\left(\frac{\partial A}{\partial x}\right)^2 m_x^2 + \left(\frac{\partial A}{\partial y}\right)^2 m_y^2} \\ &= \sqrt{18.00^2 \times 0.01^2 + 32.00^2 \times 0.02^2} \\ &= \pm 0.66\text{m}^2 \end{aligned}$$

2.5 权

2.5.1 权的概念

许多测量工作是由不同观测者用不同仪器在不同观测条件下进行的，此时，其可靠性即不同组观测值互不相同。不同权重的观测值表明了精度的差异。权数愈大，观测值精度愈高。

不等精度的观测值具有不同的可靠性，它用数字表示，这个数字称为权，通常用 p 表示。权愈大，相应的标准误差愈小。观测值 l_1，l_2，$\cdots l_n$，相应的标准差为 m_1，m_2，$\cdots m_n$，则观测值的权为 p_1，p_2，$\cdots p_n$。

为了计算具有标准差 m_i 的某观测值 l_i 的权 p_i，应用下列方程。

从上一节公式（2-23）我们知道观测值的权 p_i 与观测次数（n）成正比，这意味着权与标准差的平方成反比，即

$$p_i \propto \frac{1}{m_i^2} \tag{2-24}$$

方程（2-24）可以写为

$$p_i = \frac{\mu^2}{m_i^2} \qquad i = 1,2,\cdots,n \tag{2-25}$$

式中，p_i 为观测值 l_i 的权；m_i 为观测值 l_i 的标准差；μ 为任意常数。

例如，设观测值 l_1，l_2，l_3，相应的标准差分别为 $m_1 = \pm 2''$，$m_2 = \pm 4''$，$m_3 = \pm 6''$。观测值 l_i 的权 p_i 按公式（2-25）计算如下：

当 $\mu = m_1$ 时：$p_1 = 1 \quad p_2 = \frac{1}{4} \quad p_3 = \frac{1}{9}$

当 $\mu = m_2$ 时：$p_1 = 4 \quad p_2 = 1 \quad p_3 = \frac{4}{9}$

当 $\mu = m_3$ 时：$p_1 = 9 \quad p_2 = \frac{9}{4} \quad p_3 = 1$

从上述可以看出：权是一组比例数字，一旦 μ 值确定，观测值的权也就确定。μ 可以取不同值，观测值的权也就不同，但观测值之间权的比例关系仍不变。

权等于1的观测值称为单位权观测值；权为1的标准差称为单位权标准差。例如 $p_1 = 1$，我们称 l_1 为单位权观测值；标准差 m_1 为单位权标准差。

在水准测量中，如果每千米的精度是相等的，则水准路线高差的权与水准路线长度成反比，即

$$p_i = \frac{c}{l_i} \qquad i = 1,\ 2,\cdots n \tag{2-26}$$

式中，p_i 为第 i 条路线高差的权；c 为任意常数；l_i 为第 i 条水准路线的长度。

2.5.2 不等权观测值的平差

1. 加权平均值

一组不等权观测值的最或然值称为加权平均值，由下列公式计算：

$$x = \frac{p_1 l_1 + p_2 l_2 + \cdots + p_n l_n}{p_1 + p_2 + \cdots + p_n} = \frac{[pl]}{[p]} \tag{2-27}$$

或 $$x = x_0 + \frac{p_1\delta_1 + p_2\delta_2 + \cdots + p_n\delta_n}{p_1 + p_2 + \cdots + p_n} = x_0 + \frac{[p\delta]}{[p]} \tag{2-28}$$

式中，观测值 l_1 权为 p_1；观测值 l_2 权为 p_2；…… 观测值 l_n 权为 p_n。$[P] = p_1 + p_2 + \cdots + p_n$，$x_0$ 是 x 的近似值，以及 $\delta_i = l_i - x_0$。

如果观测值与平均值 x 的偏差为 v_i，那么 $v_i = l_i - x$。将这个方程式代入下式得

$$[pv] = [p(l_i - x)] = [pl] - [p]x$$

顾及（2-27）式，上式变为

$$[pv] = 0 \tag{2-29}$$

因此上式可作为计算的检核。

2. 一组不等权观测值的最或然值的标准差

方程（2-25）可以写为

$$\mu^2 = p_1 m_1^2 = p_2 m_2^2 = \cdots = p_n m_n^2$$

$$n\mu^2 = [pm^2]$$

$$\mu = \sqrt{\frac{[pm^2]}{n}}$$

当观测次数很大时，可以用真差 Δ 替代标准差 m

$$\therefore \quad \mu = \sqrt{\frac{[p\Delta\Delta]}{n}} \tag{2-30}$$

或 $$\mu = \pm\sqrt{\frac{[pvv]}{n-1}} \tag{2-31}$$

为了推求一组不等权观测值的最或然值的标准差，将方程式（2-27）写为

$$x = \frac{[pl]}{[p]} = \frac{p_1}{[p]}l_1 + \frac{p_2}{[p]}l_2 + \cdots + \frac{p_n}{[p]}l_n$$

设加权平均值的标准差 m_x，根据误差传播定律得出

$$m_x^2 = \frac{p_1^2}{[p]^2}m_1^2 + \frac{p_2^2}{[p]^2}m_2^2 + \cdots + \frac{p_n^2}{[p]^2}m_n^2$$

考虑权定义，即 $p_i m_i^2 = \mu^2$，上面的方程变为

$$m_x^2 = \frac{p_1}{[p]^2}\mu^2 + \frac{p_2}{[p]^2}\mu^2 + \cdots + \frac{p_n}{[p]^2}\mu^2$$

$$= \frac{1}{[p]^2}(p_1 + p_2 + \cdots + p_n)\mu^2 = \frac{\mu^2}{[p]}$$

$$\therefore \quad m_x = \frac{\mu}{\sqrt{[p]}} \tag{2-32}$$

这就是计算不等权观测值的平均值的标准差计算公式。

例 2.4 图 2-5 显示一个节点的水准测量路线。为了推求 P 点的高程，3 个已知点的高程为 $H_A = 50.148\text{m}$，$H_B = 54.032$，$H_C = 49.895\text{m}$。布设 3 条水准路线并进行水准测量。3 条水准路线长度：$l_{AP} = 2.4\text{km}$，$l_{BP} = 3.5\text{km}$，$l_{CP} = 2.0\text{km}$，3 个水准测量观测值：$h_{AP} = +1.535\text{m}$，$h_{BP} = -2.332\text{m}$，$h_{CP} = +1.780\text{m}$。

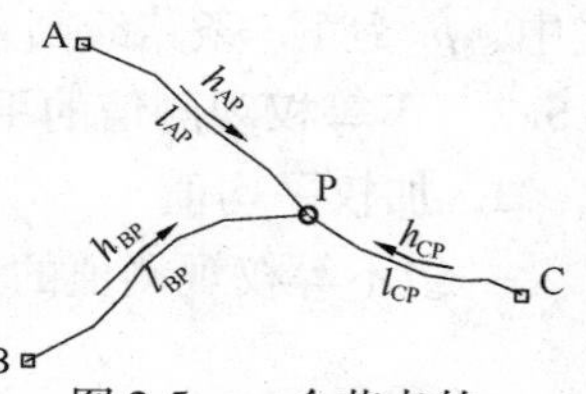

图 2-5 一个节点的水准测量路线

计算：P 点高程的最或然值（H_P）及其标准差 m_P。

【解】

表 2-2　例 2.4 列表计算

路线	高程 H_p（m）	路线长度 l_i（km）	权 $P_i=1/L_A$	v（mm）	Pv	Pvv
A－P	51.683	2.4	0.417	－0.7	－0.292	0.204
B－P	51.700	3.5	0.286	－16.3	+4.662	75.991
C－P	51.675	2.0	0.500	－8.7	－4.350	37.845
Σ	51.6837		1.203		0.02	114.040

P 点高程最或然值 H_p：

$$H_p=\frac{0.417\times 51.683+0.286\times 51.700+0.500\times 51.675}{1.203}=51.6837\text{m}$$

单位权标准差 μ：

$$\mu=\sqrt{\frac{[pvv]}{n-1}}=\sqrt{\frac{114.04}{2}}=\pm 7.6\text{mm}$$

P 点高程最或然值标准差 m_x：

$$m_x=\pm\frac{\mu}{\sqrt{[P]}}=\frac{\pm 7.6}{\sqrt{1.203}}=\pm 6.9\text{mm}$$

第3章　距离测量

3.1　概述

水平距离的测量是测量工作中实施的最基本的作业，也许也是最难做好的事。在平面测量中，每当测量地块边长或地面两点之间距离时，其距离是指水平距离。因此，线段的测量即可以直接沿着水平线测量，或间接沿着斜面测量，然后再计算相应的水平距离。在后一情况，得到的距离当然是计算的水平距离。

水平距离测量一般有下列几种方法：（1）卷尺法，（2）视距法，（3）电子测距法。

3.2　卷尺测量法

卷尺法是测量水平距离的最普通方法。当测量距离小于卷尺长度时，在点之间沿直线拉尺测定距离使尺的零点（或某个方便的分划）对着一个点把尺拉直拉紧，在另一点直接读出尺上距离。这是建筑工地上常用的方法，在那里测量距离短，倾向于用卷尺而不用全站仪，因为更便捷。

当测量距离超出卷尺长度时，必须标出相应尺长的中间点。如果总长测量要求精确，这些中间点必须形成直线。如果测量的距离较短，线段的两端相互通视的情况下不成问题。在这种情况，定线杆插在线段的两端。当第一尺长段被标志划定时，后尺手站在起始点的后面，瞄准测线两端的定线杆，然后，发信号指挥前尺手在两杆直线间左右移动直到适当的位置插下测钎。当第二尺长段和后续尺长段标志划分时，后尺手向前瞄准测线远端的定线杆，以及前尺手向后瞄准测线始端的定线杆，这两种做法都有助于使中间点保持在直线上。如果要求更高精度的作业，应在测线的一端安置经纬仪，然后在测线上设置中间点。

当测线的一端不能看到另一端，很可能应在丈量之前，在地面上标定出直线，并沿着测线中间点设置木桩（在桩顶钉大头钉）。这些木桩（如有必要，靠近木桩插定线杆）可以确保丈量在直线上。

3.3　光学法（视距法）

3.3.1　水平视线视距测量

为了视距测量，经纬仪除中心的十字丝外，还配备有与中间水平丝等间隔的较短的水平丝，称为视距丝。视距法是基于相似三角形对应边成比例的原理。图3-1显示外对光望远镜，由A和B的光线通过棱镜中心形成一对相似三角形ΔAmB与Δamb，图中AB = l，是视距尺间隔（视距间隔），ab 为视距丝间距。

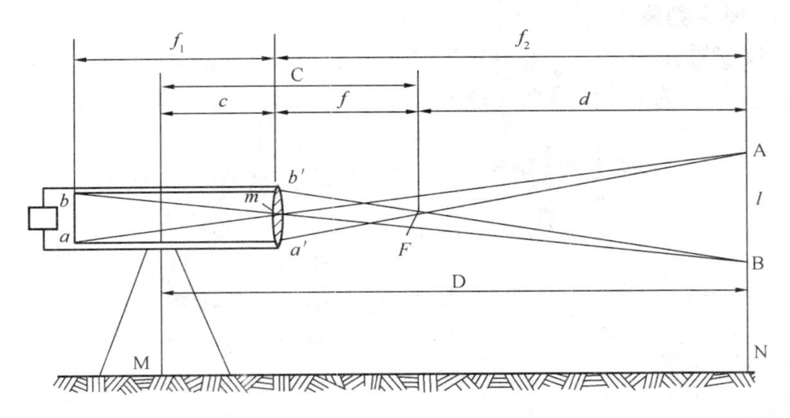

图 3-1　视距原理

在视距测量中使用的符号及其定义如下：

f = 棱镜焦距长（对任一复合物镜 f 为常数），它可通过对焦某远方目标并测量物镜棱镜中心（实际组合件的节点）和十字丝的距离来确定。

f_1 = 当望远镜对焦某固定点，物镜中心（实际组合件的节点）至十字丝平面的距离。

f_2 = 物镜中心（实际组合件的节点）至某清晰目标（当望远镜对焦该点）的距离，当 f_2 无穷大或很大时，$f_1 = f$。

p = 十字丝的间距（图 3-1 中 ab）。

$\frac{f}{p}$ = 视距间隔系数，通常为 100。

c = 从仪器中心（主轴）至物镜棱镜中心的距离。对不同视线长度物镜移进移出而有小变化，但一般认为是常数。

$C = c + f$　C 称为视距常数，尽管由于 c 有点小变化。

d = 望远镜前焦点至视距尺的距离。

$D = C + d$ 从仪器中心至视距尺的距离。

从图 3-1 的相似三角形得

$$\frac{d}{f} = \frac{l}{p} \text{或} d = \frac{f}{p}l$$

及

$$D = \frac{f}{p}l + C$$

经纬仪、水准仪及平板仪照准部中固定的视距丝，由仪器厂家仔细配置，使得视距间隔系数 $\frac{f}{p} = K$ 等于 100，则

$$D = Kl + C = 100l + C \tag{3-1}$$

当使用内对光的仪器时，望远镜的物镜仍在原位，而可移动的凹透镜调焦在物镜与十字丝平面之间改变光线的方向。适当选择 f、p 与 c 使得视距常数 $C \approx 0$，因此公式（3-1）可简化为：

$$D = 100l \tag{3-2}$$

3.3.2　倾斜视线视距测量

图 3-2 经纬仪安置在 M 点，N 点立尺。用十字丝的水平线对准 O 点读 ON（目标中心高 v），竖直角（倾斜角）为 α，仪器高（i）为经纬仪水平轴高出测站点的高度。

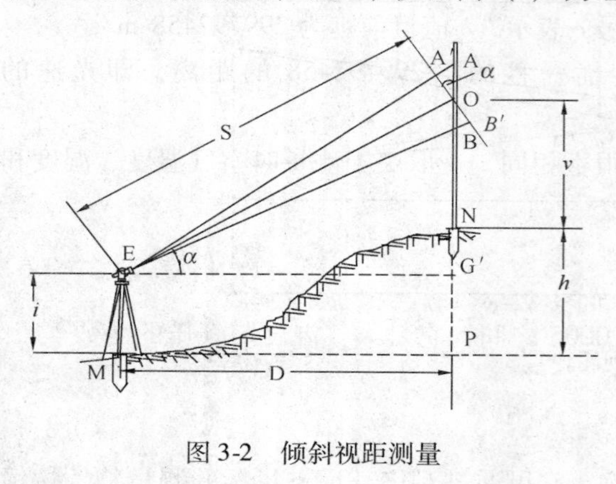

图 3-2　倾斜视距测量

设 S 代表斜距 EO，D 为水平距离 EG = MP，h 为高差 NP，从图 3-2 可得

$$D = S\cos\alpha$$

在图 3-2 中的 A，O，B 为垂直立标尺时望远镜三丝的读数。假设 AB = l。

假设能在 O 点置标尺垂直于视线，那么将得到读数 A′B′，设 $l' =$ A′B′。按照公式（3-2）我们得到

$$S = 100l'$$

在图 3-2 中，∠AA′O 可以认为直角，∠AOA′ = α，因此

$$l^1 = l\cos\alpha$$

$$S = 100l' = 100l\cos\alpha$$

$$D = S\cos\alpha = 100l\cos^2\alpha \tag{3-3}$$

从图 3-2 中看出高差 h 为

$$h = OG + GP - ON$$

$$\therefore \quad h = D\tan\alpha + i - v \tag{3-4}$$

3.4　电磁波测距（EDM）

3.4.1　原理

EDM 有两种基本测量法：脉冲法，使用发射电脉冲；相位移法，该法使用连续电磁波。

1. 脉冲法

一个强短的辐射脉冲发射到反射器，反射器立即将它沿着平行路径反射回接收器。用信号的速度乘以完成行程所需的时间计算得距离，即

$$2D = c\cdot\Delta t$$

$$D = c\cdot\Delta t/2 \tag{3-5}$$

假定脉冲离开电闸门 A 时间为 t_A，到达 B 门接收时间为 t_B，则（$t_B - t_A$）= Δt。在式（3-5）中，c 为光在介质中的速度。D 为仪器与目标间距离。

从（3-5）式可看出，距离取决于在介质中的光速和光传播时间的精度。

2. 相位移法

这项技术是使用连续的电磁波测距。虽然电磁波其性质很复杂，但它可以用最简单的周期性的正弦波形式表示，如图 3-3 所示。正弦波有下列特性：

（1）电波从波上相同点 A 移到 E 或从 D 移到 H 称为完成一个循环（或称一整波），1 秒完成整波的次数称为频率，频率用 f 表示，单位赫兹，1 赫兹为每秒传播一整波。

（2）波长是指在电波上相邻两相同点的距离，即指电波一整波传播的距离，用符号 λ，

单位用 m。

（3）周期指电波传播一整波或一个波长所需的时间，用 T 秒表示。

（4）电波介质中传播速度用 v 表示，单位用 m/s 或 km/s。

电磁波在真空中传播速度称为光速，用符号 c 表示，c 值目前定为 299792458 m/s.

有趣的是，米的定义就是光在真空中传播一秒的 1/299792458 的距离，即光速的倒数。

电磁波传播通过空气与通过真空传播速度很不相同，这取决于测量时空气温度、湿度和大气压。波速（v）可用式（3-6）表示：

$$\nu = \frac{c}{n} \tag{3-6}$$

式中，n 为大气的折射率，数值在 1.0001 至 1.0005 之间，它是大气温度与气压的函数。

上述电磁波的特性可由下面关系式表示：

$$\lambda = \frac{\nu}{f} \tag{3-7}$$

与周期波相关联的术语是波的相位。就距离测量而言，用相位更方便于鉴别整波长的一部分。表示瞬间正弦波的波幅 A 的关系式是（见图 3-3b）

$$A = A_{max}\sin\phi + A_0 \tag{3-8}$$

式中，A_{max}表示设计波的最大振幅，A_0 参考振幅，ϕ 为相位角。角度常作为相位角的单位，一个完整周期最大相位角为 360°。

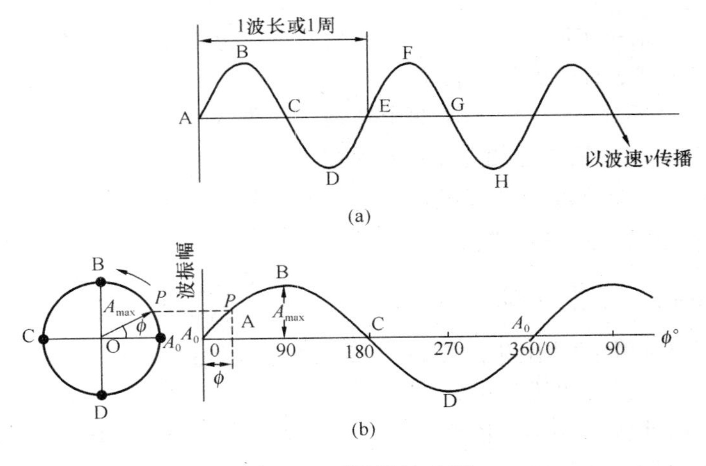

图 3-3　正弦波波动图

（a）呈现为距离或时间的函数；（b）呈现为相位角 ϕ 的函数

相位移法测定距离是通过测量发射信号与反射信号之间相位角之差。相位移通常以周期的几分之几表示。当已知波的频率与波速时，相位差可转换为距离。相位差法测量距离的方法如下：

图 3-4a 中，EDM 仪器安置在 A 点，反射器置 B 点，测量距离 AB = D。

图 3-4b 显示与图 3-4a 相同的布局图，但仅显示电磁波路径的细节。电磁波从 A 发射到 B，到 B 点立即反射（不改变相位角），在 A 点接收。为了清楚显示，图 3-4c 以相同的顺序

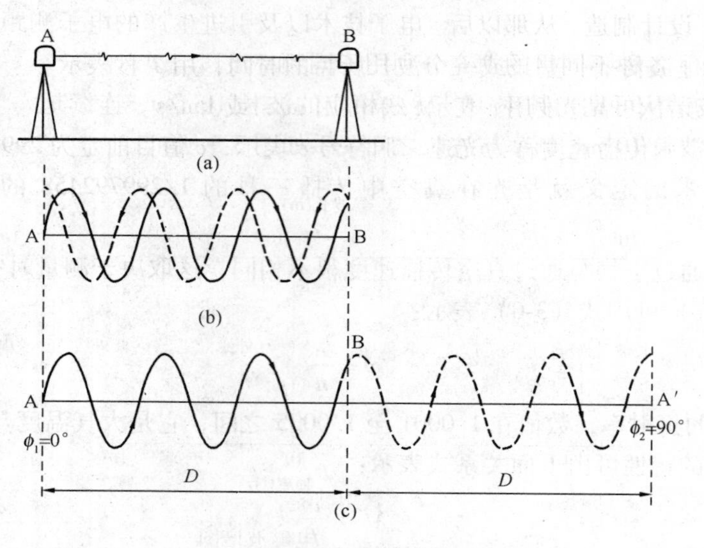

图 3-4　相位法原理

显示，返回波被展开。点 A 与 A′是同一点，因为发射器与接收器在 A 点并排在一起。

图 3-4c 显示电磁波从 A 到 A′传输经过的距离为

$$2D = n\lambda_m + \Delta\lambda_m$$

$$D = n\frac{\lambda_m}{2} + \frac{\Delta\lambda_m}{2}$$

式中，D 为 A、B 之间的距离，λ_m 为测量波的波长，n 为波长总数，$\Delta\lambda_m$ 为不足整波长的小数。因为测量双倍距离，有效测量波长为 $\lambda_m/2$。

距离 D 是由两个部分组成的，它由两个处理过程决定。

测量相位移，即 $\Delta\lambda_m$ 是通过测量相位角，在 A 点 EDM 仪器内置电子相位计或探测器测量发射的电磁波相位，设为 ϕ_1^0。同样的探测器也测量从反射镜返回波的相位 ϕ_2^0（在 A′点），比较两者给出 $\Delta\lambda_m$ 用以下的关系式：

$$\Delta\lambda_m = \frac{\lambda_m}{360°} \times (\text{以度为单位的相位移}) = \frac{\lambda_m}{360°} \times (\phi_2 - \phi_1)° \tag{3-9}$$

由于在 A' 处接收的电磁波用相位值 ϕ_2，而相位移仅提供电磁波传播超出整波数的部分。因此，需要某种方法决定未知的距离另一部分 $n\lambda_m$。这就涉及解算相位移的整波未知数的问题。有两种方法可以实现：一种是以 10 的倍数增加测量波长，直到最终测得 D 的粗值，另一种是测量距离用不同而相关的波长而形成 $2D = n\lambda_m + \Delta\lambda_m$ 形式的联立方程，解联立方程得 D。

用全站仪进行相位移测量无论使用什么技术去解算整波未知数完全自动进行，仪器按键进行测量和显示距离，没有必要按键另行计算或要求进一步的操作。

用 GPS 设备和用全站仪，虽然表面看起来有很大差别，但是它们都以同样的方法测距定位。这种情况，由卫星传输 L 波段信号，波长约 0.2m。由相位移法求得 $\Delta\lambda_m$，同样也需要解算整波未知数求得 $n\lambda_m$，比用全站仪更困难，因为到卫星的距离很长。

3.4.2　电磁波测距设备分类

第一台电磁波测距设备是微波测距仪，是雷达技术发展的结果。该仪器在 1954 年由南非大地测量厂设计制造。从那以后，电子技术以及引进生产的电子测距仪的固态设计有相当大的发展，为了各种不同目的，充分利用不同频率波段。

电磁波波谱从可见光频率 10^{14} 赫兹相应的波长 10^{-6}m，连续地一直到无线电长波频率 10^4 赫兹相应波长 10^4m，频率与波长之间关系如图 3-5 所示。

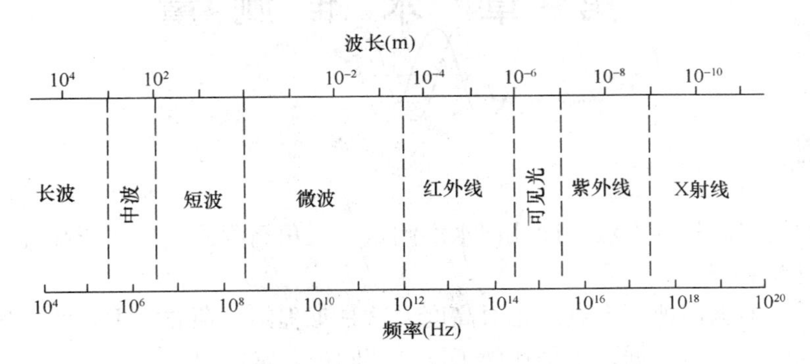

图 3-5　电磁波谱图

1. 微波测距仪

微波频率在 10^8 到 10^{12} 赫兹之间，这类仪器主要用于长距离测量。主站发射信号由远方站接收，由于信号到达远方站时变弱，所以必须放大信号，以精度相同相位返回信号。

这个系统要求两端都有操作员和通讯系统。为了使结果消除整波未知数，要使用 4 个或 5 个频段。该法适用于 30 ~ 80km 测程，获得精度为 ±(15mm ±5mm/km)。

微波测距仪用于大规模的三边测量以及需要测若干边的三角测量，因此它可作为特大土建工程项目的控制。

2. 可见光测距仪

可见光频率在 10^{14} 到 10^{16} 赫兹之间，可见光仪器适用于中等测程。远方目标站用正立方角锥棱镜，棱镜返回信号闭合于并平行于输出的信号。

光波测距仪测距经常用 3 个不同的波长以保证精度。该法适用于测程达 25km，精度为 ±(10mm ±2mm/km)的情况。

3. 红外光测距仪

红外波频率在 10^{12} 到 10^{14} 赫兹之间，红外波适用于做成短波仪器。红外光测距仪由于价格低廉因而十分普及。载波源采用砷化镓，是一种红外发射的二极管。这些二极管容易以高频调制振幅，因此提供价廉且简单方法得到调制载波。该仪器设计简约，因而紧密而轻便，可以装到经纬仪上使用。

由于红外波长接近于可见光，如同可见光一样的方式由透镜来控制。二极管主要缺点在于功率低，测程限于 2、3km，在该测程内精度为 ±(3mm ±2mm/km)。

第4章　水准测量

4.1　概述

现有几种测定高程的方法，即几何水准测量、三角高程测量、气压高程测量、流体静力学高程测量以及 GPS 等。

水准测量，即测定地面上点的相对高度，这是工程师一项非常重要的工作。无论在各类工程设计中获取必要的数据，还是在施工操作期间都非常需要。

水准测量定义的基本术语图示于图4-1。

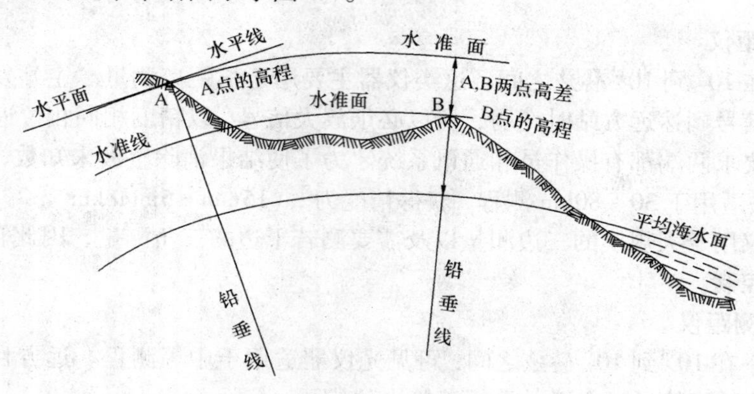

图4-1　水准测量的术语

水准面　水准面是各点都垂直于铅垂线（重力作用的方向线）的曲面。水准面其形状大致如椭球面，静止的湖面就是水准面的最好例证。

水准线　处处都位于同一个水准面上的线，因此线上各点都垂直于重力方向。

水平面　通过某点的水平面是垂直于该点的重力方向。

水平线　垂直于某点重力方向的直线，如图4-1中通过A点的水平线。

垂直线　垂直于水准面并顺沿某点的铅垂线。

垂直面　包含垂直线的某平面。

高　程　某点的高程就是沿铅垂线到基准面的高度。

归化高程　某点的归化高程是已知的或假定的基准面的高程和该点离这个基准面的高度之和。

高　差　两点的高程差就是通过该两点水准面之间的垂直距离。

基准面　基准面是涉及点的高程的参考面（例如平均海水面），参考面的高程是已知的。最普遍使用的基准面是平均海洋水准面（MSL）。

水准点（BM）　BM是已知高程的永久性的高程控制点，BM是用精密的水准测量技术和仪器装备建立的。

MSL 就是大地水准面，它是国家高程基准面并赋以高程值（用 H 表示）为 0.000m，或 0.000ft。在美国，大地水准面称为“1929 年高程基准”。在中国，青岛验潮站从 1950 年 ~ 1956 年共 7 年观测值推求与平差求得水准原点高程为 72.289m，该大地水准面称为“1956 年黄海高程系”。1987 年根据验潮站从 1952 ~ 1979 年的观测值，这个基准面进一步精化以反映长期海洋潮汐的变化，提供新的国家高程基准，1985 年水准原点高程精确定为 72.260m，这个大地水准面现在称为“1985 国家高程基准”。

工程师更关注一个点相对于另一点的相对高差，以确定两点之间的高差为目的，而不是直接推求与大地水准面的关系。因此，对于局部小地区测量方案可以选择一个完全任意的参考基准面。

4.2　水准测量原理

测量的主要工作之一是测定地面点的高程。所谓高程或高度是指高于或低于参考基准面的垂直距离。在水准测量中必要的基本设备是：

（1）能给出真正水平线的器具（水准仪）。

（2）配置分划标尺以便读取垂直高度（水准尺）。

水准测量过程中所使用的设备包括水准仪及分划标尺。装备有酒精水准器的望远镜能提供一条真正的水平视线，这就是能确定地面两点之间的高差原因所在。

测定高程最精密和最普遍使用的方法是几何水准测量，该法意味着直接测量垂直距离。由水平视线所截取的竖直分划标尺的读数差就是直接测得两个标尺站间高差。例如，根据已知高程 A 点，它高于大地水准面或任意参考基准面高程为 H_A，测定待定点 B 的高程（见图 4-2）。水准测量步骤如下：

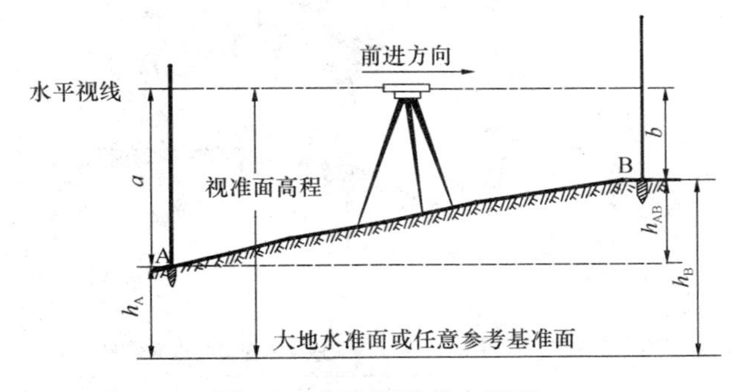

图 4-2　水准测量基本原理

首先，水准仪大约安置在 A 点与 B 点中间，水准尺分别直立在 A 点与 B 点上，如图 4-2 所示。

其次，通过望远镜的水平视线瞄准 A 点直立的水准尺并读取读数 a。这种情况，水准测量前进方向从 A 至 B，读取读数 a 水准仪要向后瞄准，所以称后视（BS）。

第三，转望远镜，通过望远镜的水平视线瞄准 B 点直立的水准尺读取读数 b。由于读取读数 b，水准仪向着前进方向瞄准，所以称前视（FS）。

从图 4-2 可以看出，A 与 B 之间的高差（h_{AB}）是

$$h_{AB} = a - b$$

即

$$\text{两点间高差} = \text{后视读数} - \text{前视读数}$$

如果已知 A 点的归化高程 RL，例如 $H_A = 100\text{m}$，那么 B 点的 RL（归化高程）为 H_B：

$$H_B = H_A + h_{AB}$$

$$H_B = H_A + a - b \tag{4-1}$$

从图 4-2 可以看出：

$$\text{HPC} = H_A + a \tag{4-2}$$

式中，HPC 称为水准仪视准轴平面高程或称仪器高程，方程（4-1）变为

$$H_B = \text{HPC} - b \tag{4-3}$$

最普遍的情况是当两点相距较远，或中间存在障碍物。因此不可能仅安置一个测站测定两点间的高差。例如图 4-3 欲测定高程的 B 点离已知点 A 很远。这种情况，水准测量实施要通过一系列测站，然后确定所求点的高程。第一测站，司仪者安置水准仪于合适的点 I_1，其位置能获得标尺的清晰读数，而不必试图在 A 与 B 的连线上。后视 A 点上后扶尺者直立的标尺，然后，前视转点 TP1 上前扶尺者直立的标尺。转点应选在合适的地点以使得前视距离，例如 I_1-TP1 大约等于它对应的后视距离，例如 $I_1-\text{A}$。最主要的要求是转点应选在稳定的物体上或使用尺垫（见图 4-11c），转点的稳定性非常重要。

第二测站，水准仪搬到超越 *TP*1 的合适位置 I_2 安置，后视 *TP*1 上直立的水准标尺，然后，前视新位置 *TP*2 直立的标尺，该标尺由后扶尺者搬来直立的（见图 4-3）。作业步骤从 *A* 至 *TP*1，从 *TP*1 至 *TP*2，从 *TP*2 至 *TP*3，最后，从 *TP*n 至 *B*。在 *TP*i 点上连续既有前视记录又有后视记录。它们是从不同仪器站得到的，这样把连续两个仪器站的视线高程联系起来以便于传递高程。这些点称为转点（*TP* 或 *CP*）。对于前视与后视在转点处的标尺必须严格安置在相同的位置，必须选择结实地点并作记号，松软的地面必须使用尺垫（见图 4-11c）。

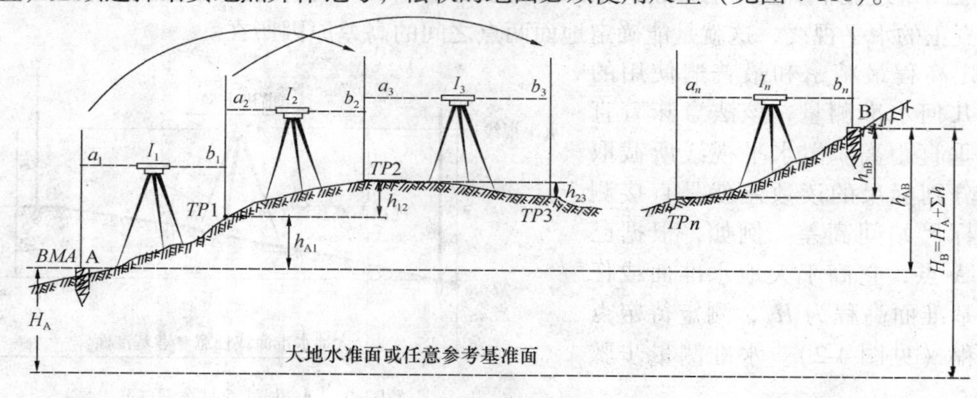

图 4-3　几何水准测量

每个站的 a_i 和 b_i 分别相加得 $\sum a_i$ 和 $\sum b_i$。最后得 A 和 B 之间的高差 $h_{AB}=\sum a_i-\sum b_i$，因此

$$H_B = H_A + h_{AB} \tag{4-4}$$

或

$$H_B = H_A + \sum a_i - \sum b_i$$

4.3　水准仪的类型及其使用

在我国水准仪按其精度可分为 DS_{05}、DS_1、DS_3 和 DS_{10} 四个等级。“D” 和 “S” 分别表示为“大地测量仪器” 和 “水准仪” 汉语拼音的第一个字母 “D” 与 “S”，其下标的数字表示 1km 水准测量高差中误差（即 ±0.5mm、±1mm、±3mm 和 ±10mm）。

按水准仪构造分可分为3种类型：微倾水准仪、自动安平水准仪和数字水准仪。

4.3.1　微倾水准仪

水准仪的基本作用在于能提供一条水平线。微倾水准仪包含4个基本部分：（1）在上部有望远镜；（2）水准管依附在望远镜旁；（3）中部有连接垂直内轴的支撑托板；（4）下部为基座，基座支撑仪器的其余部分。仪器的结构与外观显示于图4-4a与图4-4b。

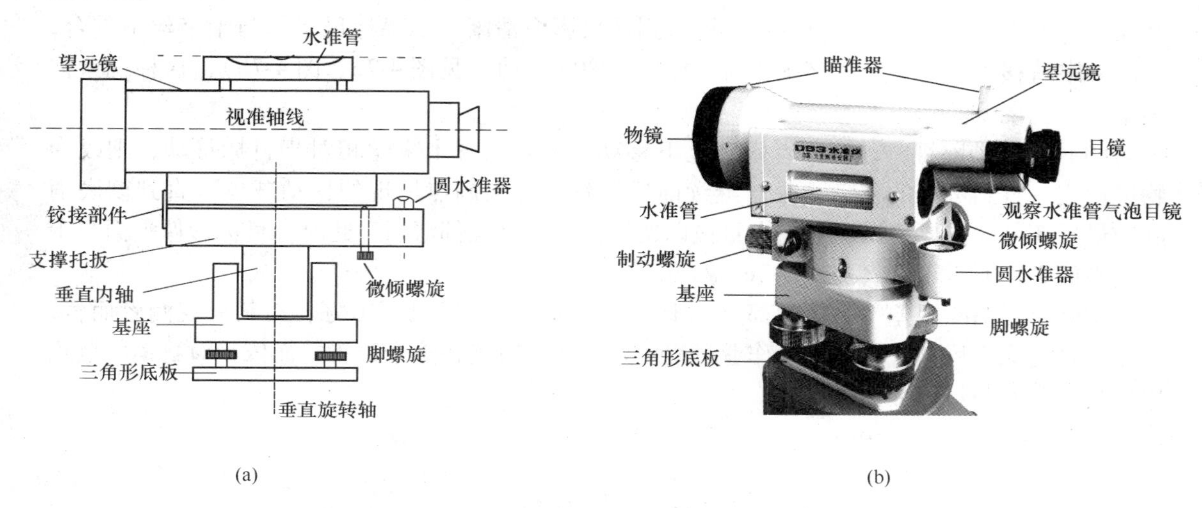

(a)　(b)

图4-4　微倾水准仪（北京测绘仪器厂DS_3）

1. 望远镜

世界上第一台望远镜是荷兰人詹·利波菲于1608年发明的，然后由伟大的数学家约翰尼斯·开普勒提议把设备改进用于测量仪器，这种类型的望远镜称为开普勒望远镜或天文望远镜。

微倾水准仪望远镜不是固定在水准仪的基座上，而是与支撑托板铰接并能在垂直面内作微小倾斜，其倾斜量由望远镜目镜下的微倾螺旋控制（见图4-4a）。脚螺旋用来居中圆水准器，因而能安置望远镜大约处在水平面上。当望远镜对标尺调焦后，使用高灵敏度的管水准器，转动微倾螺旋升降望远镜的一端使视线达到精确水平。

微倾水准仪有一个内对光望远镜，在望远镜的物镜与目镜之间附加一个凹透镜，通过使用调焦螺旋移动管内这个凹透镜进行调焦，这就能够使望远镜长度保持不变，保持密封以防灰尘进入（见图4-5）。

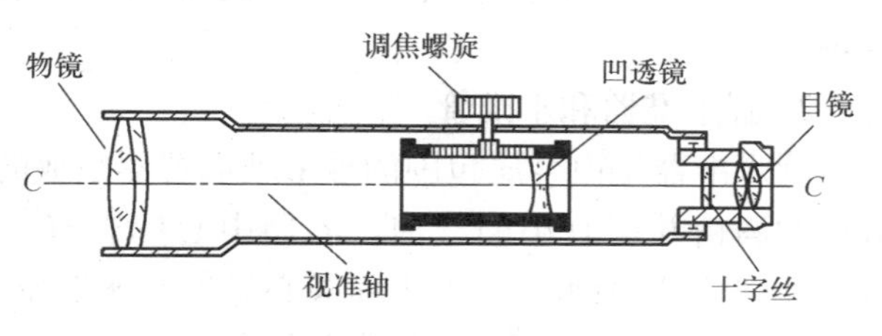

图4-5　内对光望远镜

通过旋转调焦螺旋，使双凹内调焦透镜沿着望远镜管移动直到标尺的影像聚焦到十字丝平面。十字丝被刻蚀在被称之为十字丝环的圆形薄的玻璃板上。约35倍放大率的目镜观察十字丝平面上的影像。

十字丝的不同形式如图4-6所示，从通过物镜中心到十字丝中心的直线称为望远镜的视准轴线。

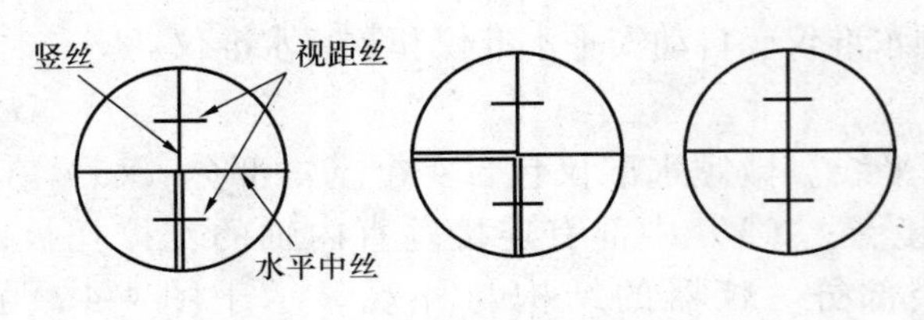

图 4-6　十字丝的不同形式

在实际观测之前，聚焦调节望远镜瞄准达到最高精度必须做下列 3 步：

（1）望远镜对准明亮无标志的目标，例如天空或白墙壁，调节目镜使十字丝清晰（即，既清又黑）。

（2）望远镜对准待测的标尺，同时眼睛保持对十字丝的聚焦，调节调焦透镜直到标尺影像清晰。如果标尺影像与十字丝不重合，观测者眼睛移动将会引起十字丝相对于标尺影像的移动（见图 4-7a、图 4-7b），这种现象称为十字丝的视差。

（3）为了消除视差，观测者眼睛上下移动。如果出现十字丝相对于目标移动，则改变物镜调焦一直到可见的移动消失。继续前后调焦，每次调焦呈现的移动减少一直到彻底消除。最后还可能需要稍微调节一点目镜以使影像与十字丝都清晰。此时，目标影像平面与十字丝平面必定非常重合（见图 4-7c）。

实际上，消除视差不必每次都要一步一步去做。当目镜已调节好，对某一观测者而言，视差一旦消除之后，一般在整个作业过程中始终保持目镜在这个位置，而依赖对物镜的调焦以使得十字丝与目标都清晰。

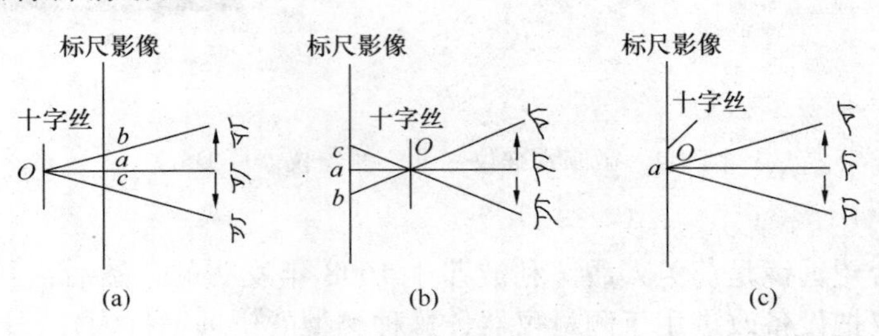

图 4-7　视差现象

（a）存在视差；（b）存在视差；（c）没有视差

望远镜质量的特性是它的放大倍率，视场大小，形成影像的亮度以及读尺的分辨率。

放大倍率是通过望远镜观察目标的大小与肉眼观察大小的比值。望远镜的分辨率或分解力是其表现细节的能力。测量望远镜在放大率上加以限制为的是能得到较好的分辨率和视场。

2. 圆水准器和水准管

圆水准器是由金属包围的球形玻璃容器组成的。它的盖内表面研磨成球状。容器内充满合成的酒精并留有小的气泡。容器中心标志有一个或多个同心圆（图 4-8）。圆水准器轴（$L'L'$）是通过圆水准器上表面圆心想象的球半径。

水准管是由金属包围的圆柱形玻璃容器组成的，它的内表面被研磨，其纵断面成圆弧形。水准管轴可想象为通过中点切于管内表面的纵向切线。当气泡居中时，水准管轴处于水平线，如图 4-9 所示。

水准管通常以 2mm 间隔注分划（图 4-9）。水准管气泡的灵敏度主要取决于它的曲率半径（R）。半径愈大，气泡的灵敏度愈高。

水准管灵敏度是 2mm 弧长所对应的圆心角（τ），如图 4-9 所示。因此

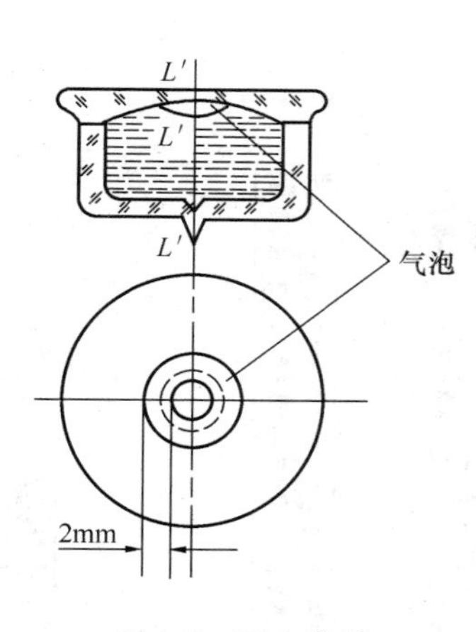

图4-8　圆水准器

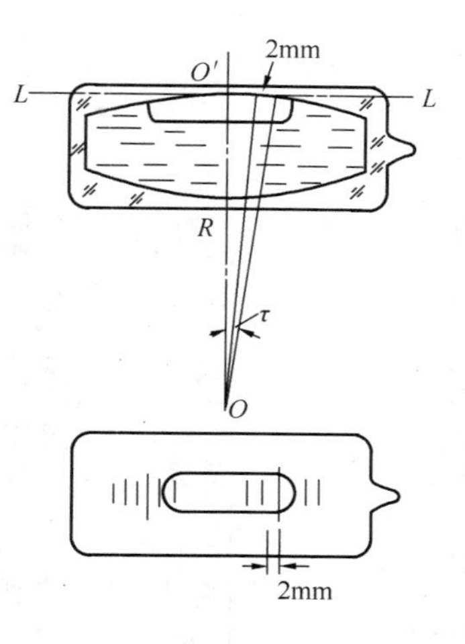

图4-9　水准管

$$\tau = \frac{2\text{mm}}{R} \times \rho'' \tag{4-5}$$

式中，R = 水准管的曲率半径；$\rho''=206265''$。

如果 $R=20.63\text{m}$ 则 $\tau=20''$。在这种情况下，气泡偏离中心一个分划间隔，就表示水准管轴线倾斜 20″。

微倾水准仪所附水准管气泡可以直接观察或借助于符合读数系统来观察（图4-10a）。符合读数系统原理图中可看到气泡两端的像，如图4-10a 和图4-10b 所示；图4-10b 显示用微倾螺旋使气泡居中影像的情况；图4-10c 显示气泡偏离中心的情况。这种方法观察气泡精度比直接观察高4倍或5倍。

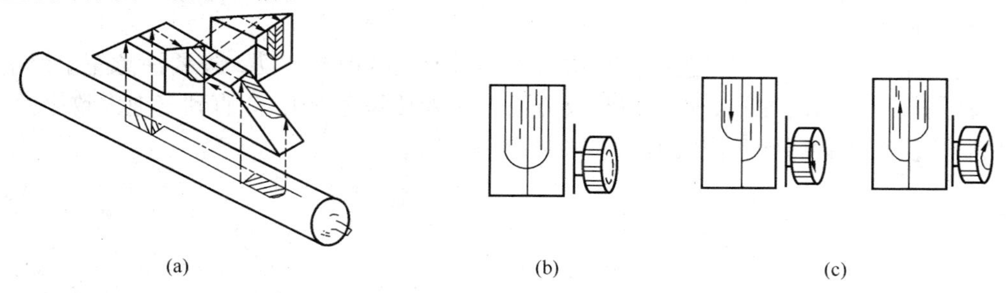

图4-10　气泡符合读数系统

（a）气泡符合读数系统的原理；（b）气泡居中图像；（c）气泡不居中图像

圆水准器作为粗略安置，而水准管用作精确测量之用。微倾水准仪有一个重要的条件，那就是视准轴线必须平行于水准管轴。

3. 水准尺与尺垫

水准尺由木材、金属或玻璃纤维制成的，并注有米与分米。大多数水准尺是伸缩性的有三、四节便于携带（图4-11a），尺上最小分划为0.01m，可估读至0.001m。米长变换通常以尺面白底黑色与红色交替显示。另外一种实心水准尺两面分划不同，一面是白底黑色（称黑面），而另一面白底红面（称红面）。黑面尺底为零，红面尺底为4.687m（1号）或为4.787m（2号）如图4-11b 所示。

图4-11c为尺垫，它是由结实的铸铁制成的，顶部呈光滑半圆球。在水准测量时，水准尺直立放置在半圆球上。

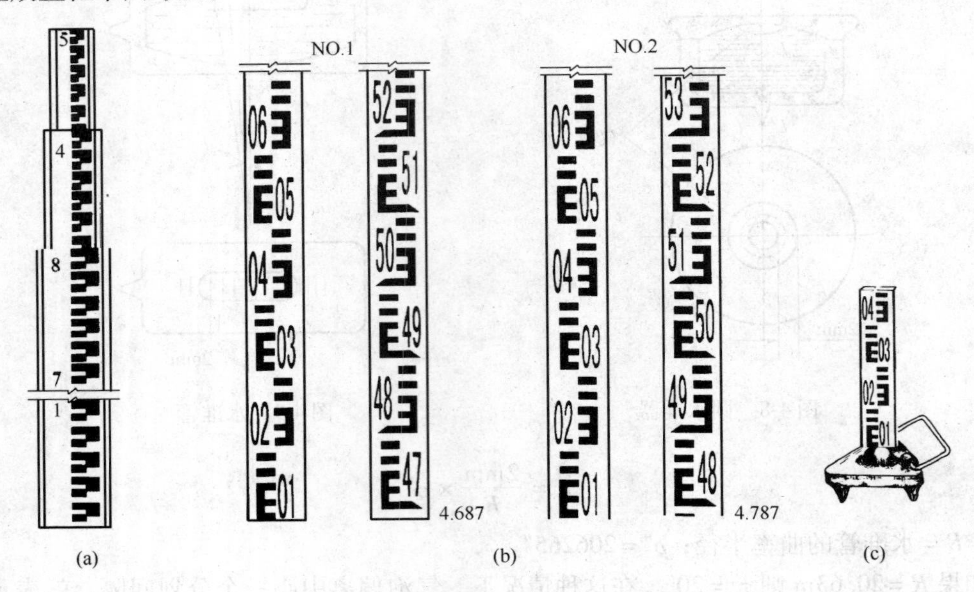

图4-11　水准测量标尺

（a）伸缩式水准尺；（b）整体式双面水准尺；（c）尺垫

4. 水准仪的使用

（1）安置仪器——张开三脚架的架腿，牢固架在地面上，把仪器安置在三脚架头上。

（2）粗平——用脚螺旋使圆水准气泡居中，其目的是使圆水准器轴大约处于铅垂线上，这就是所谓粗平。

为此，左、右手的拇指与食指夹住脚螺旋同时向内或向外转动，还要注意左手拇指运动的方向就是气泡移动的方向。首先以相反方向同时旋转脚螺旋1、2直到气泡大致停留ab线上，然后，仅旋转脚螺旋3使气泡居中（见图4-12b）。

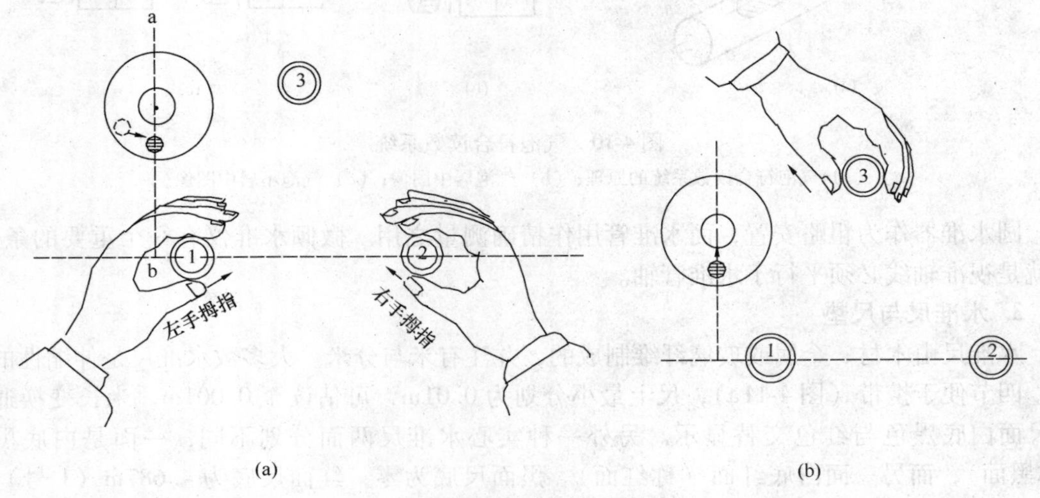

图4-12　居中圆水准器

（3）消除视差——消除视差的方法已在上述的第一部分中叙述（即“1. 望远镜”）。

（4）精平——因为水准仪并没有严密整平，因此在每次读数之前，微倾螺旋必须重新调节以使水准管气泡居中，这就是所谓精平，目的在于使视线成为真正的水平线。

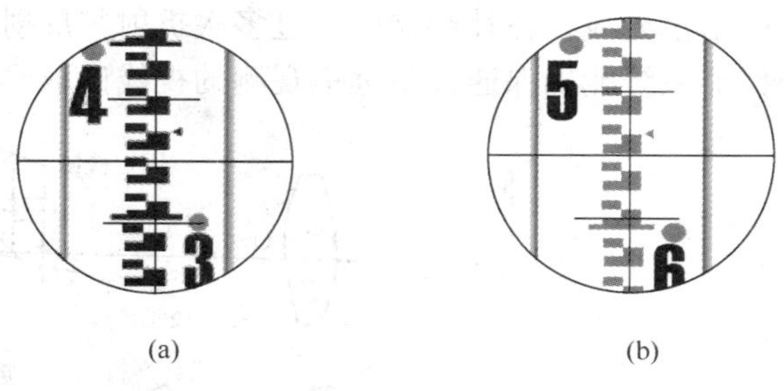

图4-13　读取标尺读数

（a）标尺数字向上增加；（b）标尺数字向下增加

（5）读数——读取水平中丝在标尺上的读数。要注意望远镜视场内标尺的数字，不论向上增加还是向下增加，读数都应该从小到大读取。图4-13a标尺读数为1.334m，图4-13b读数为1.560m

4.3.2　自动水准仪

1. 自动水准仪的特点

自动水准仪，实例如图4-14所示，不再使用酒精水准管来使视准轴水平，取代之以视准线是直接通过补偿器系统，即使望远镜光轴本身不水平，补偿器系统确保通过望远镜观察的视线读取水平读数。

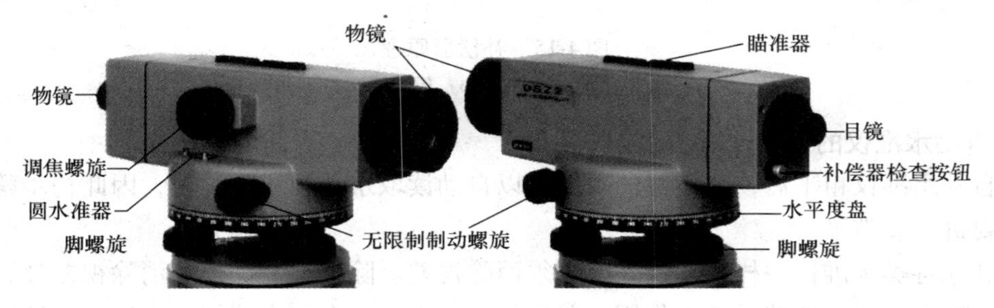

图4-14　自动水准仪（苏州光学仪器厂生产DSZ2型）

因为自动水准仪借助于低灵敏度的圆水准器仪能做到粗略整平，所以视准轴对水平面倾斜小角度 α（图4-15b），因而进入的光线射在十字丝环平面上的 b 点，位移 $Zb=f\alpha$。在K处的补偿器必须重新引导射线穿过十字丝交点Z，因此

$$f\alpha = \mathrm{Z}b = s\beta$$

和

$$\beta = \frac{f\alpha}{s} = n\alpha \tag{4-6}$$

从这里可看出，补偿器的位置对补偿过程非常重要。例如，如果补偿器固定在望远镜的中间，那么 $s\approx f/2$ 且 $n=2$，则有 $\beta=2\alpha$。补偿器的工作范围大约15′，为此要旋转脚螺旋与其相连的圆水准气泡居中。

为了补偿望远镜的微小倾斜量，补偿器需要一个固定到望远镜的反射面，需要一个受重力作用的可移动面和一个阻尼器（空气阻尼或磁阻尼）以便使移动面迅速静止，并能迅速观察标尺，这样的布置如图4-15a所示。当望远镜水平时，视线如图4-15a所示，悬吊的棱镜在重力的作用下处于某一位置。如果望远镜不水平，则视准轴在O点倾斜 α 角（见图4-15b），此时，固定的棱镜也倾斜 α 角，而悬吊的棱镜在重力的作用下处于最初的位置。进入补偿

器的水平光线，在补偿器内经过多次折射与反射，最终偏转β角并通过十字丝交点 Z。这样补偿器系统能确保通过望远镜观测的视线读取到水平读数a_0。

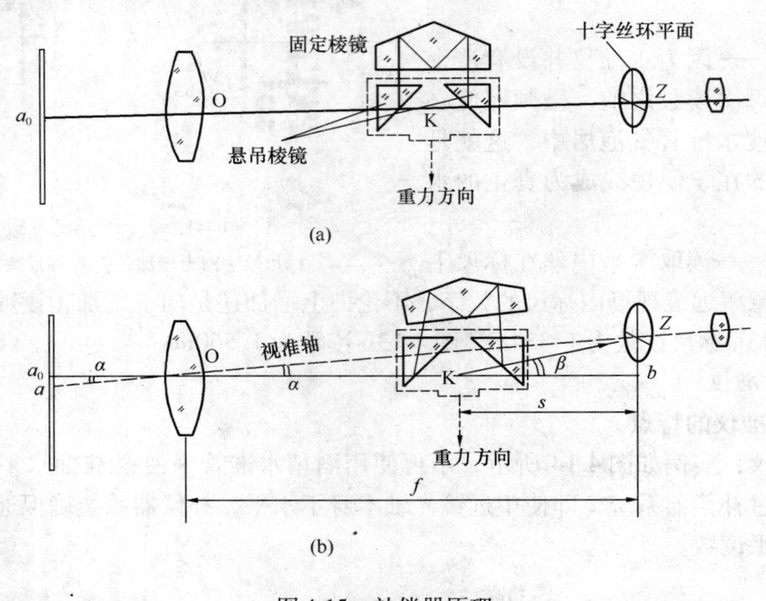

图 4-15 补偿器原理

（a）望远镜水平；（b）望远镜倾斜

2. 自动水准仪的使用

当自动水准仪粗平后，补偿器的功能可以自动读取水平视线的读数，因此初步整平以后不再需要进一步整平。

和其他各类水准仪一样，读数之前必须消除视差。除了整平操作和消除视差外，在读数开始之前还应检查补偿器是否起作用。普通的自动水准仪有一个按钮，当按一下该按钮，就会移动一下补偿器以防粘连。如果补偿器起作用，会看到水平丝移动，然后立即返回到水平视线。有些水准仪加装警示设备，当望远镜不水平时，在望远镜视场内给出可视标志。

自动水准仪的优点是，实施精密水准测量用它比用传统的水准仪具有更高的速度。归纳其原因在于它不需要居中主气泡。此外，补偿器系统消除了由于忘记居中气泡或居中不精确引起的误差。

4.3.3 数字水准仪

图 4-16 显示数字水准仪，其外观与自动水准仪、微倾水准仪相似。在使用时，安置仪器如同自动水准仪一样先把它连接到三脚架头上，调节脚螺旋使圆水准器气泡居中，然后补偿器建立水平视线，从水准标尺读取读数，所有读数与记录完全自动化。

仪器设计通过机内电脑自动读数及自动数据处理，通过显示器与键盘访问机内电脑。当水准测量时，瞄准特殊的条形码标尺（见图 4-17）进行调焦并按测量键。不必读标尺，按下测量键之后 2 ~ 3 秒显示器显示标尺读数。

仪器在光学路径上有一个电子射束分离器，电子射束分离器分离出红外射线到达光电二极管阵列，而可见光通过十字丝和目镜。当使用光电扫描模式时，扫描条形码标尺，实际上光电二极管替代观测者的人眼。光电二极管阵列转换条形码标尺图像为视频信号，然后将它放大，电子数字化后并传给微处理器进行处理。

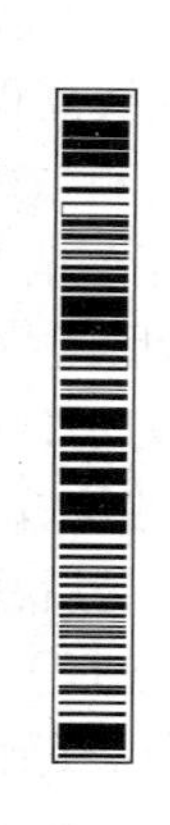

图4-16　数字水准仪（Topcon DL1－02）

图4-17　条形码标尺

捕捉的标尺条形码图像与机内存储的标尺条形码进行比较，当图像匹配时，即显示标尺读数。此外还显示仪器至标尺的距离，精度约20～25mm。所有的读数可用键盘编码。随着水准测量的进程，每个标尺的读数以及随后的所有计算都存储到水准仪的内部存储器中。

在好的条件下，数字水准仪测程可达100m。提供数字水准仪电池是标准AA电池或用可充电电池，足够一天水准测量。如果不能读取标尺电子读数（因为光线弱，有障碍物，例如树叶或者电力不足），条形码标尺的反面有通常的E形面，可以进行光学读数，用手工的方法操作仪器而取代之。

数字水准仪较传统的水准仪有许多优点，长距离观测更快，没有必要用手工读尺和记录。消除了两个难以克服的误差源——读尺错误和野外手簿记错。

数字水准仪存储的数据也可转移到移动存储卡，然后再转移到计算机，以便进一步处理。

4.4　水准测量外业施测步骤

当水准仪正确安置后，由仪器产生的视准线或视准面是重合于或非常接近于水平面的。如果这个面的高程已知，通过读竖直的水准尺读数就可求得地面点的高程。

图4-18水准仪安置在I_1，R_1与R_2为地面A与B两点竖立水准尺的读数。如果已知A点归化高程RL_A，则RL_A加标尺读数R_1就可求得仪器位置I_1视准轴的归化高程，即所谓视准轴平面的高程（HPC），或称视线高程。即

$$I_1\text{ 的视线高程} = RL_A + R_1 \tag{4-7}$$

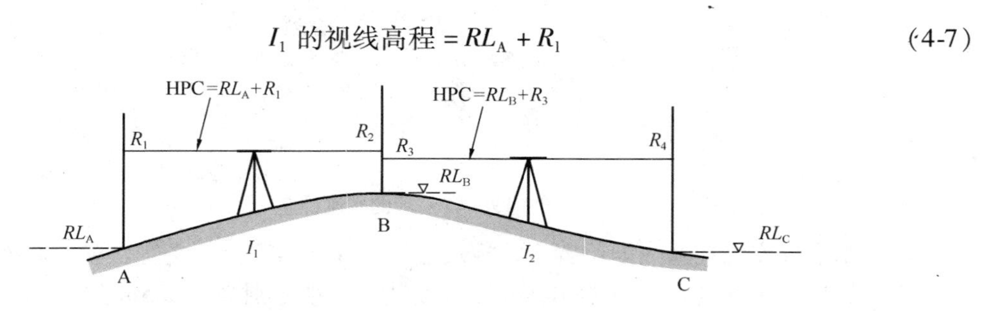

图4-18　水准测量从A到C

为了求 B 点归化高程 RL_B，必须从视线高程减标尺读数 R_2

$$RL_B = 视线高程 - R_2 = (RL_A + R_1) - R_2 = RL_A + (R_1 - R_2) \quad (4\text{-}8)$$

在这种情况下水准测量方向从 A 至 B，R_1 是水准仪面向反方向读取，故称为后视（BS），R_2 是水准仪面向 A 至 B 正方向读取，故称前视（FS）。A 与 B 之间的高差其数值与符号，由 A 与 B 两点标尺读数差给出。当 R_1 大于 R_2，$(R_1 - R_2)$ 为正，标尺底从 A 移到 B 为上升。因为 $(R_1 - R_2)$ 为正，即所谓上坡。

现在水准仪搬到新的位置 I_2 以便求 C 点高程，标尺仍留在 B 点，但尺面转向 I_2，首先要读取 R_3，在 I_2 位置它是后视，然后前视 C 点标尺读取 R_4。B 点由不同仪器站连续记录前视与后视，该点称转点（TP）。

从仪器位置 I_2 的标尺读数，计算 C 点归化高程

$$RL_C = RL_B + (R_3 - R_4)$$

由 $(R_3 - R_4)$ 求得 B 和 C 两点高差的数值与符号。在这种情况，$(R_3 - R_4)$ 为负数，因为从 B 到 C 标尺底端下降，这时标尺的读数差被看作下坡。

上述看出，不论计算上坡还是下坡，总是由后视减前视。如果此值为正，表示上坡；为负，表示下坡。

在实践中，仪器安置好之后，首先读取 BS（后视），后视总是对准水准点或已算得的归化高程。反之，前视在测站上最后读取。同一测站上在 BS 与 FS 之间读取称为中视（IS）。

更复杂的水准测量过程显示于图 4-19 的断面图与平面图，图 4-19 中两个临时水准点（*TBM*）之间的高差已测定，现求 A ~ E 各点的归化高程。用任何类型的水准仪进行读数，作业程序如下：

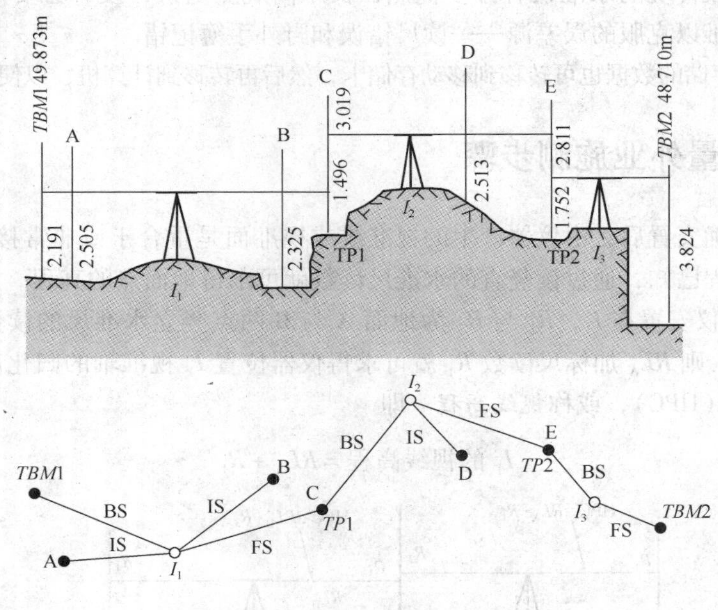

图 4-19　较复杂的水准测量

（1）水准仪安置在适当的位置 I_1，标尺底端立在 *TBM*1 上并竖直，读取后视 2. 191m。

（2）标尺依次移到点 A 和点 B，并读取读数，中视分别为 2. 505m 和 2. 325m。

（3）由于自然地面，为了测到 D 点，必须选一转点，转点选 C（*TP*1），标尺移到 C，

并读前视1.496m。

（4）在C点标尺停留时间，仪器搬到另一位置I_2，在新的仪器位置读取转点的后视读数3.019m。

（5）标尺依次移到D点和E点，读取中视（IS）2.513m和前视（FS）2.811m，这里E为另一个转点（TP2）。

（6）最后水准仪搬到I_3，对E点读取后视读数1.752m，对*TBM*2点读取前视读数3.824m。

最后标尺立在*TBM*，水准测量野外作业必须始于水准点并结束于水准点，否则，水准测量不可能检测错误。

4.5　水准测量精度的评定

在水准测量中经常使用的有3种水准路线，即

1. 附合的水准路线

实施水准测量从已知的起点*BMA*开始沿着待定的高程点测量连结到另一个已知的*BMB*点，这就是所谓附合的水准路线，如图4-20所示。理论上，路线上每段测量高差的总和（$\sum h_i$）应等于两已知水准点高程之差，即$\sum h_i = H_B - H_A$

因为测量存在误差，所以$\sum h_i \neq H_B - H_A$，由式（4-9）得水准测量闭合差：

$$f_h = \sum h_i - (H_B - H_A) \tag{4-9}$$

式中，f_h称为水准测量闭合差。

2. 闭合水准路线

实施水准测量从已知的起点*BMA*开始沿着待定的高程点测量回到起始已知*BMA*点，这就是所谓闭合的水准路线，如图4-21所示。显然水准测量闭合差f_h可以写为

$$f_h = \sum h_i \tag{4-10}$$

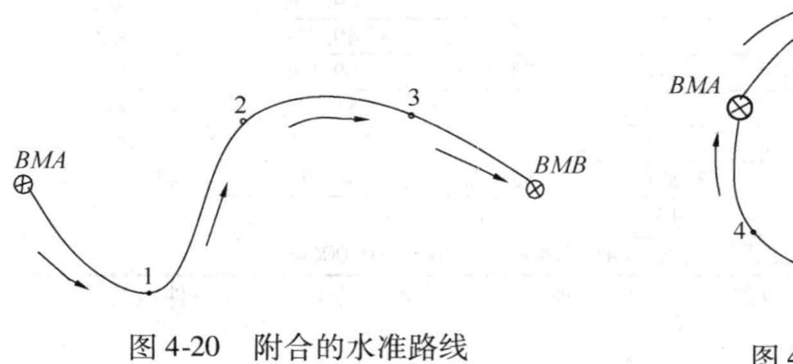

图4-20　附合的水准路线

图4-21　闭合水准路线

3. 支水准路线

由于支水准路线缺乏内部的检核条件，因此有必要往返观测未知点*P*（见图4-22），因此水准测量闭合差为

$$f_h = \sum h_{往} + \sum h_{返} \tag{4-11}$$

在4.6.1节中，按公式（4-9）求得水准测量求得闭合差为-6mm。由于闭合差的大小可以反映水准测量的质量，因此闭合差被认为是评定水准测量精度的一个标准，通常检查所

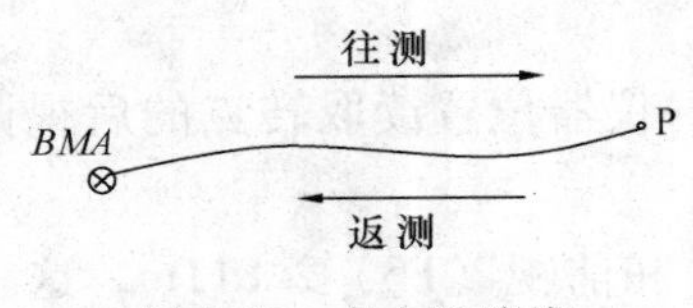

图 4-22　支水准路线

求的闭合差应小于容许闭合差的规定值。

在建筑工地及其他工程项目，实施水准测量距离较短但包含很多测站。对于这类作业，容许的水准测量闭合差由式（4-12）求得：

$$f_{h容} = \pm m\sqrt{n} \quad (4\text{-}12)$$

式中，$f_{h容}$为容许的水准测量闭合差以毫米计；m 为以毫米为单位的常数；n 为测站数。对于建筑工程水准测量，m 值常采用 ±5mm。

当从标尺读数求得闭合差与容许闭合差进行比较时，如果发现闭合差大于容许值，则该水准测量应弃去重测。如果闭合差小于容许值，则将闭合差平均分配到各测站，详见下一节叙述。

用这种方法评定水准测量的精度，现场的工程师有可能根据现场的条件选用某个 m 值。例如，土方开挖可降低要求，m 值可能选用 ±30mm；而对放样钢和混凝土结构，则 m 值可能选用 ±3mm。m 值按合同文件中的许可范围来确定。如合同文件未给出，则可由工程师根据经验简单选择。

4.6　水准测量的计算

4.6.1　视高法

在使用视高法时，视线的高程是由读取已知高程点上标尺读数进行计算。仪器移动时，视线的高程就变化。表 4-1 表示图 4-19 水准测量野外记录手簿，用视高法进行计算。视高法是基于每个仪器站计算的仪器视线高程。计算步骤列于下：

表 4-1　视　高　法

测点编号	水准尺读数			视线高程 HPC（m）	初始归化高程 RL（m）	改正值（m）	改正后归化高程 RL（m）
	后视（m）	中视（m）	前视（m）				
*TBM*1	2.191			52.064	49.873		49.873
A		2.505			49.559	+0.002	49.561
B		2.325			49.739	+0.002	49.741
C(*TP*1)	3.019		1.496	53.587	50.568	+0.002	49.570
D		2.513			51.074	+0.004	51.078
E(*TP*2)	1.752		2.811		50.776	+0.004	49.741
*TBM*2			3.824	52.528	48.704	+0.006	48.710
检核	6.962		8.131				
	f_h = 初始最后点 RL—已知最后点 RL = 48.704 − 48.710 = −0.006m						

注：日期、观测者、记簿者（如果不是观测者要单列）、测量标题、水准仪编号、天气条件以及其他有关项目如同水准尺读数都应记录。

（1）如果后视读数加临时水准点 *TBM*1 的归化高程 RL，则求得测站 I_1 的视线高程，即 49.873 + 2.191 = 52.064m，把它填入到视线高程栏，且在后视（2.191）的同一行中。

（2）为了求 A、B、C 点的初始归化高程，现从视线高程减去那些点的标尺读数，相应的计算是：

A 点的归化高程 RL = 52.064 − 2.505 = 49.559m

B 点的归化高程 RL = 52.064 − 2.525 = 49.739m

C 点的归化高程 RL = 52.064 − 1.496 = 50.568m

(3)C 点为转点，仪器移到测站 I_2，建立新的视线高程。视线高程是由 C 点后视读数加上从测站 I_1 求得的 C 点的归化高程 RL。测站 I_2 的视线高程 = 50.568 + 3.019 = 53.587m。从视线高程减去 D 与 E 的标尺读数求得它们的归化高程。

(4)继续这个过程直到 *TBM*2 的初始归化高程计算完毕。表 4-1 中初始 RL 列计算完成后，进行下列检查：

$$\sum 后视 - \sum 前视 = 最后点的初始归化高程 - 第一点的归化高程$$

即：
$$6.962 - 8.131 = 48.704 - 49.873 = -1.169$$

显然，这种情况水准测量的闭合差就是由观测值求得的最后一点的归化高程(48.704m)与已知高程(48.710m)之差。因此，高差闭合差 f_h 可写为

f_h = 计算求得的最后点的归化高程 - 最后点的已知高程 = 48.704 - 48.710 = -0.006m

进行改正常用的方法是对每个测站平均分配闭合差，但应注意仪器站闭合差的积累，改正数的符号反其闭合差的符号。

本实例有闭合差为 -0.006m，必须分配的总的调整量为 +0.006m。因为有 3 个测站，每个测站对归化高程加 +0.002m，分配的值显示在改正值纵行内，纵行内列出累计的调整值：A、B、C 高程改正 +0.002m，D 与 E 点改正为 +(0.002 + 0.002) = +0.004m。*TBM*2 改正为 +(0.002 + 0.002 + 0.002)m = +0.006m。对 *TBM*1 不需加改正，因这是临时水准点。

初始的归化高程加改正值求得改正后的归化高程列于表 4-1。

4.6.2　高差法

如果已知两个点中一个点的高程，那么另一个点的高程可由已知点高程加两点的高差求得。在一个测站水准测量时，后视读数大于前视读数，高差为正，这表示为上坡；后视读数小于前视读数，高差为负，这表示为下坡。各点的归化高程可以通过连续加上坡数（或下坡数）求得，这就是所谓高差法。

表 4-2 表示图 4-19 水准测量的所有记录，用高差法进行全部计算。

表 4-2　高　差　法

测点编号	水准尺读数			高差 h_i		改正后高差 $h_{i-改}$ (m)	改正后归化高程 RL (m)
	后视 (m)	中视 (m)	前视 (m)	上坡 (m)	下坡 (m)		
*TBM*1	2.191						49.873
					-0.314	-0.313	
A		2.505					49.560
				0.180		+0.181	
B		2.325					49.741
				0.829		+0.830	
*TP*1	3.019		1.496				50.570
				0.506		+0.507	
D		2.513					51.078
					-0.298	-0.297	
*TP*2	1.752		2.811				50.780
					-2.072	-2.071	
*TBM*2			3.824				48.710
Σ	6.962m		8.131m	1.515m	-2.684m	-1.163	-1.163
检核计算	Σ后视读数 - Σ前视读数 = 6.962 - 8.131 = -1.169m　$\sum h = \sum 上坡 + \sum 下坡 = 1.515 - 2.684 = -1.169$m 高差闭合差 $f_h = \sum h - (H_B - H_A)$ = -1.169 - (48.710 - 49.873) = -0.006m　容许的闭合差 $f_{h-容} = \pm m\sqrt{n} = \pm 5\sqrt{3} = \pm 9$mm						

注：日期、观测者、记簿者（如果不是观测者要单列）、测量标题、水准仪编号、天气条件以及其他有关项目如同水准尺读数都应记录。

（1）从 *TBM*1 至 A 有一缓慢下坡，在 *TBM*1 记录后视 2. 191m，A 点中视 2. 505m，所以从 *TBM*1 至 A 的高差为（2. 192 - 2. 505） = - 0. 314m，负号表示下坡。把 - 0. 314m 填入高差下坡栏的相应位置。

（2）重复这个步骤，A 至 B 的高差为（2. 505 - 2. 325） = + 0. 180m，正号表示上坡。把 + 0. 180m 填入高差上坡栏的相应位置。

（3）从 B 到 C 上升到第一个转点 *TP*1 高差为（2. 325 - 1. 496） = + 0. 829m。从 C 到 D 高差为（3. 019 - 2. 513） = + 0. 506m。这些数据都填入高差栏的相应位置。

（4）重复计算直到各段的高差栏计算完毕。然后一定要进行计算上的检核，这项检核是

$$\sum 后视读数 - \sum 前视读数 = \sum 上坡值 + \sum 下坡值$$

即

$$6.962 - 8.131 = 1.515 + (-2.684) = -1.169\text{m}$$

（5）用公式（4-9）计算水准测量闭合差。即

$$f_{\text{h}} = \sum h - (H_{\text{B}} - H_{\text{A}}) = -1.169 - (48.710 - 49.873) = -0.006\text{m}$$

容许的闭合差 $f_{\text{h-容}}$ 公式是

$$f_{\text{h-容}} = \pm m\sqrt{n} = \pm 5\sqrt{3} = \pm 9\text{mm}$$

由于闭合差 - 0. 006m 小于容许值，将闭合差反号平均分配到各段的高差。计算各段改正后的高差如下：

$$h_{1改} = -0.314 + 0.001 = -0.313\text{m}$$

$$h_{2改} = +0.180 + 0.001 = +0.181\text{m}$$

$$h_{3改} = +0.829 + 0.001 = +0.830\text{m}$$

$$\vdots$$

$$h_{6改} = -2.072 + 0.001 = -2.071\text{m}$$

把改正后的高差填入表 4-2 的相应位置。最后一列，从起点 *TBM*1 逐点推算从 A 至 *TP*2 各点改正后的归化高程。

例 4. 1　给出附合水准测量路线的实例，如图 4-23 所示。A 点与 E 点为两个临时水准点，其高程 $H_{\text{A}} = 89.763\text{m}$，$H_{\text{E}} = 93.504\text{m}$。测点 B、C 与 D 为待测定高程的 3 个未知点。各段测量的高差值及测站数注记于图 4-23。现要求计算未知点 B、C 与 D 平差后的高程。

【解】

野外测量手簿的成果已经整理并填入表 4-3，然后按表中的栏目进行计算。

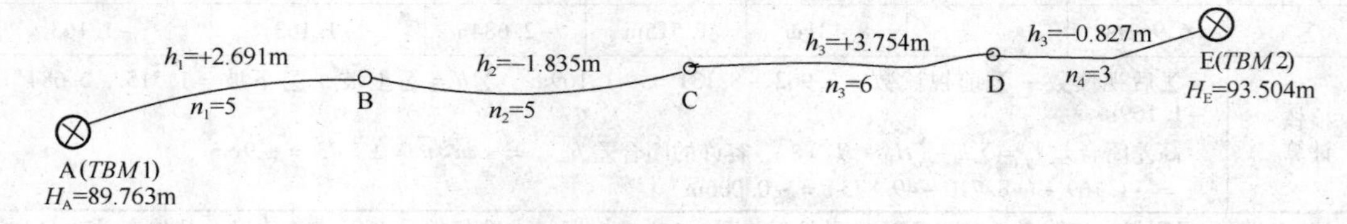

图 4-23　附合水准路线的一个实例

表 4-3　附后水准路线的计算

测点编号	测站数	测量高差 h_i (m)	高差改正数 v_i (m) v_i (m)	改正后高差 $h_{i-改}$ (m)	平差后高程 H (m)
A（*TBM*1）					89.763
	5	+2.691	−0.011	+2.680	
B					92.443
	5	−1.835	−0.011	−1.846	
C					90.597
	6	+3.754	−0.013	+3.741	
D					94.338
	3	−0.827	−0.017	−0.834	
E（*TBM*2）					93.504
Σ	19	+3.783	−0.042	+3.741	+3.741
检核计算	高差闭合差：$f_h = \sum h - (H_E - H_A) = +3.783 - (95.504 - 89.763) = +0.042$m 容许高差闭合差：$f_{h容} = \pm 12\sqrt{n} = \pm 12\sqrt{19} = \pm 52$mm 高差改正数：$v_1 = -0.0022 \times 5 = -0.011$m；$v_2 = -0.011$m；$v_3 = -0.0022 \times 6 = -0.013$m； $v_4 = -0.0022 \times 3 = -0.007$m				

（1）计算各段测量高差的总和：

$$\sum h = +2.691 - 1.835 + 3.754 - 0.827 = +3.783\text{m}$$

（2）两已知点的高差是：$H_E - H_A = 95.504 - 89.763 = +3.741$m

（3）按公式(4-9)计算高差闭合差是：

$$f_h = \sum h - (H_E - H_A) = +0.042\text{m}$$

（4）按我国公路测量容许高差闭合差是：

$$f_{h测} = \pm 12\sqrt{n} = \pm 12\sqrt{19} = \pm 52\text{mm}$$

（5）按下式计算高差改正数：

$$v_i = \frac{-f_h}{n} \times n_i \tag{4-13}$$

式中，n 为全路线测站总数；n_i 为某段测站数。

各段高差的改正数是：

$v_1 = -0.0022 \times 5 = -0.011$m　$v_2 = -0.011$m　$v_3 = -0.0022 \times 6 = -0.013$m

$v_4 = -0.0022 \times 3 = -0.007$m

（6）计算改正后高差($h_{i改}$)　　$h_{i改} = h_{i测} + v_i$

∴　$h_{1改} = h_{1测} + v_1 = +2.691 - 0.011 = +2.680$m

$h_{2改} = -1.846$m　$h_{3改} = +3.741$m　$h_{4改} = -0.834$m

（7）从 A 点开始逐点推算 B 点至 D 点的改正后高程。

4.7　仪器的检验与校正

为使仪器获得最好的结果，仪器应经常检验，如有必要，应该校正。测量仪器在建筑工地遭受连续的、过度的使用。这种情况应建立校准基地，对仪器进行一周一次检查。

4.7.1 圆水准器的检验与校正

虽然圆水准器灵敏度较低，然而在补偿器生效的功能上起着极其重要的作用。

补偿器的工作范围有限。如果圆水准器没有校正好，会导致视准轴线和竖轴过度倾斜，补偿器不能有效起作用，或当补偿器补偿时，摆锤系统大摆动致使在望远镜筒内卡住。因此，调整圆水准器非常必要。

从上述可以看出，圆水准器不但必须检测或校正，而且在使用中应严格居中。

校正的目的是确保圆水准器轴平行于仪器的竖轴。

为了调整圆水准器，首先用脚螺旋使气泡精确居中。然后，将仪器绕竖轴旋转 180°。如果气泡偏离，则旋转脚螺旋使气泡退回一半，再用校正螺丝使气泡精确居中。

4.7.2 视准轴的检验与校正

1. 微倾水准仪的检验与校正

微倾水准仪需要对视准误差进行校正。当水准管气泡居中时，如果视线不是真正水平就存在视准误差，即视线从水平面向上或向下倾斜。校正的目的是确保水准管轴平行于视准轴。

检测与校正的方法通常采用双桩法，该法实施如下，如图 4-24 所示。

（1）在地面上相隔约 60 ~ 80m 打下木桩 A 与 B，设距离为 D。

（2）水准仪安置在两木桩的中点 M 点，精确整平。水准尺依次立在两木桩上，获得水准尺读数 a_1 与 b_1，如图 4-24 所示。

因为 $AM = MB$，读数 a_1 与 b_1 中含有的误差 x 是相同的，该误差是由于视准轴误差引起的，其作用使视准轴倾斜 i 角，从图 4-24 中看出：

$a_1 - b_2 = (a'_1 + x) - (b'_2 + x) = a'_1 - b'_1 = h_{AB}$（即 A 与 B 两点的真正高差）

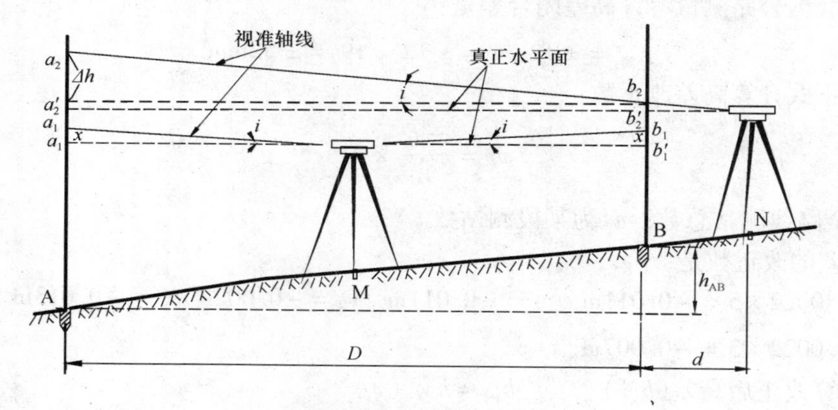

图 4-24　双桩法检测

（3）移水准仪到离 B 桩大约 3 ~ 5m 的 N 点，读 A 尺读数 a_2，读 B 尺读数 b_2，从（$a_2 - b_2$）计算 A 与 B 两点的非真正高差，即 $h'_{AB} = (a_2 - b_2)$。

（4）计算视准轴误差，即视准轴不水平，其倾斜角 i。我们设

$$\Delta h = h'_{AB} - h_{AB} = (a_2 - b_2) - h_{AB}$$
$$= [a'_2 + a_2a'_2 - (b'_2 + b_2b'_2)] - h_{AB}$$

顾及 $a'_2 - b'_2 = h_{AB}$，上述方程变为

$$\Delta h = a_2a'_2 - b_2b'_2$$

从图4-24可求得

$$i = \frac{\Delta h}{D} \times \rho'' = \frac{|h'_{AB} - h_{AB}|}{D} \times \rho'' \tag{4-14}$$

如果视准轴误差 $i < 20''$，水准仪不必校正，否则应校正。

（5）由于仪器仍在N点，我们首先推算观测A点标尺的正确读数 a'_2，由下列公式计算：

$$a'_2 = a_2 - \frac{(D + d) \times i}{\rho''}$$

最后，望远镜瞄A点的标尺，调节微倾螺旋直到视准轴线读取 a'_2。此时，视准轴线处于真正的水平面上，但是水准管气泡偏离中心。因此要调节水准管气泡的校正螺钉使气泡退回到中心。

此项校正应反复进行以确保校正成功。

2. 自动水准仪的检验与校正

关于自动水准仪，也必须做双桩检测法以确保一旦圆水准气泡居中时，补偿器能自动建立一条水平视线。

如上节所述，当推求出正确读数 a'_2 以后（见图4-24），选择两种方法之进行校正。对于大多数仪器可以用十字丝环的校正螺丝移动十字丝对准正确读数。然而，有些水准仪补偿器本身可以校正。由于操作十分精细，水准仪应送回厂家在实验室条件下进行校正。

4.8　地球曲率误差与折光差

对于短距离的水准线与水平线（通过测量员水准仪望远镜的视线）可认为是重合的，但对于长距离有必要对它们的偏差作改正。当考察水准线与水平线的偏差时，人们还必须考虑到由于地球大气使所有的视线向下折射的事实。虽然折光差的大小取决于大气的条件，但是一般认为大约是曲率误差的1/7。图4-25可看出折光差AB抵偿了部分曲率误差 AE，其结果纯误差为 $BE(c - r)$。

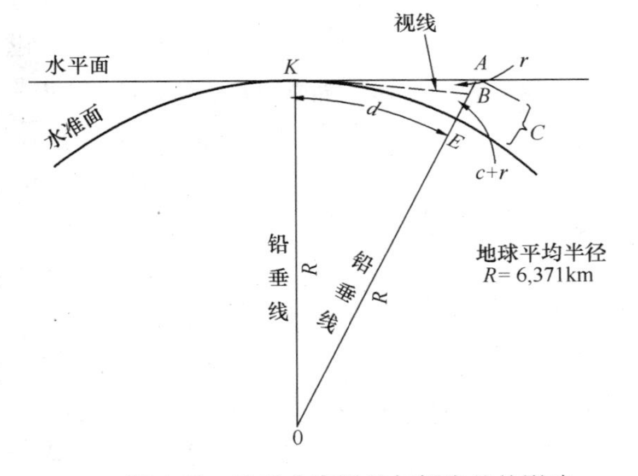

图4-25　地球曲率误差与折光差的影响

从图4-25看出曲率误差可推求如下：

$$(R + c)^2 = R^2 + D^2$$

$$R^2 + 2Rc + c^2 = R^2 + D^2$$

$$c(2R + c) = D^2$$

$$C = \frac{D^2}{2R + c} \approx \frac{D^2}{2R} \tag{4-15}$$

式中，c 为曲率误差；D 为视线长度（$D = KA$，见图4-21），R为地球的平均半径（即6371km）。

折光差受大气压、温度及地理位置的影响，但是，正如前面提到的，一般认为大约是曲率误差的1/7。曲率和折光差的共同影响可用公式（4-16）：

$$f = c - r = \frac{D^2}{2R} - \frac{1}{7} \times \frac{D^2}{2R} = \frac{3D^2}{7R} = 0.43\frac{D^2}{R} \tag{4-16}$$

从表4-4可看出，对于几何水准测量误差f相对很小，即使精密水准测量，标杆读数的距离很少超过60m，这项误差微不足道。实际上，如果相继的后视与前视从仪器到标尺的距离相等，则有效地消除了这类误差。

表4-4　不同距离（D）对应地球曲率与折光差的共同影响（f）

距离 D（m）	30	60	100	120	150	200	300	350	400	500	1000
f（m）	0.0001	0.0002	0.0007	0.001	0.002	0.003	0.006	0.008	0.011	0.017	0.068

4.9　三角高程测量

三角高程测量用于困难地区，例如山区，传统几何水准测量受阻。三角高程测量也可以用于高差很大而水平距很短的情况，例如高耸的建筑物或悬崖。测量相关两点间的竖直角及斜距，斜距测量使用电子测距仪（EDM），竖直角（或天顶距）用经纬仪测。

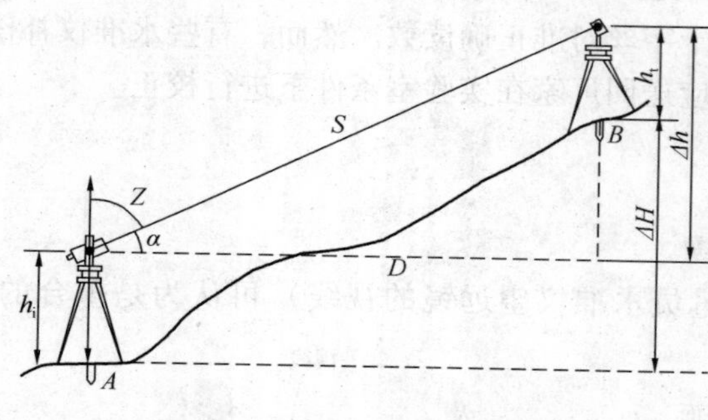

图4-26　短距离三角高程测量

全站仪包含有计算和显示水平距离及高差的算法程序。这种崭新的设备使得三角高程测量广泛运用于高差测量，包括等高线法测图。

4.9.1　短距离的三角高程测量

当测量角度时，从图4-26可看出

$$\Delta h = S\sin\alpha \tag{4-17}$$

当使用天顶距时

$$\Delta h = S\cos z \tag{4-18}$$

如果用水平距离时

$$\Delta h = D\tan\alpha = D\cot z \tag{4-19}$$

因此A与B两点的高差是

$$\Delta H = h_i + \Delta h - h_t$$

$$\Delta H = \Delta h + h_i - h_t \tag{4-20}$$

式中，h_i 为A点仪器高；h_t 为B点目标高。

这是三角测量的基本概念。仰角时，竖直角为正，俯角时，竖直角为负。天顶距总是正的，但是当天顶距大于90°，自然产生负的高差。

怎样来选定短距离是通过考虑地球曲率与折光差对照所期待的精度来推求的。对于100m地球曲率与折光差的综合影响为0.7mm；200m综合影响为3mm；300m综合影响为6mm；400m综合影响为11mm；500m综合影响为17mm（见表4-4）。

三角高程测量的基本方程（4-20）写为

$$\Delta H = S\sin a + h_i - h_t \tag{4-21}$$

取微分：

$$d(\Delta H) = \sin\alpha dS + S\cos\alpha d\alpha + dh_i - dh_t$$

转为标准误差：

$$\sigma_{\Delta H}^2 = (\sin\alpha \cdot \sigma_s)^2 + \left(S\cos\alpha \cdot \frac{\sigma_\alpha}{\rho''}\right)^2 + \sigma_i^2 + \sigma_t^2$$

假设竖直角 $\alpha = 5°$，其标准差 $\sigma_\alpha = 5''$（$= 0.000024$ 弧度），$S = 300\text{m}$，其标准差 $\sigma_s = 10\text{mm}$ 以及 $\sigma_i = \sigma_t = 2\text{mm}$，代入上面方程得：

$$\sigma_{\Delta H}^2 = 0.87^2 + 7.2^2 + 2^2 + 2^2$$

$$\sigma_{\Delta H} = 7.8\text{mm}$$

这个数值相似于300m距离地球曲率与折光差影响的大小，这说明短距离三角高程不应超过300m（参看表4-4）。这也说明：当竖直角小时，测量距离 S 的精度不很重要。但是，测量竖直角的精度却很重要，因此要求经纬仪在盘左盘右位置多次测量。

4.9.2　长距离的三角高程测量

对于长距离必须考虑地球曲率（c）及折光差（r）从图4-27可以看出A与B之间的高差（ΔH）是

$$\begin{aligned}\Delta H &= GB = GF + FE + EH - HD - DB \\ &= h_i + c + \Delta h - r - h_t \\ &= \Delta h + h_i - h_t + (c - r)\end{aligned} \tag{4-21}$$

可以看出它是短距离高差基本公式加上曲率与折光差改正。

虽然视线折向 D，但望远镜指向 H，所以测量角度从水平线起算的 α，按照 $S\sin\alpha = \Delta h = EH$，由于折射要做的改正等于 HD。

4.9.3　对向观测

对向观测是采用从A和从B的观测，最后结果取其算术平均值。如果假定两端视线是对称的且同时观测，那么，曲率与折光差的影响可以消除。例如，仰视时，$(c - r)$ 加上正的竖直角，增加了高差；俯视时，$(c - r)$ 加上负的竖直角，减少了高差，因此这两个数值取平均消除了曲率与折光差的影响。

从A观测B，方程（4－10）给出

$$\Delta H_{AB} = \Delta h_{AB} + h_{Ai} - h_{Bt} + (c - r)$$

从B观测A，方程（4－10）给出

$$\Delta H_{BA} = \Delta h_{BA} + h_{Bi} - h_{At} + (c - r)$$

对向观测的两个观测值取平均得

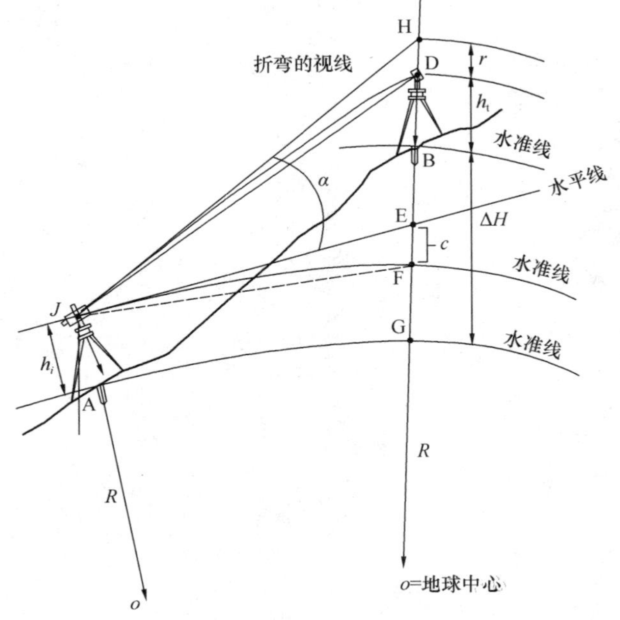

图4-27　长距离三角高程测量

$$\Delta H_{AB} = \frac{1}{2}(\Delta H_{AB} - \Delta H_{BA}) = \frac{1}{2}(\Delta h_{AB} - \Delta h_{BA}) + \frac{1}{2}(h_{Ai} - h_{Bi}) + \frac{1}{2}(h_{At} - h_{Bt}) \tag{4-22}$$

从式（4-22）可以看出，$(c-r)$ 已被消除了。这种描述不完全很客观，因为假定双方向的视线对称要求相同的地面条件、相同大气条件以及要求同时观测。

第5章　角 度 测 量

5.1　水平角与垂直角的定义

角度测量是测量及建筑业中所需的最重要工作之一。测量角度通常是用经纬仪或全站仪。

图5-1表示A、B两个目标点，经纬仪或全站仪T安置在地面点G上方的三脚架上。点A高于仪器并位于通过T水平面的上方，而点B较低，位于水平面的下方。在T点，仪器被安置在三脚架上，与G点垂直距离为h。

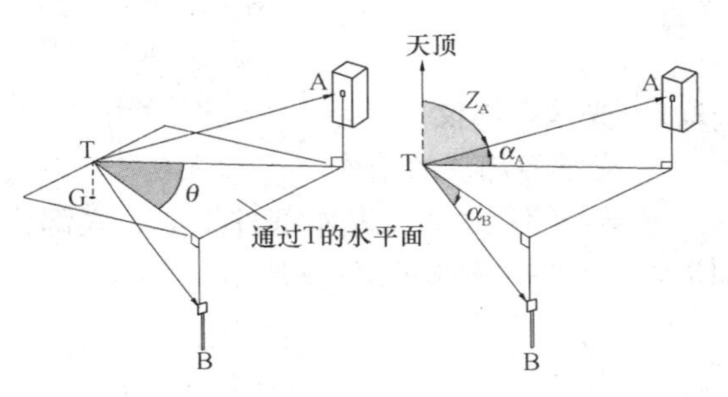

图5-1　水平角、垂直角与天顶距

在T点A与B之间的水平角不是包含A、T与B组成的斜面上的角度，而是通过T包含视线TA与TB垂直平面之间水平面的角度θ。从T瞄A的垂直角为仰角α_A，从T瞄B的垂直角为仰角α_B。

另外的角度常称为天顶角。在垂直面内，仪器上方垂直方向与视线方向之间的角度，例如，图5-1中Z_A。

水平角用于确定控制测量的方位与方向，在测图中定位细部以及对放样各种结构物都非常重要。

垂直角用于通过三角学的方法确定点的高程，还用于计算水平距离的倾斜改正。

为了测量水平角与垂直角，经纬仪或全站仪要对G点对中，必须整平以使仪器的角度读数系统处于水平面和铅垂面。虽然仪器对中、整平确保在T点测量水平角与假定仪器安置在G点测量水平角一样，但是，从T点测量垂直角与从G点测量的垂直角是不一样的。当计算高差时，必须考虑仪器高h。

5.2　经纬仪概述

经纬仪是精密仪器，经纬仪有两种基本类型，光学机械型与电子数字型，这两种仪器都能直接读到1′，20″，1″或0.1″，依仪器的精度而异。电子经纬仪能自动显示角度读数。图5-2显示BOIF（北京光学仪器厂）的光学经纬仪与电子经纬仪。

不论在工地还是其他地方，电子经纬仪是角度测量主流的仪器。但是光学经纬仪仍在使用。光学与电子这两种经纬仪本章都将介绍，但本章重点放在光学经纬仪。

在我国根据仪器的精度，经纬仪分为DJ_{07}、DJ_1、DJ_2、DJ_6和DJ_{15}五个等级，“D”与

“J”分别表示“大地测量仪器”和“经纬仪”汉语拼音的第一个字母，它们的下标表示使用仪器测量方向的精度。例如 DJ_6 表示 6″级经纬仪。

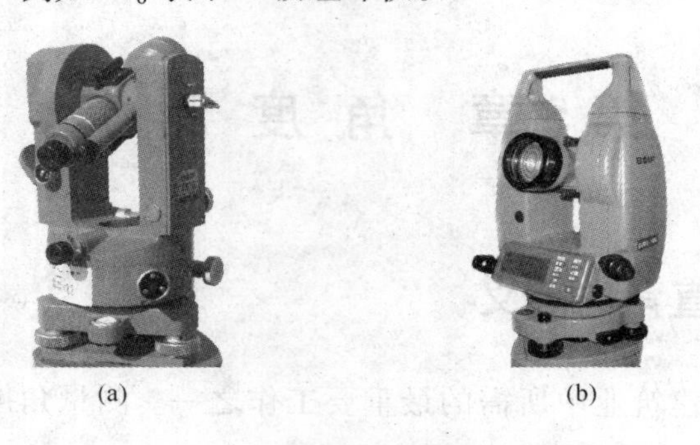

(a)　　(b)

图 5-2　BOIF 光学经纬仪与电子经纬仪

（a）光学经纬仪；（b）电子经纬仪

5.2.1　光学经纬仪

1. 光学经纬仪的结构

光学经纬仪由 3 个基本部件组成：底部为基座，中间为水平度盘（下盘）以及上部为照准部。它的结构如图 5-3a 所示。

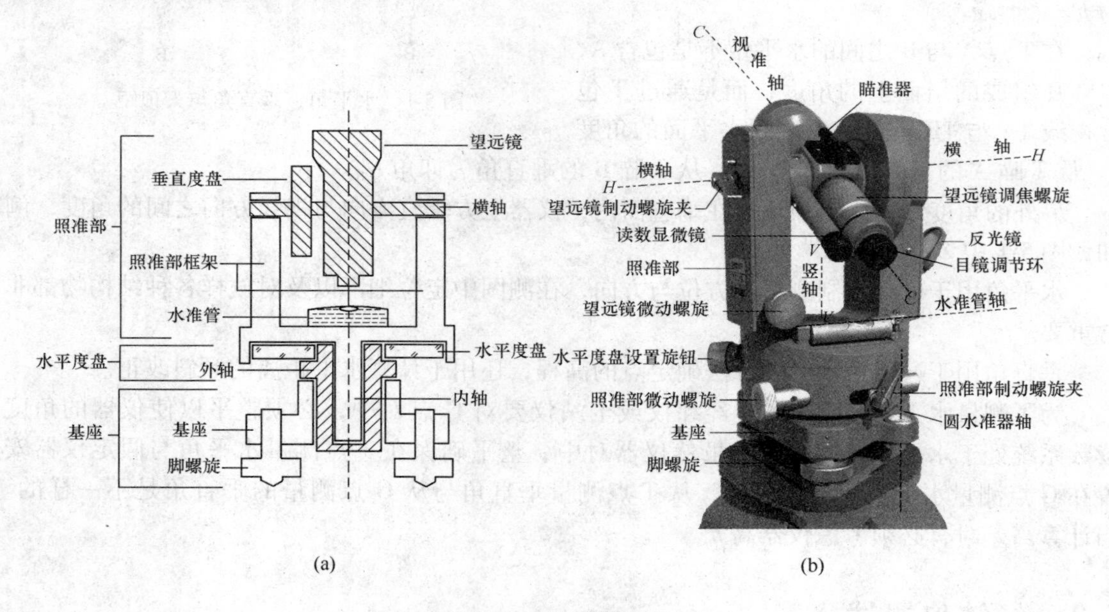

(a)　　(b)

图 5-3　光学经纬仪

（1）基座支撑仪器其他部分，基座又由脚螺旋支撑。因此，基座可在三脚架顶上独立整平。

光学经纬仪设有装置可使照准部从基座卸下。然后，用觇牌或其他设备精确在经纬仪的原位置对中。这样确保角度测量和长度测量在相同位置之间进行，从而减少误差，尤其是对中误差。

（2）水平度盘是固定在外轴套上，外轴套与内轴（即仪器的垂直轴）是同心轴。大多数经纬仪内轴是空心的，为光学对中器提供视线通道。它取代了经纬仪用垂球对地面点对中。

水平度盘与垂直度盘是玻璃制作的，其上刻蚀角度分划。经纬仪有多种类型玻璃度盘，读数精度各异，从1′到1″。然而，在工程测量中最普遍使用的是20″和1″读数的经纬仪。

现代的经纬仪没有下盘的固定螺旋夹与微动螺旋，在仪器内部有一变换水平度盘位置的装置，仅用一个控制，即所谓水平度盘设置钮或螺旋，它能转动水平度盘到任何所需的读数（见图5-3b）。

（3）照准部安装在内轴上，内轴向下装入度盘组合件的外轴内，并能在空心外轴内自由转动（见图5-3a）。度盘组合件包含一个被照准部外壳遮盖的水平度盘。外轴插入基座的空心轴筒内并能在基座轴筒内自由转动。

照准部包含直立的机架（框架），望远镜、垂直度盘、水平度盘、度盘的读数系统，上盘水准器（即水准管）、上盘制动与微动螺旋等。

2. 瞄准设备

照准部包含一个固定在水平轴（也称横轴）上的望远瞄准设备，水平轴的轴承又安放在机架上。因此，望远镜能够自由指向任何方向。具体来说，望远镜可以绕水平轴180°旋转，旋转后望远镜指向反方向，这个过程也称为望远镜纵转。

望远镜包含4个部分：（1）物镜，在望远镜的前端，它产生倒立缩小的目标影像；（2）目镜，它能放大十字丝并能根据观测者的视力对十字丝调焦；（3）十字丝环，它在望远镜筒后部提供十字丝；（4）调焦透镜，它能前后移动调焦目标的影像。

由于物镜形成的影像是倒立的，大多数经纬仪的目镜设计为直立影像。能直立影像的望远镜称为正像望远镜，而其他的望远镜称为倒像望远镜。

蚀刻在光阑上的十字线（通常称十字丝），精确形式各异，但是大多数仪器都有一条中间的水平线和两条较短的水平线，一条在中间的水平线上方，一条在下方，称之为视距丝（见图5-4）。测量垂直角，读数用中间的水平线。

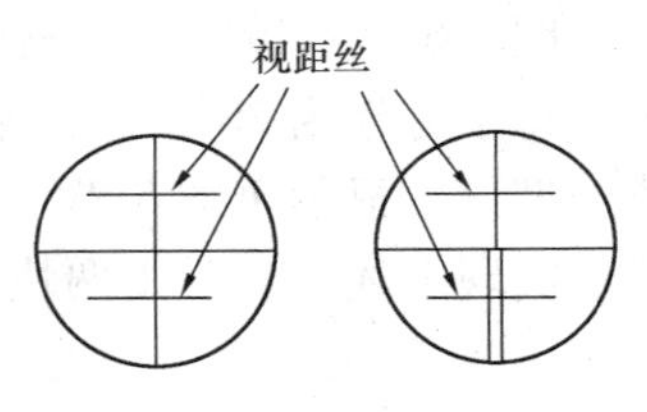

图5-4 经纬仪十字丝的式样

视准轴线是十字丝交点与物镜光学中心连线及其延长线，也可称为视线。

在开始观测之前，转目镜调焦螺旋使十字丝十分清晰。目标影像处于十字丝之前或之后而产生的十字丝视差必须消除。当通过望远镜观测时，观测者眼睛上下左右移动可以检查十字丝视差是否存在。如果目标影像与十字丝不重合，观测者眼睛移动将会引起十字丝相对于目标影像的移动。

3. 垂直度盘

垂直度盘垂直安装于水平轴，并在望远镜的一侧边，垂直度盘随望远镜一起转。

测量员面对望远镜目镜端，如果垂直度盘在望远镜的左边，则称经纬仪位于“盘左”（FL）；如果望远镜倒转，垂直度盘在望远镜的右边，称为“盘右”（FR）。

测量垂直角是倚赖垂直度盘指标，指标被设置在重力方向而言可借助于（a）高度水准器（指标水准器）或（b）自动补偿器自动安置。后一种方法在现代的经纬仪中普遍使用。

现代经纬仪测量天顶角，当盘左望远镜垂直向上瞄准，垂直度盘读数为0°。当盘左视线水平，垂直度盘读数为90°；当盘右视线水平，读数为270°。电子经纬仪没有明显的度盘，盘左与盘右通常称为位置Ⅰ与位置Ⅱ，用罗马数字印制在仪器体上。

4. 度盘读数系统

经纬仪度盘读数是借助于主望远镜旁小的辅助读数望远镜（见图 5-3b）。小的圆反光镜折射光进入用于读盘的棱镜与透镜复杂的系统。

有两种基本类型的读数系统：光学显微镜读数系统与光学测微计读数系统。

（1）光学显微镜读数系统一般用于分辨率为 20" 或更小的经纬仪。水平度盘与垂直度盘同时显示，借助于辅助读数望远镜直接读数。

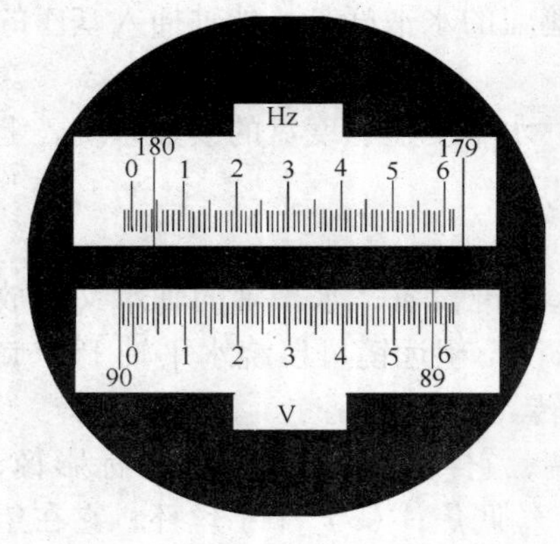

图 5-5　DJ6 光学经纬仪读数

给出直接读数的辅助望远镜可以是“线显微镜”或“盘显微镜”。

线显微镜是用一条精细的线条蚀刻在格线板作为指标，以此指标读度盘。

盘显微镜在影像平面上有一个盘，它的长度相应于度盘分划线间隔。图 5-5 说明了这种类型的读数系统，显微盘上 0′到 60′的长度等于度盘上 1°。这种类型仪器通常称为直读经纬仪，可估读到 6″。图 5-5 中水平度盘（Hz）读 180°04′24″，而垂直度盘（V）读 89°57′30″。

（2）光学测微计读数系统广泛使用线显微镜与盘显微镜相结合，和精密水准仪的平行玻璃板原理相同。

一种更精密的经纬仪，读至 1″，使用符合显微镜。它使度盘直径两边分划重合读取平均，因此平均读数消除度盘偏心误差。

图 5-6 显示借助光学测微计螺旋使度盘对径分划重合，度盘 94°和 95°之间分格数是 3，因此每个分格为 20′，所以水平度盘读 94°10′，测微盘读 2′44″，总读数为 94°12′44″。该仪器改进型读数显示于图 5-7。

实现上述过程是由两块平行玻璃板相对转动直到度盘对径分划重合。

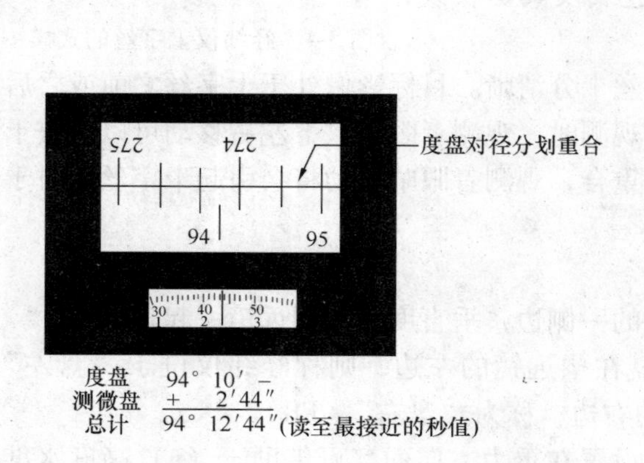

图 5-6　WildT2（老式）经纬仪读数系统

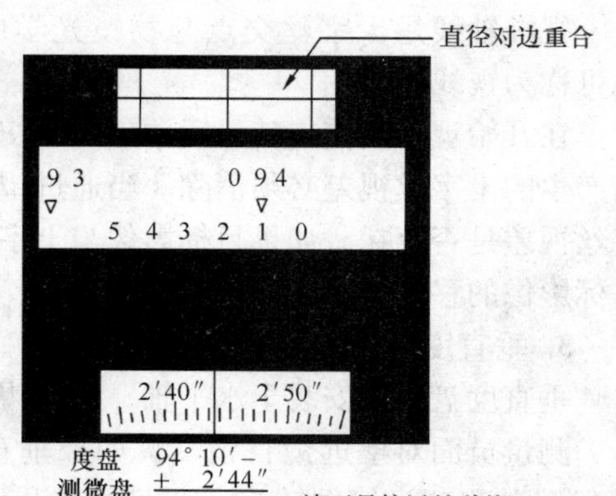

图 5-7　WildT2（新式）经纬仪读数系统

5.2.2　电子经纬仪

电子经纬仪是由精密光学器件、机械器件、电子扫描度盘、电子传感器和微处理机组成的，在微处理器的控制下，按度盘位置信息，自动以数字显示角值（水平角、垂直角）。

电子经纬仪包含有感知垂直轴及望远镜旋转的传感器，转换这些电子旋转量为水平角和垂直角，并显示在液晶显示器（LCD）或在发光二极管显示器（LED）上。读数器能把读数记录到传统的野外手簿或存储到数据收集器为以后打印或计算。仪器含有一个钟摆式的补偿器或其他某种装置以指示绝对垂直方向的垂直度盘读数。通过简单地按仪器上的按钮把度盘设置为零读数。某些电子经纬仪在后视目标之前，水平度盘读数可预设某个所要求的角度。水平度盘既可顺时针又可逆时针切换测量。当仪器转动时，度盘连续显示附近的度数。

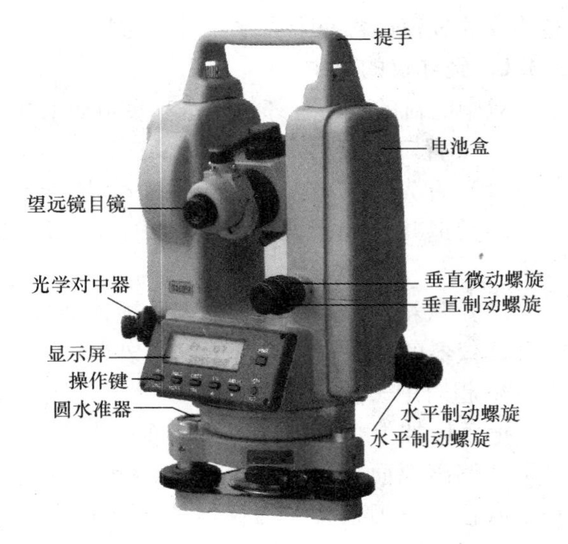

图 5-8　ET－02 激光电子经纬仪

电子经纬仪包含有数字化的度盘取代了光学经纬仪的玻璃度盘，其水平与垂直度盘实为光电二极管扫描的编码器，扫描产生的电脉冲转换为角度值以数字形式存储或显示。编码器有两种类型——增量编码系统与绝对编码系统，大多数电子经纬仪采用增量编码系统。

角度读数器读至1″，精度为0.5″～20″。测量员应检查新仪器的说明书确定其精度，有些仪器读至1″可能精度仅有5″。

图5-8显示ET－02激光电子经纬仪，它是由中国南方测绘仪器厂生产，仪器部件的名称注于图5-8上。ET－02激光电子经纬仪的键盘显示于图5-9。

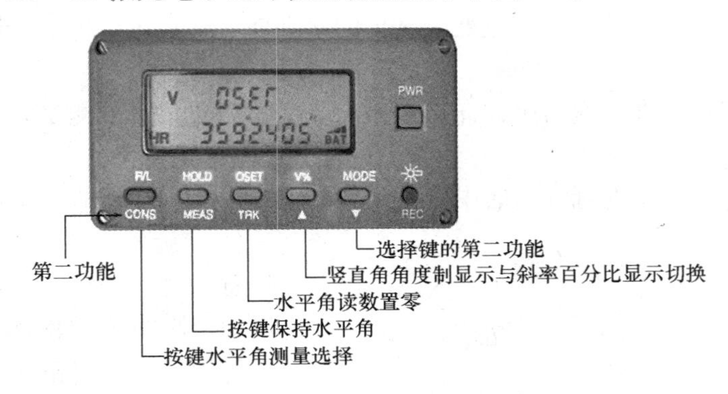

图 5-9　ET－02 激光电子经纬仪的键盘

【R/L】键：左旋测角与右旋测角转换键。右旋测角即当水平度盘度数顺时针增加时，显示屏显示HR；左旋测角即当水平度盘度数逆时针增加时，显示屏显示HL。

【OSET】键：读数归零键，连续按2次，水平度盘读数设为0°00′00″。

其他各键名称及功能详见图5-9的注释。

5.3　经纬仪的安置

安置经纬仪的目的是使经纬仪的竖轴线位于通过地面标志的中心的铅垂线方向上。安置经纬仪的过程包含两个步骤：对中与整平。

5.3.1 经纬仪的对中

对中的目的是使经纬仪的水平度盘中心位于通过地面标志的中心的铅垂线上。

1. 安置三脚架

（1）首先将三脚架安置于地面标志的上方，展开三脚架腿以适应观测者的高度。从三脚架后退几步，检查三脚架头中心是否在地面标志垂直线上方。目视三脚架头尽可能水平。

（2）从箱内取出经纬仪，牢固地连结到三脚架头上，调节三个脚螺旋达到大约相同高度（即仪器的基座大约平行于三脚架头）。

2. 粗对中——使用垂球

把垂球挂上，注意垂球与地面标志之间的偏差情况，然后通过移动架腿进行对中。如果垂球偏离地面标志不大，将三脚架腿稳固地插入地面。最后，稍松开中心连结螺旋，并稍微移动三脚架头上的仪器使垂球对准地面标志的正上方。这种方式进对中称为粗对中，其误差应在5mm范围之内。

3. 精对中——使用光学对中器

通过调节3个脚螺旋使仪器粗略整平。观测者略松开中心连结螺旋，稍微移动三脚架头上的仪器，通过对中器观察直至地面标志影像与参考标志（十字丝或一个圆圈）重合为止。由于三脚架顶面不是完全水平，应该注意仪器必须平移。如果平移时又旋转，则光学对中器的视线不再是垂直的。平移操作后，还必须检查整平与对中。

在这个阶段，经纬仪已几乎水平。但要精确整平仪器，要使用照准部的水准器。

5.3.2 经纬仪的整平

经纬仪整平的目的是使水平度盘处于真正的水平面。整平包括粗平与精平两个步骤。

1. 粗平

旋转脚螺旋使圆水准气泡居中，详见4.3.1节（看第4章水准测量）。

2. 精平

旋转脚螺旋使上盘水准管气泡居中，其步骤如下：

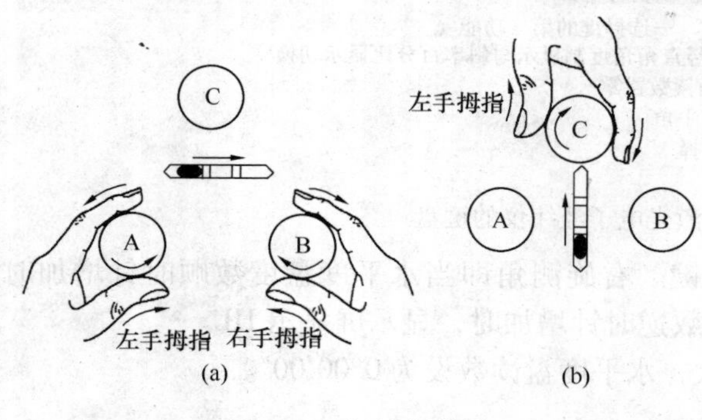

图5-10 经纬仪整平

（1）旋转经纬仪使上盘水准管平行于任意一对脚螺旋，旋转这对脚螺旋使气泡居中。脚螺旋要以相反方向同时旋转，记住气泡移动方向相应于左手大拇指的移动方向（见图5-10a）。

（2）仪器旋转90°，再居中气泡，但仅用第3个脚螺旋（见图5-10b）。

（3）重复上述过程直到这两个位置气泡都居中。

（4）现在转动仪器使得上盘水准管处于与原先成180°的位置。如果上盘水准管气泡仍然居中，则经纬仪是水平的，仪器不必调整。如果水准管气泡不居中，误差等于气泡偏离中心量的一半。例如，如果气泡向左偏离2格，误差为左偏离1格。

在粗平后有可能用电子整平经纬仪，这种情况，电子气泡显示于显示屏，使用 3 个脚螺旋去居中，如图 5-11 所示。如果仪器没有恰当地水平，电子整平会给出警示，重新粗平后电子整平才会起作用。

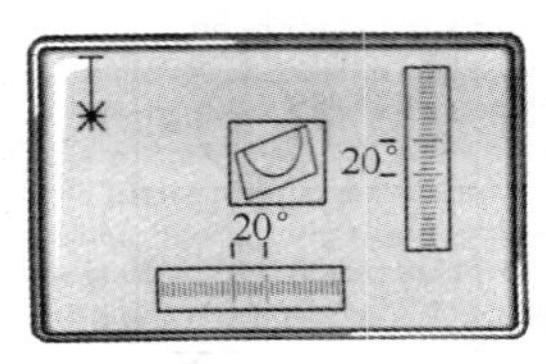

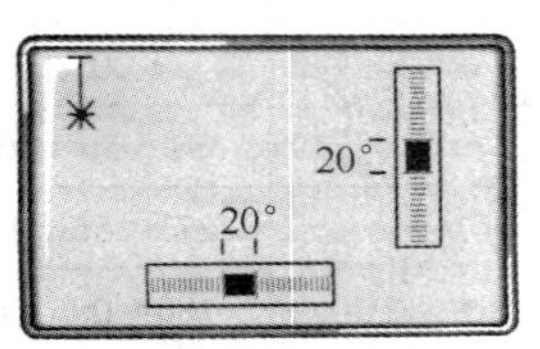

图 5-11　电子整平

在使用光学对中器或激光对中器进行对中时，关键在于使用之前，经纬仪要适当地整平。如果经纬仪不水平，对中器轴线将不垂直，尽管对中器可能显示居中，但经纬仪将是偏离中心。

如果测点占用很长时间，则有必要经常地隔一段时间检查整平与对中，特别是仪器安置在松软的地面或在强阳光之下工作。

5.4　水平角测量

假设经纬仪安置到 O 点，按上节方法精确地对中与整平。现要测定远方目标 A 与 B 之间的水平角，标杆必须设置在 A 点与 B 点，测量水平角步骤如下：

（1）经纬仪置于盘左（FL）位置，设置水平度盘为零或零附近。现代的光学经纬仪有一个水平度盘设置螺旋，它能转动水平度盘读任意所需读数。水平度盘设置之后，按一下水平度盘设置螺旋使照准部与水平度盘分离（具体操作取决于所用的仪器）。

（2）松开上盘制动螺旋夹，用光学瞄准器对准目标 A，然后固定上盘制动螺旋夹。用上盘微动螺旋使望远镜的纵丝精确对准目标 A（通常对目标的基部）。必须注意检查与消除视差，最后记下水平度盘读数“a_1”。

（3）顺时针转动仪器瞄准第二个目标 B，类似于上一步得另一读数“b_1”，因此，A 与 B 间的水平角为 β_1，$\beta_1 = b_1 - a_1$。

上述观测步骤完成上半测回。

（4）倒转望远镜使经纬仪为盘右位置（FR），开始另外半测回。但应记住：从 B 目标开始反时针旋转到 A 目标。两读数之差为第二半测回角 β_2，取 β_1 与 β_2 平均得最后水平角。

上述这些步骤完成一个测回。

（5）如果有必要，安置不同的起始读数开始，测几个测回。习惯上，沿度盘圈配置不同的起始读数以减小度盘分划不均的影响。如果测量 n 测回，每次起始读数变换 $\frac{180°}{n}$。因此，开始第二测回，瞄准 A 设置水平度盘读数为 90°。

重复（2）～（4）所述的各步完成第二测回。

每个测站至少 2 个测回，以便于角度计算时检测误差。由于每个测回独立观测，两个测回必须计算与比较之后，仪器与三脚架才能搬迁。应该注意，每次瞄准目标都要使用纵丝的中央同一部分，以便减少瞄准的误差。

表 5-1 水平角测量与计算

测站	目标	盘位	水平度盘读数 (° ′ ″)	半测回水平角 (° ′ ″)	一测回水平角 (° ′ ″)	各测回平均值 (° ′ ″)
O (第1测回)	A	L	0 00 24	91 55 42	91 56 00	91 55 50
	B		91 56 06			
	B	R	271 56 54	91 56 18		
	A		180 00 36			
O (第2测回)	A	L	90 00 12	91 55 42	91 55 40	
	B		181 55 54			
	B	R	1 56 30	91 55 38		
	A		270 00 52			

5.5 垂直角测量

5.5.1 垂直角测量的步骤

假设经纬仪安置到测站，按 5.3 节叙述方法精确对中与整平。

(1) 经纬仪置于盘左位置 (FL)，转动水平或垂直微动螺旋使十字丝的水平丝精确对准目标 A。

(2) 每次读数前，指标水准管气泡（如果附有）必须居中，然后读取垂直度盘读数“L”。

现代的经纬仪具有对垂直度盘的自动指标，读数给出相对于仪器零位的角度，可能给出相对于水平面“+角度”或“-角度”。大多数现代的仪器给出相对垂直向上为0°的角度。

(3) 倒转望远镜使经纬仪为盘右位置 (FR)，开始另外半测回。转动水平或垂直微动螺旋使十字丝的水平丝精确对准目标 A。

(4) 指标水准管气泡必须居中，然后读取垂直度盘读数“R”。

上述步骤完成一测回垂直角测量，所有的记录与计算列于表 5-2。

表 5-2 垂直角测量与计算

测站	目标	盘位	垂直度盘读数 (° ′ ″)	垂直角		指标差 (″)
				半测回 (° ′ ″)	一测回 (° ′ ″)	
O	A	L	78 18 18	11 41 42	11 41 51	+9
		R	281 42 00	11 42 00		
O	B	L	96 32 48	-6 32 48	-6 32 34	+14
		R	263 27 40	-6 32 20		

5.5.2 计算垂直角及竖盘指标差

这里有两种垂直度盘类型，一种是度盘分划从 0°到 360°顺时针增加，如图 5-12a 所示，而另一种度盘分划从 0°～360°反时针增加，如图 5-12b 所示。

当视线水平和指标水准管（如果附有）气泡居中时，理论上垂直度盘读数为90°或90°的倍数依不同仪器类型而定。实际上垂直度盘读数与其理论值之差，称为指标差 x（见图5-12）。

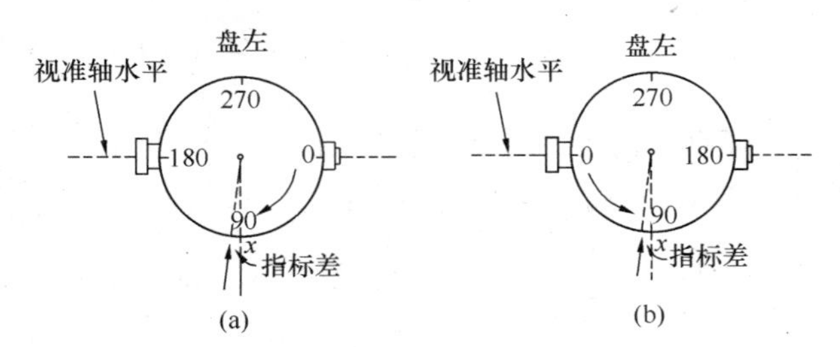

图5-12　两种类型垂直度盘
（a）度盘刻划顺时针增加；
（b）度盘刻划逆时针增加

下面将讨论顺时针增加的度盘分划的经纬仪（图5-12a）。当盘左时，垂直度读数为 L，指标差为 x，从图5-13b可得：

$$\alpha = 90° + x - L \tag{5-1}$$

设
$$\alpha_L = 90° - L \tag{5-2}$$

因此
$$\alpha = \alpha_L + x \tag{5-3}$$

式中，α 表示垂直角，α_L 表示盘左半测回垂直角。

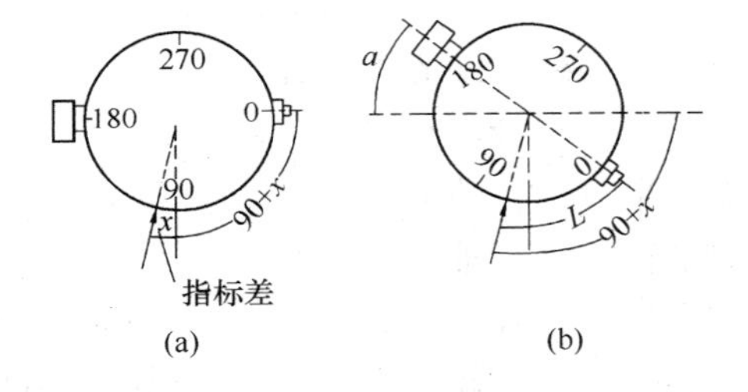

图5-13　盘左时垂直角的计算
（a）当盘左，视线水平的情况；（b）当盘左，视线倾斜的情况

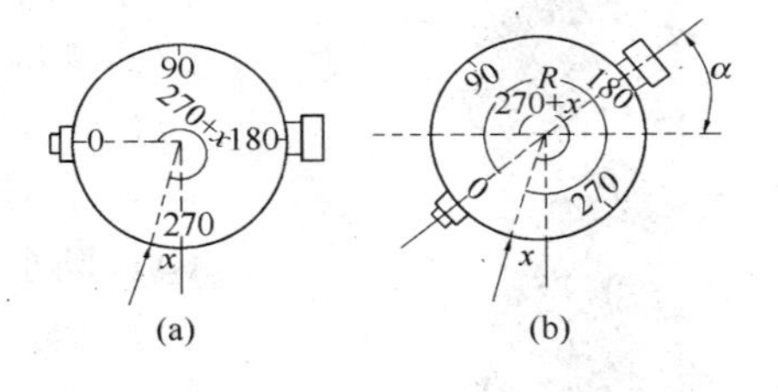

图5-14　盘右时垂直角的计算
（a）盘右视线水平时；（b）盘右视线倾斜时

与上述类似，从图5-14b可得：

$$\alpha = R - (270° + x) \tag{5-4}$$

设
$$\alpha_R = R - 270° \tag{5-5}$$

因此
$$\alpha = \alpha_R - x \tag{5-6}$$

式中，α 表示垂直角，α_R 表示盘右半测回垂直角。

（5-3）式加（5-6）式得

$$\alpha = \frac{1}{2}(\alpha_L + \alpha_R) \tag{5-7}$$

从（5-3）式减（5-6）式得

$$x = \frac{1}{2}(\alpha_R - \alpha_L)$$

将式（5-2）与式（5-5）代入上式得

$$x = \frac{1}{2}(L + R - 360°) \tag{5-8}$$

公式（5-7）与（5-8）也适用于图 5-12b 的情况，即度盘分划逆时针增加的情况，但仅是计算 α_L 和 α_R 公式不同，它们是：

$$\alpha_L = L - 90°$$

和

$$\alpha_R = 270° - R$$

所有的记录与计算列于表 5-2。

5.6　经纬仪的检验与校正

5.6.1　经纬仪各轴的几何关系

经纬仪各轴线的布局如图 5-15 所示。最重要的几何关系如下：

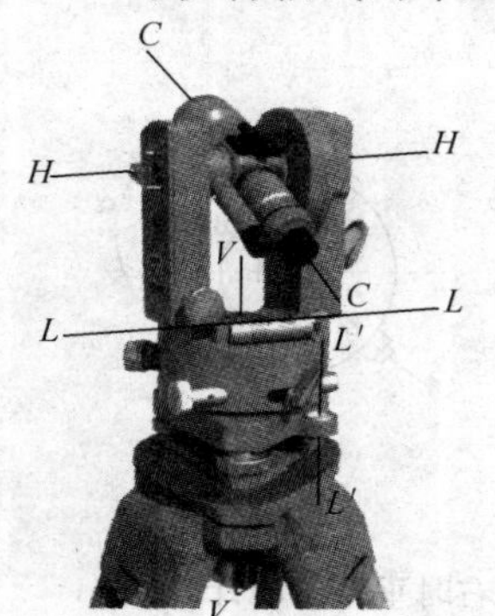

图 5-15　经纬仪的轴线

（1）水准管轴应处于垂直于垂直轴（经纬仪的旋转轴）的平面上，即 $LL \perp VV$。

（2）视准轴应垂直于水平轴，即 $CC \perp HH$。

（3）十字丝纵应垂直于水平轴。

（4）水平轴应垂直于垂直轴，即 $HH \perp VV$。

（5）圆水准器轴应平行于垂直轴，即 $L'L'//VV$

（6）垂直度盘指标应处于正确位置，即当视准轴水平和指标水准管（如果附有）气泡居中，理论上垂直度盘读数是 90°或 0°依经纬仪垂直度盘类型而定。

经纬仪调整的目的是使其主要轴线处于正确的几何关系。

5.6.2　长水准管的检验与校正

安装在照准部的盘水准器，它为高灵敏度的管状水准器，通常也称长水准管。当长水准管气泡居中时仪器的竖轴线必须真正垂直。仪器厂家制造的仪器竖轴是严格地垂直于安装有长水准管的水平度盘。为了确保仪器竖轴真正垂直，有必要调准长水准管轴线平行于水平度盘。

检验：设水准管轴线不平行于水平度盘而有角度误差 e。水准管平行于一对脚螺旋，粗略整平，然后转 90°，用第三个脚螺旋再整平。现在退回原先的位置，用一对脚螺旋精确整平，将出现图 5-16a 情况。现在仪器转 180°将出图 5-16b 情况，气泡偏歪量表示仪器误差的 2 倍（$2e$）。

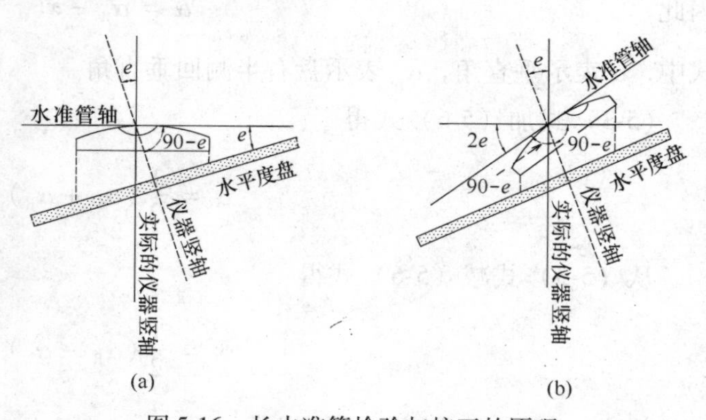

图 5-16　长水准管检验与校正的原理

校正：旋转一对脚螺旋使气泡

严格退回一半，使仪器轴转动 e 角，因而它真正垂直了，即使那里没有可用的校正工具，这个阶段仪器可以用了，气泡偏离相应于 e 的量，现在调节其校正螺丝抬高或降低水准管的一端使气泡居中。

5.6.3 视准轴的检验与校正

检验的目的是确保视准轴垂直于横轴。如果视准轴不在它的正位确置，那么，当望远镜绕水平轴旋转时，视准轴线扫描出圆锥面而不是垂直平面。

检验：安置仪器并精确整平，望远镜瞄准约50m外与仪器高同高位置A点精细标志（图5-17）。如果视线垂直于水平轴，那么当望远镜倒转180°，在视线的方向大约50m且与仪器高大约同高位置放置的水平尺得记号 A_1。但是，假设视线与水平轴交角（$90°-c$），因而，当盘左望远镜瞄目标A后，倒转望远镜在水平尺上给出记号 A_L（图5-17a）。现在仪器处于盘右的位置，重新瞄准A，倒转望远镜后在水平尺上给出记号 A_R（图5-17b），从图中明显看出：A_LA_R 距离代表4倍仪器误差（$4c$）。

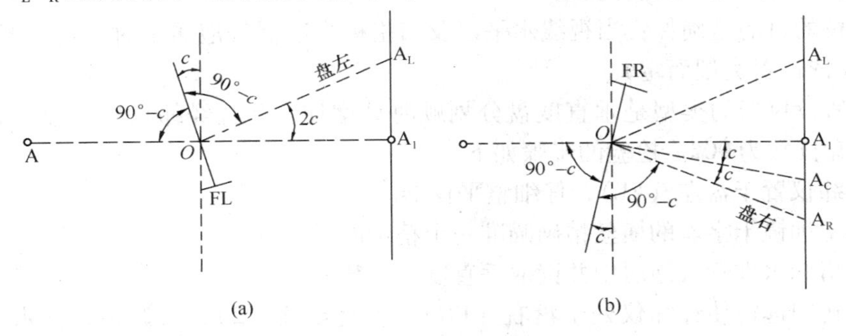

图5-17 视准轴的检验与校正的原理

（a）盘左；（b）盘右

校正：调节十字丝环两边的调节螺丝使十字丝按方位移动从 A_R 移到 A_1 与 A_R 的中点 A_C，这就是 A_LA_R 距离的1/4。

十字丝的移动可能影响纵丝相对于横轴的位置，即它应该垂直于横轴。这对通过望远镜对细圆点垂直上下移动来检测。如果纵丝离开该圆点，则说明纵丝不垂直于横轴，应用调节螺丝改正。

5.6.4 十字丝环的检验与校正

在进行视准轴检校时，移动了十字丝环。这可能会影响纵丝垂直于横轴的设置。

假定视准轴的检校刚刚完成，应采用下列步骤：

（1）重新仔细地整平仪器，以任一盘位瞄准A。

（2）在瞄准A时上下移动望远镜。如果纵丝仍然在A点，则设置是正确的。

（3）如果有必要校正，应转动十字丝环一直到在望远镜上下转动时而纵丝不离开A点为止。

检测5.6.3与5.6.4是独立进行，两者检测同时着手一直到每一项都获得满意的结果。

仪器厂家制造十字丝环已使得纵丝与横丝相垂直，设置纵丝垂直因而也设置水平丝水平。

5.6.5 横轴的检测

这项检测的目的是为了安置横轴垂直于竖轴。当仪器整平时，则横轴也水平。如果横轴不水平，望远镜不能给出铅垂面，这将产生不正确的垂直角与水平角。

应该指出的是，由于现代仪器的优秀构造，这项校正可以不做，因此，大多数仪器不提供这种校正。仪器厂家宣称这项误差在现代的设备中不会出现。然而，在实践中通过盘左与盘右的平均读数将获得满意的结果。

5.6.6　圆水准器的检验与校正

这项检校的目的是确保圆水准器轴平行于仪器的垂直轴。否则，即使圆水准器气泡居中，仪器竖轴不是处于大致铅垂的位置。此外，如果圆水准器没有适当地整平，电子水准器将不起作用。

圆水准器检校很简单。当长水准管按5.6.2节所述方法安置和检校之后，必须把仪器严格地整平。此时，如果圆水准气泡偏离中心，那么通过或多或少调节圆水准器下面的3个校正螺丝使圆水准器气泡居中即可。

5.6.7　垂直度盘指标的检验与校正

这项检校的目的是确保：当视线水平以及当指标水准器气泡居中时，垂直度盘读90°或90°的倍数（依仪器类型而定）。

检验：假设仪器的类型是垂直度盘分划顺时针增加，并且经纬仪置于盘左，视线水平时，垂直度盘读数为90°。检验的步骤如下：

（1）经纬仪置于盘左（FL），仔细整平仪器。

（2）用望远镜十字丝的横丝精确瞄准一个精细的目标点。

（3）使指标水准管气泡居中并读取垂直度盘读数 L。

（4）纵转望远镜使经纬仪处于盘右（FR），重复步骤（2），再调整指标水准管气泡并且读取垂直度盘读数 R。

例如，检验结果：盘左垂直度盘读数 $L=78°18'18''$，盘右垂直度盘读数 $R=281°42'00''$。按公式（5-8）求得

$$x=\frac{1}{2}(78°18'18''+281°42'00''-360°)=+9''$$

校正：经纬仪位于盘右，望远镜视线仍然在瞄准目标点。

（1）方程（5-4）说明：当经纬仪位于盘右时，垂直度盘的正确读数应为（$R-x$），即（$R-x$）$=281°42'00''-09''=281°41'51''$。通过旋转指标水准管水平螺旋对准这个读数。这导致指标水准管气泡偏离中心。

（2）调节指标水准管的校正螺丝使指标水准管气泡居中。

对于自动垂直指标的经纬仪，应查阅厂家仪器手册取得正确的校正方法。

5.6.8　光学对中器的检验与校正

光学对中器的视准轴必须与仪器的竖轴重合。有两种检验的方法，视所用仪器类型而定。

（1）如果光学对中器是安装在照准部并能绕着竖轴旋转（图5-18a）。

首先把仪器架好，并精平。然后在仪器下面地面固定一张纸，光学对中器轴线与纸相交处做一标志。转动照准部180°做第二标志。如果两个标志重合，光学对中器不必调整。如果不重合，两个标志的中点就是光学对中器轴的正确位置。查阅仪器使用手册调节对中器的

十字环或调节对中器的物镜。

（2）如果光学对中器安装在基座上，它不能旋转（图5-18b）。

经纬仪的一边安放在工作台上，经纬仪底座面向某墙面，标出通过光学对中器在墙上的交点。转动基座180°再在墙上做标志。如果两个标志重合，对中器不必调整。如果不重合，对中器十字丝环应调节使其交于两标志的中点。

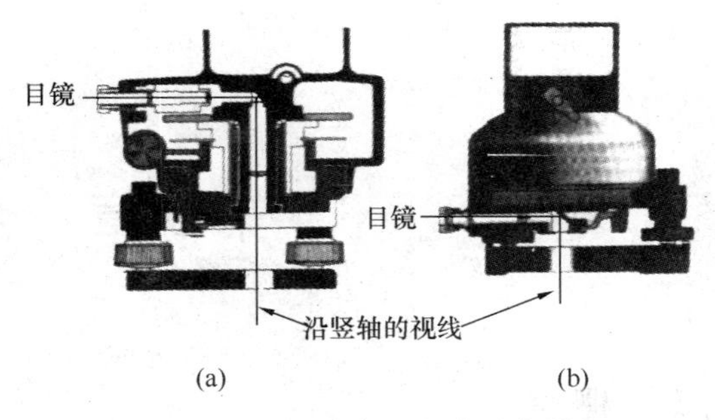

图5-18　两种类型光学对中器

（a）光学对中器安装在照准部；

（b）光学对中器安装在基座

第6章　全站仪

6.1　概述

今日的测量，全球定位系统的使用不断增加，然而最普遍使用的仪器还是全站仪。

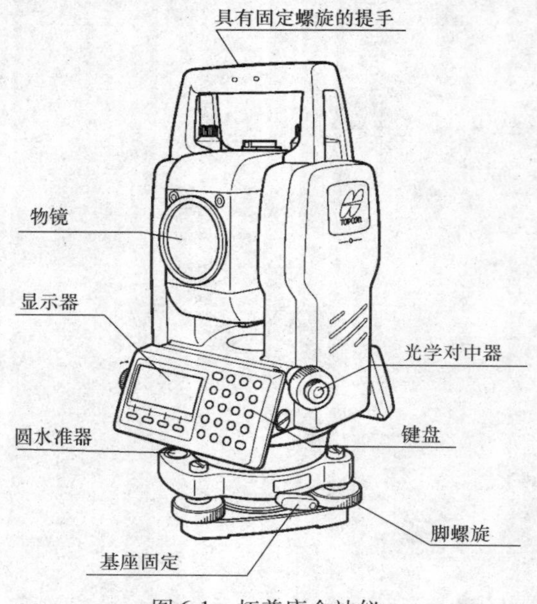

图6-1　拓普康全站仪

全站仪是集成电子经纬仪和电子测距仪的电子仪器。一台典型的全站仪如图6-1所示。它的外观很像电子经纬仪。全站仪把测距与测角结合一起放在同一单元。用全站仪进行测量更容易、更快、精度更高。

全站仪能完成各种不同的测量任务，并能存储大量的数据。它能平均多次角度测量，平均多次距离测量，并能测定坐标、高程、面积，计算大气及仪器的改正等等。全站仪的所有功能由微处理器（或电脑）控制，通过键盘存储与显示。

使用全站仪，仪器置于待测线的一端，某种形式的反射器安置于另一端，并使仪器与反射器之间视线无障碍物，如图6-2所示，图中反射器是装在细部杆上的棱镜，望远镜瞄准这个棱镜。测量顺序是先初始化，然后从测站对着反射器发射信号，部分信号返回测站。这个过程几秒钟，给出斜距以及水平角和垂直角。

全站仪也可用于无反射器模式，这就有可能测量遥远点位或悬崖绝壁的区域。甚至危险

图6-2　用全站仪测量

图6-3　机器人全站仪在使用

的区域，例如滑坡的区域，用这个方法都能安全有效地测量。

有些仪器有电动机驱动，使用自动化目标识别寻找并锁定棱镜，这个过程完全自动化不需要人工干预。有些全站仪可受细部杆控制，由一个人操纵测量，如图 6-3 所示。

6.2 全站仪的特点

1. 望远镜视准轴与 EDM 测距发射光轴同轴

全站仪系统包含电子经纬仪及袖珍的 EDM 装置以使得 EDM 测距发射光轴和望远镜的视准轴为同轴的光学系统。因此测量时瞄准目标棱镜中心就能同时测定目标点的水平角、垂直角和斜距。

2. 全站仪能执行各种操作，能监测仪器的状态并具有广泛的多种内置程序

全站仪能读并记录水平角度与垂直角度以及斜距，全站仪的微处理器能执行各种操作：例如平均多次角度观测值，平均多次距离测量值，确定坐标 X，Y，Z，气象改正和仪器改正等等。

全站仪能自动监测仪器的状态（例如电池状态、水平轴与垂直轴状态、返回信号强度等等）。

典型的全站仪内置程序包括测定点位、设置自由站、方位角计算、悬高测量、偏距测量、放样点位和面积测量等等。所有这些将在 6.5 节讨论。

图 6-4 尼康 DTM750 全站仪

3. 全站仪具有大容量的数据存储能力和数据传输能力

许多现代的全站仪都有机载数据存储器，因而可省去手工数据采集器。例如，托普康 GTS300 全站仪数据能存储到仪器内存中，其容量约 1300 点。从全站仪经 RS-232 电缆把数据传输到电脑。

有些全站仪有内存卡（每卡大约存 2000 点）。有些厂家使用 PCMCIA 卡，它由 PCMCIA 读卡器直接读入计算机。另外一些全站仪通过电脑同仪器（或它的键盘）连结把数据可直接下载到电脑。

图 6-4 显示尼康 DTM750 全站仪设置有读卡器，仪器上方读卡器插应用程序卡，下方读卡器插数据存储卡。数据存储卡储满后，取出用 PCMCIA 读卡器（现代笔记本电脑的标准配置）读入电脑。全站仪操作软件兼容 MS-DOS，允许用户定义附加应用软件。

4. 全站仪有双轴液晶补偿器

全站仪引入双轴液晶补偿器。内置的双轴倾斜传感器不间断地监测垂直轴在两个方向的倾斜量，即在视准轴方向（即 x 轴）和水平轴方向（即 y 轴）的倾斜量，它计算补偿值并自动改正水平角和垂直角。索佳 1000 全站仪的双轴补偿器说明显示于图 6-5。

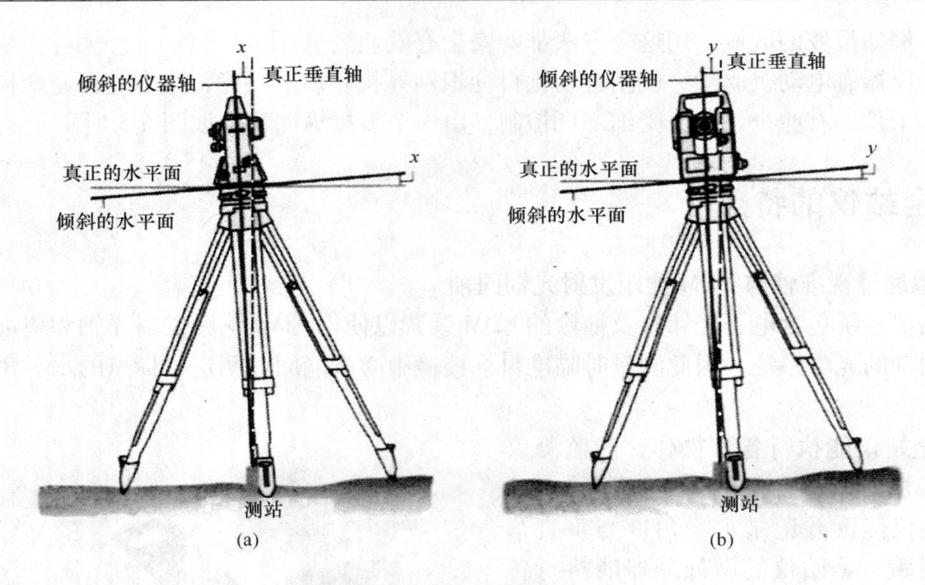

图 6-5　双轴补偿器示意图
（a）侧面图；（b）正面图

6.3　全站仪的结构及其组件

6.3.1　全站仪的结构

全站仪的结构包含 3 个部分：测距单元或 EDM，测角设备或电子经纬仪以及机载微处理器。大多数 EDM 单元发射红外载波束，单棱镜测程 300 ~ 3300m（1000 ~ 11000 英尺），三棱镜测程 900 ~ 11000m（3000 ~ 36000 英尺）。

角度测量单元实际上是一台电子经纬仪。水平盘与垂直盘采用某种特殊的编码，由光电二极管读取。刻划度盘可以 2 种方式编码——增量编码系统和绝对编码系统。大多数全站仪具有增量编码。无论采用哪一种，瞄准目标以后都可以指定 0°或任意所需的度数。显示度盘分辨率 0.5 ~ 20″。

全站仪有机载微处理器，可监测仪器状态，并对测量的数据进行改正。微处理器控制角度与距离的获取，然后计算水平距离、高差、坐标等等。

全站仪结构包括垂直轴、横轴和视准轴，这些轴线应相互垂直。其他部分包括有脚螺旋的基座，带显示屏的键盘和安装在机架上的望远镜，它可绕横轴旋转。同经纬仪一样进行整平，调节脚螺旋使盘水准器或电子水准器气泡居中。望远镜能翻转成为盘左（或盘Ⅰ）和盘右（或盘Ⅱ）的位置。全站仪绕垂直轴水平转动由水平制动夹和微动螺旋控制。望远镜绕横轴旋转由垂直制动夹和微动螺旋控制。这些螺旋由无限制的摩擦驱动装置替代，不需要固定螺旋夹。有些全站仪引入双速粗驱动可以快速定位目标和精细准确定位目标。所有的全站仪都有测角用的水平度盘与垂直度盘，用度分秒显示或以 gon 显示。全站仪的角度精度依仪器而异，用在施工测量的多数仪器是 1 ~ 10″。所有的全站仪都有光学对中器或激光对中器等。

现行的全站仪大概分为 3 类。用于建筑与施工用的仪器测程较短，角度技术规格较低，但它们制造更坚固耐用，具有防水和高度防尘。证据显示，所有的工地中95%的距离测量在500m以内，10″角度精度已满足大多数放样作业。索佳10系列和托普康GPT 3005是这一类仪器的实例，如图6-6所示。第二类全站仪包含那些测量应用型的全站仪，有更高的角度与距离技术要求，更多功能，更高数据存储和处理的能力。索佳030R系列和天宝3600DR全站仪是这一类仪器的实例，如图6-7所示。第三类包含机动化全站仪，它技术规格最优，价格最贵。徕卡TPS1200和天宝S6全站仪是这一类仪器的实例，如图6-8所示。这些全站仪使用自动目标识别（ATR）技术，需要全站仪安装ATR传感器。

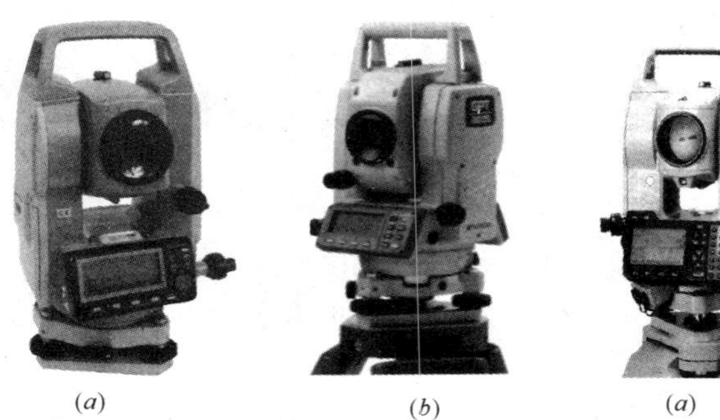

图6-6 建筑与施工用全站仪
（a）索佳10系列；（b）托普康GPT3005

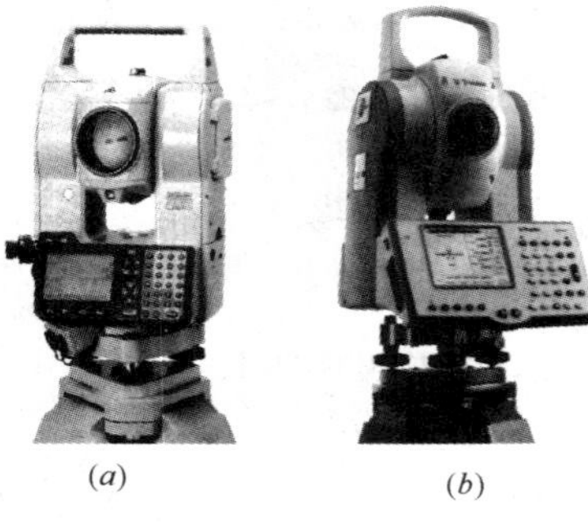

图6-7 测量应用型全站仪
（a）索佳030R；（b）天宝3600DR

虽然角度测量与距离测量可分别进行，但多数是共同使用，结合起来在控制测量、绘图及放样中界定位置使用。

同全站仪一样，工地测量越来越多使用GPS设备。预言这种趋势还在继续，尽管GPS的应用不断增长。但全站仪在工地测量中还是主导的仪器，将延续某一定时期，最终将发现两者相互补充而不是抗争。

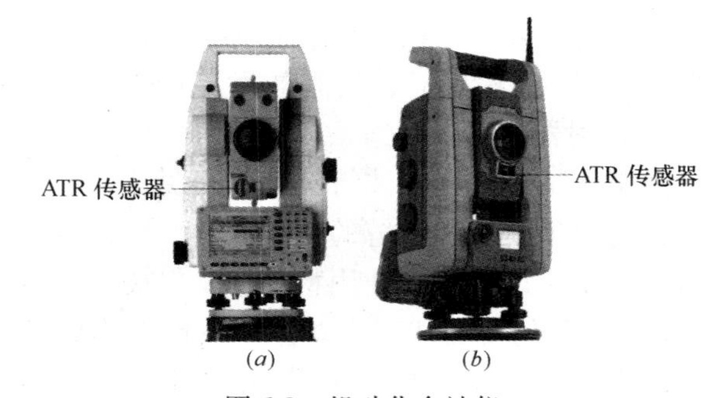

图6-8 机动化全站仪
（a）徕卡TPS1200；（b）天宝S6

6.3.2 全站仪的组件及附件

1. 键盘与显示屏

全站仪通过控制板来激活，控制板包含一个键盘与多行的液晶显示屏（LCD）。全站仪显示屏能防潮并可照亮，有的加入对比度调节以适应不同角度观察。许多仪器有两个控制板更便于使用。键盘可供用户选择和实施不同的测量模式，能改变仪器参数，允许访问特定的软件功能。有些键盘加入执行特定任务的多功能键，而另一些键盘用键去激活并显示菜单系统，使全站仪可作为电脑使用。

全站仪电子化记录角度与距离，以数字形式记录原始数据（斜距、垂直角与水平角）。为了绘图，从键盘输入代码定义正在观测的地物。当数据下载到室内电脑与绘图仪进行处理就会更快。在数字键盘上，代码仅由数字表示，而在字母的键盘上，代码可由数字和（或）字符表示，它给出多种用途，扩大了应用范围。现在很多全站仪有大的图形显示屏，在工地上有可能进行数据编辑。

图 6-9 给出键盘与显示屏的实例。有些全站仪的键盘与显示屏可以从仪器上卸下，与其他全站仪或 GPS 接收机互换，这就是所谓集成化测量，应用简单的界面使不同仪器与系统之间数据共享。这些综合的键盘与显示屏不仅控制仪器，而且还成为数据存储的设备。

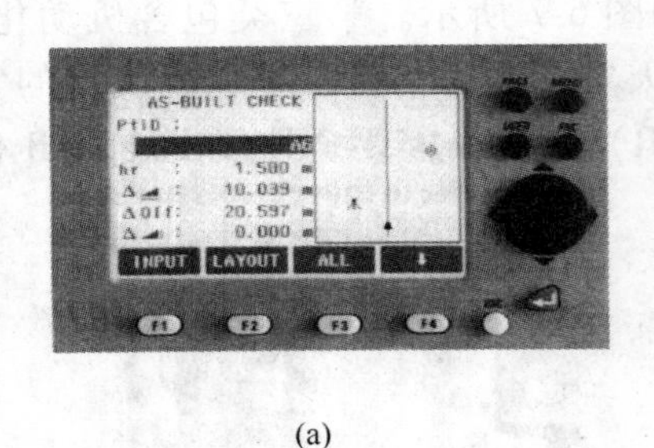

(a)

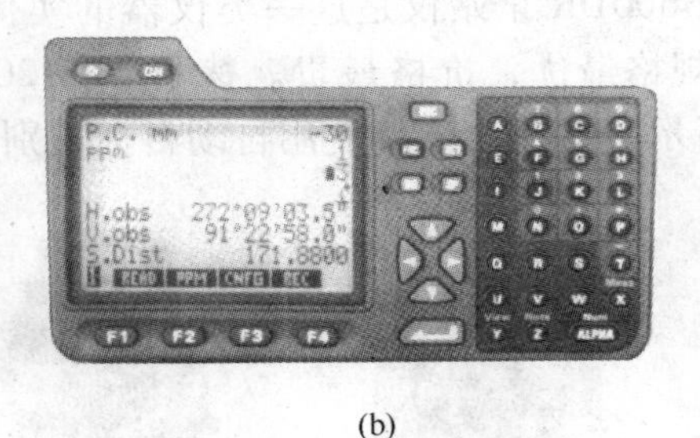

(b)

图 6-9　全站仪键盘与显示屏的实例

（a）徕卡 TCR410C 具有工地所需的基本功能；（b）索佳 030R 具有更强的测绘功能

2. 电源

用于测量仪器有 3 种类型充电电池：镍合金氢化物（NiMh）电池，镍镉（NiCad）电池和锂离子电池。镍合金氢化物（NiMh）电池使用最普遍，因为它符合标准的录像机电池并且容量比镍镉电池大。然而，镍镉电池已流行很多年仍广泛使用，比起镍合金氢化物（NiMh）电池更多循环充电。锂离子电池，它具有易充电和好维护的优点。作为替代充电电池，一些仪器可用 AA 碱性电池。大多数全站仪都有电源指示器，有些有自动节能特点，当仪器不使用一段时间后，仪器自动关机或进入休眠待命模式。

3. 反射器

由于全站仪发射电波与脉冲是可见光或红外光（它传播像光，但不可见）。平面镜原本可作为反射器，但遗憾的是这需要反射镜精确定向，因为发射电波与脉冲传播很窄。为了解决这个问题，总是使用特殊的反射棱镜即所谓正立方角锥棱镜。它是用一个与立方体面夹角为 45°的平面切割正立方体玻璃而构成的立方角锥，如图 6-10a 所示。这样的棱镜返回的电波总能精确地平行于入射波的路径，除非对于棱镜前面的垂直线形成的入射角大约超过 20°范围，如图 6-10b 所示。由此看出，棱镜的定向要求不高，因此在现场能做到快速安置。

现有各种测程的反射棱镜，有适用于短距离测量（小棱镜）和长距离测量（大棱镜和组合棱镜）。可固定于三脚架的单棱镜和三棱镜组（即为附有光学觇标的组合棱镜），如

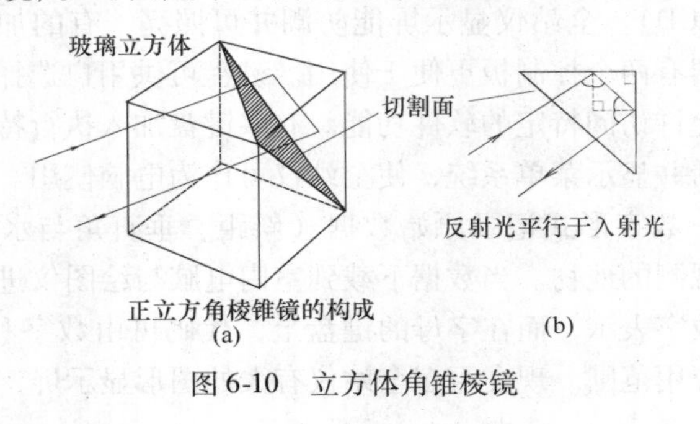

图 6-10　立方体角锥棱镜

图 6-11　单棱镜与三棱镜组

图6-11所示。

与反射棱镜相联系的是棱镜常数——这是棱镜的有效中心与对中点铅垂线之间的距离。由于玻璃折射的特性，当载波穿过棱镜时减慢速度，因此其有效中心通常位于物理中心的后面，如图6-12所示。棱镜常数典型的是 -30mm 或 -40mm，这个值键入全站仪作为每次测量距离的自动改正值。如果忽视或应用不正确，则在所有距离测量中将产生系统误差，应用外业任何方法都无法消除。因此，使用棱镜进行全站仪各种测量，验明正确的棱镜常数非常重要。即使对反射片测量和采用无反射镜模式也有必要输入棱镜常数。

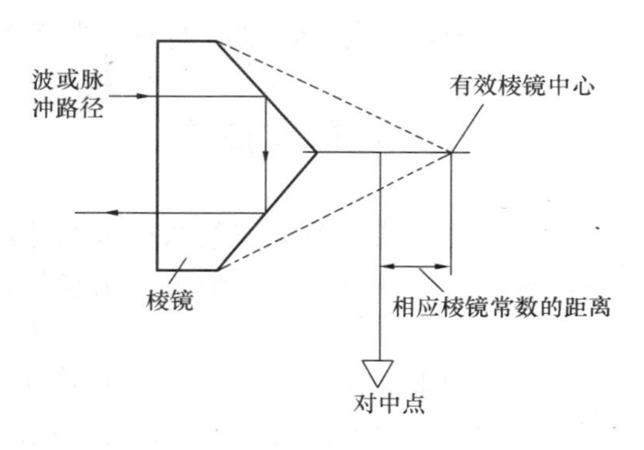

图6-12　棱镜常数

6.4　全站仪的使用

全站仪开始工作时通常需要对测站点进行对中与整平，对中用光学对中器，整平用盘水准器或用仪器显示器上的电子气泡。大多数全站仪都允许用户随时观察仪器偏歪情况，如图6-13所示。当仪器不水平超出某规定范围，全站仪会发出警报并停止工作。

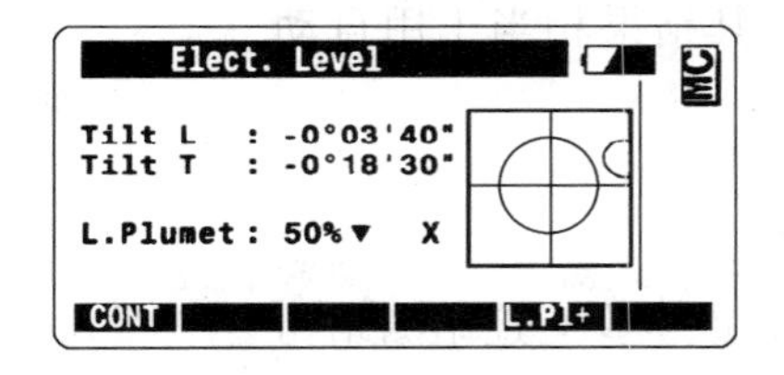

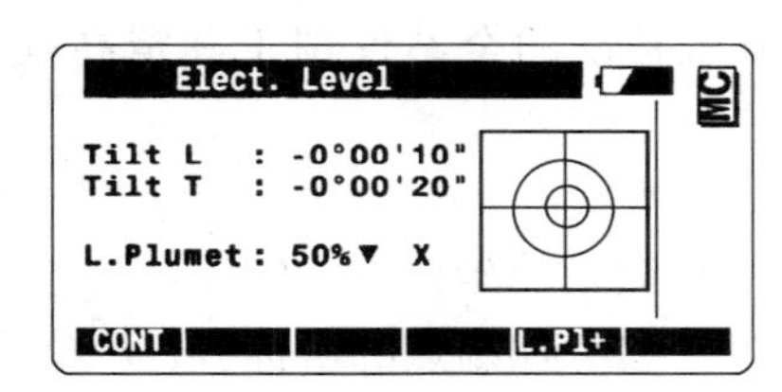

图6-13　徕卡公司全站仪显示倾斜误差

当仪器在测站上整平和对中之后，首先打开电源并初始化系统，通常转望远镜通过天顶距90°。其次，某些参数通过控制面板输入。这些参数主要有下列：

（1）距离测量和角度测量的单位，距离以米或英尺和角度以度或哥恩（gon）分别输入；

（2）温度，华氏度（℉）或摄氏度（℃），大气毫米汞柱或英寸汞柱（有些系统有内置温度和气压传感器）；

（3）棱镜常数；

（4）地球曲率及折光改正设置；

（5）角度或距离测量重复数（计算平均值）。

当安置全站仪并输入必要的参数初始化之后，现在可以开始各种测量工作。

例如，测量水平角，望远镜瞄准后视站的反射镜和目标并按零设置键通过键盘设水平度盘为0°或预设的方位角。转动仪器并指向第二点，水平角自动显示在显示屏上并记录在野外手簿中，或存储在机内存储器或数据采集器，以供今后使用。

全站仪测量仪器至反射器的斜距以及垂直角与水平角，然后仪器内的微处理器计算斜距的分量即水平距和高差。另外微处理器用这些计算分量及直线的方位角确定直线的南北分量与东西分量和新点的坐标，这些新点的坐标存入内存储器。

6.5　全站仪的基本操作

全站仪及（或）其附属的数据采集器已程序化实现多种测量功能。所有全站仪的作业方案需要仪器站以及至少一个参考站参与，以便使其后所有相联系的各测站都可以用坐标 X，Y，Z 来确定。因为测量方案要求在开始数据采集之前，计划好测站并标志好后视站，计划的测站的坐标 X，Y，Z 必须上传或用手工的方法输入到全站仪的微处理器；另外，后视站的坐标也必须上传或用手工的方法输入至全站仪的微处理器，至少测站至参考站的方位角必须输入，然而，如果后视站的坐标已上传或手工输入，那么至后视站的必要方位角很容易用存储的数据反算求得。仪器安置之后，在进行仪器定向之前，仪器与棱镜的高度必须量测与记录。

6.5.1　点位测定

仪器正确定向以后，任何视点的坐标（北、东、高程）就能确定。显示与记录格式如下：NEZ 或 ENZ（见图 6-14）。选择格式反映了软件程序处理野外数据的格式的要求。在此时，给视点加编号及属性代码（点的描述），这些与定位数据一同记录。这个方案广泛应用于地形测量。这里应指出：具有双轴补偿的全站仪用于确定高程，其结果相当于用自动或数字水准仪。

点位测定方案：

已知：（1）测站的坐标 N，E 及 Z。

（2）参考控制点的坐标 N，E 及 Z，或至少有连结测站与控制点连线的方位角。

在外业之前，上述测站与参考站信息从计算机上传至全站仪或手工输入。

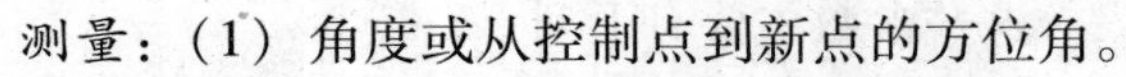

测量：（1）角度或从控制点到新点的方位角。

（2）从测站至新点的距离。

计算：（1）新点的坐标 N，E 及 Z。

（2）测站至新点的连线的方位角及其距离。

图 6-14　点位测定

6.5.2　自由站

这项技术允许测量员可以在称之为自由站任何方便的位置安置全站仪。然后，通过瞄准事先已知坐标的参考点来确定仪器位置的坐标及高程（见图 6-15）。

当仪瞄准两个已知点时，则有必要记录测站至两参考点之间的距离及角度。

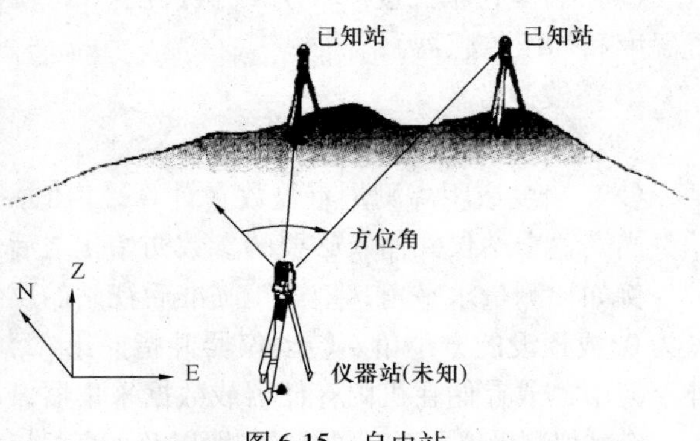

图 6-15　自由站

当瞄准多个（3个或更多）已知点，仅需测量点之间的角度。强调这一点很重要，大多数测量采用多次读数，而不是采用求解的最小必要读数。这些多余的观测提高了测量精度并检核了成果。一旦测定了测站的坐标，仪器现在可以进行定向，测量员就能用本节叙述的其他测量技术继续测量。

交会法方案

已知：（1）控制点1号坐标 N，E 和 Z。

（2）控制点2号坐标 N，E 和 Z。

（3）追加的照准点坐标 N，E 和 Z（总数多达10个），这些可以手工输入或由电脑上传依全站仪容量而定。

测量：（1）照准点之间的角度。

（2）如果仅有两个控制点被瞄准，则从测站至控制点间的距离也要测量。测量的量（角度和距离）愈多，自由站的精度愈好。

计算：测站的坐标 N，E 及 Z。

6.5.3 放样（或测设）点位

放样点位的坐标与高程上传至全站仪以后，放样或测设软件将能使测量员定位任何点。启动放样软件，输入放样点的编号，仪器显示屏上显示每一个所要求的点位上放置棱镜必须左右、前后、上下移动的量（见图6-16）。这种能力在房产及施工放样中很有用。

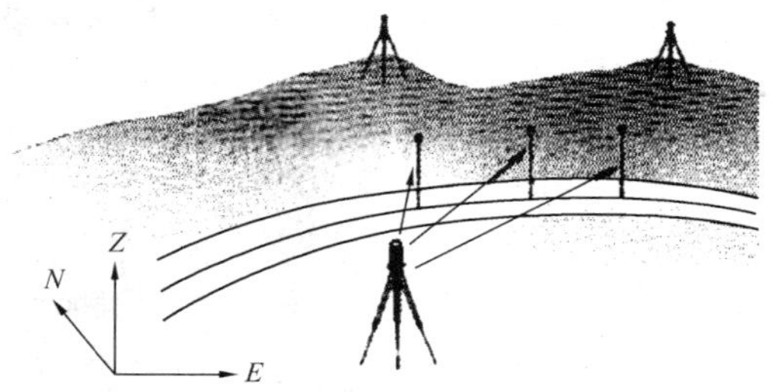

图6-16 放样或设置点位

放样点位测量方案：

已知：（1）测站的坐标 N，E 及 Z。

（2）参考点的坐标 N，E 及 Z，或至少有测站至控制点的连线的方位角。

（3）计划放样点的坐标 N，E 及 Z（手工输入或由计算机上传）。

测量：（1）在仪器显示屏上显示从控制点至放样点的角度或方位角，以及控制点至放样点的距离。角度可手工旋转仪器，或如果仪器装有马达驱动则可自动化。

（2）当棱镜真正移到所要求的放样点位时重新测量距离（水平距及垂直距）。

6.5.4 悬高测量

测定不可到达点的高程（例如电线、桥梁部件）可以简单地采用瞄准该目标下方直立安置的杆装棱镜。当瞄准目标本身，目标的高度立即显示出来（棱镜高首先要输入到全站仪），见图6－17。

悬高测量方案：

已知：（1）测站的坐标 N、E、Z。

（2）参考点的坐标 N，E 及 Z，或至少有测站至控制点的连线的方位角。

测量：（1）测量从参考点至目标的水平角或方位角，以及测站至目标下方（或上方）安置的棱镜的距离。

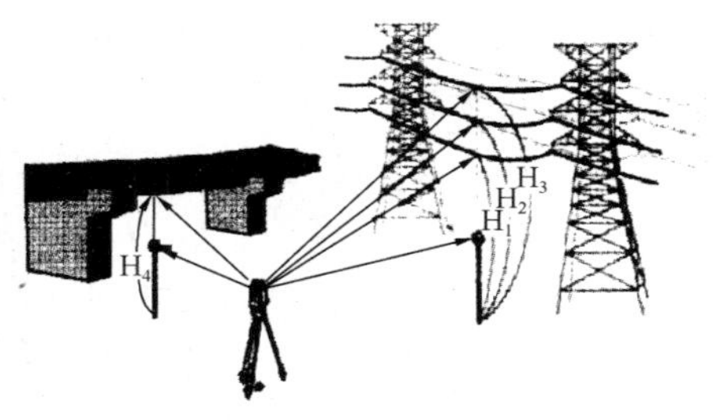

图6-17 悬高测量

（2）测量到棱镜和到目标点的垂直角。

（3）测站仪器高 h_i。

计算：从地面到目标点的距离（和它的坐标，如果需要的话）。

6.5.5　偏心测量

当目标被隐藏，全站仪看不见，必须采用偏心测量法。棱镜要安放在适当的位置以便测定目标的位置。这里有两种情况叙述如下：

距离偏心测量方案（见图 6-18）：

已知：（1）测站的坐标 N、E、Z。

（2）参考点的坐标 N，E 及 Z 或方位角。

测量：（1）测站至偏心点的距离。

（2）从偏心点到待测点（与仪器视线成直角）的距离。

计算：（1）被隐藏待测点的坐标 N，E，Z。

（2）测站至隐藏待测点的方位角与距离。

角度偏心测量方案（见图 6-19）：

已知：（1）测站的坐标 N，E，Z。

（2）参考点的坐标 N，E 及 Z 或方位角。

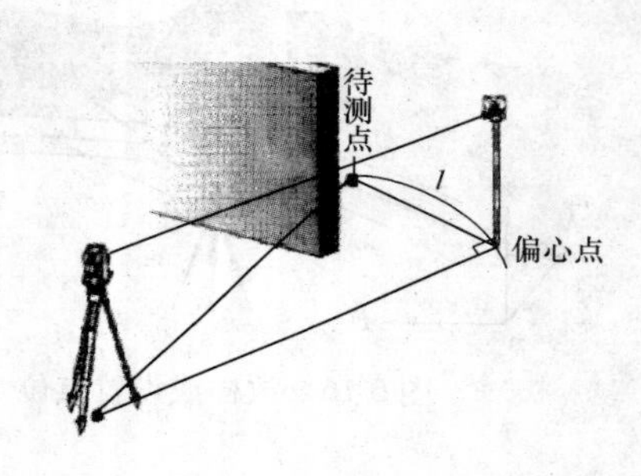

图 6-18　距离偏心测量

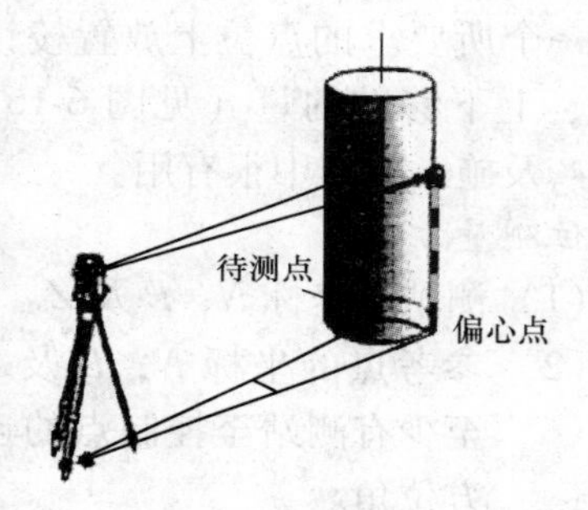

图 6-19　角度偏心测量

测量：（1）从测量点任一边的棱镜到目标中心的角度。（如果测量两旁，棱镜必须安置两旁与测站相等距离）。

计算：（1）被隐藏待测点的坐标 N、E、Z。

（2）测站至隐藏待测点的方位角与距离。

6.5.6　面积测量

当选择这个方案时，处理器将计算测量点所包围的面积。处理器首先按前述的方法测定每个站的坐标，然后用各点的坐标计算面积（见图 6-20）。

面积测量方案：

已知：（1）测站的坐标 N，E 及 Z。

（2）参考点的坐标 N，E 及 Z，或至少有测站至参考点的连线的方位角。

测量：（1）从控制点到各新点的角度或方位角。

（2）测站至新点的距离。

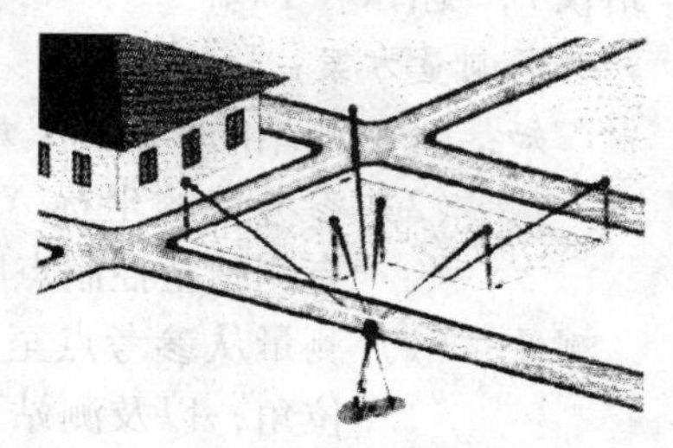

图 6-20　面积测量

计算：（1）面积边界点的坐标 N，E 及 Z。

（2）坐标点所包围的面积。

6.6 全站仪的仪器误差与距离测量精度

6.6.1 全站仪的仪器误差

在第5章，图5-15显示经纬仪各轴线的正确配置，这同样适用于全站仪。实际上，各轴线关系难以达到完善的要求，致使全站仪产生如同经纬仪一样的仪器误差。有一些在5.6节已经叙述。大多数高级的全站仪都能对水平及垂直视准误差、横轴倾斜以及补偿器指标误差进行量测和改正，叙述于以下各节。

1. 水平视准轴误差（或视线误差）

由于视线不垂于横轴引起视准轴的误差，它影响所有水平度盘的读数，并随视线的倾斜而增加，但是，用双盘位观测可以消除。对于单盘位观测，用全站仪机内校准功能测定 c 值（见图6-21），即实际视线与横轴垂直线之间的夹角。然后，全站仪自动对所有水平度盘的读数进行改正。如果 c 值超过规定范围，应退回厂家进行调整。

2. 横轴误差

全站仪的横轴不垂直于仪器竖轴引起横轴误差。当望远镜水平时没有影响，望远镜倾斜时它的误差会影响水平度盘读数，尤其是对大倾角影响更大。同视准轴误差一样，横轴误差可以通过两个盘位观测予以消除。图6-22显示横轴误差 α 可以用校准功能测定，然后对水平度盘读数施加改正。如果 α 太大，仪器应返回厂家进行调整。

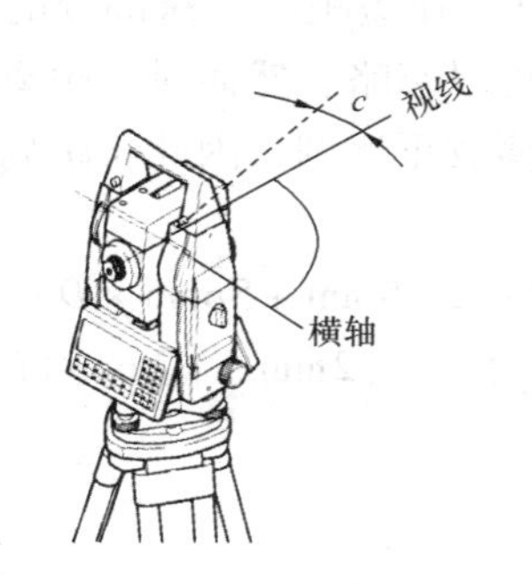

图6-21 水平方向视准轴误差

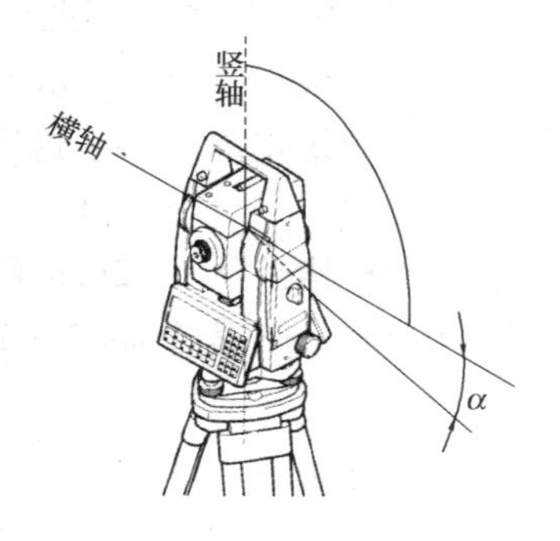

图6-22 横轴误差

3. 补偿器指标差

经纬仪或全站仪没有仔细整平产生的误差不能通过两个盘位观测予以消除。如果仪器安装补偿器并打开开关，那么补偿器能测量仪器残留的倾斜并对水平角与垂直角进行改正。然而，所有的补偿器都有一个沿视线方向的纵向误差 l 和垂直于视线方向的横向误差 t（见图6-23），称为零点误差。用机内的校准功能测定 l 与 t 值以便对测量的角度进行改正。

4. 垂直视准误差（垂直指标差）

当视线水平时，如果垂直度盘的0°～180°线与仪器竖轴线不垂直，则全站仪存在垂直指标差。在所有的垂直度盘读数中都包含有这个垂直指标的误差。如同水平视准轴误差一样，可以通过两个盘位的读数加以消除，或者通过测定 i，如图 6-24 显示的指标差，使用另外的校准程序。在这种情况下，误差 i 可以直接对垂直角施加改正。

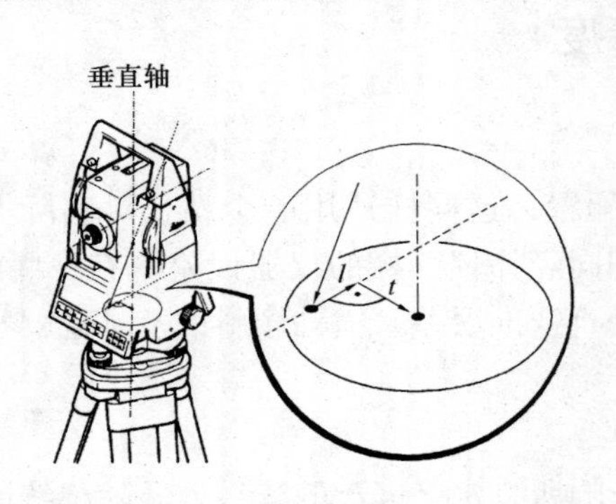

图 6-23　补偿指标误差

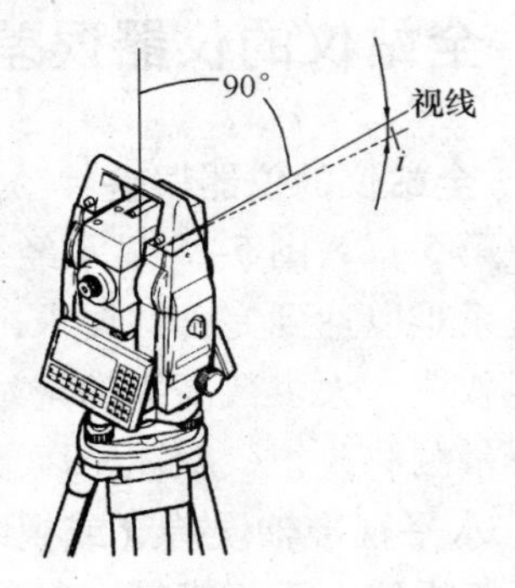

图 6-24　垂直指标差

6.6.2　全站仪距离测量精度

各种全站仪距离测量精度都可以用下列形式表示：

$$\pm(a\quad \text{mm} + b\text{ppm} \times D)$$

式中常数 a 独立于被量测的距离，它是由非用户所能控制的仪器内部的各种误差组成的。它主要包含零位误差（即全站仪和反射器它们的机械中心、电子中心及光学中心位置之差）和周期误差（由于全站仪内部信号的产生和处理遭受到有害的干扰造成的）。

系统误差 b 与所测距离成正比例，式中 ppm（每百万分之几）相当于每千米误差 ±1mm。它取决于测量时的大气条件和振荡器频率位移。在短距离，这部分的距离误差与仪器误差和对中误差相比较很小，对于一般的测量工作可以忽略。然而对长距离测量，大气条件是电子测距的最重要的误差源。由于这些误差与距离成正比例，因此长距离施加计算改正要特别注意记录气象条件。

全站仪的典型技术规格是从 $\pm(2\text{mm}+2\text{ppm}\times D)$ 至 $\pm(5\text{mm}+5\text{ppm}\times D)$。现取 $\pm(2\text{mm}+2\text{ppm}\times D)$ 为例，100m 距离误差为 ±2mm。1.5km，则为 $\pm(2\text{mm}+[2\text{mm/km}\times 1.5\text{km}])=\pm 5\text{mm}$。

第7章　控 制 测 量

7.1　概述

控制测量是在测区中建立控制网，以便求得控制点的坐标。控制测量工作主要包含控制点的踏勘选点、观测和计算。这些已在1.5节中叙述了。对于各种工程项目，实施控制测量是为了确定测图、施工放样以及其他三维空间定位工作所需的参考点的位置。

在大规模的测量工作中，例如某大工业区或某城镇的测量，首先要做的事是建立控制网。采用从整体到局部测量工作的组织原则，即选择少量的基本控制点（第一级）形成覆盖全测区的高等级精度的控制网。如有必要，在主测区内划分为更小的覆盖局部区域的控制网图形建立二级控制。如果还有必要，则建立第三级控制。虽然建筑工地面积通常很小，但是要求精度很高，因此，和大面积测量一样，建筑工地的测量也要建立高精度的控制网。

控制测量一般分为两种：水平控制及高程控制。覆盖待测区域布置成网状形式。布设控制点，其点位要便于其他测量能方便和精确地与它们相连结。

水平控制实施可用精密的导线测量、三角测量、三边测量以及这些方法的某种组合。采用哪种方法应考虑地形条件、现有的设备、必要的资料和经济因素。

对于大多数工程项目，水平控制点的位置是以平面直角坐标来确定（相当于数学上的X，Y坐标）。这在建筑工地十分通用，因为当使用直角坐标进行放样和其他三维作业，测量工作大大简化，较少发生错误。

使用导线测量技术，确定控制点的相对位置是通过测量在每个点上相邻站之间的水平角并且测量相邻两个站间的水平距离。导线测量的所有外业工作及计算将在本章讨论。

与水平控制相随的是，所有工地要建立某种形式的高程控制。工地上高程控制多数采用水准测量的方法。用此法在工地周围建立一系列的水准点。当然也可使用全站仪的三角高程测量方法。对于大型建筑和工程项目来说，测定控制点的位置必须考虑地球曲率。

7.2　方向的概念

7.2.1　子午线

在测量中，叙述某线的方向是通过该线与某基准线（或方向）所形成的水平角。此事涉及到确定被称为子午线的参考线。这条参考线通常选择真子午线或磁子午线。真子午线是通过地理北极、南极和观测者位置的平面与地球表面的交线如图7-1所示。真子午线方向的是通过瞄准太阳或已知天文位置的星体之一求得的（一般是测定北极星）。

磁子午线是通过磁北极、磁南极和观测者位置的平面与地球表面的交线。磁子午线的方

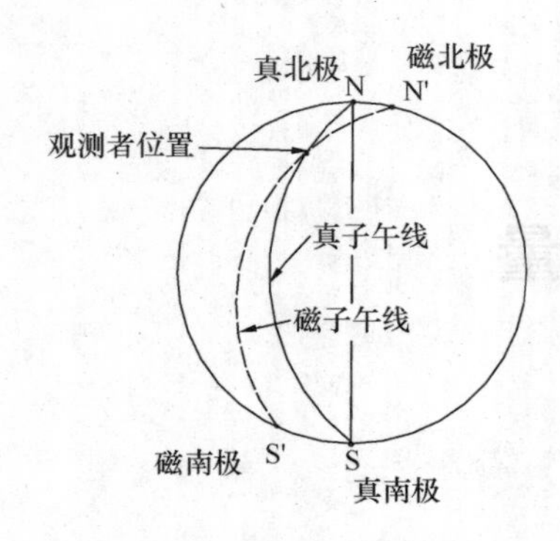

图 7-1　真子午线与磁子午线

向是观测者位置的罗盘仪磁针所指的方向。如果既没有真子午线也没有磁子午线可以利用，观测者可利用假定子午线，即仅为方便起见而采取的任意方向线。

真子午线适用于大范围地区的各种测量，实际上更适用于土地边界的各种测量，它们不会随时间而改变，几十年以后还可以重建。磁子午线存在受多种因素影响的缺点，其中就有随时间变化。

有时在有限范围内测量使用另外类型的子午线。通过具体区域的某一点的直线被选为基准子午线，在区域内所有其他的子午线都假定平行于这条子午线，这就是所谓坐标子午线。为了适应工地条件，坐标子午线可以是沿某地物定线（例如大桥的中心线或沿某建筑物线）。坐标子午线的使用不必考虑区域内不同点子午线收敛角的问题。

7.2.2　方位角与象限角

1. 方位角

指定某线方向普遍使用的是方位角。某线方位角是从子午线北端顺时针量测至该线的角度。方位角的大小从 0°～360°。AB、AC、AD 的方位角如图 7-2a 所示，大小分别为 60°、152°和 294°。

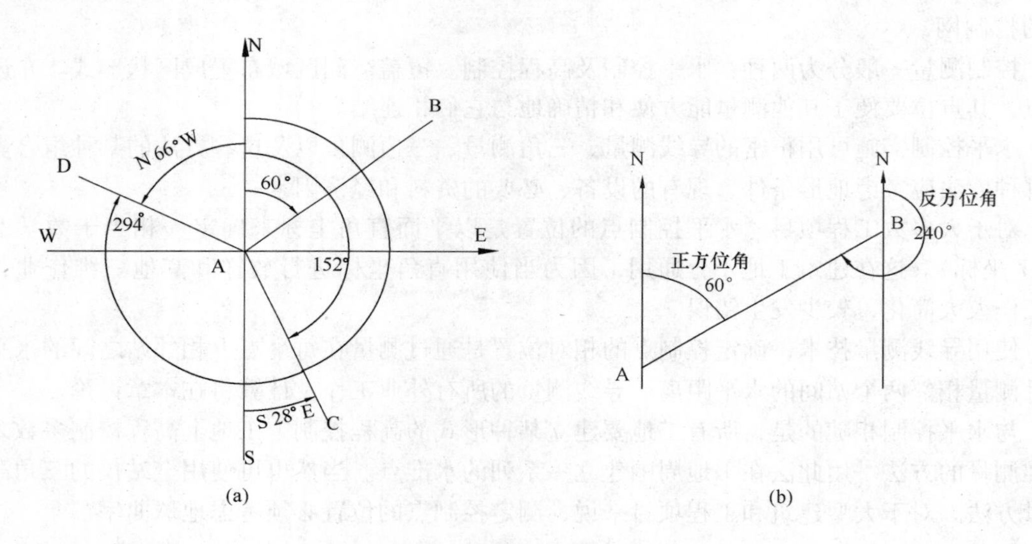

图 7-2　方位角与象限角

（a）方位角与象限角；（b）正反方位角

可以认为每条线有两个方向，如图 7-2b 所示有 AB 方向和 BA 方向。正方向与反方向的指定通常可以任意选择，假定从 A 点至 B 点的方向为正方向，那么从 B 点至 A 点的方向为反方向。因此，AB 的方位角 A_{AB} 为正方位角，而从 B 到 A 的方位角 A_{BA} 称为 AB 的反方位角。例如，AB 线的正方角 $A_{AB}=60°$，则反方位角 $A_{BA}=A_{AB}+180°=240°$（见图 7-2b）。正方位角 $A_{正}$ 与反方位角 $A_{反}$ 的关系可写为

$$A_{反}=A_{正}\pm 180° \tag{7-1}$$

式中，如果 $A_{正}<180°$，则 $+180°$；如果 $A_{正}>180°$，则 $-180°$。

方位角分为真方位角、磁方位角及坐标方位角，取决于所用的子午线。故使用哪一种子午线应说明清楚。

2. 象限角

表示直线方向的另一种方法是它的象限角。某线段的象限角是指该线段所处的象限并且与那个象限内的子午线形成的锐角，象限角由子午线的北端或南端量算。象限角从 0°到 90°，它不能大于 90°。因此，图 7-2a 中 AB 的象限角读为北 60°东或写 N60°E，AC，AD 和 AE 象限角分别为 S28°E，S50°W，和 N66°W。如果某线段的方向平行于子午线并且向北，象限角可写为 N0°或正北；如果垂直于子午线并向东，象限角可写为 N90°E 或正东。

7.3　导线测量

工程测量的原理之一是必须建立水平控制和高程控制以便于测量细部和工程放样。导线是一系列连续的线，其长度及相邻线之间的角度由外业测量确定。导线测量是指建立导线点及所作的必要测量工作，主要是包括角度测量和距离测量。角度测量用经纬仪或全站仪，距离测量用电子测距仪（EDM）或全站仪，有时也用钢卷尺。

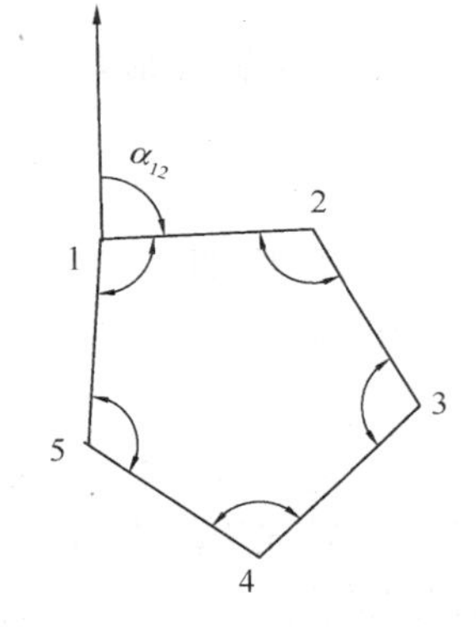

图 7-3　闭合导线

导线有三种类型：闭合导线、附合导线和支导线。

7.3.1　闭合导线

图 7-3 中导线从 1 站开始，经过 2、3、4 站又回到 1 站。站 1 为已知点或为某一假定的点。由于它本身闭合回来，所以这种导线称为闭合导线或多边形导线。

在外业工作中，角度用经纬仪或全站仪测量。对于 n 边闭合多边形，内角和应等于 $(n-2)\times180°$。因此，在野外通过对角度观测值总和与理论值的比较就可检查角度的观测，这是闭合导线的主要优点。

在图 7-3 中，如果给出第 1 点的坐标及第 1 条边的方位角，则所有连续的各点的坐标就可计算出。

7.3.2　附合导线

图 7-4 为典型的附合导线，从精确的坐标点 B 开始闭合到精确的坐标点 C。点 A、B、C、D 是现有精确坐标控制网的控制点。导线 B123C 称为附合导线。在野外应测量导线 1、2、3 的角度，并同时测量导线边与高级边的连结角（即 φ_B 与 φ_C）。

附合导线由于起始于控制点并又结束于控制点，因而具有对观测值的外部检核条件。

7.3.3　支导线

支导线是一种自由的、张开的导线，它起始于已知点而结束于未知点，它不闭合于任何已知点，如图 7-5 所示。因此，测量技术要

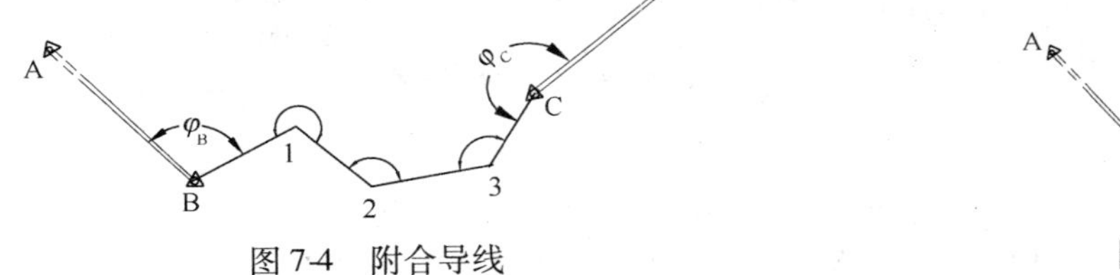

图 7-4　附合导线

图 7-5　支导线

求精细并且必须重复测量以防止发生错误。导线边至少测 2 次（往返各 1 次），角度也至少测 2 次。支导线经常用于路线测量。

7.4　导线测量外业工作

7.4.1　踏勘与建立导线点标志

踏勘是任何测量项目极为重要的部分，其目的是决定导线点的最佳位置。踏勘工作概述如下：

（1）导线的相邻点必须通视，以便于进行观测。

（2）测站应位于坚固而平整的地面，以便在测站进行角度观测时，能稳定地支撑经纬仪和三脚架。

（3）如果控制网的目的仅仅为了测量地形细部，那么测站点的位置应能提供观测地形有最好的视野，从每个站能测量最大量的细部。

（4）道路和管线工程需要布设附合导线，由于这些场地狭长，道路或管线的形状指明了导线的形状。

（5）如果导线是用来进行放样，例如放样道路中心线，那么导线点的布置应能提供最好的位置来放样道路交点和切线点。

踏勘完成时，为了测量期间或更长期间的需要，应建立导线点标志。

导线点标志的构建及类型根据测量目的而定。对于一般目的的导线，采用木桩，把木桩打入地面，使桩顶与地面大致齐平，在桩顶上打入钉子以表示导线点的精确位置。更多永久性导线点的标志要求构筑混凝土。

7.4.2　角度测量

导线点在地面一旦选定好之后，野外工作的第二步就是用经纬仪测量导线边的夹角。

多数情况，由于测点的标志不能直接看到，有必要在观测站上设置瞄准标志。瞄准标志必须垂直立在测站标志之上，否则会产生对中误差。

瞄准标志应是完全垂直并精确对准地面的标志。如果瞄准标志不直立，即使瞄准标志的底部已精确对中，也将产生对中误差，如图 7-6a 所示。

从图 7-6a 看出，观测点越低，对中误差越小。因此在测角时，应总是观测目标标志的最低可视点。

一些简单类型的瞄准标志如图 7-6b 所示，对它们的使用给出以下一些建议：

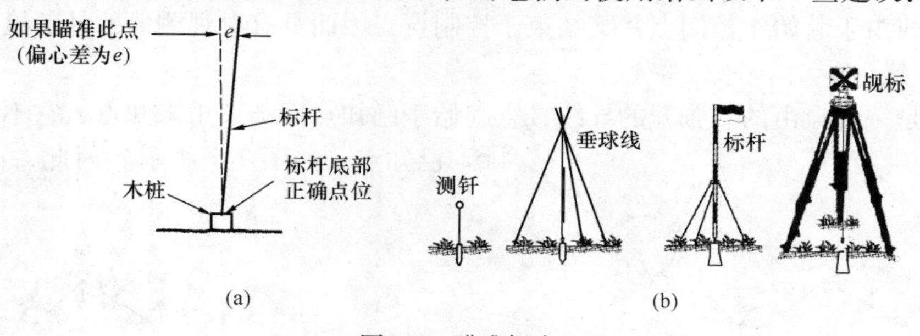

图 7-6　瞄准标志

（a）瞄准标志不垂直；（b）多种瞄准标志

（1）当导线边长很短时，瞄准标志可采用木桩顶上打入一钉子。

（2）如果地面标志不能直接看到，但通视条件良好，可以在木桩顶上插一根测钎。

（3）用3根竹杆捆扎成三脚架以悬挂垂球，观测垂球线。

（4）对于较长的边，最好使用三脚架支撑的标杆或觇标。

在外业工作中，每个站测量左角或测量右角，必选其一。测量左角观测顺序是（见图7-7）：（1）盘左观测后站；（2）盘左观测前站；（3）盘右观测前站；（4）盘右观测后站。

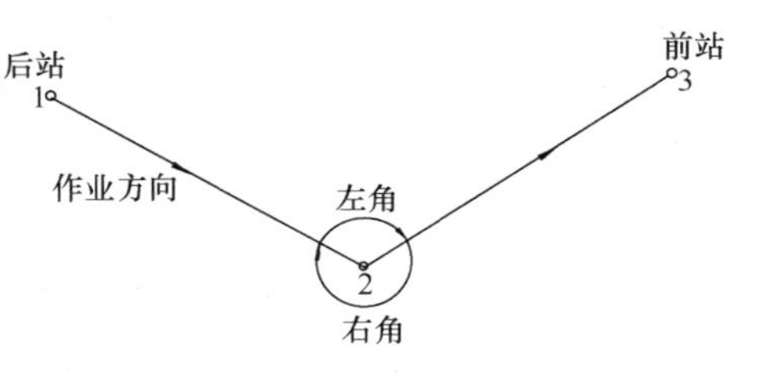

图7-7　左角与右角

上述观测顺序可称为："后—前—前—后"。相反地，测量右角观测顺序是："前—后—后—前"。

计算左角或右角的公式如下：

（1）左角 = 前视度盘读数 - 后视度盘读数

（2）右角 = 后视度盘读数 - 前视度盘读数

7.4.3　距离测量

导线的距离测量通常是用钢尺或EDM仪器，我国导线距离测量的技术要求列于表7-1和表7-2。

导线边长测量通常采用满足工程要求的最简单、最经济的方法，经常使用电子设备和卷尺，并能获得高等级的精度。当使用EDM时，这种导线测量称为电子导线测量。

表7-1　我国光电测距导线的技术规格

分类	导线全长（km）	平均边长（m）	测距中误差（mm）	测角中误差（″）	导线全长相对闭合差
三等	15	3000	≤±18	≤±1.5	≤1/60000
四等	10	1600	≤±18	≤±2.5	≤1/40000
一级	3.6	300	≤±15	≤±5	≤1/14000
二级	2.4	200	≤±15	≤±8	≤1/10000
三级	1.5	120	≤±15	≤±12	≤1/6000

表7-2　我国图根控制的技术规格

导线长度	导线相对闭合差	边长	测角中误差		DJ6测回数	方位角闭合差	
			一般	首级控制		一般	首级控制
≤1.0M	≤1/2000	≤1.5测图的最大视距	30″	20″	1	$\pm 60''\sqrt{n}$	$\pm 40''\sqrt{n}$

注：M是测图比例尺分母；n是导线角数。

7.5　平面直角坐标系

平面直角坐标系定义如图7-8，它分成4个象限，用典型的数学习惯，即向北向东的轴为正的，向南向西为负。4 个象限Ⅰ，Ⅱ，Ⅲ和Ⅳ按顺时针排列。

在纯数学里，坐标轴定义为 x 轴和 y 轴，角度是从 x 轴以逆时针方向量测。在测量中，x 轴被称为东西轴（E），y 轴被称为南北轴（N），α 角从 y 轴（南北轴的北端）顺时针量测。

另外一种定义法：南北方向定义为 x 轴，垂直于南北方向的东西方向定义为 y 轴，α 角从 x 轴的北端顺时针量测。

对于各种测量及工程，坐标原点可选在区域的西南角，因而所有的点其坐标均为正。

从图 7-8 可以看出，求 B 点的坐标，需要 A 点的坐标及 AB 两端的坐标差。我们有

$$\left.\begin{aligned} E_B &= E_A + \Delta E_{AB} \\ N_B &= N_A + \Delta N_{AB} \end{aligned}\right\} \tag{7-2}$$

或

$$\left.\begin{aligned} Y_N &= Y_A + \Delta Y_{AB} \\ X_B &= X_A + \Delta X_{AB} \end{aligned}\right\} \tag{7-3}$$

式中，ΔE_{AB}或 ΔY_{AB}称为横坐标增量（或称东西坐标增量）；ΔN_{AB}或 ΔX_{AB}称为纵坐标增量（南北坐标增量）。

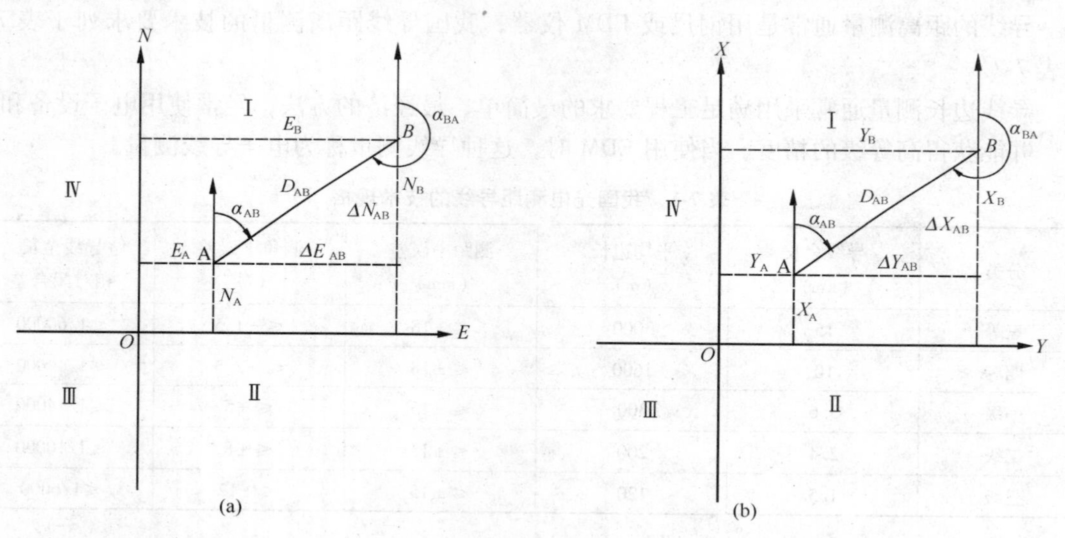

图 7-8　平面直角坐标系

从图 7-8 很容易求得坐标增量的公式：

$$\left.\begin{aligned} \Delta X_{AB} &= D_{AB}\cos\alpha_{AB} \\ \Delta Y_{AB} &= D_{AB}\sin\alpha_{AB} \end{aligned}\right\} \tag{7-4}$$

公式（7-4）中，ΔX_{AB}和 ΔY_{AB}的符号依 α_{AB}所在的象限而定。

如果已知 A、B 两点坐标 $A(x_A, y_A)$ 和 $B(x_B, y_B)$，从图 7-8b 便得到：

$$\tan\alpha_{AB} = \frac{\Delta Y_{AB}}{\Delta X_{AB}} \tag{7-5}$$

$$D_{AB} = \frac{\Delta Y_{AB}}{\sin\alpha_{AB}} = \frac{\Delta X_{AB}}{\cos\alpha_{AB}} \tag{7-6}$$

$$D_{AB} = \sqrt{\Delta X^2 + \Delta Y^2} \tag{7-7}$$

根据公式（7-5），（7-6）和（7-7），方位角与边长可求得，但应注意公式（7-5）通常不能直接给出正确的方位角，应根据 ΔX_{AB}和 ΔY_{AB}的符号决定其方位角属哪个象限进行判断处理。

7.6 导线测量的计算

7.6.1 闭合导线的计算

1. 准备数据

当导线外业完成之后，应计算平均角度与平均边长。最好画导线草图，列上所有数据，这样有助于以后计算并减少错误的发生。

野外数据如图7-9所示，在导线图上填写角度与边长。各点坐标及测量数据填入表中，见表7-1。

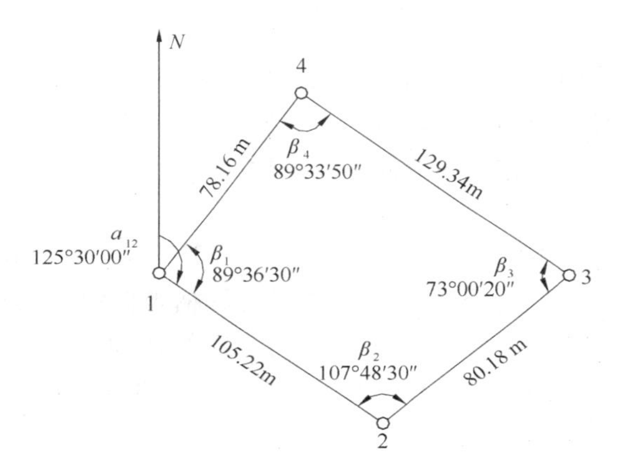

图7-9 闭合导线示意图

2. 角度闭合差的计算与调整

闭合导线观测的角度既可以是内角也可以是外角，通过比较角度观测值总和与下列理论值推求角度闭合差。

（1）角内角总和 = $(n-2) \times 180°$

或　（2）各外角总和 = $(n+2) \times 180°$

式中，n 是导线角数或边数。

设$\sum\beta_{测}$ 代表观测角总和，f_β 代表角度闭合差，则

$$(1)\ f_\beta = \sum\beta_{测} - (n-2) \times 180°$$

或

$$(2)\ f_\beta = \sum\beta_{测} - (n+2) \times 180°$$

$f_{\beta容}$代表角度闭合差容许值，不同等级导线不相同。对于测图控制，容许的角度闭合差如下：

$$f_{\beta容} = \pm 60''\sqrt{n} \tag{7-8}$$

式中，n 是导线角数或边数。

当角度闭合差（f_β）与它的容许值（$f_{\beta容}$）比较可能出现两种情况：

（1）角度闭合差小于容许值，这成果可以验收。

关于角度闭合差的调整，可以采用平均分配的方法，因为每个角度以相同的方式测量，每个角度误差产生闭合差的机率是相同的。角度闭合差反其符号平均分配到每个观测角，即每个角的改正值是

$$v_\beta = \frac{-f_\beta}{n} \tag{7-9}$$

本例是 $v_\beta = \dfrac{-(-50'')}{4} = 12.5''$，改正值凑整至秒，即为两个 $-12''$，另外两个 $-13''$。

改正后角　$$\beta_{改} = \beta_{测} + v_{\beta} \tag{7-10}$$

改正后角度计算之后，应检查改正后导线角总和是否等于理论值，即

$$\sum\beta_{改} = (n-2)\times180°(当观测内角时)$$

或

$$\sum\beta_{改} = (n+2)\times180°(当观测外角时)$$

(2)闭合差大于容许值，这成果不能验收，角度应重测。

闭合导线角度闭合差的计算及角度调整列于表 7-1。

3. 方位角的计算

闭合导线第 1 边方位角已知或假定，用 α_{12}表示见图 7-9a，从图可得

$$\left.\begin{aligned}\alpha_{23} &= \alpha_{12} + 180° \pm \beta_2\\ \alpha_{34} &= \alpha_{23} + 180° \pm \beta_3\\ \alpha_{41} &= \alpha_{34} + 180° \pm \beta_4\\ \alpha_{12} &= \alpha_{41} + 180° \pm \beta_1\end{aligned}\right\} \tag{7-11}$$

式中　β_1、β_2、β_3、β_4，分别表示 1、2、3、4 各点的角度。在使用上面的公式时，应注意符号。当测量左角时，应给导线角“+”号，当测量右角时，应给导线角“-”号。应特别注意：在上述方程中应使用改正后的导线角，即 $\beta_{改}$，最后应做检查计算，即

$$\alpha_{41} + 180° \pm \beta_1 = \alpha_{12}(检核)$$

本例：　$$215°53'17'' + 180° + 89°36'43'' = 125°30'00''$$

4. 计算纵坐标增量(Δx)和横坐标增量(Δy)

按照公式(7-4)计算每条边的 Δx 和 Δy，填入表 7-1。之后计算纵坐标增量总和$\sum\Delta x$ 与横坐标增量总和$\sum\Delta y$。从图 7-10a 可以看出$\sum\Delta x=0$ 和$\sum\Delta y=0$。

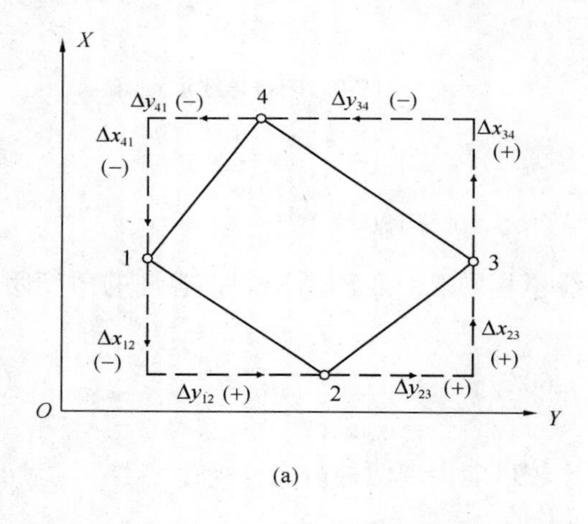

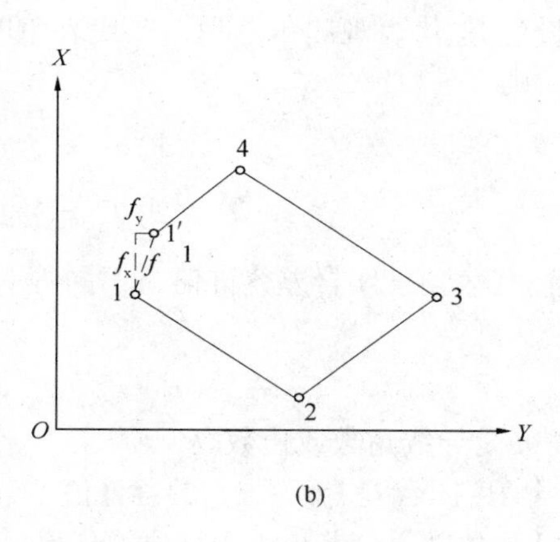

图 7-10　闭合导线的平差

(a)理论导线图；(b)实际导线图

由于测角和测边存在误差，导致$\sum\Delta x\neq0$，$\sum\Delta y\neq0$。设

$$\left.\begin{aligned}f_x &= \sum\Delta x\\ f_y &= \sum\Delta y\end{aligned}\right\} \tag{7-12}$$

式中，f_x 称为纵坐标增量闭合差，f_y 称为横坐标增量闭合差。

理论上，$f_x=0$，$f_y=0$，从图 7-10a 可明显得到

$$f_x = \sum \Delta x = (-\Delta x_{12}) + (-\Delta x_{23}) + \Delta x_{34} + \Delta x_{41} = 0$$
$$f_y = \sum \Delta y = \Delta x_{12} + \Delta x_{23} + (-\Delta x_{34}) + (-\Delta x_{41}) = 0$$

由于外业测量误差，导线终点为1′而不是1(图7-10b)，闭合差全长为f，由下式求得

$$f = \sqrt{f_x^2 + f_y^2} \tag{7-13}$$

本例 $$f = \sqrt{0.09^2 + (-0.07)^2} = \pm 0.11\text{m}$$

为了求得导线的测量精度，将闭合差与导线全长($\sum D$)作比较求得全长的相对闭合差K，即

$$K = \frac{f}{\sum D} \tag{7-14}$$

本例 $$K = \frac{0.11}{392.90} = \frac{1}{3571}$$

已知容许的相对闭合差为$\frac{1}{2000}$，而实际的相对全长闭合差小于容许值，因此导线外业作业合格。

按边长成正比进行坐标增量的调整，即

$$\left.\begin{aligned} v_{x_i} &= \frac{-f_x}{\sum D} \times D_i \\ v_{y_i} &= \frac{-f_y}{\sum D} \times D_i \end{aligned}\right\} \tag{7-15}$$

式中，v_{x_i}=第i边纵坐标增量改正数，v_{y_i}=第i边横坐标增量改正数。显然坐标增量的改正数应满足下列公式：

$$\left.\begin{aligned} \sum v_x &= -f_x \\ \sum v_y &= -f_y \end{aligned}\right\} \tag{7-16}$$

现在开始计算改正后纵坐标增量Δx与改正后的横坐标Δy，即

$$\left.\begin{aligned} \Delta x_{i改} &= \Delta x_i + v_{x_i} \\ \Delta y_{i改} &= \Delta y_i + v_{x_i} \end{aligned}\right\} \tag{7-17}$$

所有改正后的坐标增量计算之后，检查改正后坐标增量总和应等于零，即

$$\left.\begin{aligned} \sum \Delta x_{i改} &= 0 \\ \sum \Delta y_{i改} &= 0 \end{aligned}\right\} \tag{7-18}$$

5. 计算导线各点坐标

对于闭合导线，为了计算导线各点的坐标，起始点的坐标必须是已知的，起始点的坐标也可以假定。坐标计算过程如下：

$$\left.\begin{aligned} x_2 &= x_1 + \Delta x_{12} \\ x_3 &= x_1 + \Delta x_{23} \\ &\vdots \\ x_1 &= x_n + \Delta x_{n,1} \end{aligned}\right\} \text{（检核）} \tag{7-19}$$

表 7-3 闭合导线的计算

点号	水平角 (° ′ ″)	改正数 v (″)	改正后水平角 (° ′ ″)	方位角 α (° ′ ″)	边长 D (m)	坐标增量 Δx (m)	坐标增量 Δy (m)	改正后坐标增量 Δx (m)	改正后坐标增量 Δy (m)	坐标 x (m)	坐标 y (m)
1	2	3	4 = 2 + 3	5	6	7	8	9	10	11	12
1										500.00	500.00
				125 30 00	105.22	−2 −61.10	+2 +85.66	−61.12	+85.68		
2	107 48 30	+13	107 48 43							438.88	585.68
				53 18 43	80.18	−2 +47.90	+2 +64.30	+47.88	+64.32		
3	73 00 20	+12	73 00 32							486.76	650.00
				306 19 15	129.34	−3 +76.61	+2 −104.21	+76.58	−104.19		
4	89 33 50	+12	89 34 02							563.34	545.81
				215 53 17	78.16	−2 −63.32	+1 −45.82	−63.34	−45.81		
1	89 36 30	+13	89 36 43							500.00	500.00
				125 30 00							
总和	359 59 10	+50	360 00 00		392.90	−0.09	+0.07	0.00	0.00		

$f_\beta = -50''$

$f_x = +0.09 \quad f_y = -0.07$

$f_{\beta容} = \pm 60''\sqrt{n} = \pm 60''\sqrt{4} = \pm 120''$

全长闭合差 $f = \sqrt{0.09^2 + (-0.07)^2} = \pm 0.11\text{m}$

容许全长相对闭合差 $K_{容} = \frac{1}{2000}$

全长相对闭合差 $K = \frac{0.11}{392.90} = \frac{1}{3571}$

7.6.2　附合导线的计算

在高级控制点 B 与 C 之间布置附合导线，如图 7-11 所示。

导线两端的控制点的精确坐标如下：

B 点坐标：$X_B = 509.580\text{m}$

$Y_B = 675.890\text{m}$

C 点坐标：$X_C = 529.000\text{m}$

$Y_C = 801.540\text{m}$

已知 AB 起始的方位角 $\alpha_{AB} = 127°20'30''$，CD 最终的方位角 $\alpha_{CD} = 34°26'00''$，这些数据填入表 7-4。

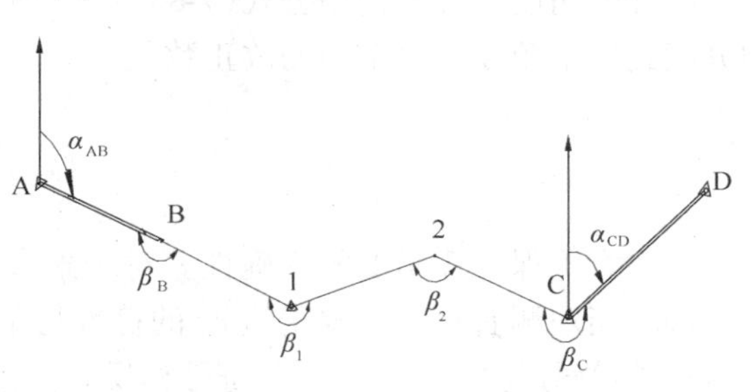

图 7-11　附合导线示意图

附合导线的计算与闭合导线有许多相同，仅是角度闭合差的计算和坐标增量闭合差的计算不同，因为附合导线两端要连接到高级点。以下将讨论它这两项的计算。

1. 角度闭合差的计算与调整

在图 7-11 中，A、B、C、D 为高等级已知控制点，β_B 与 β_C 为已测量的连结角，β_1 与 β_2 为已测量导线右角。从 7-11 图中可看出：

$$\left.\begin{aligned} \alpha_{B1} &= \alpha_{AB} + 180° - \beta_B \\ \alpha_{12} &= \alpha_{B1} + 180° - \beta_1 \\ \alpha_{2C} &= \alpha_{12} + 180° - \beta_2 \\ \alpha_{CD} &= \alpha_{2C} + 180° - \beta_C \end{aligned}\right\} \tag{7-20}$$

上述 4 个方程相加得：

$$\alpha_{CD} = \alpha_{AB} + 4 \times 180° - \sum(\beta_B + \beta_1 + \beta_2 + \beta_C)$$

上面方程写成普遍公式如下：

$$\alpha_{终} = \alpha_{始} + n \times 180° - \sum \beta_{右} \tag{7-21}$$

或

$$\alpha_{终} = \alpha_{始} + n \times 180° + \sum \beta_{左} \tag{7-22}$$

我们用测量的右角值 $\beta_{右测}$ 代替理论的右角值 $\beta_{右}$，计算得最终的方位角用 $\alpha'_{终}$ 表示，则公式(7-21)变为

$$\alpha'_{终} = \alpha_{始} + n \times 180° - \sum \beta_{右测} \tag{7-23}$$

同理

$$\alpha'_{终} = \alpha_{始} + n \times 180° + \sum \beta_{左测} \tag{7-24}$$

本例

$$\alpha'_{CD} = 127°20'30'' + 4 \times 180° - 822°54'06'' = 34°26'24''$$

角度闭合差 f_β 是测量值与理论值之差，即

$$f_\beta = \alpha'_{终} - \alpha_{终}$$

本例　　　　　　$f_\beta = \alpha'_{CD} - \alpha_{CD} = 34°26'24'' - 34°26'00'' = +24''$

容许的角度闭合差仍用公式(7-8)计算。如果$f_\beta < f_{\beta容}$，则角度测量合格。角度闭合差平均分配到每个角度，每个角的改正数是

$$v_\beta = \frac{f_\beta}{n} \tag{7-25}$$

要注意确保计算角度的正确性。对于附合导线，观测左角与观测右角改正值v_β计算是不同的。当观测右角时，改正值v_β的符号与角度闭合差f_β符号相同，但是观测左角时，改正值v_β的符号与角度闭合差f_β符号相反。

本例附合导线测量导线右角，改正值v_β的符号与角度闭合差f_β符号相同，$f_\beta = +24''$，则$v_\beta = \frac{+24''}{4} = +6''$，见表7-4。

2. 计算纵坐增量(Δx)和横坐标增量(Δy)

对于附合导线，每边的坐标增量总和应等于终点坐标与始点坐标之差，即

$$\sum \Delta x_{理} = x_{终} - x_{始} \tag{7-26}$$

$$\sum \Delta y_{理} = y_{终} - y_{始} \tag{7-27}$$

由于测角与测边存在误差，因此$\sum \Delta x_{测}$与$\sum \Delta y_{测}$不等于理论值，导致坐标增量闭合差，即

$$f_x = \sum x_{测} - \sum x_{理} \tag{7-28}$$

$$f_y = \sum y_{测} - \sum y_{理} \tag{7-29}$$

$$f = \sqrt{f_x^2 + f_y^2} \tag{7-30}$$

本例$f_x = -0.031\text{m}$，$f_y = -0.033\text{m}$，$f = +0.045\text{m}$，$K = \frac{0.045}{178.670} = \frac{1}{3953}$。导线全长相对闭合差小于容许值，因此观测值合格。附合导线坐标增量闭合差的调整方法同闭合导线，计算过程列于表7-4。

表 7-4　附合导线的计算

点号	水平角观测值 (° ′ ″)	改正数 (″)	改正后水平角 (° ′ ″)	方位角 α (° ′ ″)	边长 D (m)	坐标增量 ΔX (m)	坐标增量 ΔY (m)	改正后坐标增量 ΔX (m)	改正后坐标增量 ΔY (m)	坐标 X (m)	坐标 Y (m)
1	2	3	4	5	6	7	8	9	10	11	12
A											
				127 20 30							
B	128 57 32	+6	128 57 38							509.580	675.890
				178 22 52	40.510	+7 −40.494	+7 +1.144	−40.487	+1.151		
1	295 08 00	+6	295 08 06							469.093	677.041
				63 14 46	79.040	+14 +35.581	+15 +70.579	+35.595	+70.594		
2	177 30 58	+6	177 31 04							504.688	747.635
				65 43 42	59.120	+10 +24.302	+11 +53.894	+24.312	+53.905		
C	211 17 36	+6	211 17 42							529.000	801.540
D											

$f_\beta = +24''$

$\sum D = 178.670$; $f_x = -0.031$; $f_y = -0.033$; $f = +0.045$; $K = 1/3970$

第 8 章　地 形 测 量

8.1　概述

地形测量是测定地球表面或其局部的外形（地貌）即三维特征，并且定位在其上的自然地物与人工地物。自然地物包括树木、河流、湖泊等等；人工地物如公路、桥梁、水坝、房屋等等。这些地物与地貌的特征表示在地形图上。地形图是借助于各种线条和惯用符号图解表示某区域。

地形图是一种表示地物与地貌水平位置与垂直位置的图。它与平面图的区别是附加了可量测形式的地貌。地形图的显著特征是地球表面形状由等高线来描绘。

地形图，工程师用来决定公路、铁路、桥梁、房屋、运河、管道、传输线、水库及其他工程的最合理、最经济的位置；地质学家用来调查矿藏、石油、水及其他资源；建筑师用于房屋建筑与园林设计；农学家用于土壤调查。

地形测量包括：（1）建立测图区域的水平与垂直控制系统，该系统要包括经由高精度测量所连结的关键测站；（2）测量细部，包含选择地面细部点并从控制点用较低精度方法进行测量。

实施地形测量可用航空摄影测量法、地面测量法或数字化测图法。当今地形测量主要用航空测量技术，用现代计算机化的摄影测量或激光图像建立平面及数字高程模型（DEM）。较小面积的地形测量常用电子设备来完成，例如全站仪。

然而，地面测量方法仍然经常使用，尤其是小面积大比例尺测图。即使在使用摄影测量法，仍有必要进行地面测量来建立控制，野外检查已测绘的地物以确定精度都需要地面测量。本章重点叙述地面测量法并介绍几种测定地形的外业方法。

8.2　等高线

表示某一特定地区地形最普遍的方法是使用等高线。等高线可以定义为水平面与地面相切而得的曲线，因此等高线上的每个点都有相同的高程，即假定横切面的高程。图 8-1 表示一个山头被高于基准面 320m、330m、……350m 的一系列水平面所切情形。事实上，静止的湖面的边缘就是一条等高线，如果湖面下降或上升，湖面新位置的边缘将代表另外的等高线。这些等高线显示在地形图和平面图上表示了该地区的地形。

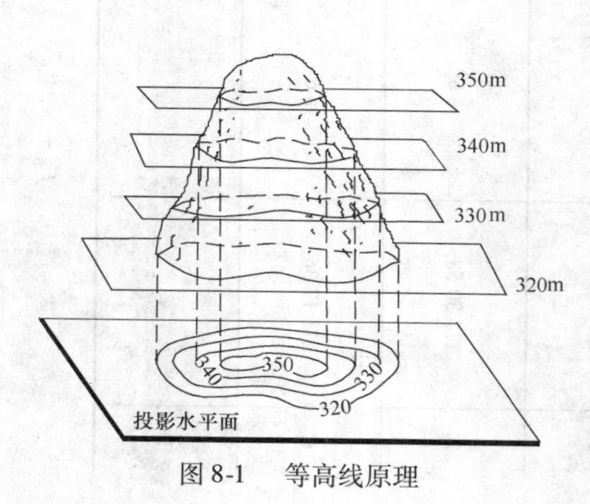

图 8-1　等高线原理

地形图等高距是指两条等高线之间的垂直距离。等高距的确定要根据测图的目的和地面情况（山地或平地）而定。普通的地形图，等高距从2～20m。但对于平坦的地区可能小到0.5m，山区大到20～50m。

等高距的选择是个重要的课题，它取决于测图例尺和测区地貌情况。在起伏大的地区，等高距通常较大，以避免图上太多等高线影响读图。表8-1表示我国大比例地形图等高距的规定。

表8-1 我国大比例尺地形图的等高距的规定（m）

地形类别＼比例尺	1：500	1：1000	1：2000
平 地	0.5	0.5	0.5或1
丘陵地	0.5	0.5或1	1
山 地	0.5或1	1	2
高 地	1	1或2	2

地形图有4种类型等高线：基本等高线（首曲线）、间曲线、助曲线及计曲线。

基本等高线是指那些按大比例尺测图规定（见表8-1）的等高线，如例中100m、102m、104m、108m、110m等等（见图8-2）。

间曲线是二分之一的基本等高线，这些等高线在图上描绘成长虚线，如例中的103m，这就是所谓半距等高线。

助曲线是四分之一的基本等高线，这些等高线在图上描绘成短虚线，如例中的108.5m，这就是所谓四分之一等高线。

计曲线是那些较粗的等高线，为读图方便，每第5条等高线加粗并注数字，这样的曲线称为计曲线（见图8-2），如例中100m、110m。由计曲线的高程帮助用户估计出相邻等高线高程。

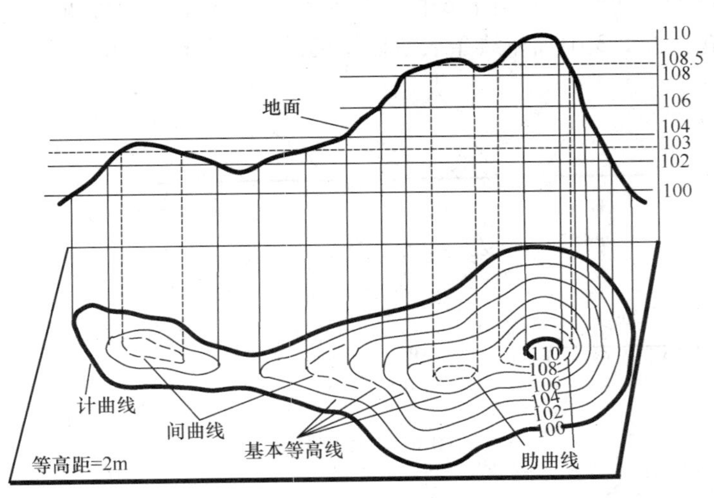

图8-2 等高线的种类

8.3 细部测量的方法

8.3.1 细部点的定义和选择

地形控制建立之后，地形测量的第二步就是从控制点测定细部点。这些细部点包含地貌的特征和所有天然地物和人工地物。要注意必须有足够的地物点和独立高程点才有可能画等高线。

细部点可分为下列4个主要类别：

硬细部是指清晰可辨的地物。主要指各种构筑物，例如建筑物、道路、围墙等等，具有明显边缘或能确定精确位置的都可归于这一类。

软细部是指不能明确界定的地物，这些主要指天然地物，例如河岸、灌木丛、树木和其他植被。没有明显边缘或不能确定精确位置的都可归于这一类。

架空细部是指地面上方的地物，例如电力线、电话线。

地下细部是指地面以下的地物，例如水管与下水道。

许多类型的符号用于表示细部，普遍采用标准的格式。推荐使用那些容易理解的那些符号和缩写词。然而，用于工程测量的大比例尺图，许多地物能按实际形状画出，因此没有必要用符号表示。

细部点的选择非常重要。天然和人工地物，例如河流、道路及建筑物，细部点应选地物轮廓的转折处，比如建筑物的角点，道路的交点和转折点等等。对于地貌，细部点应能代表山脊、山谷、山顶、洼地以及鞍部。

地形测量采用的仪器与方法取决于测量的目的，必要的精度等级，地面的类型，测图比例尺及等高线间距。如果要求高等级的精度，方位角测定用陀螺经纬仪，水平距离测定用EDM设备或全站仪，高程测定用水准仪。

8.3.2 野外确定点的基本方法

图8-3图示说明野外确定点P的6种方法，这一切基于水平控制。头4种方法的每一种必须已知一条线（AB距离）。采用第6种方法必须已知3个点的位置或能在图中确认，这就是所谓三点问题。图8-3中加粗线条表示已知的距离。在相应的图中要测量的量是：

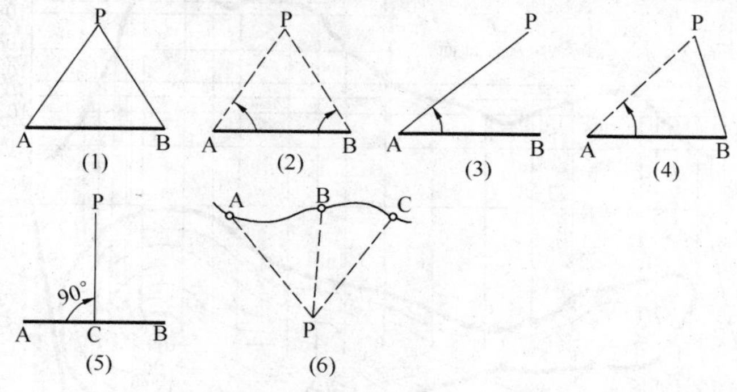

图8-3　定位P点

1. 两个距离；
2. 两个角度；
3. 一个角及一条相邻边的距离；

4. 一个角及一条相对边的距离；

5. 一个距离和一个垂直支距；

6. 在待定点上的两个角。

最经常使用的是第 3 种方法，但有经验的人员主要考虑使用哪种方法最适合实际情况，必须考虑野外与室内（计算与绘图）的要求。

8.3.3　传统的地形测量法

传统的地形测量一般使用经纬仪和视距法。当使用该法时，水平距离及高差是用经纬仪对水准尺测量尺间隔和垂直角来确定的。

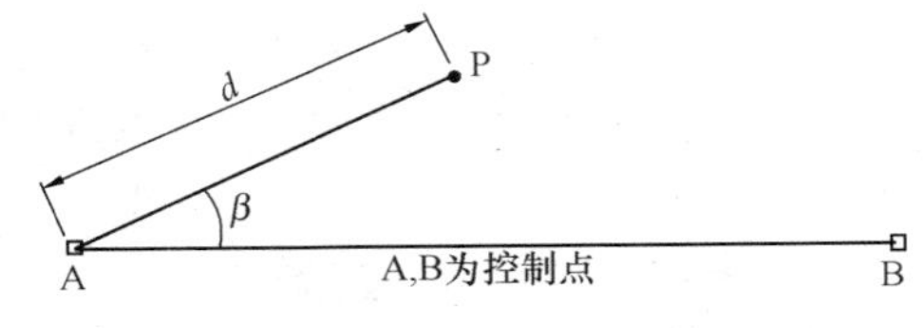

图 8-4　极坐标法测定细部

视距测量技术是用极坐标法（辐射法）测定细部点位置，其基本原理如图 8-4 所示，在野外用视距法测量 d，用经纬仪水平度盘读数求得 β，借助于量角器与比例尺来确定 P。

视距测量技术有效用于等高线测量，特别是在没有明显细部点的开阔地区。视距法非常适合于细部点采集。一个测站的细部测量的步骤叙述如下：

（1）安置仪器于控制点 A，按通常的方法对中与整平。测量并记录横轴高于测站标志的高度（hi）。对于细部测量标准的做法是仅使用一个盘位测量水平角与垂直角，因此经纬仪应校正好。

图 8-5　视距法测量细部的原理

（2）选择一个适当的控制点作为基准目标（RO），例如图 8-5 中 B 点。对着 RO 方向水平度盘置零，测量 AB 与测站至细部点 1 之间水平角 β。细部点 1 是建筑物的拐角点，在细部点 1 上应竖直立标尺。

（3）读取细部点 1 上的标尺的尺间隔 l 及视准轴的倾斜角 α，计算 A 点和 1 点之间的水平距离（d_{A1}）及高差（h_{A1}）。碎部点 1 的高程由 $H_1 = H_A + h_{A1}$ 求得。

（4）为了画细部点，需要量角器和比例尺。根据要求的比例尺在图纸上展绘 1 点，其高程注在点位旁。

其他细部点以同样方法测量并将其展绘在图上。

（5）根据测量点绘图并内插等高线

如果采用手工绘制等高线，通常使用目估内插法，其步骤如下：

当画等高线时，首先用铅笔画出地性线，山脊线画虚线，山谷线画实线，然后用内插法标出等高线通过的点。由于细部点是选在坡度变化处，所以两相邻高程点之间地面坡度可以认为是均匀的，即两点间的高差同其水平距成正比。在两相邻碎部点连线上，用铅笔内插并标志所需等高线的位置，其方法就是所谓目估内插法。图 8-6a 中，设两个地面细部点 A 与 C，它们的高程分别为 207.4m、202.8m。如果测图的等高距 1m，则必定有 5 条等高线，它们的高程为 203m、204m、205m、206m 和 207m。根据平距与高差成正比例的原理，先目估定出高程为 203m 的 m 点和高程为 207m 的 g 点（见图 8-6a），然后将 mq 的距离四等分，定出高程为 204m、205m、206m 的 3 个点，即 n、o、p。同法定出其他相邻两碎部点间等高线应通过的那些点。当所有的等高线位置都确定后，具有相同高程点的各点用圆滑的曲线相连形成图 8-6b。

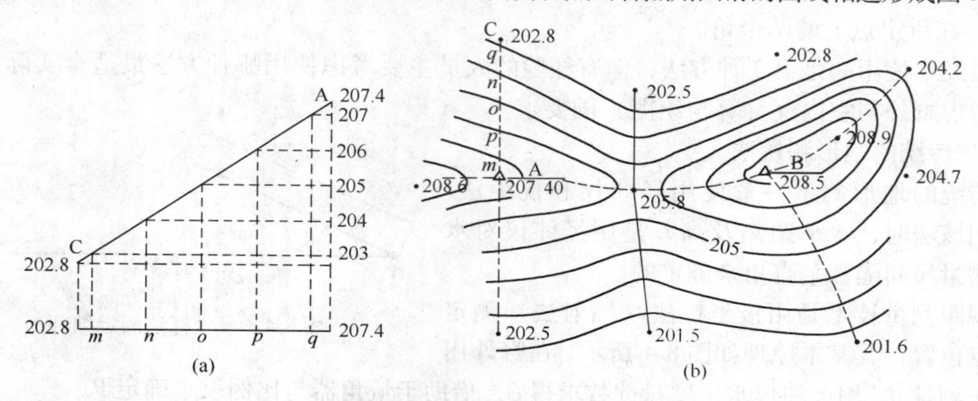

图 8-6　画等高线

8.4　全站仪细部测量

8.4.1　辐射法原理

在第 6 章全面的介绍了全站仪，这里叙述全站仪细部测量。全站仪与固定在细部杆上的棱镜联合使用，应用辐射法定位细部点。

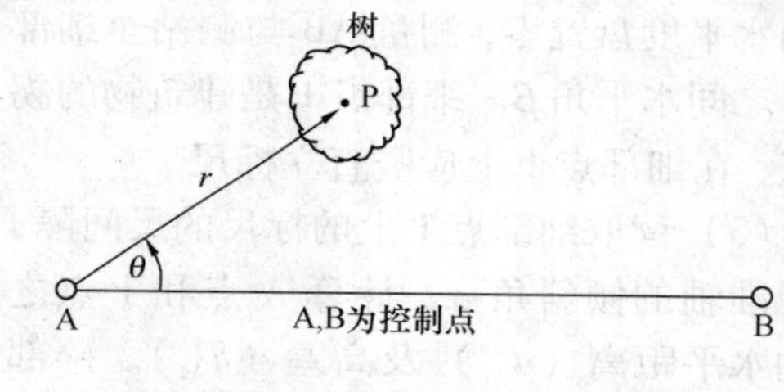

图 8-7　辐射法的原理

辐射法的原理如图 8-7 所示，从测线 AB 测定待定点树 P 的位置。测定树的平面位置通过测量 A 至 P 的水平距离 r 及测线 AB 和 A 点至 P 所夹的水平角 θ，因此辐射法包含测量水平角和水平距。另外，细部点上立细部杆，通过对棱镜测量垂直角，并结合棱镜高和全站仪的仪器高可求得点的归化高程。因此，测点的三维数据可以求得，即平面位置和它的归化高程。

使用全站仪的辐射法非常有效进行等高线图测量，尤其是在开阔地区没有很明显的细部点。另外由于全站仪的高精度性能，能用来收集任何类型的细部，无论是硬细部点或软细部点。传统的方法用全站仪测量细部需要两个人，一人操作全站仪，另一个人扶细部杆。然而，在最近几年全站仪机动化得到发展，用自动目标识别技术能搜索与跟踪细部杆上的棱镜。这就是所谓机器人全站仪，一人就可以操作。这个系统仪器按常规放置于控制点上无人操作，操作员操控细部杆，实际选择并测定细部点。

虽然这种机器人全站仪使用在不断增加，但是双人全站仪细部测量仍然是常态。因此本节叙述双人全站仪操作，一人看仪器，另一人持细部杆。

8.4.2　全站仪细部测量外业

外业期间，全站仪依次安置在控制网的每一点上，由控制点测定细部点。因为全站仪固有的高精度，它就有可能用这种方法测定远距离的细部点。然而，如果距离太远，两人工作彼此有效的联系变得很困难。如果靠声音联系，建议视线距离不要超过 50m。大于这个距离建议使用收发两用的无线电设备。如果使用自动控制系统，当然不成问题。

有几种可能的观测方法，以下仅叙述一般用途的方法。假定已知控制点的归化高程，图 8-8 显示从控制点 A 向细部点 P 所作的细部测量。

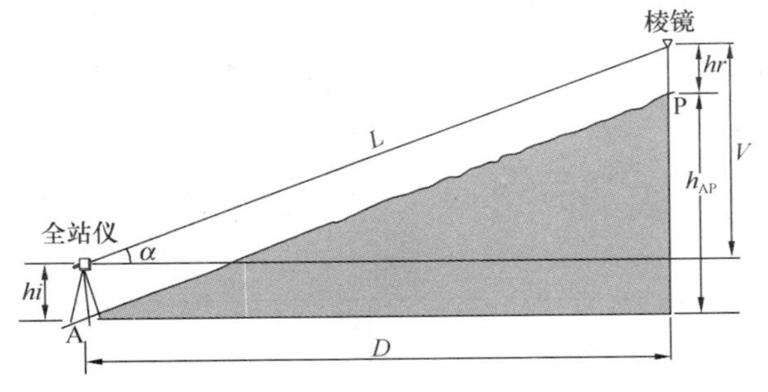

图 8-8　用全站仪进行细部测量

（1）仪器安置在控制点 A 测站标志上方，按通常方法对中与整平。细部测量标准的方法是仅用一个盘位测量水平角与垂直角，因此全站仪应在作业前校正好。

（2）测量并记录仪器高（*hi*），选择附近的控制点作为基准目标（RO），瞄准这个点记下水平度盘的读数。如果这个 RO 点不能直接看到，有必要在 RO 点上直立一个目标。在辐射范围内的细部点都相对于这个选择的方向来确定。有些工程师更喜欢沿着 RO 方向设置水平度盘读数为零，尽管这不是必需的。

（3）每个测站实际观测操作法依所用仪器类型手工记录或自动记录而异。最简单的情况是记录水平度盘读数，垂直度盘读数以及斜距 L。在仪器上直接给出水平距离 D 和斜距的垂直分量 V，这两个值应同水平度盘读数一起记录。如果某仪器能在显示屏上计算与显示点坐标（X，Y）及归化高程（*RL*），那么直接记录这些数值。

（4）如果采用手工记簿和绘图，应在记录表格上注明细部点的类型。如果采用自动化记录和绘图，对细部点应统一编号，对地物赋予地物代码以便于软件处理数据和绘图识别。细部杆的棱镜（反射镜）的中心高于细部杆底部的高度（*hr*）必须记录。由于细部杆是可伸缩的，其高度按需要设置。如果采用手工记簿和计算，将 *hr* 设置等于测站仪器高 *hi* 可简化计算，*hi* 被 *hr* 抵消了，斜距的分量 V 就等于 A 和 P 两点的高差。如果观测值记录到数据采集器和存储卡内，则 *hr* 可设为任何方便之数。然而一旦设置之后，除非有绝对的必要棱镜高不要轻易改动。由于每次 *hr* 的改变，新的值要用手工键入数据采集器或仪器的键盘，这样既费时又容易忘记而导致出现错误。

（5）P 点定位之后，移到第 2 点重复操作直到观测完成。从实践来看，细部点选择以顺时针的方向为好，这样可减少扶尺员走路。但是，如果观测采用计算机绘图软件包，在移动到另一地物之前，最好是编排好确定具体地物所必需的全部碎部点，例如相同建筑物的各角点，沿相同路边石的所有点，野外所有独立高程点等等。

（6）在测站结束之前，最后应再瞄准基准目标（RO）以便检查观测过程中水平度盘是否有变动。如有则所有的读数不可靠应重测。因此，在全站仪一系列读数期间应经常后视 RO，比如说每测 10 个细部点。

8.5　数字测图技术

近年来，计算机技术的发展使得许多功能强大的桌面电脑可利用带有先进的外围设备。使用这些，许多测量机构和大型民用工程承包商为电脑绘制测量平面图研制了他们自己的内部系统。此外，制造业体系也利用自动绘制测量平面图系统。一个自动测量绘图系统的原理以框图的形式显示在图 8-9 中。各个阶段所涉及的内容在下面的各节中叙述。

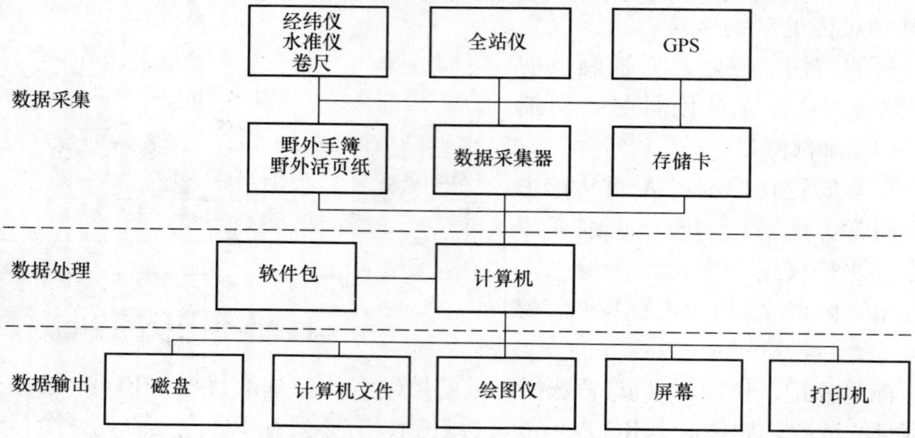

数据采集：经纬仪 水准仪 卷尺；全站仪；GPS；野外手簿 野外活页纸；数据采集器；存储卡

数据处理：软件包；计算机

数据输出：磁盘；计算机文件；绘图仪；屏幕；打印机

图 8-9　计算机辅助绘图系统

8.5.1　数据的采集

原始数据采集通常用各种测量设备如图 8-9 所示。传统的方法把所有的观测值手工记录到野外用纸上或野外手簿上。现如今，观测值用电子记录到数据采集器或存储卡中，该卡可插入到全站仪，或使用 GPS 接收机。电子源的数据能很快地通过准标的 RS232 接口或 USB 接口传输到电脑。如果使用野外纸张，观测值要用手工仔细键入，这样很费时且存在潜在的误差。

对于野外测量的每个点，用代码来定义细部点的类型，在图上细部点处加注释（如有必要时）和其他信息，例如地物的名称和尺寸。实际上，代码替代了野外的草图，但是我们建议画草图，这有助于后续绘图。

全站仪能记录角度与距离，并赋予它们适当代码直接存入数据存储装置，其内容可以从存储装置传输到计算机以便于后续的处理。这种数据的传送需要适当的接口。

经纬仪与 EDM 系统相结合也可以配以数据存储单元使用，但在很多情况下采集数据属于半自动性质，其原因是因为经纬仪和 EDM 的读数，以及相关的代码必须手工输入到数据存储装置。

用传统的方法，所有的野外观测记录在野外用纸上，并附随草图，数据加代码通过键盘输入到计算机。

数据采集是整个过程的最重要部分。野外观测的质量和地物代码非常影响测量的结果。如果采集数据阶段，正确地实施了对各点给予正确的代码，那么后续的处理与输出阶段无疑会很快。

8.5.2　数据的处理

所有数据输入计算机后，以野外观测文件形式存储。借助专门设备软件，计算机操作员检查所有的读数，然后自动计算野外测量的细部点的三维坐标。这些信息连同每个点的代码存为坐标文件。在这个阶段开始编辑坐标文件以确保最后标在图上的信息位置正确，要求附注正确并且符合野外测量员的描述。用软件进行编辑过程常称为交互图形例程。交互式绘图能使所选任一小区域的坐标文件显示在图形屏幕上，观察到的信息可以在图形屏幕上用光笔或电子光标按需要进行更改、移动或删除。当点改变时，就计算新的坐标并改变相应的代码，这些信息经过坐标文件显示在数字监视器上。不同层的信息或几个层的信息联合都可在屏幕上显示以便编辑。使用点代码的坐标文件编辑这些图层。图层包含这样的数据，如特征、轮廓线图形、符号、建筑物、独立高程点、道路、水文要素、地下设施、架空线等等。

坐标文件也用于计算机产生等高线。由于等高线不能穿过某些地物，例如岸堤、路堑、沟渠、建筑物等等，应告知计算机不能通过这些地物画等高线。在坐标文件中加标志，即岸堤等边缘加上适当标志。等高线的信息通常放在单独的信息层。

8.5.3　数据的输出

绘图文件一旦建立以后，数据就可以以各种方式绘制。数据绘在高分辨率的图形屏幕上，可以检查绘图的完整性，精度等等。如果利用交互绘图，绘制的地形可以删除、增强、改正、画交叉阴影线、作标志、三维量度等等。在这个阶段，屏幕显示的硬拷贝既可打印在简单的打印机上，也可输出到喷墨打印机上。绘图文件可直接绘制在数字打印机上，类似于图8-10显示的那样。按所要求比例尺绘制成果图，仅受图纸大小限制。有些绘图仪仅有一两支笔，现有绘图仪带有4～8支笔。具有各种颜色的绘图笔，各种粗细的线条。用数字绘图仪在施工工地上绘制平面图和断面图变得越来越普遍。

图8-10　惠普8支笔绘图仪

8.6　数字地形模型（DTM）

局部地球自然表面的数学表示常用的名称为数字地形模型，即DTM，由于数据存储以数字形式，即X，Y和RL。然而，用于这种表示还有许多其他的名称，例如，数字高程模型（DEM）和数字地面模型（DGM）。但是对工程测量应用最多的是数字地形模型（DTM），这是因为大多数人是考虑平面的数据（X，Y）和地貌数据（RL，地理要素和自然地物，例如河流，山脊线等等）。

为了形成DTM，对要建DTM的区域进行细部测量。由于自然表面形状差异的随机性，代表地面形状的各测点组成的网通常将是随机的图形，它包含平面坐标及高程。

很多情形，涉及航空测量的摄影测量法用来提供地面信息制作数字地面模型，这些方法非常适合获取广大区域的三维信息，广大区域地面测量那将是非常费力的。

用下列方法之一从野外数据生成DTM。

1. 基于方格的地形建模

该技术在工地区域建立有规则的方格网，方格顶点的高程RL是由外业观测的数据点内插求得。方格大小的选择要使得基于它表示的不规则的地面有足够的精度。该法的一个缺点是方格顶点不同实际野外数据点重合，这使得不规则的地表面变圆滑了，导致方格生成的等高线小于真实的地面。更多的缺点在于在工地上一系列地貌已按一定形式仔细测量的任何山脊以及斜坡上附加的变化不能在方格网上精确地再现，这也影响生成等高线的形状。

2. 基于三角形的地形建模

该法是用实际的野外测量点作为 DTM 的节点。软件联结所有的数据点形成一系列互不重叠连续的三角形，这些三角形每个节点为一个数据点产生不规则的三角形网（TIN）。

这样一种技术没有上述基于方格法的缺点，能更真实地表示测量的地面。由于仔细测量的一些地貌作为限制条件，例如路堤的顶部和底部，就可真实再现地貌并能考虑是否要生成等高线。这样也就有可能设置地表某个区域没有等高线。这一点在基于方格系统是不可能做到。

DTM 除用于生等高线外，在工程测量中还有许多应用。DTM 可以从不同角度去观察，DTM 呈现为丝网构架的景观，对感兴趣的区域可以加亮显示。图 8-11a 与图 8-11b 显示了基于方格网与基于三角形网相同地区丝网构架的景观。一旦建立了 DTM，地貌的体积，例如湖泊、矸子山、储料堆和采石场等的体积很容易求得，并且可以迅速生成纵断面图与横断面图。

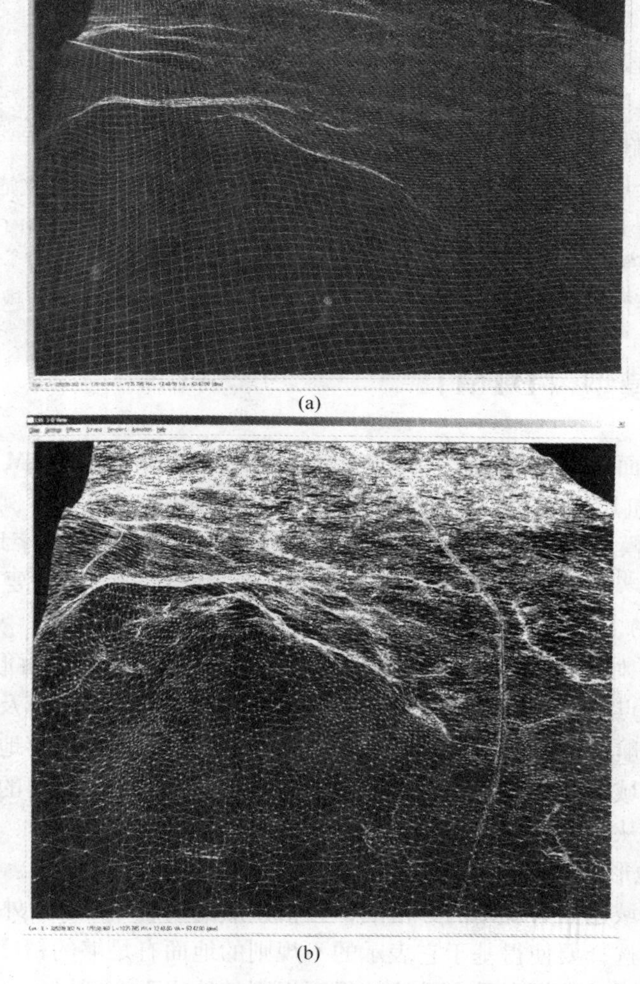

(a)

(b)

图 8-11　数字地面模型（DEM）的实例

（a）由方格法生成 DTM 丝网构架，网格大小 50m；

（b）由三角形法（TIN）生成 DTM 丝网构架

第 9 章　公路曲线测量

9.1　圆曲线

在公路和铁路的设计中，把道路的直线段与固定半径或变化半径的曲线相联结，如图 9-1 所示。曲线的用途是使道路转弯，通过两直线之间的角度 Δ 使道路转折，因而 Δ 称为转折角。

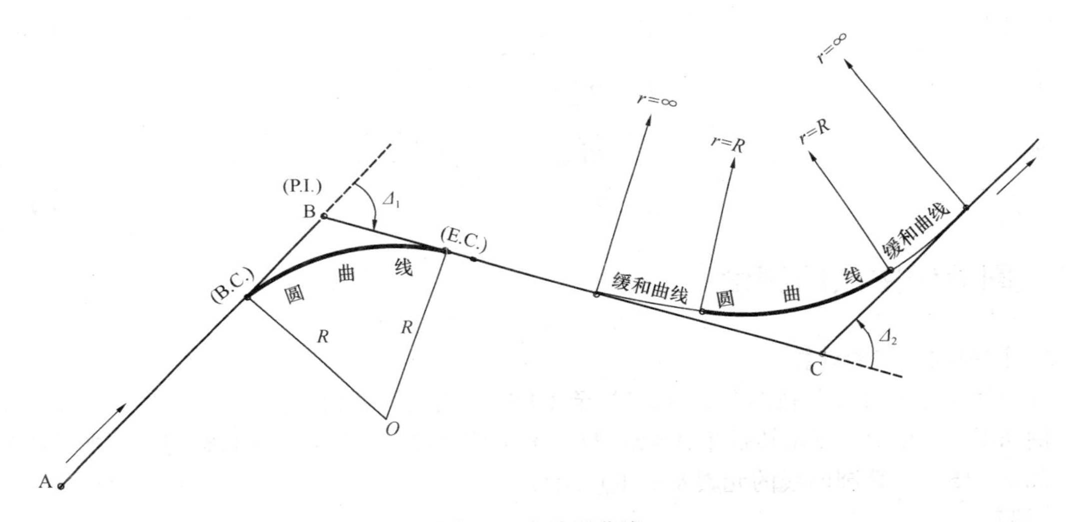

图 9-1　公路平曲线

如图 9-1 所示的曲线为公路平曲线，由于在设计和施工的测量都认为在水平面上。平曲线有两种主要类型：

（1）圆曲线，它的半径是常数，如图 9-1 左半部分所示。

（2）缓和曲线，它的半径是变化的，如图 9-1 右半部分所示。

一种简单的圆曲线由固定半径的一段圆弧组成，如图 9-2 所示。

圆曲线元素如图 9-2 所示，切线的交点（P. I. ）也称顶点（V）；曲线起点（B. C. ）和曲线终点（E. C. ），也称为切曲点（T. C. ）和曲切点（C. T. ）；曲线中点（M. C. ）是圆曲线的中点；B. C. 、M. C. 与 E. C. 称为圆曲线的三主点。

从 B. C. 到 P. I. 的长度和从 P. I. 到 E. C. 的长度称为切线长（T）；曲线长（L）是从 B. C. 到 E. C 沿着曲线弧长测量的距离，外距是从顶点到曲线沿辐射线测量的距离；两条切线方向的改变称为转折角 Δ，它等于圆心角。

圆曲线元素及其关系显示如下：

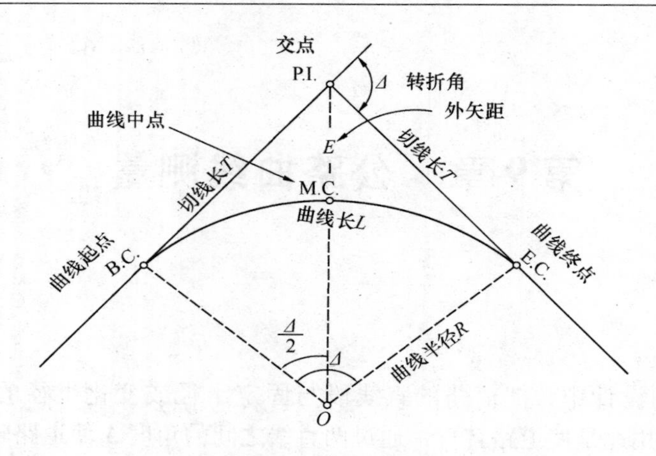

图 9-2　圆曲线元素

切线长　$$T = R \cdot \tan\frac{\Delta}{2} \tag{9-1}$$

曲线长　$$L = R \cdot \frac{\Delta}{\rho} \tag{9-2}$$

外距　$$E = R\left(\sec\frac{\Delta}{2} - 1\right) \tag{9-3}$$

切曲差　$$D = 2T - L \tag{9-4}$$

9.2　圆曲线主点的测设

9.2.1　圆曲线主点的计算

圆曲线主点的计算是按照公式（9-1）至（9-4），给出算例如下：

例 9.1　已知野外测量转折角 $\Delta = 39°27'$，P. I. 里程桩编号 = K5 + 178. 64，圆曲线半径 $R = 120\text{m}$。要求计算圆曲线的元素及三主点编号。

【解】

$$T = 120 \cdot \tan\frac{39°27'}{2} = 43.03\text{m}$$

$$L = 120 \cdot \frac{39°27'}{3437.75'} = 82.62\text{m}$$

$$E = 120\left(\sec\frac{39°27'}{2} - 1\right) = 7.48\text{m}$$

$$D = 2 \times 43.025 - 82.624 = 3.44\text{m}$$

交点桩里程实质上是从道路起点测量的水平距离，因此

$$\left.\begin{aligned} &\text{B. C. 里程编号} = \text{P. I. 里程编号} - T \\ &\text{M. C. 里程编号} = \text{B. C. 里程编号} + L/2 \\ &\text{E. C. 里程编号} = \text{M. C. 里程编号} + L/2 \end{aligned}\right\} \tag{9-5}$$

为了避免计算的错误，用下式进行检核计算：

$$\text{E. C. 里程编号} = \text{P. I. 里程编号} + T - D \tag{9-6}$$

本例，P. I. 里程是 K5 + 178. 64，计算步骤如下：

P. I. 里程编号	K5 +178. 64
$-T$	43. 03
B. C. 里程编号	K5 +135. 61
$+L/2$	41. 31
M. C. 里程编号	K5 +176. 92
$+L/2$	41. 31
E. C. 里程编号	K5 +218. 23

用公式（9-6）作检查计算：

E. C. 里程编号 = K5 +178. 64 +43. 03 −3. 44 = K5 +218. 23

两次计算 E. C. 里程编号完全相同，可以确信上述的计算是正确的。

9. 2. 2　圆曲线主点的测设步骤

1. 测设圆曲线的 B. C. 与 E. C.

（1）经纬仪安置在转折点 P. I. 上，按以前叙述的方法对中与整平。

（2）后视道路的起点（A），如图 9-1 所示，或后视后面的一个转折点，从转折点 P. I. 沿着视线方向量切线长 T，然后把提前写好 B. C. 点编号的木桩打入地面，这就是 B. C. 点。

（3）同样地，前视下一个转折点 C（见图 9-1），从转折点 P. I. 沿视线方向量切线长 T，然后把提前写好 E. C. 点编号的木桩打入地面，这就是 E. C. 点。

2. 测设圆曲线中点 M. C.

（1）经纬仪安置在 B 点上不要移动，但必须注意检查对中与整平的情况（见图 9-1）。

（2）前视下一个转折点 C，并设置水平度盘读数为 0°00′00″。

（3）旋转经纬仪的照准部，设置水平度盘读数为 $\frac{180^\circ-\Delta}{2}$（参见图 9-1 与图 9-2），这时望远镜的视准线处于∠CBA 角度平分线上，沿角平分线从交点 P. I. 量外距 E，然后把提前写好 M. C. 点编号的木桩打入地面，这就是 M. C. 点。

9. 3　圆曲线细部的测设

圆曲线细部点测设的目的是要在地面上每隔一定距离间隔打木桩建立起圆曲线曲线的中心线。一般情况，当圆曲线长度小于 40m 时，测设圆曲线 3 主点能够满足工地施工的要求。然而，当地形起伏不平，或曲线较长，或较短半径的圆曲线，这种情况，圆曲线的细部测设要求每隔 10m 或 20m 打桩。

在中线测量时，中心桩的编号必须选择整数桩，因此，靠近曲线 B. C. 的第 1 个曲线中心桩的编号变成整数桩编号。请看下列实例：

已知某公路曲线 B. C. 的里程编号为 K8 +720. 56，E. C 的里程编号为 K8 +767. 18。如果要求曲线上每隔 10m 打桩，那么靠近 B. C. 第 1 点 P_1 桩号应是 K8 +730，第 2 点 P_2 是 K8 +740，第 3 点 P_3 是 K8 +750，第 4 点 P_4 是 K8 +760。

测设圆曲线细部点主要有下列 3 种方法

1. 切线支距法（直角坐标法）

切线支距法是基于直角坐标系，它的坐标原点为曲线 B. C. 点或曲线 E. C. 点，过曲线点的切线为 x 轴，原点的半径方向为 y 轴。在野外测设之前应首先计算细部点的坐标 x，y。测设原理见图 9-3，计算公式如下：

$$\left.\begin{aligned} \varphi_i &= \frac{l_i}{R}\left(\frac{180^\circ}{\pi}\right) \\ x_i &= R\sin\varphi_i \\ y_i &= R(1-\cos\varphi_i) \end{aligned}\right\} \tag{9-7}$$

式中，l_i为细部点 i 到曲线点 B. C. （或 E. C. ）的弧长；φ_i 为弧长 l_i 对应的圆心角；x_i 为沿切线的纵坐标（见图 9-3b）；y_i 为支距长，即垂直于 x 轴的横坐标 y（见图 9-3b）。

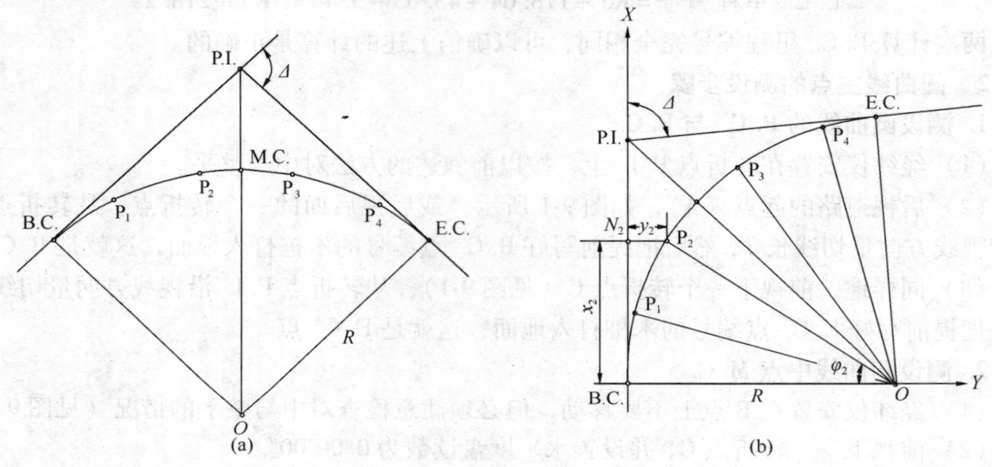

图 9-3　切线支距法

例 9.2　设 P. I. 的里程编号为 K8 +745. 72，转折角 $\Delta = 53°25'20''$。曲线半经设计为 $R = 50$m。要求每隔 10m 钉中心桩。试计算曲线元素及所有的测设数据。

【解】　从公式（9-1）至（9-4）求得：

$T = 25.16$m，$L = 46.62$m，$E = 5.97$m，$D = 3.70$m。

B. C 里程编号 = K8 +720. 65，M. C. 里程编号 = K8 +743. 87，E. C 里程编号 = K8 +767. 18。

不同的弧长代入公式（9-7），其结果列于表 9-1。

表 9-1　切线支距法测设曲线计算

主点	细部点		弧长 l_i	x（m）	y（m）	弦长（m）
B. C. K8 +720. 56				0. 00	0. 00	
	P_1	+730	9. 44	9. 38	0. 89	9. 43
	P_2	+740	19. 44	18. 95	3. 73	9. 98
M. C. K8 +743. 87			23. 31	22. 47	5. 34	3. 87
	P_3	+750	17. 18	16. 84	2. 92	6. 13
	P_4	+760	7. 18	7. 16	0. 51	9. 98
E. C. K8 +767. 18				0. 00	0. 00	7. 17

测设曲线步骤如下：

（1）在地面上从曲线点 B. C. 沿切线方向量 x 并插一测钎。

（2）垂直于切线设置支距，用光学直角棱镜或其他工具。

（3）一旦垂直方向线设定，沿直线丈量支距长 y，并在地面打木桩确定曲线的细部点。

（4）重复上述方法，用其他的支距长度得到更多的点，一直到曲线长度的一半。

（5）当支距长度变得很长时，最好从另外一个曲线点测设另一半曲线，例如从 E. C. 点，这样做可以避免由于垂直线偏差引起误差。

2. 偏角法测设曲线

测设曲线最普遍使用的方法是偏角法。偏角是在 B. C.（或 E. C.）上从主切线至曲线上某一点量测的角度，图 9-4b 中，角度 Δ_1、Δ_2、Δ_3 和 Δ_4 分别表示 P_1、P_2、P_3 和 P_4 各点的偏角。

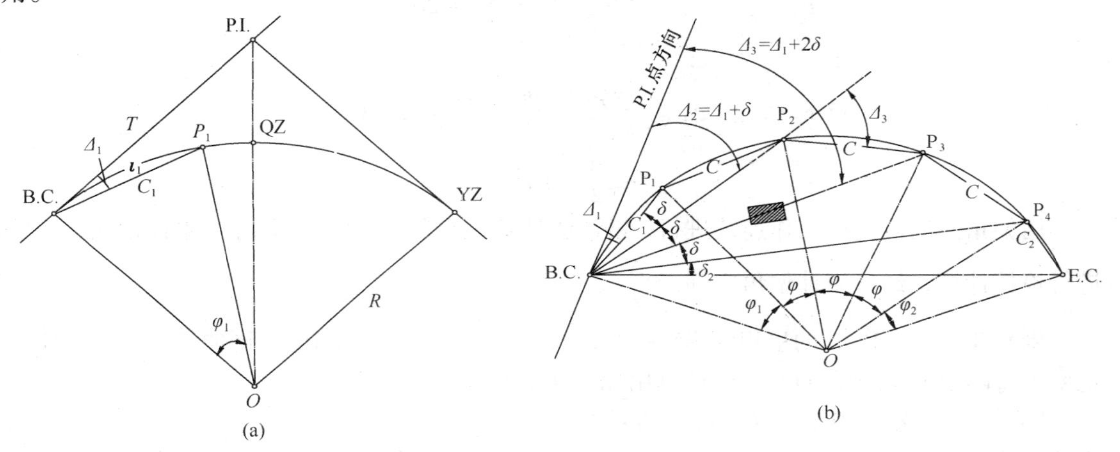

图 9-4　偏角法

偏角法通常测设曲线上的整桩。曲线段的里程桩测设密度要大于直线段，根据 5m、10m、20m 等不同规定而定。为了使曲线上的里程桩为整数，则曲线头尾两段弧长通常不是整数。第 1 段分弧定名为 l_1，尾段分弧定名为 l_2，相应的弦长分别定名为 C_1 和 C_2。曲线中间相邻两点间整弧定名为 l_0，它们相应的弦长为 C。

在图 9-4 中，第 1 段分弧是从 B. C. 点至 P_1 点，从图 9-4a 可求得 P_1 点的测设数据，相应弧长 l_1 的圆心角 φ_1 可按下列公式计算：

$$\varphi_1 = \frac{l_1}{R}\left[\frac{180^\circ}{\pi}\right] \tag{9-8}$$

相应于第 1 段分弧 l_1 的偏角 Δ_1 以及弦长 C_1 分别按下列公式计算：

$$\Delta_1 = \frac{\varphi_1}{2} = \frac{l_1}{R}\left[\frac{90^\circ}{\pi}\right] \tag{9-9}$$

$$C_1 = 2R\sin\Delta_1 \tag{9-10}$$

相应于尾段分弧 l_2 的圆心角为 φ_2，尾分弧对应的圆周角 δ_2 和相应的弦长 C_2 分别按下列公式计算：

$$\Delta_2 = \frac{\varphi_2}{2} = \frac{l_2}{R}\left[\frac{90^\circ}{\pi}\right] \tag{9-11}$$

$$C_2 = 2R\sin\delta_2 \tag{9-12}$$

圆曲线的中间部分，相邻两点之间整弧长等于 l_0。相应于整弧长的圆心角为 φ 与相应于整弧长的圆周角 δ 和弦长 C 分别由下列公式计算：

$$\delta = \frac{\varphi}{2} = \frac{l_0}{R}\left[\frac{90^\circ}{\pi}\right] \tag{9-13}$$

$$C = 2R\sin\delta \tag{9-14}$$

各细部点的偏角列如下：

点 P_1　　Δ_1

点 P_2　　$\Delta_2 = \dfrac{\varphi_1 + \varphi}{2} = \Delta_1 + \delta$

点 P_3　　$\Delta_3 = \dfrac{\varphi_1 + 2\varphi}{2} = \Delta_1 + 2\delta$

⋮　　⋮

点 E. C.　　$\Delta_{E.C.} = \dfrac{\varphi_1 + n\varphi + \varphi_2}{2} = \Delta_1 + n\delta + \delta_2$

$= \dfrac{\Delta}{2}$（检核）

用偏角法测设曲线工作连续进行，累计法计算每点的偏角，即所谓“累计偏角”。为了进行检核计算，最后累计偏角应等于 $\dfrac{\Delta}{2}$。

例 9.3　已知 P. I. 的里程为 K5 +135.22，转折角 $\Delta = 40°21'10"$，圆曲线半径 $R = 100\text{m}$，现要求每隔 20m 钉细部点桩，试计算用偏角法的放样数据。

【解】　从公式(9-1)～(9-4)求得：$T = 36.75\text{m}$，$L = 70.43\text{m}$，$E = 6.54\text{m}$，$D = 3.07\text{m}$。用偏角法测设圆曲线细部点按公式(9-8)～(9-14)计算，计算结果列于表 9-2。

表 9-2　偏角法测设曲线计算

主点	细部点	弧长（m）	转折角 Δ（°　′　″）	累计转折角 Δ（°　′　″）	弦长 C（m）
B. C. K5 +098.47				0　00　00	
		1.53	0　26　18		1.53
	P_1　+100			0　26　18	
		20.00	5　43　46		19.97
	P_2　+120			6　10　04	
		20.00	5　43　46		19.97
	P_3　+140			11　53　50	
		20.00	5　43　46		19.97
	P_4　+160			17　37　36	
		8.90	2　32　59		8.90
E. C. K5 +168.90				20　10　35	

测设步骤如下：

（1）安置经纬仪于 B. C.（或 E. C.），然后松开照准部瞄准交点 P. I.，并且设置水平度盘读数为 0°00′00"。

（2）转动照准部设置水平度盘读数为偏角 Δ_1（0°26′18"），然后沿着视线方向丈量弦长 C_1（1.53m），并在地上钉一木桩确定了曲线上的第 1 点，即 P_1（K5 +100）。

（3）经纬仪不要移动，转动照准部设置水平度盘读数为所需的累计偏角值 Δ_2（6°10′04"）。一人握住尺子前端并使尺读数 C（即弦长 19.97m）对准 P_1 点，另一人在 P_2 点附近拉紧尺子的后端（零分划）并摆动直至当与经纬仪的视线相交为止，这时立即在零分划处插一测钎，这就是曲线上的 P_2点（桩号 K5 +120）。

这个过程继续进行，一直钉到 E. C. 点。

（4）理论上，这样测设的最后点应与已钉的 E. C. 点重合，但是由于误差的积累或测量中其他缺陷，产生了闭合差。如果闭合差不合格，应检查所有的观测并进行调整直到最终获得满意的桩位。

偏角法是一种精度高、实用性强、灵活性大的曲线细部测设常用方法，它可在曲线上的任意一点或交点 P. I. 处设站。但由于距离是逐点连续丈量的，前面点的点位误差必然会影响到后面点的测量精度，点位误差是逐渐累积的。因此，对于长曲线前半段曲线由 B. C. 来测设，后半段曲线由 E. C. 来测设，产生的小误差可在曲线的中部进行调整。

3. 极坐标法（使用全站仪测设）

当使用全站仪测设时，如图 9-5 所示，应计算每个点的偏角（Δ_1、Δ_2、Δ_3 等等）以及从 B. C. 至每点的距离（C_1、C_2、C_3等等），因此首先必须计算如表 9-3 所需的各种数据。

关于偏角 Δ_1 和弦长 C_1 在上节中已推导，即公式（ 9-9 ）和（ 9-10 ）：

$$\Delta_1 = \frac{\varphi_1}{2} = \frac{l_1}{R}\left[\frac{90°}{\pi}\right]$$

$$C_1 = 2R\sin\Delta_1$$

$$\delta = \frac{\varphi}{2} = \frac{l_0}{R}\left[\frac{90°}{\pi}\right]$$

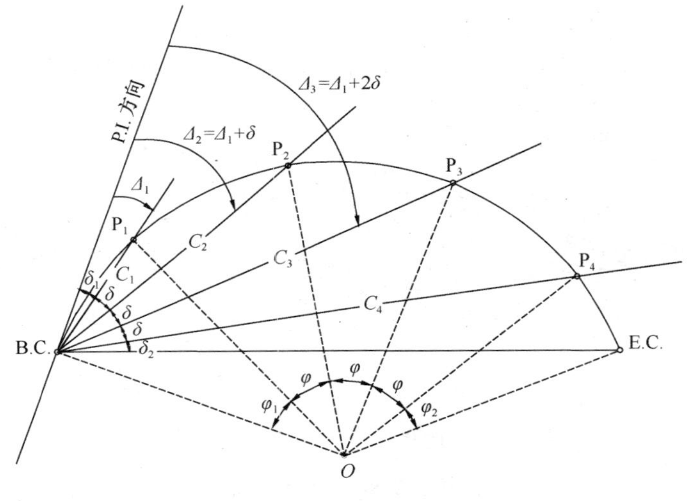

图 9-5　极坐标法

从图 9-5 可以看出：

$$\Delta_2 = \Delta_1 + \delta \qquad C_2 = 2R\sin\Delta_2$$

$$\Delta_3 = \Delta_1 + 2\delta \qquad C_3 = 2R\sin\Delta_3$$

$$\Delta_4 = \Delta_1 + 3\delta \qquad C_4 = 2R\sin\Delta_4$$

$$\vdots \qquad\qquad \vdots$$

例 9. 4　所有原始数据与例 9-3 相同，用极坐标法测设圆曲线计算列于表 9-3。

表 9-3　用极坐标法测设计算

主点	细部点	弧长 (m)	转折角 Δ (° ′ ″)	累计转折角 Δ (° ′ ″)	总弦长 C (m)
B. C. K5 +098. 47				0 00 00	
		1. 53	0 26 18		1. 53
	P_1 +100			0 26 18	
		20. 00	5 43 46		
	P_2 +120			6 10 04	21. 48
		20. 00	5 43 46		
	P_3 +140			11 53 50	41. 23
		20. 00	5 43 46		
	P_4 +160			17 37 36	60. 56
		8. 90	2 32 59		
F. C. K5 +168. 90				20 10 35	68. 98

当测设曲线时，全站仪安置在 B. C. 点，视线向前瞄准 P. I. 点并且水平度盘设为 0°00′00″，向右转偏角 0°26′18″，并测量弦长 1. 53m 定 P_1 点（见图 9-5）。然后，向右转偏角 6°10′04″，并测量总弦长 21. 48m 定 P_2 点。用类似的方法通过设置偏角并且测量总弦长来定下其他各点。当图 9-5 完成测设 P_4 时，我们应丈量 P_4 与 E. C. 之间距离，看它是否等于它们两点间的理论弦长。目视检查曲线应均匀圆滑。显然，这个方法要求从 B. C. 到各点的视线清晰可见。

9. 4　缓和曲线

9. 4. 1　概述

缓和曲线是一种半径不断变化的曲线。如果用于联结直线至半径为 R 的曲线，那么缓和曲线开始的半径等于直线半径（∞），最后的半径是曲线的半径 R（见图 9-6）。

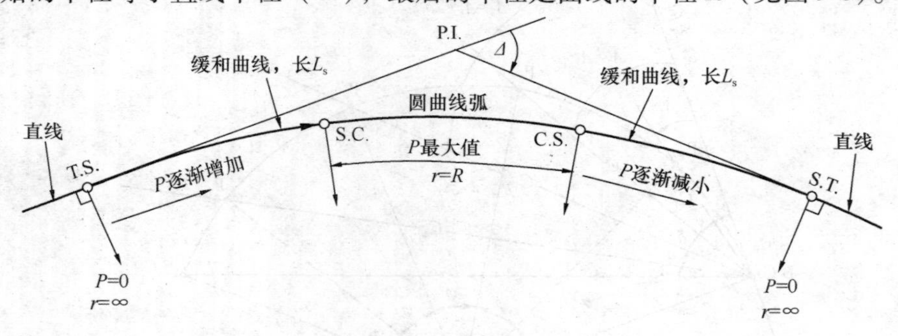

图 9-6　缓和曲线

设车辆沿直线行驶车速为 v，作用于车辆的力就是它的重量 w，作用力向下，等量的反作用力通过车辆向上。当车辆在切线点 T. S. 进入半径为 R 的曲线，一个附加的离心力 P 作用于车辆，已知 $P = wv^2/gr$，式中 w 是车辆的重量，g 是重力加速度 $g = 9.8\text{m/sec}^2$ 以及 r 是某点的曲率半径。如果 P 大，车辆有一个力指向曲线外侧，可能使车辆打滑或翻倒。

在直线道路上，r = 无穷大，则 $P = 0$。

在半径为 R 的圆线上，$r = R$，则 $P = wv^2/gR$。

9. 4. 2　螺旋线的特征方程

道路曲线通常由两条缓和曲线和一条圆曲线组成（图 9-6）。车辆行驶从 T. S. 到 S. T.，

力 P 从零逐渐增加到圆曲线上的最大值，然后又逐渐降低至零。这样就大大地减少了车辆打滑，降低乘客的不舒适感。

对于固定车速 v，作用于车辆的力 P 是 wv^2/gr。由于设计道路曲线针对特定的车速和车辆的重量，故车速与车重可以认为是常数，因此得出 $P \propto \frac{1}{r}$。

然而，如果这个力允许沿着曲线均匀地增加，那么这种力 P 还必须同 l 成正比，即 $P \propto$。式中 l 就是从切线点到所关注点的曲线长。

综合上述，这两个要求可以得出 $l \propto \frac{1}{r}$，即 $rl = K$，这里 K 为常数。如果 L_S 为每边缓和曲线总长度，R 为圆曲线的半径，则 $RL_S = K$。

缓和曲线的基本要求在于曲率半径 r 必须随 l 的长度（从缓和曲线起点开始计算）不同而不同。从切线点沿曲线测量的长度与那个点的曲率半径的乘积是常数。

满足缓和曲线基本要求的标准曲线是回旋螺旋线。这个曲线在实际使用中有些改进，简化为立方形螺旋线或抛物线。

螺旋线如图 9-7 所示，T. S. =切线至螺旋线的点，S. C. =螺旋线至圆曲线的点，C 为从 T. S. 起算的螺旋线长度为 l 的任意点。在 S. C. 点切于曲线的切线交切线于 M 点，r 是 C 处螺旋线的半经，β 是在 C 点切线与后一边的切线 AX 的夹角，β_S 为偏转角或螺旋线角，即螺旋线终点的切线与后一边的切线 AX 组成的夹角。R 圆曲线半径，L_S 缓和曲线长。从缓和曲线基本要求，我们得

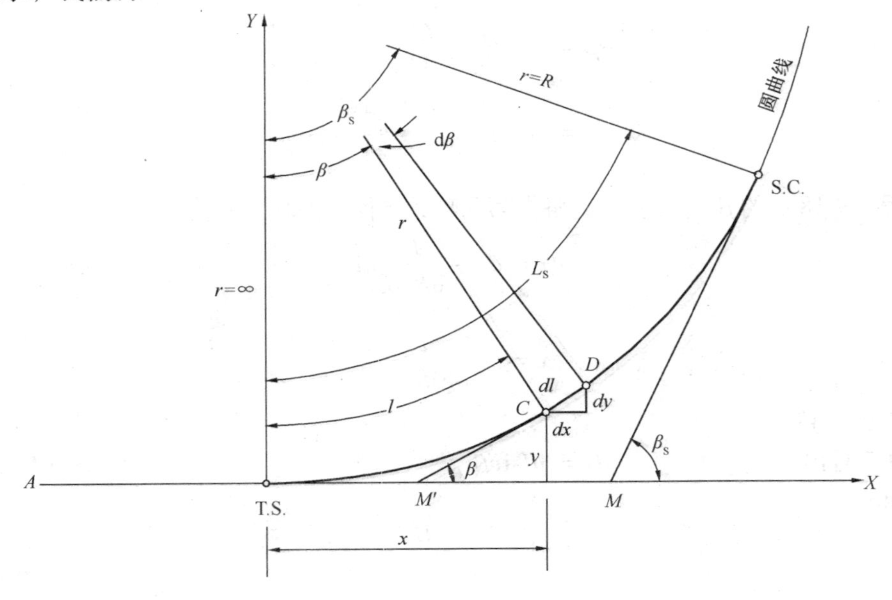

图 9-7　理想的缓和曲线——回旋螺旋线

$$l\,r = L_S\,R \tag{9-15}$$

螺旋线任一点的曲率半径 r 由上式可求得，即

$$r = \frac{RL_S}{l} \tag{9-16}$$

从图 9-7 看出，设想在螺旋线任意点取一微分段 dl，对应微分段 dl 的中心角为 $d\beta$，我们得

$$r d\beta = \mathrm{d}l \tag{9-17}$$

$$d\beta = \frac{\mathrm{d}l}{r} = \frac{l}{R}\frac{\mathrm{d}l}{L_S}$$

积分后求得

$$\beta = \frac{l^2}{2RL_S} \tag{9-18}$$

这就是螺旋线的特征方程。

当 $l = L_S$，方程（9-18）变为

$$\beta_S = \frac{L_S}{2R} \tag{9-19}$$

9.4.3　曲线上点的直角坐标

从切线点 T. S. 很容易测设曲线，为此要求建立曲线点的直角坐标系。参见图 9-7，T. S. 是坐标原点，C 点的坐标为（x，y），任意点 D 点的坐标为（$x+dx$）和（$y+dy$），dl 是 C 与 D 沿曲线的距离，β 是 C 点切线与 AX 组成的角，$\beta + d\beta$ 是 D 点切线与 AX 组成的角。

C 点坐标 x：

从图 9-7 看出

$$dx = dl\cos\beta \tag{9-20}$$

展开 $cos\beta$

$$\cos\beta = 1 - \frac{\beta^2}{2!} + \frac{\beta^4}{4!} - \cdots$$

∴

$$dx = dl\left(1 - \frac{\beta^2}{2!} + \frac{\beta^4}{4!} - \cdots\right) \tag{9-21}$$

将方程（9-18）的 β 代入上式，略去第 3 项，方程（9-21）变为

$$dx = dl - \frac{l^4}{8R^2L_S^2}dl$$

积分得

$$x = l - \frac{l^5}{40R^2L_S^2} \tag{9-22}$$

C 点的坐标 y：

从图 9-7 看出

$$dy = dl\sin\beta \tag{9-23}$$

展开 $sin\beta$ 得

$$\sin\beta = \beta - \frac{\beta^3}{3!} + \frac{\beta^5}{5!} - \cdots$$

∴

$$dy = dl\left(\beta - \frac{\beta^3}{3!} + \frac{\beta^5}{5!} - \cdots\right) \tag{9-24}$$

将方程（9-18）的 β 代入上式，略去第 3 项，方程（9-24）变为

$$dy = dl\left(\frac{l^2}{2RL_S^2} + \frac{l^6}{48R^2L_S^2}\right)$$

积分得

$$y = \frac{l^3}{6RL_S} - \frac{l^7}{336R^2L_S^2} \tag{9-25}$$

从这些方程求得直角坐标 x 和 y。

9.4.4 缓和曲线的计算与测设

1. 缓和曲线的计算

图 9-8 图解说明圆曲线如何向内移动（向圆心），在变短圆曲线的两端留下空间插入螺旋线。圆曲线从主切线向内移动量称为 p，移动的结果，圆心 O 离主切线距离（$R+p$）。

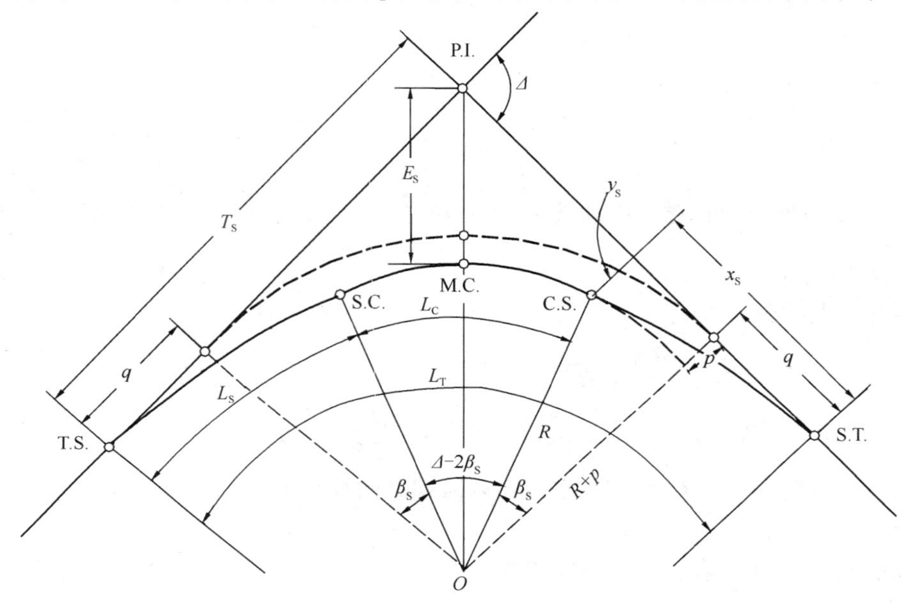

图 9-8　螺旋线的几何形状与符号

图 9-8 可以看出内移值 p

$$p = y_S - (R - R\cos\beta_S) \tag{9-26}$$

展开 $\cos\beta$ 并代入方程（9-26）得

$$p = \frac{L_S^2}{24R} \tag{9-27}$$

图 9-8 可以看出切线增长值 q

$$q = x_S - R\sin\beta_S \tag{9-28}$$

展开 $\sin\beta$ 并代入方程（9-28）得

$$q = \frac{L_S}{2} - \frac{L_S^3}{240R^2} \tag{9-29}$$

从方程（9-29）可以看出，当圆曲线半径足够大时，公式中的第 2 项很小可以略去。切线的增长值大约是缓和曲线的一半，即 $q = \frac{L_S}{2}$。

缓和曲线的元素及它们之间的关系如下所述：

（1）切线长　$$T_S = (R+p)\tan\frac{\Delta}{2} + q \tag{9-30}$$

(2) 主曲线（圆曲线弧）长　$L_C = R(\Delta - 2\beta_S)\frac{\pi}{180°}$　(9-31)

(3) 曲线总长　$L_t = L_C + 2L_S$　(9-32)

(4) 外距　$E_S = (R + p)\sec\frac{\Delta}{2} - R$　(9-33)

(5) 切曲差　$D_S = 2T_S - L_T$　(9-34)

为了计算的方便，对方程 (9-30)、(9-31)、(9-32)、(9-33) 与 (9-34) 做一些小的改变得

切线长 T_S：

$$T_S = (R + p)\tan\frac{\Delta}{2} + q$$

$$T_S = R\mathrm{tg}\frac{\Delta}{2} + \left(p\tan\frac{\Delta}{2} + q\right) = T + t \quad (9\text{-}35)$$

式中，圆曲线长为 $T = R\tan\frac{\Delta}{2}$，尾加数为 $t = p\tan\frac{\Delta}{2} + q$。即，缓和曲线切线长 T_S 等于圆曲线的切线长 T 加尾加数 t。

主曲线（圆曲线弧）长 L_C：

$$L_C = R(\Delta - 2\beta_S)\frac{\pi}{180°}$$

$$= R\Delta\frac{\pi}{180°} - 2R\beta_S\frac{\pi}{180°} = L - 2R\left(\frac{L_S}{2R}\frac{180°}{\pi}\right)\frac{\pi}{180°} = L - L_S \quad (9\text{-}36)$$

即，主曲线（圆曲线弧）长 L_C 等于圆曲线长 L 减螺旋曲线长 L_S。

总曲线长 L_T：　$L_T = L_C + 2L_S = (L - L_S) + 2L_S = L + L_S$　(9-37)

即，总曲线长 L_T 等于圆曲线长 L 加螺旋曲线长 L_S。

外距 E_S：

$$E_S = (R + p)\sec\frac{\Delta}{2} - R$$

$$= \left(R\sec\frac{\Delta}{2} - R\right) + p\sec\frac{\Delta}{2} = E + e \quad (9\text{-}38)$$

式中，圆曲线外距为 $E = R\sec\frac{\Delta}{2} - R$，尾加数为 $e = p\sec\frac{\Delta}{2}$。即，缓和曲线外距 E_S 等于圆曲线外距 E 加尾加数 e.

切曲差 D_S 为两倍切线长与曲线长之差，即

$$D_S = 2T_S - L_T \quad (9\text{-}39)$$

圆曲线半径 R 和缓和曲线长 L_S 的确定要根据公路的等级及地面的条件。测量转折角 Δ，按照上面所列公式求出测设曲线所需的数据。

缓和曲线各点里程桩公式如下：

T. S. 里程 = P. I. 里程 − 切线长 T_S
S. C. 里程 = T. S. 里程 + 螺旋曲线长 L_S
M. C. 里程 = S. C. 里程 + 主曲线长度 $L_C/2$　(9-40)
C. S. 里程 = M. C. 里程 + 主曲线长度 $L_C/2$
S. T. 里程 = C. S. 里程 + 螺旋曲线长 L_S

为了避免计算的错误，用下列公式作核计算：

$$\text{S. T. 里程} = \text{P. I. 里程} + T_S - D \tag{9-41}$$

例 9.5　已知公路设计行车速度 120km/h，交点桩 P. I. 的里程为 K9 +658.86，转折角 $\Delta = 20°18'26''$（右偏），设计曲线半经 $R = 600$m，设计缓和曲线长 $L_S = 100$m。要求计算曲线元素以及缓和曲线主点的里程。

计算列于表 9-4，最好是使用笔记本电脑进行编程计算，计算效率将大大提高。

表 9-4　缓和曲线测设计算表

工程项目：××××× 　　地点：××××××

已知数据		交点桩号：P. I. 8　　编号：K9 +658. 86 转折角 $\Delta = 20°18'26''$（右偏） 圆曲线半经 $R = 600$m　　缓和曲线长 $L_S = 100$m。
缓和曲线特征参数的计算		切线增长值　$q = \frac{l_h}{2} - \frac{l_h^3}{240R^2} = 50.00$m
		圆曲线内移值　$p = \frac{l_h^2}{24R} = 0.69$m
		回旋螺旋线角　$\beta_h = \frac{l_h}{2R} \times \frac{180°}{\pi} = 4°46'29''$
缓和曲线元素的计算	T_S	圆曲线的切线长　$T = R\tan\frac{\Delta}{2} = 107.46$m
		切线长尾加数　$t = p\tan\frac{\alpha}{2} + q = 50.12$m
		加缓和曲线后的切线长　$T_S = T + t = 107.46 + 50.12 = 157.58$m
	L_T	圆曲线长　$L = R\Delta\frac{\pi}{180°} = 212.66$m
		缓和段曲线长　$L_S = 100$m
		总曲线长　$L_T = L + L_S = 212.66 + 100 = 312.66$m
	L_C	主曲线长　$L_C = L - L_S = 112.66$m
	E_S	圆曲线的外距　$E = R\left(\sec\frac{\Delta}{2} - 1\right) = 9.55$m
		外距尾加数　$e = p\sec\frac{\Delta}{2} = 0.70$m
		加缓和曲线后的外距　$E_S = E + e = 9.55 + 0.70 = 10.25$m
	D_S	切曲差　$D_S = 2T_S - L_T = 2.26$m
缓和曲线各主点编号的计算	直缓点　$TS = PI - T_S = K9 + 501.28$	略图
	缓圆点　$SC = TS + L_S = K9 + 601.28$	
	曲中点　$MC = SC + \frac{L_C}{2} = K9 + 657.61$	
	圆缓点　$CS = MC + \frac{L_C}{2} = K9 + 713.94$	
	缓直点　$ST = CS + L_S = K9 + 813.94$	
校　　核	$ST = PI + T_S - D_S = K9 + 813.94$	

（1）缓和曲线参数的计算

按公式（9-19）计算螺旋线角

$$\beta_S = \frac{L_S}{2R} \times \frac{180°}{\pi} = \frac{100 \times 180}{2 \times 600 \times \pi} = 4°46'29''$$

按公式（9-27）计算内移值 p

$$p = \frac{L_S^2}{24R} = \frac{100^2}{24 \times 600} = 0.69\text{m}$$

按公式（9-29）计算切线增长值 q

$$q = \frac{L_S}{2} - \frac{L_S^3}{240R^2} = \frac{100}{2} - \frac{100^3}{240 \times 600^2} = 50\text{m}$$

按公式（9-22）与（9-25）分别计算缓和曲线上任意点的直角坐标 x 和 y，即

$$x = l - \frac{l^5}{40R^2L_S^2} \text{ 和 } y = \frac{l^3}{6RL_S} - \frac{l^7}{336R^2L_S^2}$$

当 $l = L_S$，把 L_S 代入上列公式得 S. C. 点的坐标 x_S 与 y_S，即

$$x_S = L_S - \frac{L_S^3}{40R^2} = 100 - \frac{100^3}{40 \times 600^2} = 99.93\text{m}$$

$$y_S = \frac{l^2}{6R} = \frac{100^2}{6 \times 600} = 2.78\text{m}$$

（2）计算缓和曲线主要元素

切线长：$T_S = T + t$

$$T = R\tan\frac{\Delta}{2} = 600 \times \tan\frac{20°18'26''}{2} = 107.46\text{m}$$

$$t = p\tan\frac{\Delta}{2} + q = 0.69 \times \tan\frac{20°18'26''}{2} + 50.00 = 50.12\text{m}$$

$$\therefore \quad T_S = T + t = 107.46 + 50.12 = 157.58\text{m}$$

总曲线长 L_T：

$$L = R\Delta\frac{\pi}{180°} = 600 \times 2°18'26'' \times \frac{\pi}{180°} = 212.66\text{m}$$

$$\therefore \quad L_T = L + 2L_S = 212.66 + 100 = 312.66\text{m}$$

主曲线长 L_C：$L_C = L - L_S = 212.66 - 100 = 112.66\text{m}$

外距：$E_S = E + e$

$$E = R\left(\sec\frac{\Delta}{2} - 1\right) = 600\left(\sec\frac{20°18'26''}{2} - 1\right) = 9.55\text{m}$$

$$e = p\sec\frac{\Delta}{2} = 0.69 \times \sec\frac{20°18'26''}{2} = 0.70\text{m}$$

$$\therefore \quad E_S = E + e = 9.55 + 0.70 = 10.25\text{m}$$

切曲差：$D_S = 2T_S - L_T = 2 \times 157.58 - 312.66 = 2.50\text{m}$

（3）缓和曲线主点里程的计算

P. I. 里程	K9 + 658.86
$-T_S$	157.58
T. S. 里程	K9 + 501.28
$+L_S$	100.00

S. C. 里程	K9 +601. 28
$+L_C/2$	56. 33
M. C. 里程	K9 +657. 61
$+L_C/2$	56. 33
C. S. 里程	K9 +713. 94
$+L_S$	100. 00
S. T. 里程	K9 +813. 94

检核：S. T. 里程 = P. I. 里程 + $T_S - D$ = K9 +658. 86 +157. 58 −2. 5 = K9 +813. 94

检核通过，所以上述计算过程是正确的。

2. 缓和曲线主点测设步骤

（1）经纬仪安置在交点 P. I. 上，对中整平。后视道路中线并从交点 P. I. 沿道路中线测量切线长 T_S = 157. 58m，并把预先写好编号（K9 +501. 28）的木桩打入地面。

同样地，前视道路中线，从交点 P. I. 沿中线测量切线长并把预先写好编号（K9 +813. 83）的木桩打入地面。

（2）仪器不要移动，前视瞄准 S. T.，设置水平度盘读数为 0°00′00″。然后旋转经纬仪照准部，设置水平度盘读数为 $\frac{180° - \Delta}{2}$，这时望远镜视线就处于角平分线上，由交点 P. I. 沿角平分线测量外距 E_S，并把预先写好编号（K9 + 657. 55）的木桩打入地面。

（3）假定以 T. S. 点为坐标原点，T. S. 至 P. I. 的切线方向定义为坐标系的 x 轴，垂直于切线方向定义为 y 轴。用切线支距法测量 x_S =99. 93m，y_S = 2. 78m 得到 S. C. 点，它的里程编号为 K9 +601. 28。

（4）同样地，假定以 S. T. 点为坐标原点，S. T. 至 P. I. 的切线方向定义为坐标系的 x 轴，垂直于切线方向定义为 y 轴。用切线支距法测量 x_S = 99. 93m，y_S = 2. 78m 得到 C. S. 点，它的里程编号为 K9 +713. 83。

（5）打木桩时，应该注意在木桩顶上钉小钉子以标志点的精确位置。

9. 5　竖曲线

公路两坡度线之间，在竖直平面方向提供平稳缓和变换的曲线称为竖曲线。它的作用是确保沿曲线行驶车辆运输安全，旅客舒适，以及确保使相会的车辆能有足够视线距离以便于车辆能够安全停靠或安全通过。

当道路两坡道形成土丘，该曲线称为凸形竖曲线，如图 9-9a 所示。当两坡度线形成谷

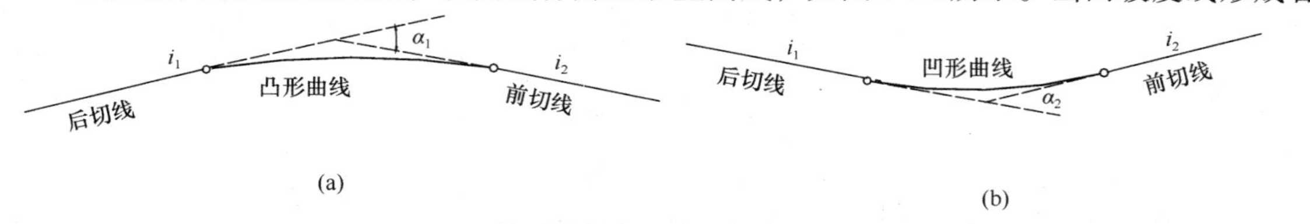

图 9-9　两种类型的竖曲线

（a）凸形竖曲线；（b）凹形竖曲线

地，生成凹形竖曲线，如图 9-9b 所示。假设后切线的坡度为 i_1，前切线的坡度为 i_2，坡度通常以百分比表示。相邻两坡度线的坡度代数差用 α 表示，即

$$\alpha = i_1 - i_2 \tag{9-42}$$

上式中，α 也可称为总坡度差。正的 α 表示竖曲线为凸形，而负的 α 表示竖曲线为凹形。

图 9-10 显示竖曲线使用的术语。当汽车沿道路行驶，最先遇到的坡度线称后切线。另一条坡度线称前切线。从后切线的坡度改变为前切线的坡度要求圆滑且逐渐变化，所以有必要选择某种曲线。圆曲线是适用的。两坡度线的交点缩写为 V，竖曲线的起点缩写为 BVC，竖曲线的中点缩写为 MVC，竖曲线的终点缩写为 EVC。

如同水平的圆曲线一样，竖曲线的元素及其关系可用下列公式表示：

从图 9-10 可求得切线长：　$T = R\tan\dfrac{\alpha}{2}$

由于 α 很小，所以 $\tan\dfrac{\alpha}{2} = \dfrac{\alpha}{2}$．

$$\therefore \quad T = \frac{1}{2}R(i_1 - i_2) \tag{9-43}$$

从图 9-10 可求得竖曲线长：　$L = R(i_1 - i_2) = 2T$　　(9-44)

图 9-10 中，△CAO 为直角三角形，因此

$$T^2 + R^2 = (E + R)^2 = E^2 + 2ER + R^2$$

$$T^2 = E(E + 2R)$$

上式中外距 E 与竖曲线的设计半径 R 相比较是非常之小，用 $2R$ 代替（$E + 2R$）产生的误差极小，所以外距 E 可以写为

$$E = \frac{T^2}{2R} \tag{9-45}$$

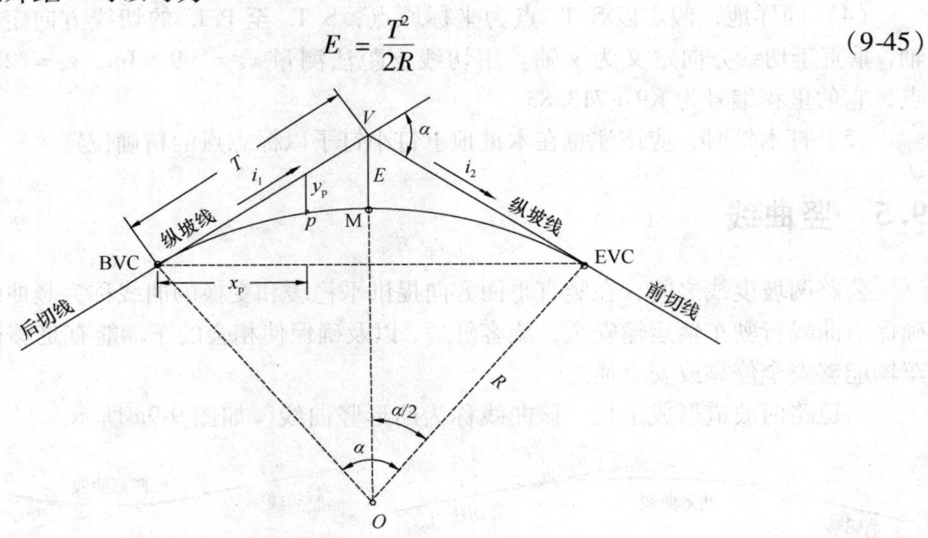

图 9-10　测设竖曲线

为了满足施工以及土方量计算的需要，必须计算竖线上各点的高程改正数，即计算从地面坡度线至竖曲线的垂直距离。从图 9-10 可看出，竖曲线中点 M 的高程改正数就是外距 E。

为了计算竖曲线上其他各点的高程改正数有必要建立起平面直角坐标系，以竖曲线的起点 BVC 或终点 EVC 为坐标原点，过坐标原点的水平方向为 x 轴，竖直方向为 y 轴。因此，竖曲线上任一点 p 的高程改正数可用下式计算：

$$y_p = \frac{x_p^2}{2R} \tag{9-46}$$

式中，x_p 为竖曲线上任意点 p 至竖曲线起点 BVC 或终点 EVC 的水平距离；

y_p 为竖曲线上任意点 p 的高程改正数，也可称为切线的支距（在垂直方向从切线至曲线的距离）。

因此，竖曲线上任意点 p 的设计高程可写为

$$H_p = H'_p \pm y_p \tag{9-47}$$

式中，H_p 为竖曲线上任意点 p 的设计高程；

H'_p 为相应于任意点 p 在纵坡线上的实际高程。

公式（9-47）计算时，对于凸形竖曲线第 2 项应取负号，而对于凹形竖曲线应取正号。

例 9.6　假设某公路变坡点的里程桩号为 K6 + 144，变坡点的高程为 $H_0 = 44.50$m，两相邻坡段的坡度分别设计为 $i_1 = +0.6\%$，$i_2 = -2.2\%$。凸形竖曲线的设计半径为 $R = 3000$m。要求在竖曲线上每隔 10m 打桩，试计算竖曲线元素及所有测设数据。

【解】

（1）计算竖曲线元素

总坡度差　　$\alpha = i_1 - i_2 = 0.006 - (-0.022) = 0.028$

切线长　　$T = \frac{1}{2}R(i_1 - i_2) = \frac{1}{2} \times 3000 \times 0.028 = 42\text{m}$

曲线长　　$L = 2T = 84\text{m}$

外距　　$E = \frac{T^2}{2R} = 0.29\text{m}$

（2）竖曲线主点的里程与高程的计算

BVC 里程：K6 + 144 − T = K6 + 102　　高程：44.50 − 0.6% × 42 = 44.25m

MVC 里程：K6 + 102 + $L/2$ = K6 + 144　　高程：44.50 − 0.29 = 44.21m

EVC 里程：K6 + 102 + L = K6 + 186　　高程：44.50 − 2.2% × 42 = 43.58m

（3）竖曲线细部点计算

全部计算列于表 9-5：

表 9-5　竖曲线的计算

主点	桩点编号	桩点至 BVC 或 EVC 的距离 x（m）	高程改正数 y（m）	纵坡线上实际高程 H'（m）	竖曲线上设计高程 H（m）
竖曲线起点 BVC	K6 + 102	0	0.00	44.25	44.25
	+ 112	10	0.02	44.31	44.29
	+ 122	20	0.07	44.37	44.30
	+ 132	30	0.15	44.43	44.28
变坡点 V	K6 + 144	42	0.29	44.50	44.21
	+ 156	30	0.15	44.24	44.09
	+ 166	20	0.07	44.02	43.95
	+ 176	10	0.02	43.80	43.78
竖曲线终点 EVC	K6 + 186	0	0.00	43.58	43.58

第 10 章　面积与体积的计算

10.1　面积计算

面积的计算基于平面图或断面图，或直接从野外获得的数据。下面几节将讨论面积计算最普遍使用的方法。

10.1.1　透明纸法

为了测量不规则曲线所包围的面积，可以使用带有方格的透明纸。将透明纸覆盖在平面图上。在面积的边界内，计数充满的方格数，沿着面积边界数方格就要判断一下。一般来说，正好精确半个方格要数，大于半格视为一格，小于半格弃去，这就是所谓平衡补偿（见图 10-1）。

$$总面积 = 方格数 \times 方格面积 \times M \qquad (10\text{-}1)$$

式中，M 为平面图比例尺分母。

10.1.2　三角形法

直线边的图可以分成形状较好的三角形（见图 10-2）每个三角形的面积可用下列公式计算：

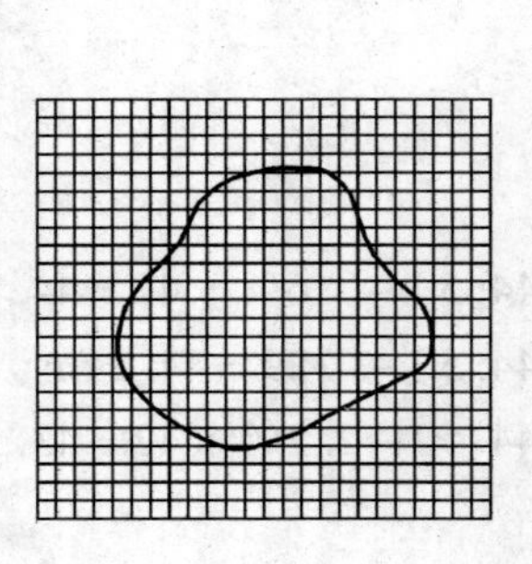

图 10-1　透明纸法

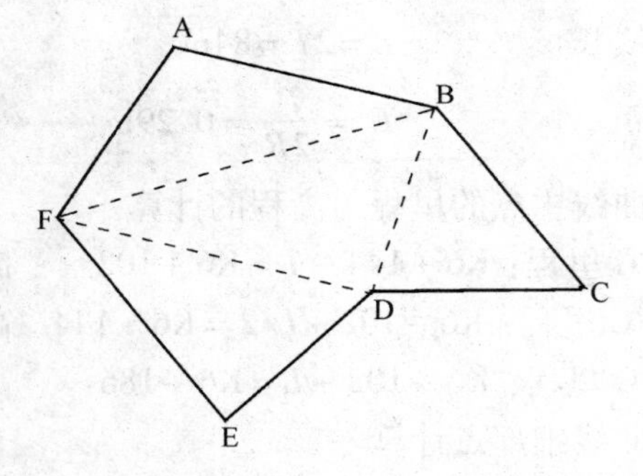

图 10-2　三角形法

（1）$A = \sqrt{s(s-a)(s-b)(s-c)}$

式中，a，b，c 为三角形的三条边，s 为

$$s = \frac{1}{2}(a+b+c).$$

（2）$A = \frac{1}{2}$（三角形底 ×三角形高）

（3）$A = \frac{1}{2}ab\sin C$

式中，C 为 a 和 b 两条边所夹的角。

任意直线边的面积可分成若干三角形计算，然后把各三角形面积相加。

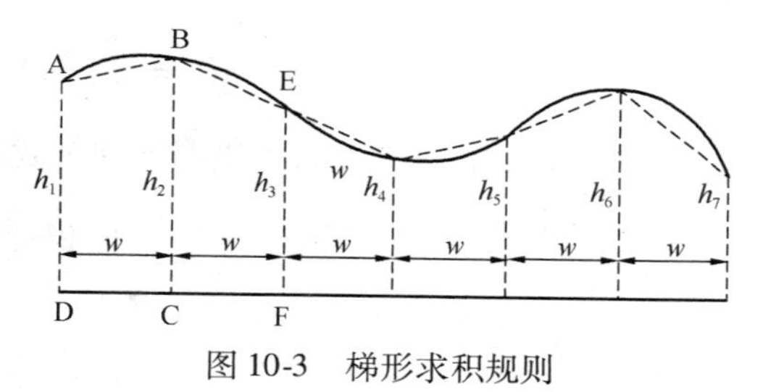

图 10-3　梯形求积规则

10.1.3　梯形求积规则

在图 10-3 中，第 1 个梯形 ABCD 的面积 $=\frac{h_1+h_2}{2}\times w$

第 2 个梯形 BEFC 的面积 $=\frac{h_2+h_3}{2}\times w$ 等等

总梯形面积 = 各梯形面积之和

$$=A=w\left(\frac{h_1+h_7}{2}+h_2+h_3+h_4+h_5+h_6\right) \qquad (10\text{-}2)$$

注意：（1）如果第一条或最后一条的纵坐标为零，仍包含在公式中。（2）这个公式表示曲线边界下虚线所包围的面积，如果的边界线在外边，则计算的面积偏小，反之偏大。

10.1.4　辛普生规则

在辛普生的规则中，在两个纵距之间的边界曲线被假定为抛物线，因此也称抛物线规则，如图 10-4a 所示。推求面积的公式每次使用三个连续的纵距和两段。因此，公式要求偶数段和奇数的纵距。

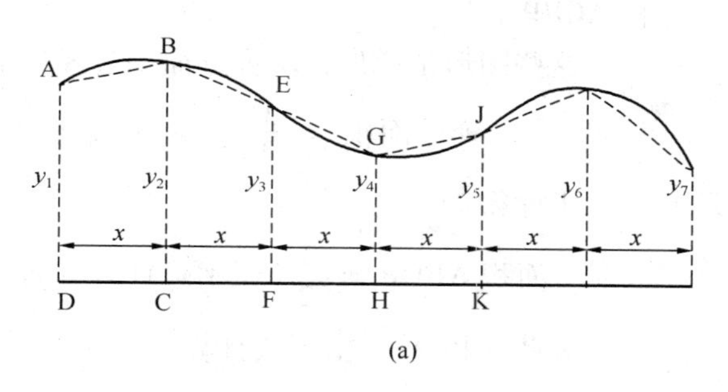

(a)

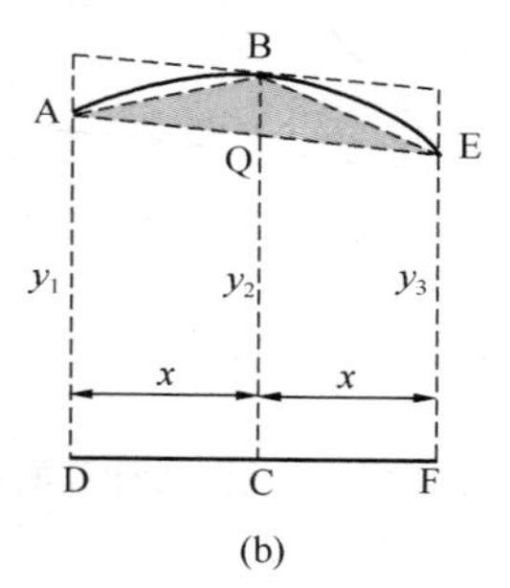

(b)

图 10-4　辛普生规则

（a）边界线按抛物线；（b）两段间的面积

让我们设想面积的两个连续的段，如图 10-4b 所示，连结第 1 纵距点 A 和最后的纵距点 E 而形成的 DAQEFCD 是一个梯形，它可用公式 $\left[\frac{1}{2}(y_1+y_3)\right]\times 2x$ 计算，式中 y_1 与 y_3 为两边的纵距，x 是两个纵距之间的固定间隔。总面积包含这个梯形的面积和显示阴影的抛物线的面积。这个抛物线的面积是封闭的四边形面积的三分之二。这个四边形的底可认为 $2x$，高是 BQ，BQ = BC − QC，BC $=y_2$ 与 QC $=\frac{1}{2}(y_1+y_3)$，因此，阴影的面积如下：

$$\text{阴影的面积}=\frac{2}{3}\left[y_2-\frac{1}{2}(y_1+y_3)\right](2x)$$

第 1 纵距和第 3 纵距之间的总面积是上述这两个面积的总和，即

$$\text{ABEFCDA 的面积}=\frac{1}{2}(y_1+y_3)(2x)+\frac{2}{3}\left[y_2-\frac{1}{2}(y_1+y_3)\right](2x)$$

$$= \frac{1}{3}(x)[y_1 + 4y_2 + y_3]$$

在图 10-4a 中间两段的面积用下列公式计算：

$$\text{EGJKHFE 的面积} = \frac{1}{3}(x)[y_3 + 4y_4 + y_5]$$

各部分面积相加得总面积

$$\text{总面积} = \frac{1}{3}(x)[y_1 + y_7 + 2(y_3 + y_5) + 4(y_2 + y_4 + y_6)] \tag{10-3}$$

辛普生规则表述如下：首尾纵距相加，加上 2 倍第 3、5、7 等纵距之和，再加 4 倍第 2、4、6 等纵距之和，最后把这个总和乘以 $x/3$，这里的 x 为纵距之间的距离。

注意：（1）这个规则假定边界模式为横跨偶数段的抛物线，因此比梯形求积更精确。（2）公式要求奇数个纵距，因而有偶数段。

10.1.5　坐标计算面积公式

在导线测量，三角与三边测量计算中，直线边图形各边的交会点其坐标已计算，这就有可能用坐标来计算控制网各边包围的面积，用直角坐标法计算面积。

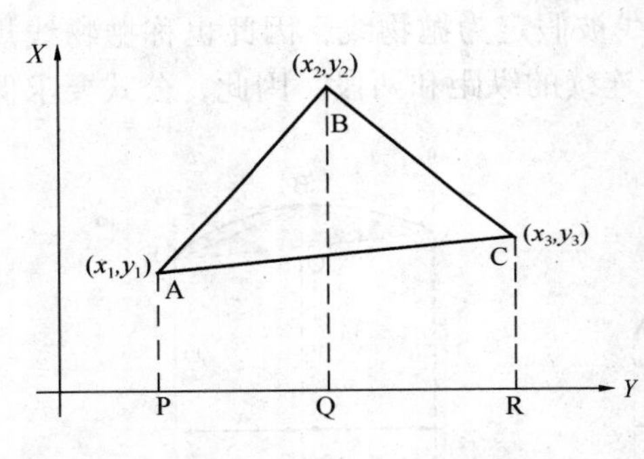

图 10-5　坐标法计算面积

图 10-5 显示 3 点顺时针排列的控制网 ABC，求 ABC 的面积。

面积 ABC = 面积 ABQP + 面积 BCRQ − 面积 ACRP

这些图形是梯形，求梯形面积公式如下：

$$\text{梯形面积} = \frac{1}{2}(\text{高} \times \text{宽})$$

由此得

$$\text{面积 ABQP} = \frac{1}{2}(x_1 + x_2)(y_2 - y_1)$$

因此面积 ABC 由下式计算：

$$\text{面积 ABC} = \frac{1}{2}(x_1 + x_2)(y_2 - y_1) + \frac{1}{2}(x_2 + x_3)(y_3 - y_2) - \frac{1}{2}(x_1 + x_3)(y_3 - y_1)$$

因此

$$2 \times \text{面积 ABC} = x_1y_2 - x_1y_1 + x_2y_2 - x_2y_1 + x_2y_3 - x_2y_2 + x_3y_3 - x_3y_2 - x_1y_3 + x_1y_1 - x_3y_3 + x_3y_1$$

重新整理上式得

$$2 \times \text{面积 ABC} = x_1y_2 + x_2y_3 + x_3y_1 - (y_1x_2 + y_2x_3 + y_3x_1)$$

虽然本例仅 3 边的图形，上式能应用于 n 边图形，普遍的公式如下：

$$2\text{倍面积} = 2 \times A = x_1y_2 + x_2y_3 + \cdots + x_{n-1}y_n + x_ny_1 - (y_1x_2 + y_2x_3 + \cdots y_{n-1}x_n + y_nx_1)$$

使用符号 $\sum$ 表示，上式变为

$$A = \frac{1}{2}\left(\sum_{i=1}^{n} x_iy_{i+1} - \sum_{i=1}^{n} y_ix_{i+1}\right) \tag{10-4}$$

当使用这个公式时必须注意 x 与 y 的下标。对于闭合多边形，公式（10-4）中编号如果 $i = n$，则 $y_{i+1} = y_{n+1} = y_1$ 且 $x_{i+1} = x_{n+1} = x_1$。用这个公式有可能很容易计算闭合多边形内的面积，其步骤如下：

（1）为了方便计算，各点坐标按下列方式排列如图 10-6 所示，注意第 1 点坐标在最后重复排列上。

（2）从 x_1 到 y_2，从 x_2 到 y_3，……等等，画实线；然后从 y_1 到 x_2，从 y_2 到 x_3，……等等，画虚线。

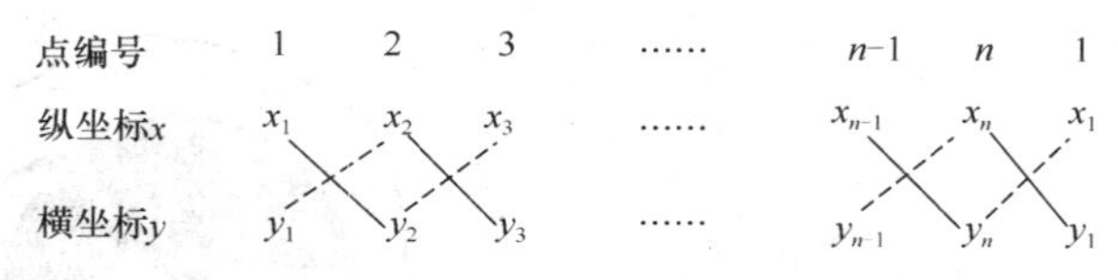

图 10-6　各点坐标排列图

（3）实线相连两坐标的乘积之和减去虚线相连两坐标的乘积之和等于二倍多边形面积，即

$$2A = 实线两坐标乘积之和 - 虚线两坐标乘积之和$$

应用坐标法，如果计算得面积为负，负号应略去。

例 10-1　已知某四边形四个角的坐标已标注于图 10-7 上，试计算四边形的面积。

【解】

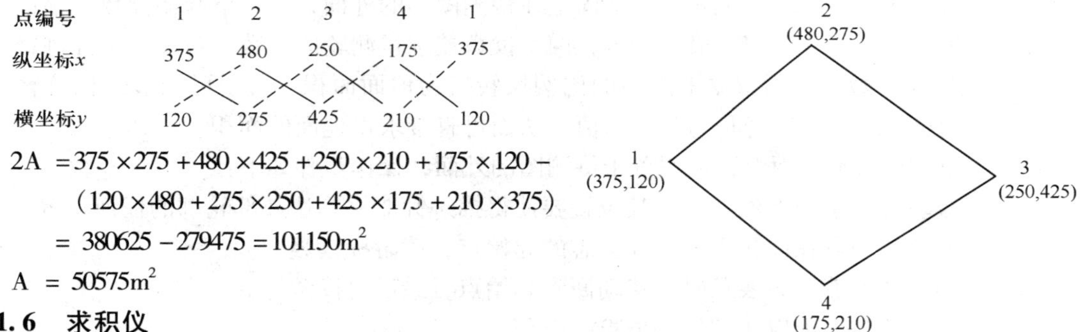

$$2A = 375 \times 275 + 480 \times 425 + 250 \times 210 + 175 \times 120 - (120 \times 480 + 275 \times 250 + 425 \times 175 + 210 \times 375)$$

$$= 380625 - 279475 = 101150\text{m}^2$$

$$A = 50575\text{m}^2$$

图 10-7　四边形

10.1.6　求积仪

1. 极点机械求积仪

求积仪是测定不规则平面图形面积的机械装置。它能取得高精度。图 10-8 显示极点求积仪的主要特点。

仪器包含两个臂，极臂 JP 和描迹臂 JT，它们通过铰接点 J 并通过极块的铰钉固定在平面图于 P 点彼此相对自由移动。M 是有分划的测轮，T 为描迹点。当 T 沿着面积周围移动时，由于描迹点的动作不同，测轮在图上部分旋转，部分滑动（图 10-8）。测轮被分为 10 个大格，每个大格又分为 10 小格。因此，测轮能直读一周的1/100，而游标紧靠着测轮可读至测轮转动的 1/1000。测轮与水平计数圆盘齿轮传动连接，计数圆盘显示转数，测轮转 1 周相应圆盘转 1 格。因此读数包含 4 个数字：

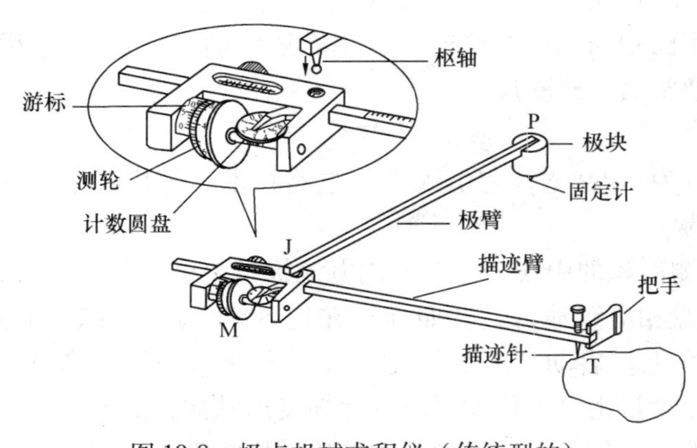

图 10-8　极点机械求积仪（传统型的）

圆盘指针读数，测轮的 1/10 与 1/100 数，游标读至测轮的 1/1000 数。

使用极点求积仪有两种方式：

（1）把极块置于待测图形的外面（见图 10-9）；

（2）把极块置于待测图形的内部（见图 10-10）。

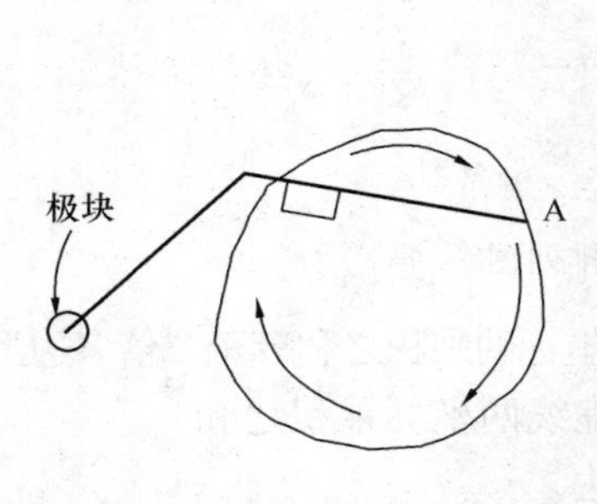

图 10-9　极块在图形之外

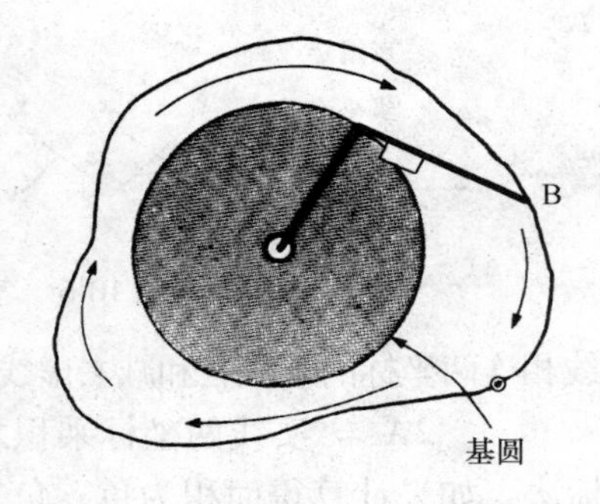

图 10-10　极块在图形之内

在通常情况下，极块（固定针）P 固定在待测图形的外面，先记下起始读数，描迹针仔细沿着面积的周围移动并记下最后读数，两个读数差求得测轮的转数。对于固定描迹臂的仪器直接求得 mm^2 数，然后根据平面图的比例尺转换为地面面积。对于可调节的描迹臂，根据图的比例尺可设置描迹臂长为某特殊值，从而可直接求得地面的面积。

平面图放在水平桌面上，极块置于待测图的外面，操作步骤如下：

（1）极块放置于图形的外面，其位置要使得描迹针能够到达外部轮廓的任何位置。

（2）安放描迹点在面积界线上并标志的起始点，读游标读数。

（3）描迹点沿着边界线顺时针移动回到起始点，最后再读游标。

（4）两个读数差乘以比例尺分母求得面积。

（5）重复操作直到取得 3 个相符合的值，最后取平均值。

极块置于待测图形内部（见图 10-10）：

测法与极块放在待测图形外面一样，但是应加一个常数，如图 10-10 所示，图中阴影的面积就是基圆面积，在这种情况下，两臂互相垂直，测轮垂直于移动的路径，没有滚动只有滑动，因此产生读数零变化。该常数附在求积仪上，必要时须转换。

无论使用哪一种方法，求积仪应使用已知面积进行检核，如果发现不符值应计算改正因子，施用于所有读数。为此目的求积仪通常提供检验尺。

2. 数字求积仪

图 10-11 显示索佳电子求积仪，型号为 KP-90N，它引入集成电路技术，扫描臂上带有透镜和描迹点，极块与极臂被滚轴所取代。

开始量测时，描迹臂大约安放在待测面积的中线处。镍—镉电池提供电源。打开开关，测量单位可选，例如 m^2，比例尺通过键盘相应键输入。按照前述在边界线上选择开始的参考点或作一个标志，然后描迹点定位在图上。启动“开始”键，显示器显示为零，描迹点沿边界顺时针移动回到参考点，系统的移动量由电子传动译码器感知发出脉冲，经电子处理以数字形式显示面积。仪器提供两种特殊的比例尺，即不同的水平比例尺与垂直比例尺，还

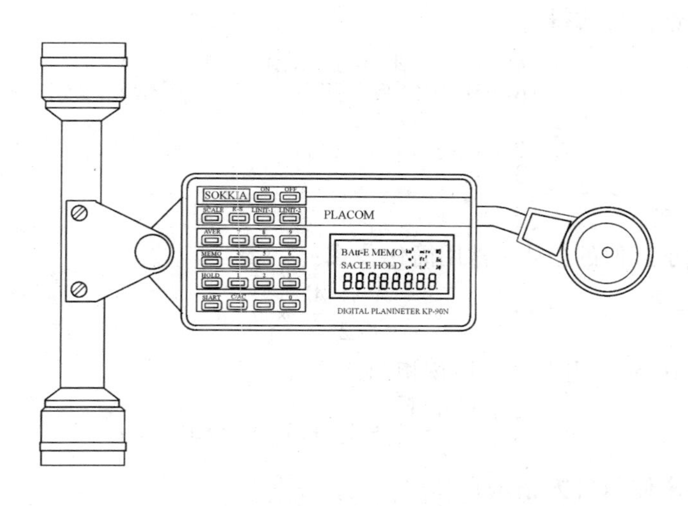

图 10-11　数字求积仪（索佳 KP-90N 电子求积仪）

有其他的功能。

10.2　体积的计算

10.2.1　体积的计算公式

1. 平截头棱锥体公式

平截头棱锥体（图 10-12）可以定义为由两个平行的平面图形和一条直线沿着这两个图形边缘移动产生的面所包围界定的固体。平截头棱锥体公式的证明可以在立体几何课本上找到，其公式表示如下：

$$V = \frac{h}{6}(A_1 + 4M + A_2) \tag{10-5}$$

式中，V 为平截头棱锥体的体积；h 为两个约束平面或基面之间的垂直距离；A_1 为一个基面面积；A_2 为另一个基面的面积；M 为中间截面面积，即位于两基面中间且平行于基面并切固体形成的截面。

平截头棱锥体公式的应用：

用求积仪描迹每一层等高线求得面积 A_1 和 A_2，h 为等高线间距。但是中间的截面积无法求得，然而可以估算。如果假定相邻等高线所包围的面积是相似的图形，并假定中间截面积相似于这两个图形，那么，中间截面的线性尺度等于两个基面线性尺度的平均。由于相似图形线性尺度是图形面积的平方根，所以

$$\sqrt{M} = \frac{\sqrt{A_1} + \sqrt{A_2}}{2}$$

$$M = \frac{A_1 + 2\sqrt{A_1 A_2} + A_2}{4}$$

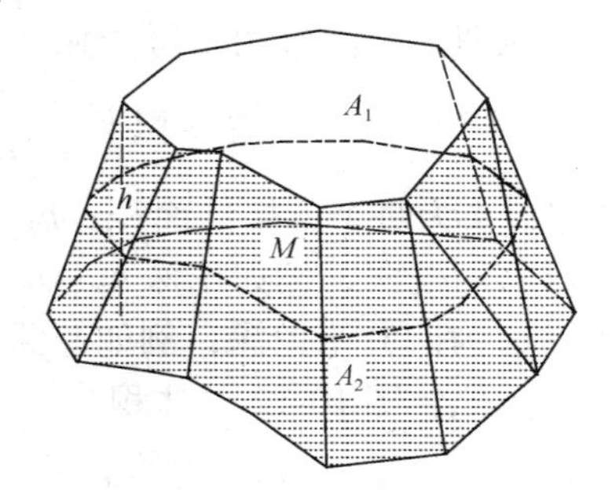

图 10-12　平截头棱锥体

上式代入平截头棱锥体公式得

$$V = \frac{h}{6}\left(A_1 + 4\frac{A_1 + 2\sqrt{A_1A_2} + A_2}{4} + A_2\right)$$
$$= \frac{h}{3}(A_1 + \sqrt{A_1A_2} + A_2) \tag{10-6}$$

2. 两端面积公式

两端面积公式是近似的公式，常用简化公式：

$$V = \frac{A_1 + A_2}{2} \times h \tag{10-7}$$

式中符号同方程（10-5），当有几个体积相加，公式变为

$$V = h\left(\frac{A_1 + A_2}{2} + A_2 + A_3 + \cdots + A_{n-1}\right) \tag{10-8}$$

10.2.2　格点高程求体积（方格法）

此法用于求深挖土方的体积，例如地下室、地下水箱等等的体积。

在地面建立方格网、矩形网或三角形网并求格网交点的高程。较小格网计算体积精度较高，但野外工作量增加，所以通常采用折中的办法。

每格网点的地形高程必须已知，要计算每个网格点现有地形至设计高程的挖深。

图 10-13 显示 10m 的方格，方格顶点注明挖深。查看含有 $h_1\ h_2\ h_6\ h_5$ 的方格，显示于图 10-14。

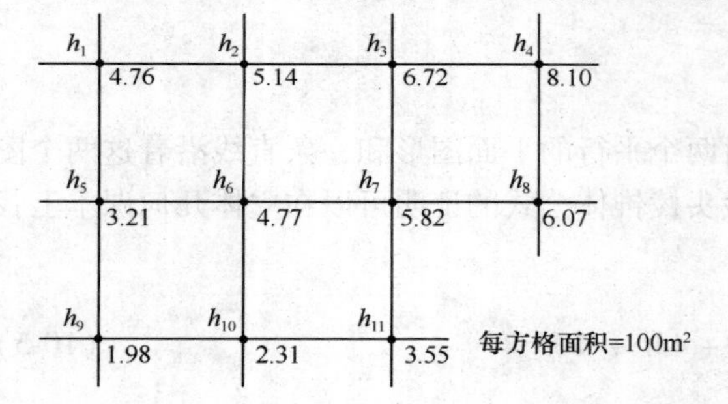

图 10-13　方格网

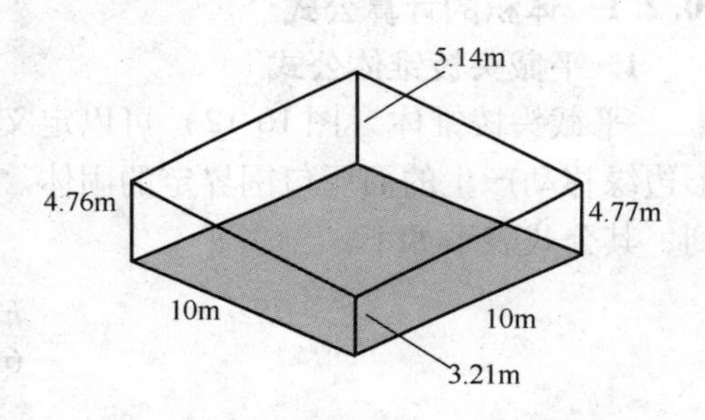

图 10-14　一个独立方格

假设方格顶点之间地面斜坡均匀，方格体积公式是

$$体积 = 平均高 \times 平面面积$$
$$= \frac{1}{4}(4.76 + 5.14 + 4.77 + 3.21) \times 100 = 447\text{m}^3$$

相似的方法应用于每个独立方格，因而导出适用方格网或矩形网格的普遍公式：

$$总体积\ V = \frac{A}{4}\left(\sum h_S + 2\sum h_D + 3\sum h_T + 4\sum h_Q\right) + \delta V \tag{10-9}$$

式中，A 为每个方格的面积；h_S 为独立挖深，例如挖深 h_1、h_4、h_8、h_9 与 h_{11} 仅用 1 次；h_D 为双共用挖深，例如挖深 h_2、h_3、h_5 与 h_{10} 用了 2 次；h_T 为三共用挖深，例如挖深 h_7 用了 3 次；h_Q 为四共用挖深，例如挖深 h_6 用了 4 次；δV 为方格外的所有体积应分别计算。

因此，图 10-12 总体积为

$$总体积\ V = \frac{100}{4}[4.76 + 8.10 + 6.07 + 1.98 + 3.55 + 2(5.14 + 6.72 + 3.21$$

$$+2.31)+3\times5.82+4\times4.77]$$

$$=25(24.46+34.76+17.46+19.08)=2394\text{m}^3$$

这个结果仅是近似的，因为假定方格顶点之间地面坡度是均匀的。

如果用三角形格网，普遍公式按下面方法修改：

（1）$\frac{A'}{3}$替换，$\frac{A}{4}$，式中 A' 为每个三角形的面积。

（2）挖深会出现包含有 5 个和 6 个三角形共用。

10.2.3　等高线图计算体积

这个方法特别适合于计算非常大的体积，例如水库、土坝、矸子山等等的体积。

采用这个系统要计算由每条等高线所包围的面积，然后把它看作横截面积。等高线的间隔提供了横断面的间距，用平截头棱锥体公式或两端面积公式计算体积。如果用平截头棱锥体公式，要求等高线数为奇数。每条等高线所包含的平面面积可用求积仪或其他方法。

成果的精度很大程度上取决等高线的间距，但通常情况不需要很高的精度，例如计算库容，体积计算通常凑整至 1000m³ 就足够了，考察下面的例子。

图 10-15 显示设计水库及坝墙图，图显示通过水库的横切面积和每条等高线与坝墙所包围的面积。

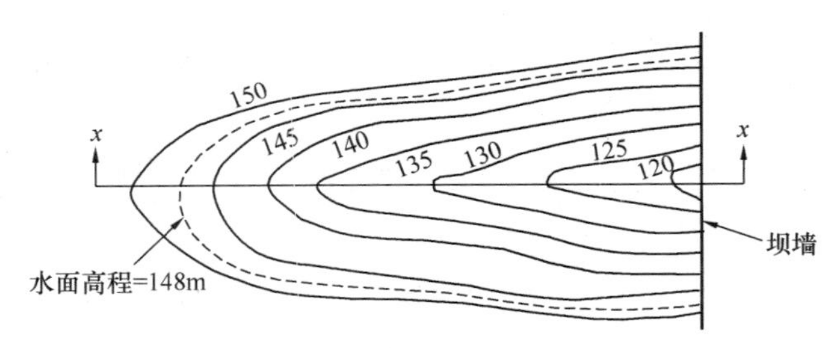

图 10-15　设计水库图

等高线之间储水的体积参看图 10-16，等高距 5m，水库水面高程 148m，要求计算库容量。

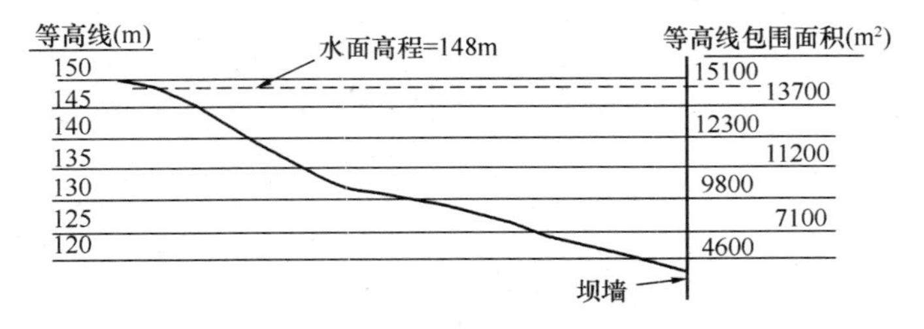

图 10-16　水库垂直断面

总体积 = 等高线 148m 与 145m 间的体积 + 等高线 145m 与 120m 间的体积 + 低于 120m 等高线的小体积。

等高线 148m 与 145m 间的体积用公式（10-7）计算，即

$$=\frac{15100+13700}{2}\times3=43200\text{m}^3$$

等高线 145m 与 120m 间的体积由公式（10-8）计算，即

$$= \frac{5}{2}\left[13700 + 4600 + 2 \times (12300 + 11200 + 9800 + 7100)\right] = 247750\text{m}^3$$

低于120m等高线的小体积，由于等高线间隔减小，比如说1m，用两端面法或截棱锥公式计算任选一种。如果非常小，则可忽略。设这个体积 = δV，因此

$$\text{总体积} = 43200 + 247750 + \delta V \quad (10\text{-}10)$$
$$= (290950 + \delta V)\,\text{m}^3$$

（通常凑整至1000m^3）。公式（10-10）的第2项的计算是用两端面法计算等高线145m和120m之间的体积，换一种方法用截棱锥公式，须用等高线145m和125m之间的体积（因要保持等高线为奇数）和等高线125m与120m间用端面法计算。如果这样做等高线145m和120m之间的体积为248583m^3。

第 11 章　施 工 测 量

11.1　概述

施工测量，可称为建筑物放样或测设，它是利用测量仪器和技术将平面图上的建筑物转移到实地的过程，因而它与测量相反。换言之，施工测量涉及到图上的尺寸规格转换到实地，以便按正确位置施工。这种类型测量有时称为定线和测设高程与坡度。建筑工程测量员的工作通常称为放样工作。

施工测量为各种工程，例如公路、街道、管线、桥梁和建筑物提供放样的准线和坡度高程。施工放样标志拟建工程应有水平位置（线）和垂直位置即高程（或坡度）。建筑承包人可以从测量的标志量测待建设施每个部件的精确位置。

在工程开始之前，首先必须进行地形测量与绘图以表示拟建区域的一切物体的位置。在施工过程中，测量员应测设构造物和实施其他必要的任务。建筑完工后，要进行竣工测量以提供实际工程的位置与尺寸。

施工测量作为所有建筑及其局部的基础已变得越来越重要，据估计测量工时的 60% 花费在定线和测设高程定位类型的测量工作。

建筑物的各个部分应安放到正确的位置。为此在实际放样测量之前，施工测量员应建立参考线或基线，这项内容将在 11.2 节中详细讨论。

在建筑物放样测量中，首先确定从控制点定位建筑物点的方向和距离的必要的数据，然后由室内电脑直接输入到仪器。因此，测量员可指导持棱镜者沿着计算的方向线移动，一直到定位点的距离等于计算的值。使用不同的技术完成这个目标，最普遍使用的技术称为自由站。允许测量员在任何方便的位置安置全站仪，然后通过瞄准已知坐标的参考站，确定测站的坐标与高程。仪器安置在测站（控制点）并正确定向以后，就能显示出从控制点至放样点的方位角与距离。现在很多全站仪都有这个功能，放样点的坐标及高程上传到全站仪，仪器的显示器能显示每个所要求的点位安置棱镜必须左/右，前/后，上/下的移动量。

11.1.1　施工测量的任务

施工测量的任务至少包括下列 4 个方面：

（1）测量工地的现有情况，包括地形，现有建筑物和基础设施，有可能测量的地下基础（例如管底标高、检修口管直径）。

（2）放样参考点和设置指导新建筑施工的标记。

（3）在施工期间检查建筑物的定位。

（4）工程完成以后要做竣工测量，以提供工程完工后的位置与尺寸。这些测量不仅提供修建的记录，而且提供一种检查看工程是否按照设计图。

施工放样包含 3 个方面的问题，无论放样什么建筑物都必须做到：

（1）定位位置正确；

（2）高程正确；

（3）确保结构的垂直性。

11.1.2　放样建筑物

在定位建筑物时，带有平头钉的木桩应打入地面以标志房角，这样的桩仅是暂时的，因为在基础挖掘施工时必然要移动。为了在基础施工时提供许多房角的标志，每个房角附近设置“龙门板”。

龙门板由打入地面的几个木桩和以某特定高程的水平方向钉在桩上的木板组成。龙门板设置在每个角，离每角方向线1～2m，在水平板的适当位置钉小钉（或刻V痕），以便在钉子（或V口）上缠线从一个角到另一个角拉线。房屋角定义为两条线相交处，拉线也提供参考高程。为维护建筑的平整性从参考高程所作的高程测量，参看图11-1和图11-2。图11-1显示一个角的透视图，图11-2显示平面图，C、D、J和I为简易房屋的4个角。

龙门板应牢固建立在地上，为了防止移动，必要时要支撑加固。拉线可以取消，在以后施工时放回。例如基槽需挖多深必须拉线向下测量决定。然后把拉线移开，在槽沟内进行人工或机械挖掘。在做基础砌石时拉线作为参考线。

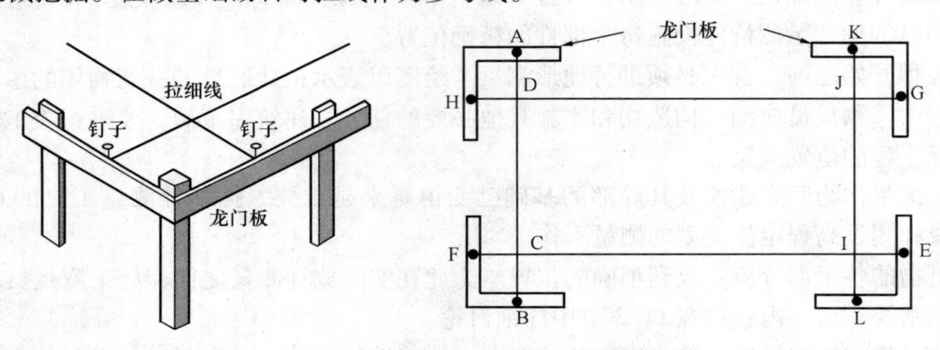

图11-1　龙门板　　　　图11-2　龙门板平面图

设置龙门板步骤如下：

（1）在地上测设建筑物的每个角桩，采用带平头钉的临时木桩。

（2）在每个角附近打龙门桩，按要求的高程水平钉龙门板，用水准仪及水准尺来测设所需的高程。

（3）定义房屋每一边的线要钉小钉（例如图11-2的A点与B点），通过所要求的角点（图11-2中的C点与D点）拉线，拉紧A点与B点之间的线。

（4）经纬仪安置在角点C，瞄准A，转直角，在E点钉一小钉使∠ACE＝90°。

（5）从E点通过经纬仪下C点延长线拉紧，在龙门板上确定F点，在F点钉小钉，缠线于F小钉，EF与AB交点确定角点C。

（6）设置房屋尺寸CD，精确建立角点D。

（7）搬经纬仪到D点，瞄准B点，转直角，在G点钉小钉使∠GDB＝90°。

（8）从G点通过D点延长确定H点，在H点钉小钉，从G到H拉紧线并固定，AB与GH交线确定角点D。

（9）设置房屋尺寸CI与DJ，建立角点I与J。

（10）精确通过角点 I 与 J 拉紧直线确定 K 与 L 两点，在 K 与 L 处钉小钉并拉紧线，KL 与 FE 相交定角点 I，KL 与 HG 相交定角点 J。

如果这些步骤都做得完美，那么建筑物测设正确。测量 IJ 可以确认测设是否正确，也可以测量对角线看是否等于正确的长度。

以上讨论涉及直角的极简单建筑物测设。当然，许多建筑物更为复杂，有的非常复杂。尽管如此，以上所述的一般程序给出了所需的基本步骤。

11.2　基线

在实际放样测量工作开始之前，有必要仔细建立参考线或基线。基线是两个已知点连接的线。基线也可以是现有相邻建筑物的建筑线的延长线。为了放样工程所需的基线应由测量员在图上做出设计。

野外的主要控制点，例如，图 11-3 中导线点 E 和 F，通过测量角度 α 和距离 l 来建立基线 AB，如图 11-3 所示。然后从基线的每一端点测直角丈量辅助支距长，以确定建筑物 Z 的两角点 R 和 S。一旦 R 和 S 打桩后，测量 RS 水平距离与其设计值比较检查。如果在容许的范围内，用点 R 和 S 作为基线测设角点 T 和 U。

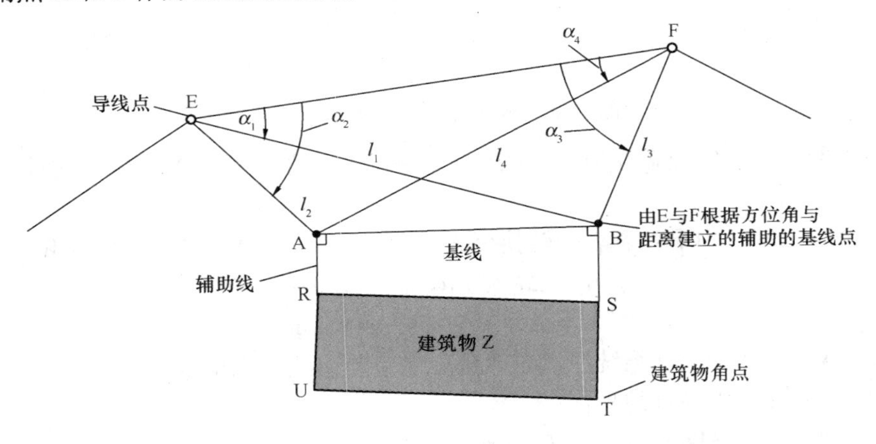

图 11-3　基线

在某些情况，设计者可能规定两个现有建筑物点之间连线作为基线。然后从该基线测直角与量支距，或从基线量距离放样设计点。这种方法的精度主要取决于基线建立的精度。

11.3　格网的使用

土木工程许多结构是由钢或钢筋混凝土柱支撑的水泥地板组成。由于这些柱子照常布置的互相垂直，使用格网形式，格网的交点确定了柱子的位置，因而非常便于放样。放样中使用几种不同的格网如下：

1. 测量网：就是原始地形测量与绘图所建立的直角坐标系（图 11-4）。

2. 施工网：施工网为设计者所用。

施工网定义了工程的主要建筑线的位置与方向，如图 11-4 所示。施工网最佳位置可以

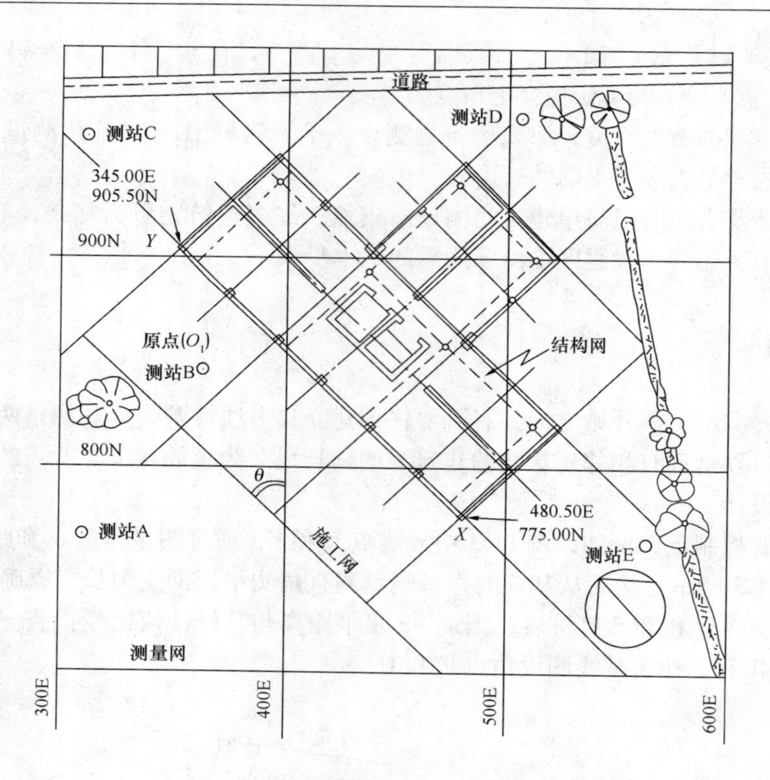

图 11-4　测量网、施工网与结构网

通过绘有施工网的透明纸覆盖在原平面图纸上，进行简单的移动使其上主要设计单元方位合适。

为了测设施工网，用下列著名的转换公式将施工控制网坐标转换为测量网的坐标：

$$E = \Delta E + E_1\cos\theta - N_1\sin\theta$$

$$N = \Delta N + N_1\cos\theta + E_1\sin\theta$$

式中　ΔE, ΔN——两个坐标系统原点东西坐标增量与南北坐标增量；

E_1, N_1——施工网点的坐标；

θ——两格网相对旋转角；

E, N——测量网点的坐标。

因此，选择的点，例如 X 点和 Y 点（图 11-4）要将施工坐标转换为测量坐标，以便从测量控制点用极坐标法或交会法进行测定。现在用 XY 作为基线，用经纬仪和钢尺测设施工网，所有角度用正倒镜观测，格网间隔用钢尺在标准拉力下仔细丈量确定。

3. 结构格网：结构格网是围绕具体建筑物或构造物来建立的，它包含许多细部，例如柱子。这些点位从施工格网去测设不具有足够的精度。结构格网通常是用施工格网来建立，并以物理的方法建在混凝土的楼板上（见图 11-4），采用与施工格网相同的坐标。

11.4　控制垂直度

控制垂直度有多种方法，最普遍使用的方法如下：（1）使用垂球，（2）使用经纬仪，（3）使用光学对中器。

11.4.1　使用垂球

对于低层建筑物可以使用5～10kg的垂球，如图11-5所示。如果外墙是完全垂直的，当垂球线与木桩中心重合，则顶上水平尺距离 d 将等于基础木桩的支距。这个概念可用于内部及外部提供有用的洞和缺口的情况。垂球应较重，比如5kg，垂球和绳线需保护以避风吹。垂球浸入水桶中可以减弱垂球的摆动。类似于决定矿井的垂直度一样但很少有危险性。为便于从底层到顶层的传递，在各中间层预留直径约为0.2m的孔洞，这可能需要与建筑设计师协商。

11.4.2　使用经纬仪

使用该法的前提是经纬仪已完全调整好。当望远镜绕横轴旋转时，视线描绘出垂直平面。该法叙述如下：

经纬仪要依次安置在第一层楼板标志的每条参考线的延长线上。望远镜瞄准已被转换的特殊的线，抬高望远镜对准所要求的楼层地板，视线对准地板的点作一个标志。对所有的四个角重复这个过程，传递全部8个点，如图11-6所示。一旦8个标志转递完毕，把它们相连结，测量它们间之距离，并测量对角线长度作校核。

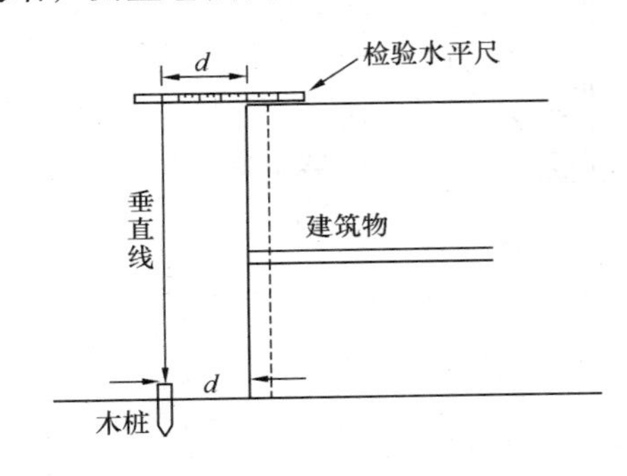

图11-5　垂球线控制垂直度

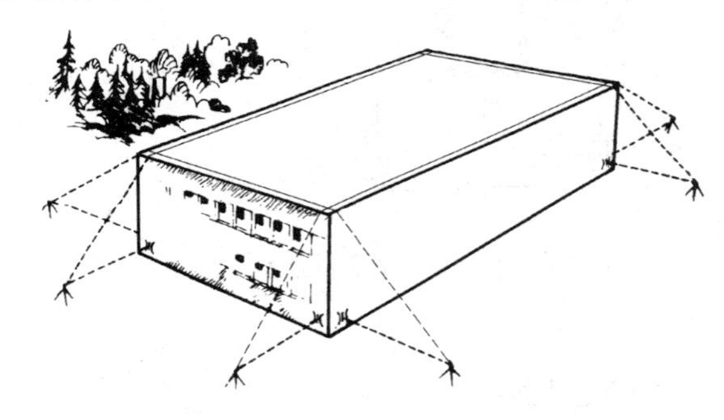

图11-6　多层建筑控制点传递

如果建筑物的中心线已建立，该法的变种是依次安置经纬仪于每条中心线上，传递4个点而不传递8个点，如图11-7所示。在每个楼板上所作的测量建立起互相垂直的两条线。

如果经纬仪并未完全调节好，传递点位必须用两个盘位，并取其平均位置。另外，由于涉及到大高度角，要求经纬仪必须仔细整平，并且需要利用附件转折目镜以使操作者通过望远镜寻找目标。

11.4.3　使用光学对中器

对于高层建筑最普遍使用的是自动光学对中器。现有各种类型的光学对中器都具有向上和向下瞄准建立垂直线，这些由工厂生产，它能与经纬仪交换使用三脚基座。图11-8显示了这样的仪器，它有两个望远镜，一个是低功率的向下定位地面标志，一个是高功率的向上对准目标。向上瞄准的望远镜装有相同类型的补偿器，类似于以前在自动安平水准仪使用的补偿器，因而确保视线自动垂直，即使仪器倾斜几分也能保证视线垂直。仪

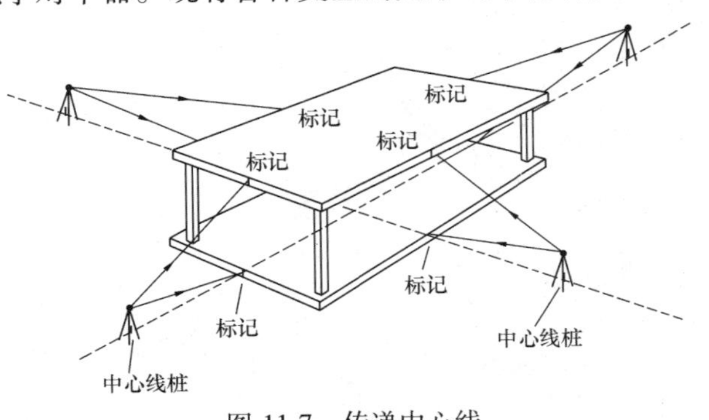

图11-7　传递中心线

器与三脚安平基座相连，基座固定在三脚架上，对准地面点进行对中操作。由于向下瞄准的望远镜没有补偿器，仪器整平实质上是用平行于望远镜轴安置的长水准管来完成的。圆水准器安装仪器顶上。

这种仪器提供竖直的视线精度达 ±1 秒（200m 误差 1mm）。为了确定它与垂直线的偏差，仪器绕 90°旋转并在 4 象限位置观测加以改正，得到的 4 个标记形成正方形，正方形对角线交点就是正确的投射中心点。

在地面建立了一个基本图形，依此进行定位测量。如果这个图形随着工程的进展垂直向上传递，那么从各层相同的图形进行定位测量就能确保建筑物的垂直度（图 11-9）。

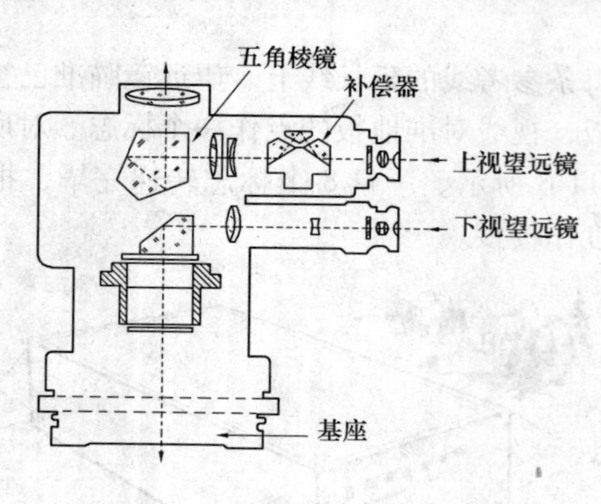

图 11-8　光学自动对中系统

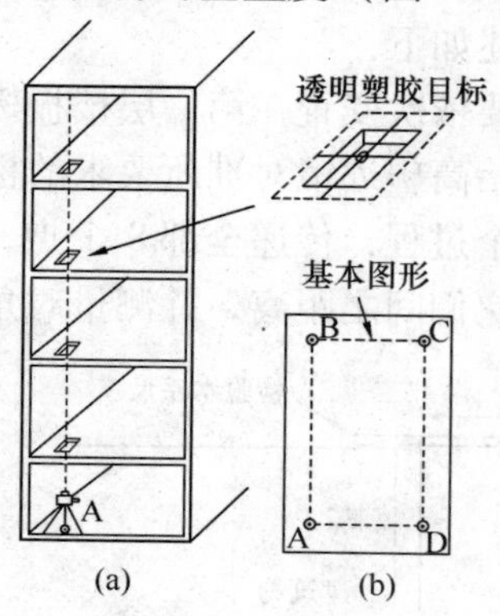

图 11-9　检测建筑物的垂直度

（a）正视图；（b）平面图

为了确定上层楼面上基本图形的点位，用一种透明塑胶目标板安放在上层楼板预留的小孔洞上（图 11-9b），按上述的方法确定中心点位。

使用哪种基本图形的形状取决于建筑物设计形状。对于长矩形建筑物简单的基线就足够了，但也使用 T 形状和 Y 形状。

第 12 章　全球定位系统

12.1　概述

全球定位系统（GPS）是基于卫星的系统，它能用于测定地球上任意点位的位置，它是由美国国防部于 20 世纪 70 年代初研发的。GPS 在全球范围内提供连续地（每日 24 小时）、实时地、三维定位、导航与授时。持有 GPS 接收机的任何人都能访问这个系统，因此可用于需要定位坐标的任何应用。

卫星定位概念始于 1957 年 10 月苏联发射第一颗人造地球卫星，接着美国海军研发了海军卫星导航系统（NNSS），这个系统一般称为海军导航系统，但是该系统测定点位需要很长时间，短距离测定相对点位精度低，其应用局限于大地测量和低等动态导航。1973 年美国国防部开始研发 NAVSTAR（Navigation System Timing And Ranging）全球定位系统（GPS），并于 1978 年发射了第一颗卫星。

GPS 卫星每日在轨道上很精确地围绕地球 2 圈并向地球发射信号。GPS 接收机接收信号并用三角测量法计算用户的位置。实质上，GPS 接收机能比较卫星发射的信号的时间和它接收的时间，这时间差告知 GPS 接收机现在离卫星有多远。现在，从多个卫星进行距离测量，接收机就能确定用户的位置，并将它显示在接收机的电子地图上。

卫星系统的最初目的是为了能够快速测定飞机、船舰和其他军事部门的大地位置。尽管该系统的研发是为了军事目的，但大大有益于其他群体，例如国家大地测量、私人专业测量和非专业的大众。GPS 系统可用来完成传统测量技术所做的任何事情。

现在，GPS 完全开放经营，相对定位精度达到几个毫米，观测时间短到几分钟。GPS 测量距离大于 5km 通常比 EDM 精度高，因此 GPS 广泛应用于工程测量。GPS 的引进在实际工程测量中的影响超过了 EDM 的影响，除了高精度外，GPS 还有下列优点：

（1）通常称为基线（简单直线）测量成果来看，不仅产生两站之间的距离，而且还给出测站的坐标 $X/Y/Z$，或东坐标/北坐标/高程，或纬度/经度/高程。

（2）无需通视，与其他传统测量系统不同，测站之间视线不要求通视，但是每个测站天空视野必须清晰以便能“看见”相关的卫星。

（3）大多数卫星测量设备配备有防风防雨装置，这样的系统在任何天气下都能进行观测，无论在白天、夜间、浓雾天都不影响观测。

（4）卫星测量可以一人操作，大大节省了时间与劳力。

（5）可在陆地、海洋及飞机上进行定位测量。

（6）可连续不断进行测量，从而大大改进变形监测。

因为卫星环绕全球，卫星定位的坐标系统是全球的而不是局部地方的。因此，如果坐标要求局部基准面或某种投影，则必须了解局部投影与基准面和卫星坐标系统之间的关系。

12.2　GPS 系统

GPS 系统由 3 部分组成：（1）空间部分：GPS 卫星本身；（2）控制系统，由美国军事部门操控；（3）用户部分，包括军事用户和民用用户以及他们的 GPS 接收设备。

12.2.1　空间部分

空间部分由卫星组成（见图 12-1），现在至少有 24 颗卫星环绕地球，几乎成圆形轨道，高度 20200km。6 个轨道平面等间隔（见图 12-2），每个条轨道上有 4 颗卫星。每个轨道面与赤道面倾斜 55°，轨道周期 12 小时。这意味着卫星每天 24 小时完成 2 圈。24 颗卫星组成 GPS 星座。卫星独立出现在地平线以上有 5 个小时。系统的设计使得至少有 4 颗卫星出现在地平线至少 15°以上。

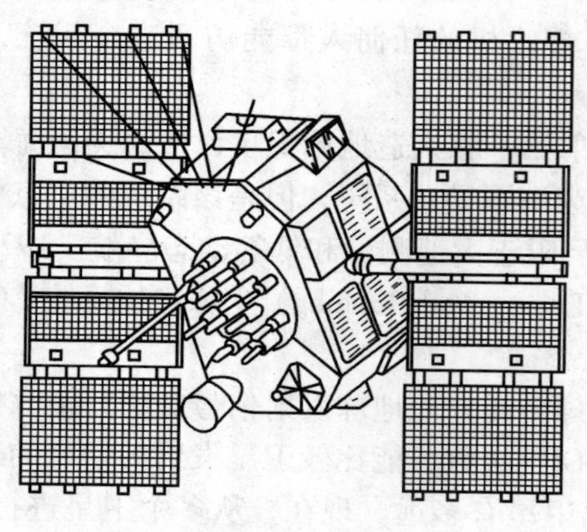

图 12-1　GPS 卫星

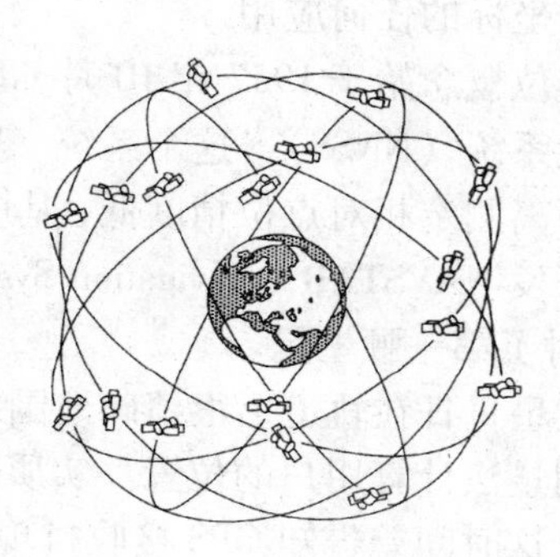

图 12-2　GPS 星座轨道

GPS 卫星设计寿命为 7.5 年。单翼太阳能电池阵列展开面积为 7.2m^2，太阳能电池板面保持垂直于太阳。在进入阴影期时由 3 个镍镉蓄电池供电。反冲式叶轮控制卫星在空间的定向与定位。卫星发射无线电信号给用户，每个卫星带有两个铷和两个铯原子钟以确保精确授时。

12.2.2　控制部分

GPS 的控制部分由 1 个主控制站，3 个注入站和 5 个监测站组成（见图 12-3）。

主控站（MCS）位于美国科罗拉多斯普林斯附近的施里佛空军基地的统一空间管理中心（CSOC）内。为了能用 GPS 精确定位，必须知道每个卫星在任意时间的精确位置。尽管卫星位置非常精确，但是卫星的时钟有漂移，所以必须保持 GPS 时间与 CSOC 定义的时间同步。当卫星围绕地球转动时，卫星受到地球的不同引力，太阳和月亮的吸引力，以及太阳的辐射压，所有这些都会引起卫星轨道随时间的变化，必须用工具进行测量和预测。为此有 13 个跟踪站组成的网连续不断地监测全部视野中 GPS 卫星，与此同时每个卫星的时钟也得到监视。比较 GPS 时间计算改正值以便保持卫星时钟与 GPS 时间同步。在主控站预测卫星轨道位置（这就是所谓历书预测）和计算卫星时钟改正值。

监测站设在阿森松群岛（位于南美洲与非洲中间大西洋中部），迪戈加西亚（位于印度洋），卡瓦加兰（位于太平洋），科罗拉多斯普林斯和夏威夷。监测站是遥控的无人值守站。

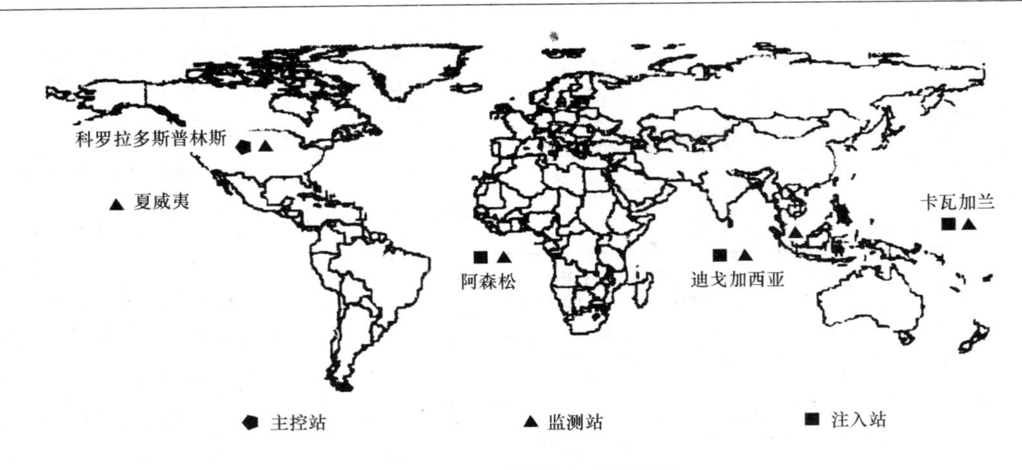

图 12-3　GPS 系统的控制部分

每个站都有 GPS 接收机、时钟、气象传感器、数据处理器和通讯设备。它们的功能是观测卫星广播导航信息和卫星时钟误差和漂移。

监测站自动收集和处理数据并传送到主控站。主控站比较各监测站传来的数据，计算当时导航信息和卫星时钟的误差。因此，能计算改正的导航信息为下一步传播。

注入站设在美国阿松森、迪哥加西亚和卡瓦加兰。主控站周期性发送改正后位置信息和时钟数据给注入站，然后由注入站上传这些数据给每个卫星。最后卫星用改正后的信息传播给用户端。

12.2.3　用户部分

GPS 接收机系统包含天线、接收机本身、命令输入及显示装置以及电源。图 12-4 显示 GPS 接收机系统，接收机与天线容纳于盒中，盒安装在三脚架上。命令显示装置固定到三脚架一条腿上，但为方便观测者使用，也可以重新安装。电源盒装在三脚架另一条腿上。天线接收卫星信号并转换为电能便于接收机使用。接收机在微处理器的控制下，处理信号转换为伪距并计算接收机的概略坐标。大多数接收机系统有机内的数据存储装置或外部输出接口允许与其他计算机相连。

图 12-4　GPS 接收机系统

GPS 接收机通常有一个或多个通道，通道包含硬件和必要的软件以便搜索卫星信号代码和（或）同时进行载波相位测量。接收机有 4 个或更多的专用通道，每次能同时搜索不同的卫星。这样的接收机，从多个卫星可获得伪距和相位数据，可以立即测定接收机的位置及接收机的时钟误差。

12.3　GPS 信号结构

各种 GPS 信号与基本频率的关系如图 12-5 所示。

GPS 星座每个卫星连续播放用于无线电通讯的 L 段的两种电磁波信号。卫星中，精密的

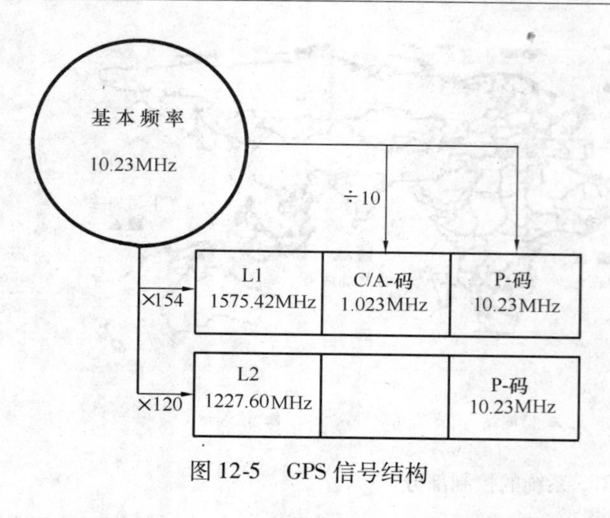

图 12-5　GPS 信号结构

原子钟有基本钟率或频率 f_0 = 10.23MHz 和由原子钟导出分配给 GPS 的两个 L 段频率：L1 = 154 × f_0 = 1575.42MHz 和 L2 = 120 × f_0 = 1227.60MHz。L1 和 L2 都起载波作用。每颗卫星传输导航信息，它是低速率 50bps（比特/秒）数据流，它调制在 L1、L2 载波上。完整的导航信息包含 25 帧，每帧长 1500 比特，即总长 37500 比特（即完整导航信息需 750s，即 12.5min 传播）。每帧分为 5 个子帧，每子帧 300 比特。导航信息连同其他信息包含有：作为时间函数（广播星历）的 GPS 卫星坐标、卫星钟改正参数、卫星健康状况、卫星历书以及大气改正参数。

和导航信息一样，两个二进制代码连续地调制在 L 段信号上，他们被称为粗/搜索码，即 C/A 码和精码，即 P 码。但是，粗码和精码与导航信息包含数据不相同，不由地面控制段上传给卫星，而是由每个卫星连续广播。

C/A 码和 P 码都由数字“0”和“1”组成，每个卫星传播唯一的代码。代码的结构十分周密，但是听起来代码像电子随机“噪声”，所以叫做伪随机码 PRN。虽然它表现为随机性，但它们是由某数学法则产生的并具有可复制的序列。C/A 码调制在 L1 载波上，而 P 码调制在 L1 和 L2 这两个载波上。C/A 码在 L1 载波上速率为每秒 1.023 百万字节（Mbps 或频率 $f_0 \div 10 = 1.023$MHz），即每毫秒重复 1 次，因为这给出码元宽度（相当 1 字节）293m，这意味着 C/A 码测距使用波长 293m。

精确的 P 码作为导航的主要代码。P 码的码速率就是基本频率的码速率，即 10.23MHz（频率 f_0 = 10.23MHz，码元长 29.3m）。P 码是非常长的二进制数字序列，266 天以后重复一次。它比 C/A 码快 10 倍。取 P 码循环 266 天乘它的速率 10.23Mbps（兆比特每秒）即得 2.35×10^{14}bit（比特）。把 266 天的长度分割为 38 段，每段长为一个周（星期）。其中 32 段分配给各卫星，即每个卫星传播 P 码中独特的一个周段，它在周六至周日午夜初始化。

C/A 码提供 GPS 标准定位服务（SPS），授权为民用用户服务。精密定位服务（PPS），它既可访问 C/A 码，又可访问 P 码，P 码的设计主要为军事部门用户。

20 世纪 80 年代美国国防部使该系统可供民用并实行选择可用性（SA）和反电子欺骗（AS）两种限制性的政策。设计 SA 降低信号的精度，导致水平定位精度为 100m。从 2000 年 5 月开始永久性解除 SA。这项政策的改变，提高 GPS 定位精度 5 倍（例如误差从 100m 降到 10～20m）。限制民用的另一种方法是反电子欺骗，将 P 码加密，改变为保密码 Y 码。使用差分定位技术，AS 及大部分的天然误差和其他误差可以消除。

12.4　定位基本原理

12.4.1　基本概念

GPS 的概念相当简单。如果已知地球上某个点（GPS 接收机）至 3 个 GPS 卫星的距离及卫星的位置，那么接收机的位置可应用简单交会法原理确定。但我们如何求得至卫星的距离及卫星的位置呢?

正如前述，每个卫星连续传播无线电信号，该信号由两个载波，两个代码和一条导航信息组成。当 GPS 接收机开关打开时，它能通过接收机天线接收 GPS 信号，一旦接收机得到 GPS 信号，机内的处理软件就能处理它，信号处理的部分结果包含有由数字代码（即所谓伪距）表示的是至 GPS 卫星的距离以及由导航信息表示的卫星坐标。

理论上，仅需要有同时跟踪 3 颗卫星测距离即可，此时接收机将是 3 个球的交会点，每个球以接收机至卫星的距离为半径并以对应的卫星为球心（图 12-6)。然而，从实际观点出发，有必要规测第 4 个卫星以便计算接收机的钟差。

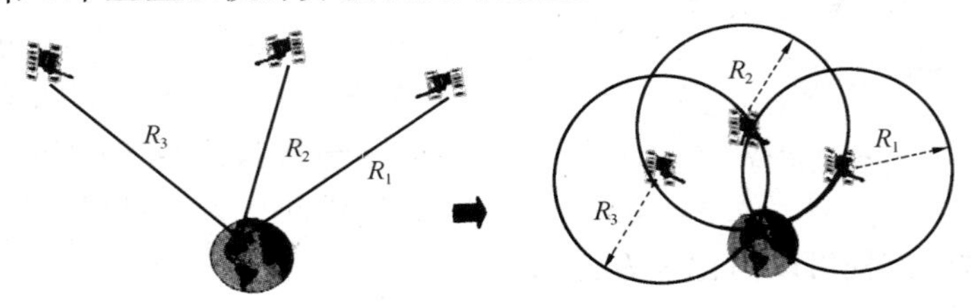

图 12-6　GPS 定位基本原理

定位测量通常分为两类：代码测量和相位测量。民用代码测量限制使用 C/A 码，当用于点位定位，仅提供精度 10 ~ 15m。应用各种差分技术，点位精度能达亚米至 5m。P 码测量提供军事部门具有更高的精度。使用一台 GPS 接收机点位定位的技术是跟踪卫星代码信号以便直接测定接收机站的坐标。

12.4.2　代码测量

为了定位，GPS 接收机测定至若干个 GPS 卫星的距离，然后用这些计算接收机天线的坐标。在 GPS 中，存在有用于测定距离的伪距和载波相位两种观测值。大多数的接收机能用代码测距法测定伪距，因此用手持的绘图级别接收机能提供低精度的、一般即时的点位。使用测地型接收机则用载波相位测量能得到高精度的定位。

*代码测距法*是 GPS 定位最简单的方法，用一台接收机就可以了。为了测定至卫星的距离用代码测距法，接收机测量 C/A 码从卫星至接收机运行时间，然后计算两者之间的距离 $R = c \times \Delta t$，这里 c 为信号的速度（光速)，Δt 为信号从卫星至接收机传播的时间。但是使用这些观测值有些困难，因为测量过程是单向的，必须测量信号离开卫星的时间和 PRN 信号到达接收机时间，这两个观测值是与 GPS 时间同步的，否则，测量通过的时间有误，距离就不正确。

为了解这些问题，对 GPS 代码测量研发了以下的技术：所有的卫星用它们机上的原子钟产生独立的 C/A 码，它与 GPS 时间的关系是已知的。当接收机锁定卫星时，进来的信号激发接收机产生 C/A 码等同于由卫星传输的信号。然后，由接收机产生的复制代码与卫星代码作比较，用所谓自相关技术处理。接收机代码在时间上位移直到与卫星代码同相。接收机产生代码位移的量等于卫星至接收机传播的延迟，将它乘以光速求得卫星和接收机的距离。

上面已经叙述，所有卫星配备有原子钟，它的 GPS 时间偏差通过导航信息精确知道，这使得从卫星传播的 RPN 代码精确相关于 GPS 时间。实际上，因为价格太贵接收机没有配备原子种，代之配备电子钟（振荡器)，其精度低于原子钟，但远高于日常的时钟。因此，接收机的时钟与 GPS 时间不经常同步，接收机产生的复制码将包含有钟差，由于该原因接收机测定的距离有偏差，故称为伪距。实质上是钟差导致观测值产生的偏差。

与伪距测量的同时接收机也要处理导航代码，由此计算观测时的卫星位置。

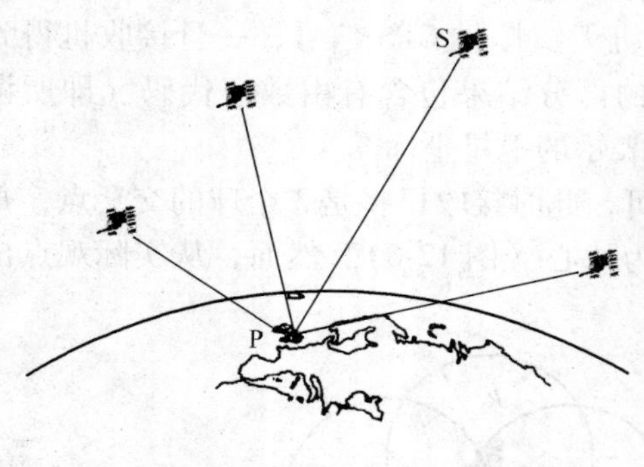

图 12-7　GPS 单点定位

计算点位把卫星看作已知坐标的控制点，接收机为未知点 P 对卫星测定距离。图 12-7 中 P 点至任意卫星 S 的距离可以定义为

$$R_{SP} = \rho + c(dt - dT) \tag{12-1}$$

式中　R_{SP}——从卫星 S 至 P 的几何距离；

ρ——从卫星 S 至 P 的伪距；

dt——卫星 GPS 时间的钟差；

dT——接收机 GPS 时间的钟差。

由于卫星的钟差 dt 可以用导航代码所包含的数据进行模拟，这个方程可以简化为 $R_{SP} = \rho - cdT$。

几何距离 R_{SP} 可写为

$$R_{SP} = \sqrt{(X_S - X_P)^2 + (Y_S - Y_P)^2 + (Z_S - Z_P)^2} \tag{12-2}$$

式中，X_S，Y_S，Z_S 为卫星的已知坐标；X_P，Y_P，Z_P 为 P 点的未知坐标。

如果同时观测 4 个卫星，可以得到下列定位方程：

$$\left.\begin{aligned}
&\text{卫星 1}\quad \sqrt{(X_{S1} - X_P)^2 + (Y_{S1} - Y_P)^2 + (Z_{S1} - Z_P)^2} = \rho_1 - cdT\\
&\text{卫星 2}\quad \sqrt{(X_{S2} - X_P)^2 + (Y_{S2} - Y_P)^2 + (Z_{S2} - Z_P)^2} = \rho_2 - cdT\\
&\text{卫星 3}\quad \sqrt{(X_{S3} - X_P)^2 + (Y_{S3} - Y_P)^2 + (Z_{S3} - Z_P)^2} = \rho_3 - cdT\\
&\text{卫星 4}\quad \sqrt{(X_{S4} - X_P)^2 + (Y_{S4} - Y_P)^2 + (Z_{S4} - Z_P)^2} = \rho_4 - cdT
\end{aligned}\right\} \tag{12-3}$$

这 4 个方程可以解算 4 个未知数：接收机的坐标 X_P、Y_P、Z_P 和接收机钟差 dT。

上述看出，在 GPS 测量中至少必须搜索并测量 4 个卫星。然而这不能保证一定成功，有时需要观测更多卫星。

代码测距法也称为导航解决方案，这达到了 GPS 原始的目的。因为在这个过程中接收机钟差很容易测定。GPS 接收机的时钟不必是精确的原子钟，这样降低接收机成本适合野外使用。在任意给定的导航解决方案中，使用计算机多次解算方程。软件安装在接收机内，操作者无需执行任何计算，仅需读取接收机实时显示的位置或作电子化记录。

12.4.3　载波相位测量

测量至卫星距离的另一种方法是通过载波相位。GPS 接收机测量相位延迟用比较来自卫星的信号同由接收机产生的相似信号。这种相位测量是采用 L1 与 L2 的载波（并非调制的 C/A 码与 P 码），即使用某种技术去掉调制在 L1 与 L2 上的调制的伪随机码 PRN。因此，GPS 接收机到卫星的距离简化为接收机与卫星之间的整载波个数和接收机上不足整波个数之和再乘以载波波长（见图 12-8）。用载波相位测定距离精度大大高于伪距测量。由于信号处理技术能够提高观测的分辨率大约为载波波长的百分之一，因此载波相位测量可能达到精度 1.9mm（用 L1 信号波长 19cm）和 2.4mm（用 L2 信号波长 24cm）。

但是，这里存在一个问题：载波是纯的正弦波，这意味着所有的周波看起来都一样。因此，GPS 接收机没法区分是哪一个。换句话说，当接收机开关打开后不能决定卫星与接收机之间有多少个整波，唯一能很精确测量的是不足一个整波的部分（小于 2mm），而内部完整波数仍然不知道，或称整周未知数。幸好接收机开关打开后，接收机具有保持搜索相位变化的能力。这意味着只要信号没有丢失或没有出现周跳，则初始的整周未知数不会随时间流逝而改变。

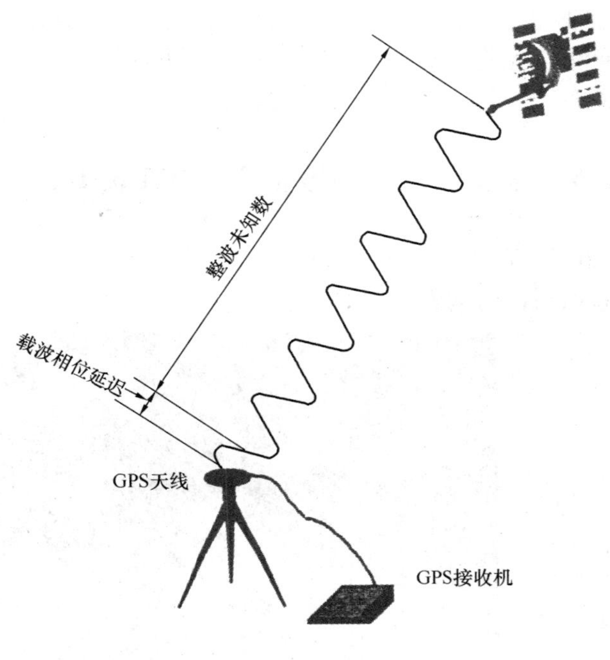

图 12-8　载波相位测量

很显然，如果解决了整周未知数，就可得到高精度距离测量值，达到高精度的定位。高精度定位习惯上是通过所谓相对（或差分）定位技术，包括实时和事后处理模式。遗憾的是都需要 2 台接收机同时搜索天空中相同的卫星。

为了既迅速又稳定求解载波整周未知数已提出了许多技术措施。一种简单的办法是首先用接收机按代码测距法求得接收机概略位置，然后进行载波相位测量。解决整周未知数参数更多不同的定位技术和方法已超越本书的范围。

12.5　相对定位

GPS 相对定位也称差分定位，使用 2 台（或多台）GPS 接收机同时搜索相同的卫星以确定它们的相对坐标（见图 12-9）。其中一台作为参考站，或基站，它在某地点保持固定，已知精确坐标（即预先测量）。其他接收机所谓流动站或遥测站，其坐标未知。用两台接收机同时观测以确定相对于基站的位置。流动站采用什么作业模式根据 GPS 操作类型而定。

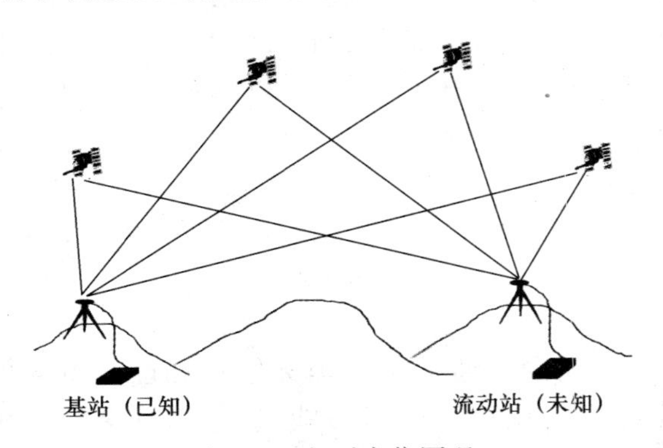

图 12-9　相对定位原理

相对定位最少要有 4 颗共同卫星，然而，多于 4 颗卫星可以提高 GPS 定位的精度。载波相位测量或伪距测量都可用于相对定位。各种定位技术都可提供实时定位信息或延后（即后处理）定位信息。一般来说，相对定位精度高于独立定位的精度。这要看相对定位时是用伪距测量还是载波相位测量，可以分别获得几米到几毫米的精度等级。这主要是因为同时搜索 2 个或多个卫星或多或少包含相同的误差，两台接收机之间距离越短，误差越小。因此，如果我们取两台接收机观测值之差（因此又称为差分定位），共同的误差将被消除，并且空间相关的误差将减少，取决于基站接收仪与流动站接收机之间的距离。

12.6　实时动态 GPS 测量（RTK GPS）

RTK 是“Real Time Kinematic”的缩写。实时动态定位测量是基于载波而使用两台或多台接收机同时搜索相同的卫星进行的相对定位的技术（图 12-10）。

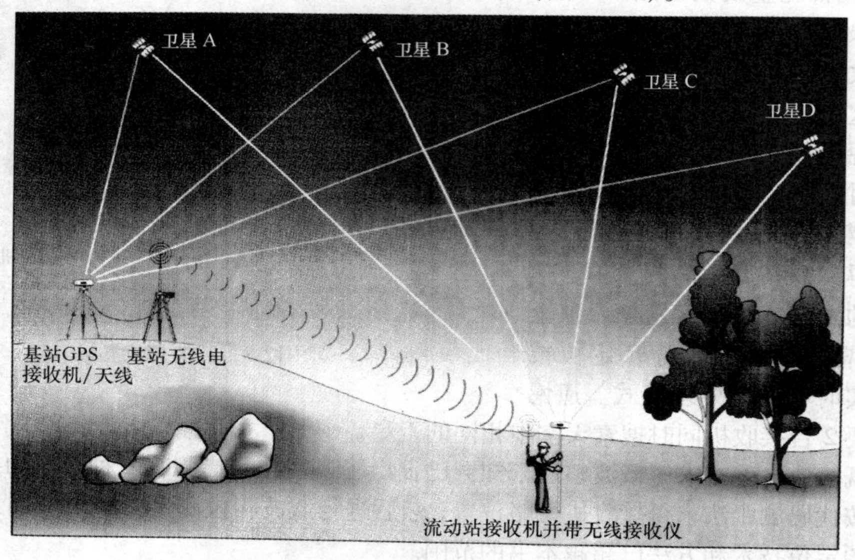

图 12-10　实时动态 GPS 测量

使用这种方法时，基站的接收机放在已知点保持固定，并与无线电发射机相连(图 12-10)。流动站的接收机装入背包（或固定在某移动物体上）并与无线电接收仪相连结。RTK 要求基站测量卫星，处理基线的改正值，然后经过无线电通讯把改正值传输给多个流动站接收机。

基站传输代码和载波相位数据到流动站接收机。流动站接收机用这些数据以帮助解决整周未知数问题，解决基站与流动站坐标差的变化问题。

无线电传播载波相位数据从基站到流动站接收机传输距离，使用无线电放大器可以延伸距离大约达 10km，更长的距离要求商业注册。有些国家要求商业注册距离更短。

该法适用于：（1）已知点附近要测量大量的未知点（在 15 ~ 20km 以内）；（2）未知点的坐标必须实时提供；（3）两台接收机之间传播路径相对无障碍。

12.7　大地坐标系

12.7.1　大地坐标系概念

坐标系是定义点位（也称为坐标）的一套规则。通常包括坐标原点和已知方向的参考线（或称轴），图 12-11 显示某三维坐标系，使用三条参考线（X，Y 和 Z）交坐标系的原点 C。

坐标系可分为一维、二维与三维共 3 种，根据需要识别点位的坐标数而定。例如，识别高于海平面的点的高程需要一维坐标系。

坐标系也可以根据参考面、轴的方向及原点来分类。三维大地坐标系也称地理坐标系，参考面选择椭球体面，轴的方向与原点由两个面来确定：即通过南北极即 Z 轴的子午线面（子午面是通过北极与南极的平面）和椭球体的赤道面（见图 12-11 的细节）。

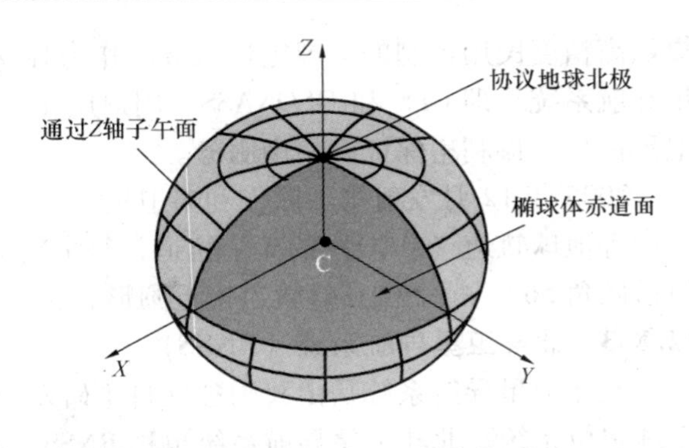

图 12-11　三维坐标系

对 GPS 用户特别重要的是三维大地坐标系，在这个系统中点的坐标由大地纬度（φ），大地经度（λ）和参考面的高度（h）来确定。图 12-12 显示这些参数。大地坐标（φ、λ 与 h）很容易转换为笛卡尔直角坐标（X，Y 与 Z），为此，必须已知椭球体的参数（a 与 f）。

12.7.2　WGS-84 大地坐标系

1984 年世界大地坐标系（WGS-84）是三维、地心参考系。WGS-84 系统的坐标原点是地球质量中心，Z 轴指向由国际时间局（BIH）定义 1984 年 1 月 1 日 0 时的协议地球极（CTP）；X 轴指向（BIH）1984.0 的子午面和 CTP 的赤道面的交线；Y 轴与 Z，X 轴形成右手规则坐标系。相应于 WGS-84 大地坐标系有一个 WGS-84 椭球体。WGS-84 椭球体及有关常数采用国际大地测量协会和国际大地测量与地球物理联合会第 17 届大会的推荐值。该系统图解说明见图 12-12。

WGS-84 椭球体定义参数如下：

长半径　$a = 6378137.0\text{m}$

短半径　$b = 6356752.3142\text{m}$

扁 率　$f = \dfrac{a-b}{a} = 1/298.257223563$

角速度　$\omega = 7.292\,115 \times 10^{-5}$ rad/sec

谐波系数　$C_{2.0} = -484.16685 \times 10^{-16}$

重力常数　$G_{\text{M}} = 398\,600.5\text{m}^3/\text{sec}$，

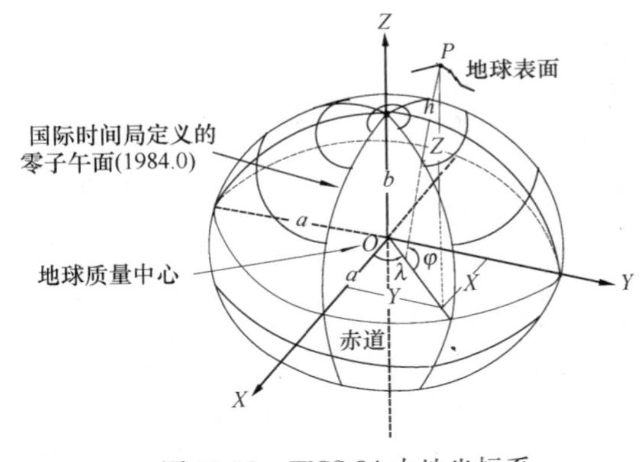

图 12-12　WGS-84 大地坐标系

12.8　其他卫星导航系统

12.8.1　格洛纳斯（GLONASS）系统

俄罗斯政府研制的类似于 GPS 系统，称为 GLONASS。1982 年 10 月发射第 1 颗 GLONASS 卫星，GLONASS 系统包含 24 颗使用卫星，3 个接近圆形轨道安排相隔 120°，高度 19100km，与赤道倾斜角 64.8°，每颗卫星绕地球是 11 小时 15 分。GLONASS 每颗卫星用相同的代码和不同的频率，而 GPS 每颗卫星用 2 种频率和不同的代码。有些 GPS 接收机兼容接收 GPS 和 GLONASS 信号，这就提高了卫星的可利用性和复合系统的完善性。

12.8.2　伽利略（GALILEO）系统

伽利略系统是由欧盟（EU）和欧洲宇航局（ESA）研制的全球导航系统（GNSS），它

提供高精度民用控制的全球定位服务。作为独立的导航系统，伽利略能同时共同使用其他卫星导航系统，即 GPS 和 GLONASS。用户用相同的接收机在混合系统中能接收来自任何一种卫星信号。伽利略保证高精度服务。

2005 年 12 月发射第 1 颗伽利略卫星。伽利略星座由 30 颗（27 颗在用，3 颗备用）分布于中等地球轨道（MEO）的 3 个轨道，每个轨道平面有 9 颗等间隔的卫星加一颗备用卫星，轨道倾角 56°，每个伽利略轨道接近圆形，长半径 29600km，运行周期约 14 小时。

12.8.3　北斗卫星导航系统（BNSS）

北斗卫星导航系统是由我国独立自主研发，目前是区域性的，将逐步扩展为独立运行的全球定位系统。北斗卫星导航系统也称 BNSS，第一代北斗系统由 3 个卫星组成的（两颗工作的，一颗备用的）分布在地球同步轨道上作为北斗导航实验系统。两颗工作卫星先后分别在 2000 年 10 月与 12 月发射的，备用卫星是 2003 年 5 月发射的。

自从 2007 年 4 月以来，中国发射了一系列北斗-2 导航卫星，到 2012 年 10 月已经发射了 16 颗北斗-2 导航卫星，建成了第二代北斗导航系统。北斗导航系统为中国及周边地区提供导航及通讯服务。北斗系统有地面控制段包含一个中心控制站和一个地面改正站，用户段包含一个地面接收/发射终端。

中国规划在 2020 年左右建成全球定位系统，它包含 5 颗地球同步轨道卫星和 30 颗非地球同步轨道卫星。

参 考 文 献

[1] McCormac, Jack C. Surveying [M] // Hoboken, N. J. Wiley, 2012.

[2] Hofmann-Wellenf. B. GNSS-Global navigation Satelite GPS, GLONASS, Galileo and More [M] // Wien, New York: Springer, c 2008.

[3] R. Subramanian. Surveying and Levelling [M] // New York Oxford university press c 2007.

[4] Xu, Guochang 1953 -. GPS: theory algorithms and application [M] 2nd ed. // Berlin: Springer, c 2007.

[5] W. SchoFild and M. Breah. Engineering Surveying [M]6th ed. // Oxford; Boston: Butterworth-Heinemann, 2007.

[6] J. Uren W. F. Price. Surveying for Engineers [M] 4th ed. // Basingstocke: Macmillan 2006.

[7] Ahmed EL-Rabbany. Introduction to GPS the Global Positioning System [M] 2nd ed. // Boston, MA: Artech House, c 2006.

[8] Johnson, Aylmer, 1951 -. Plane and Geodetic Surveying [M] Lond New york: Spon Press, 2004.

[9] Field Harry L. , 1949 -. Ladscape Surveying [M] // Clifton Park, N Y: Thomson/Delmar Learning, 2004

[10] Barry F. Kavangh. Surveying with Construction Application [M] // Upper Saddle River, N. J. Pearson Pretice Hall, c 2004.

[11] Kavangh, Barry F. Surveying: Principle and Application [M] 6th ed. // Upper Saddle River, N. J. Pratice hall, c 2003.

[12] Schof ield, W. (Wifred). Engineering Surveying: theory and examination problems for student [M] 5th ed. // Oxford; Boston: Butterworth-Heinemann, 2001.

[13] Jack McCormac. Surveying [M] 4th ed. // New York: John Wiley and Sons, Inc. , c1999.

[14] Arther Bannister, Stanly Raymond, Raymona Baker. Surveying [M] 7th ed. // Harlow, Essex, England, Adson Wesley Longmand Ltd. , 1998.

[15] Heribert Kahmen, Wolfgand Faig. Surveying [M] // Walter de Gruyter & Co. , Berlin30. 1998.

[16] Francis, H. Moffitt, John D. Bosstter. Surveying [M] 10th ed. // An imprint Wesley Longman, Inc. 1998.

[17] Moffit, Framcis H. Surveying [M] 10th ed. // Melo Park, Calif: Addison-Wesley, c1998.

[18] Anderson, J. M. (James McMurry),1926 -. Surveying theory and pratice [M] 7th ed. // Boston: WCB/McGraw-Hill, c 1998.

[19] French Gregory T. Understanding the GPS [M] // Bethesda, MD: GroResearch, c 1996.

[20] William Irvine. Surveying for Construction [M] // London, McGraw-hill Book Co. , c1995.

[21] Claney, John, 1936. Site Surveying and Levelling [M] 2nd ed. // Lomold, 1991.

[22] Leick, Afred. GPS Satellite Surveying [M] // New York:, Wiley, c 1990.

[23] Kahmen, Heribert, 1940 -. Surveying [M] // Berlin; New york: W. de Gruyter, 1988.

[24] Rusell C. Brinker, Roy Minick. The Surveying handbook [M] // New York: Van Nostrand Reinhold, c 1987.

[25] Philip Kissan. Surveying for civil engineers [M] 2nd ed. // New York: McGraw-Hill c1981.

[26] Hevubin Charless, 1931 -. Principle of Surveying [M] // Reston Va, Reston Pub. Co. , c1978.

[27] Russell C. Brinker Paul R. Wolf. Elementary Surveying [M] // Thomas Y. Crowell Company, Inc. 1977.

[28] 阎浩文,杨维芳.测量学原理与方法[M].北京:科学出版社,2009.

[29] 尹晖. 测绘工程专业英语 [M]. 武汉: 武汉大学出版社,2004.

[30]　陈学平．实用工程测量［M］．北京：中国建材工业出版社,2007.
[31]　全国科学技术名词审定委员会．测绘学名词(第三版)［M］．北京:科学出版社,2010.
[32]　严勇,王颖．英汉·汉英 测绘学词汇手册［M］．上海:上海外语教学出版社,2009.
[33]　英汉测绘词汇编辑组编．英汉测绘词汇［M］．北京：测绘出版社,1980.
[34]　Tom McArthur,Beryl Atkins．邱述德译,许孟雄审校．英汉双解简明英语短语动词词典［M］．北京:外语教学与研究出版社,1987.